Analog Function Circuits

Analog Function Circuits

Fundamentals, Principles, Design and Applications

K.C. Selvam

CRC Press
Taylor & Francis Group
Boca Raton London New York

CRC Press is an imprint of the
Taylor & Francis Group, an **informa** business

First edition published 2022
by CRC Press
6000 Broken Sound Parkway NW, Suite 300, Boca Raton, FL 33487-2742

and by CRC Press
2 Park Square, Milton Park, Abingdon, Oxon, OX14 4RN

CRC Press is an imprint of Taylor & Francis Group, LLC

Library of Congress Cataloging-in-Publication Data
Names: Selvam, K. C., author.
Title: Analog function circuits : fundamentals, principles, design and applications / K.C. Selvam.
Description: First edition. | Boca Raton : CRC Press, [2022] | Includes bibliographical references and index.
Identifiers: LCCN 2021025429 (print) | LCCN 2021025430 (ebook) | ISBN 9781032081601 (hardback) | ISBN 9781032117652 (paperback) | ISBN 9781003221449 (ebook)
Subjects: LCSH: Operational amplifiers. | Analog integrated circuits.
Classification: LCC TK7871.58.O6 S445 2022 (print) | LCC TK7871.58.O6 (ebook) | DDC 621.39/5--dc23
LC record available at https://lccn.loc.gov/2021025429
LC ebook record available at https://lccn.loc.gov/2021025430

ISBN: 978-1-032-08160-1 (hbk)
ISBN: 978-1-032-11765-2 (pbk)
ISBN: 978-1-003-22144-9 (ebk)

DOI: 10.1201/9781003221449

Typeset in Times
by SPi Technologies India Pvt Ltd (Straive)

Dedicated to my loving wife
S. Latha

Contents

PART C Design of Function Circuits

PART E Miscellaneous Function Circuits

PART F Applications of Function Circuits

Preface

When I joined the Indian Institute of Technology (IIT) Madras in 1987, I used to work with operational amplifiers. I got my first function circuit in 1993. During 1993–1995, I published 9 papers on function circuits. From 1996 to 2010, I suffered a setback, as I was diagnosed with Bipolar disorder disease, and I was away from the research. In 2011, I made a comeback and started my research after a long hiatus. From 2011 to 2016, I published 34 papers on function circuits. In 2017–2018, I got concepts for 118 multiplier circuits and published a book titled, *Design of Analog Multipliers with Operational Amplifiers* published by CRC Press/Taylor & Francis. During 2019, I got concepts for 127 multiplier-cum-divider circuits and published another book titled, *Multiplier-Cum-Divider Circuits: Principles, Design, and Applications*, also published by CRC Press. In 2020, I got concepts for more than 500 function circuits and, as a result, I decided to publish this book.

I am highly indebted to my mentor Prof. Dr. V.G.K. Murti who taught me about function circuits, Prof. Dr. P. Sankaran who introduced me to IIT Madras, Prof. Dr. K. Radha Krishna Rao who taught me about operational amplifiers, Prof. Dr. Bhaskar Ramamurti who motivated me to do this work, Prof. Dr. V. Jagadeesh Kumar who guided me in the proper way of the scientific world, Dr. M. Kumaravel who trained me to do experiments with op-amps, Prof. Dr. Enakshi Bhattacharya who encouraged me to work on this book, Prof. Dr. Devendra Jalihal who kept me in a happy and peaceful atmosphere, and Prof. Dr. David Koil Pillai who supervised all my research work at IIT Madras.

I am also indebted to psychiatry Drs. S. Sathiyanathan and R. Rajkumar, and Dr. V. Vasantha Jayaram who gave me medical treatment during 1996–2010. I am also indebted to Drs. Shiva Prakash and D. Saraswathi, who continue my medical treatment.

I am thankful to Dr. T.S. Rathore, former professor of IIT Bombay who reviewed all my papers, and Prof. Dr. Raj Senani, former editor of *IETE Journal of Education*, who published most of my papers.

I thank my father, Venkatappa Chinthambi Naidu; my mother, C. Suseela; my wife, S. Latha; older son, S. Devakumar; and younger son, S. Jagadeesh Kumar, for providing a happy and peaceful home life.

My sincere gratitude also goes to Gauravjeet Singh Reen, Senior Commissioning Editor at Taylor & Francis who showed a keen interest in publishing all my concepts on function circuits. His hard work in bringing this book to reality is commendable.

Further thanks go to my apprentice trainee batchmates and friends, T. Padmavathy, V. Selvaraju, Dr. C.R. Jeevandos, and P.V. Suguna, with whom I started my career at the IIT Madras Central Electronics Centre in 1987.

Last but not least, I would like to thank my friends, Prof. Dr. K. Sridharan, Prof. Dr. R. Sarathi, Dr. Balaji Srinivasan, Dr. T. G.Venkatesh, Dr. Bharath Bhikkaji, Dr. Bobey George, Dr. S. Anirudhan, and Drs. R. Aravind and Anantha Krishnan for their constant encouragement throughout my research work; also, other staff, students, and faculty of the Electrical Engineering Department at IIT Madras for their immense help during the experimental setups, manuscript preparation, and proofreading.

Useful Notations

V_1	First input voltage
V_2	Second input voltage
V_3	Third input voltage
V_O	Output voltage
V_R	Reference voltage/peak value of first sawtooth waveform
V_T	Peak value of first triangular waveform
V_P	Peak value of second triangular wave/sawtooth wave
V_C	Comparator 1 output voltage in the first sawtooth/triangular wave generator
V_M	Comparator 2 output voltage by comparing sawtooth/triangular waves with one input voltage
V_N	Lowpass filter input signal
V_{S1}	First generated sawtooth wave
V_{S2}	Second generated sawtooth wave
V_{T1}	First generated triangular wave
V_{T2}	Second generated triangular wave
V_S	Sampling pulse
V_1'	Slightly less than V_1 voltage
V_2'	Slightly less than V_2 voltage

Abbreviations

MPDD	Multiplexing Peak Detecting Divider
MPDM	Multiplexing Peak Detecting Multiplier
MPDMCD	Multiplexing Peak Detecting Multiplier-Cum-Divider
MPRD	Multiplexing Peak Responding Divider
MPRM	Multiplexing Peak Responding Multiplier
MPRMCD	Multiplexing Peak Responding Multiplier-Cum-Divider
MPSD	Multiplexing Peak Sampling Divider
MPSM	Multiplexing Peak Sampling Multiplier
MPSMCD	Multiplexing Peak Sampling Multiplier-Cum-Divider
MTDD	Multiplexing Time Division Divider
MTDM	Multiplexing Time Division Multiplier
MTDMCD	Multiplexing Time Division Multiplier-Cum-Divider
PDD	Peak Detecting Divider
PDM	Peak Detecting Multiplier
PDMCD	Peak Detecting Multiplier-Cum-Divider
PPDD	Pulse Position Detecting Divider
PPDM	Pulse Position Detecting Multiplier
PPDMCD	Pulse Position Detecting Multiplier-Cum-Divider
PPRD	Pulse Position Responding Divider
PPRM	Pulse Position Responding Multiplier
PPRMCD	Pulse Position Responding Multiplier-Cum-Divider
PPSD	Pulse Position Sampling Divider
PPSM	Pulse Position Sampling Multiplier
PPSMCD	Pulse Position Sampling Multiplier-Cum-Divider
PRD	Peak Responding Divider
PRM	Peak Responding Multiplier
PRMCD	Peak Responding Multiplier-Cum-Divider
PSD	Peak Sampling Divider
PSM	Peak Sampling Multiplier
PSMCD	Peak Sampling Multiplier-Cum-Divider
SPDD	Switching Peak Detecting Divider
SPDM	Switching Peak Detecting Multiplier
SPDMCD	Switching Peak Detecting Multiplier-Cum-Divider
SPRD	Switching Peak Responding Divider
SPRM	Switching Peak Responding Multiplier
SPRMCD	Switching Peak Responding Multiplier-Cum-Divider
SPSD	Switching Peak Sampling Divider
SPSM	Switching Peak Sampling Multiplier
SPSMCD	Switching Peak Sampling Multiplier-Cum-Divider
STDD	Switching Time Division Divider

STDM	Switching Time Division Multiplier
STDMCD	Switching Time Division Multiplier-Cum-Divider
TDD	Time Division Divider
TDM	Time Division Multiplier
TDMCD	Time Division Multiplier-Cum-Divider

Introduction

Analog multipliers, dividers, multipliers-cum-dividers, squarers, square rooters, square root of multiplication, vector magnitude circuits, multifunction circuits, and phase sensing circuits are called "analog function circuits" (AFCs).

An analog multiplier is a function circuit which accepts two input voltages V_1 and V_2 and produces an output voltage $V_O = \frac{V_1 V_2}{V_R}$, where V_R is the proportional constant, and usually it will be a constant reference voltage. The symbol of a multiplier is shown in Figure 0.1.

An analog divider is a function circuit which accepts two input voltages V_1 and V_2 and produces an output voltage $V_O = \frac{V_2}{V_1} V_R$, where V_R is a constant reference voltage. Figure 0.2 shows the symbol of an analog divider.

A multiplier-cum-divider (MCD) is a function circuit which accepts three input voltages V_1, V_2, and V_3 and produces an output voltage $V_O = \frac{V_2 V_3}{V_1}$. Its symbol is shown in Figure 0.3.

An analog squarer is a function circuit which accepts an input voltage V_1 and produces an output voltage $V_O = \frac{V_1^2}{V_R}$. Its symbol is shown in Figure 0.4, where V_R is constant reference voltage.

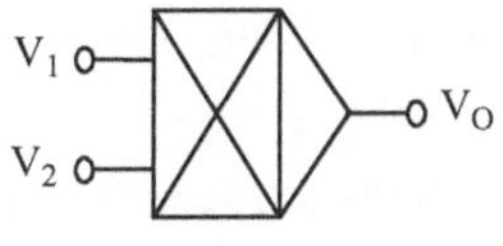

FIGURE 0.1 Multiplier symbol.

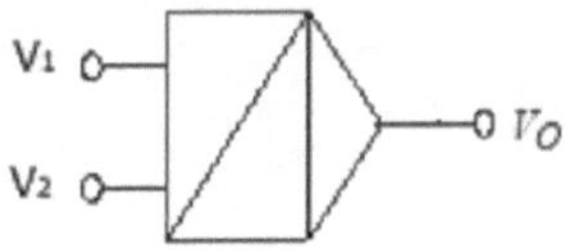

FIGURE 0.2 Divider symbol.

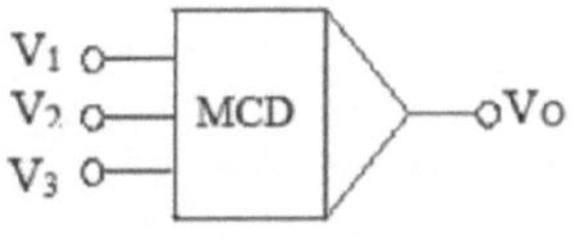

FIGURE 0.3 Multiplier-cum-divider symbol.

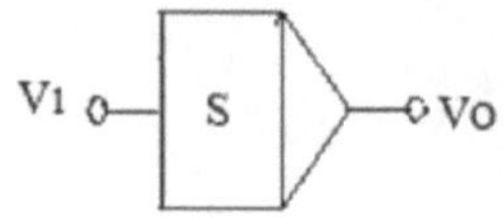

FIGURE 0.4 Squarer symbol.

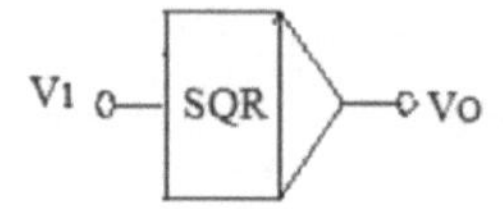

FIGURE 0.5 Square rooter symbol.

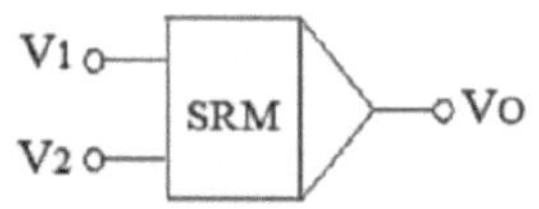

FIGURE 0.6 Square root of multiplication symbol.

An analog square rooter is a function circuit which accepts an input voltage V_1 and produces an output voltage $V_O = \sqrt{V_1 V_R}$, where V_R is a constant reference voltage. Figure 0.5 shows the symbol of an analog square rooter.

A square root of multiplication (SRM) is a function circuit which accepts two input voltages V_1 and V_2 and produces an output voltage $V_O = \sqrt{V_1 V_2}$. Its symbol is shown in Figure 0.6.

A squarer and divider (SAD) is a function circuit which accepts two input voltages V_1 and V_2 and produces an output voltage proportional to $V_O = \frac{V_2^2}{V_1}$. The symbol of SAD is shown in Figure 0.7.

A vector magnitude circuit (VMC) is a function circuit which accepts two input voltages V_1 and V_2 and produces an output voltage $V_O = \sqrt{V_1^2 + V_2^2}$. The symbol of VMC is shown in Figure 0.8.

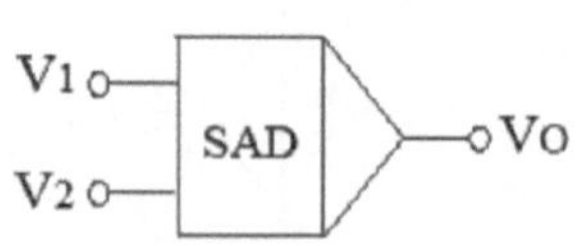

FIGURE 0.7 Squarer and divider symbol.

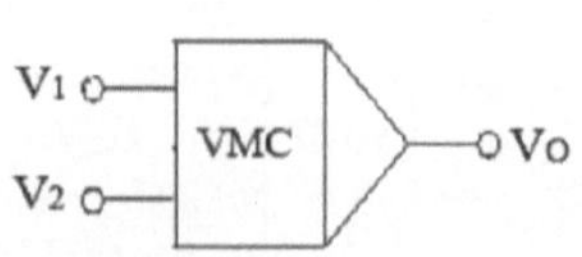

FIGURE 0.8 Vector magnitude circuit symbol.

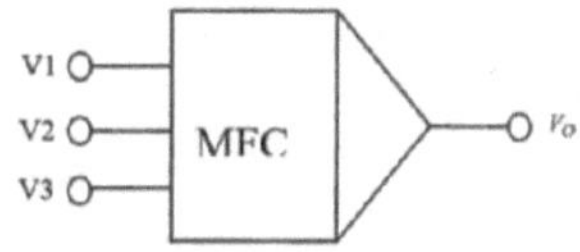

FIGURE 0.9 Multifunction converter symbol.

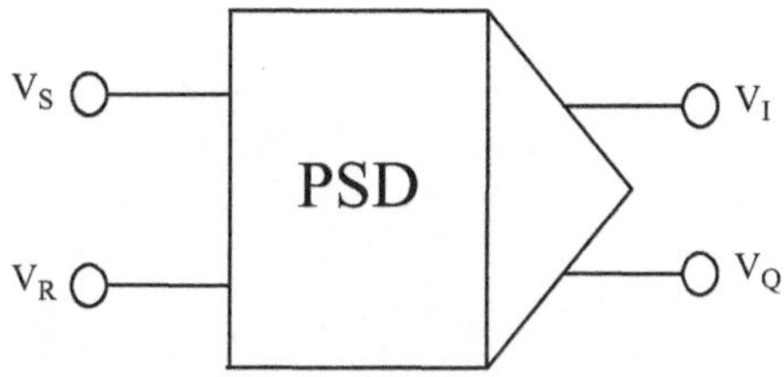

FIGURE 0.10 Phase sensitive detector symbol.

A multifunction converter (MFC) is a function circuit which receives three input voltages V_1, V_2, and V_3 and produces an output voltage $V_O = V_2\left(\frac{V_3}{V_1}\right)^m$ where m is an integer. Figure 0.9 shows the symbol of MFC.

A phase sensitive detector (PSD) is a function circuit which accepts two sinusoidal signals V_S and V_R of the same frequency with phase difference Ø and provides two DC output voltages in which one DC output voltage V_Q is proportional to the quadrature component SinØ, and the another DC output voltage V_I is proportional to the in-phase component CosØ. Figure 0.10 shows the symbol of a PSD.

Part A

Fundamentals of Function Circuits

1 Components of Function Circuits

Analog function circuits such as multipliers, dividers, multipliers-cum-dividers, square rooters, and vector magnitude circuits are designed with the following components: (i) transistors, (ii) field effect transistors (FETs), (iii) switches, (iv) multiplexers, and (v) operational amplifiers (op-amps). Transistors were invented by Bell Telephone Laboratories in 1948. Electronics changed in a major way after transistors were invented. A metal-oxide-semiconductor field effect transistor (MOSFET) is the most widely used electronic device, and complementary metal-oxide-semiconductor (CMOS) technology is the technology of choice in the design of integrated circuits. Transistors are a preferable device in very demanding analog circuit applications, both integrated and discrete. FETs and MOSFETs can be used as switches and multiplexers. These switches and multiplexers play an important role in the design of analog function circuits with op-amps. The components of function circuits are described in this chapter.

1.1 TRANSISTORS

A transistor is a semiconductor device with three layers. There are two types of transistors: nρn (n type – ρ type – n type sequence) and ρnρ (ρ type – n type – ρ type sequence). The symbols of transistors are shown in Figures 1.1(a) and 1.1(b). The transistor has three terminals, Base (B), Collector (C), and Emitter (E), and it has two PN junctions.

A single PN junction has two modes of operation: forward bias and reverse bias. The transistor with two PN junctions has four possible modes of operation, depending on the bias condition of each PN junction. The basic principle of a transistor is that the voltage between two terminals controls current through a third terminal. The current in a transistor is due to flow of both electrons and holes, hence it is called a "bipolar transistor."

In a transistor, the emitter current (I_E) will be the addition of base current (I_B) and collector current (I_C), i.e.,

$$I_E = I_B + I_C \tag{1.1}$$

Transistors are used for both current and voltage amplifications. The base current is very small.

$$I_C = \alpha I_E$$

$$\alpha = \frac{I_C}{I_E} \tag{1.2}$$

DOI: 10.1201/9781003221449-2

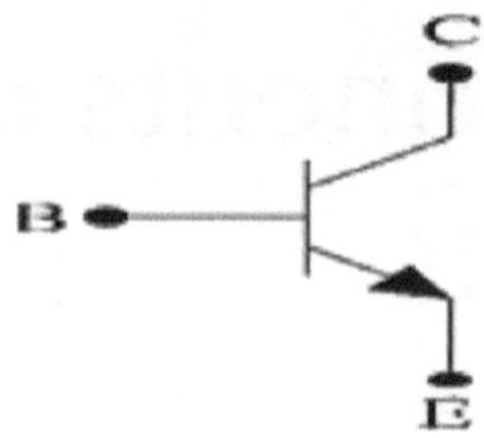

FIGURE 1.1 (A) npn transistor.

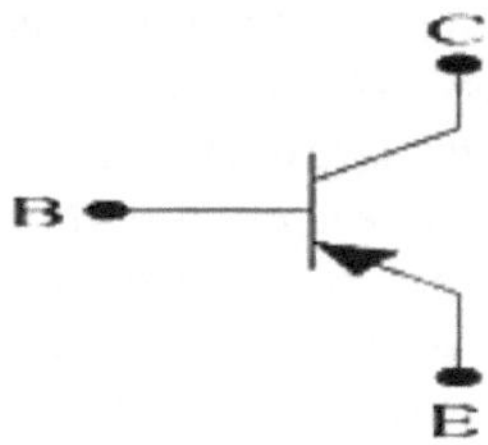

FIGURE 1.1 (B) pnp transistor.

where α is the emitter to collector gain or ratio of collector current to emitter current. The value of α will be normally 0.96–0.995. The collector current is almost equal to emitter current.

$$I_C = \beta I_B$$

$$\beta = \frac{I_C}{I_B} \tag{1.3}$$

where β is the base collector current gain or the ratio of collector current to base current.

The typical values of β are ranges from 25 to 300.

$$\beta = \frac{\alpha}{1-\alpha} \tag{1.4}$$

$$\alpha = \frac{\beta}{1+\beta} \tag{1.5}$$

1.2 TRANSISTOR BIASING

Transistors can be biased by (i) base, (ii) collector-to-base, and (iii) voltage divider bias methods.

The transistor base bias circuit is shown in Figure 1.2. From Figure 1.2

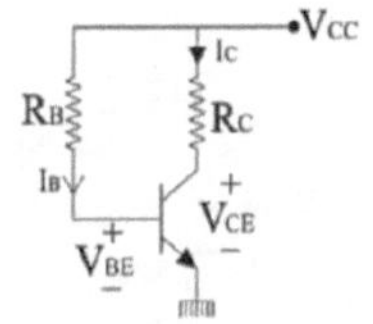

FIGURE 1.2 Transistor base bias.

$$I_B = \frac{V_{CC} - V_{BE}}{R_B} \tag{1.6}$$

The V_{BE} is 0.7 V for silicon transistors and 0.3 V for germanium transistors

$$I_C = \frac{V_{CC} - V_{CE}}{R_C} \tag{1.7}$$

$$V_{CE} = V_{CC} - I_C R_C$$

The transistor collector-to-base bias circuit is shown in Figure 1.3. From Figure 1.3.

$$V_{CE} = V_{BE} + I_B R_B$$

$$I_B = \frac{V_{CE} - V_{BE}}{R_B} \tag{1.8}$$

$$V_{CE} = V_{CC} - R_C\left(I_C + I_B\right)$$

$$V_{BE} + I_B R_B = V_{CC} - R_C\left(I_C + I_B\right)$$

$$V_{BE} + I_B R_B = V_{CC} - R_C\left(\beta I_B + I_B\right) \tag{1.9}$$

The above equation can be simplified as

$$I_B = \frac{V_{CC} - V_{BE}}{R_B + R_C\left(1+\beta\right)} \tag{1.10}$$

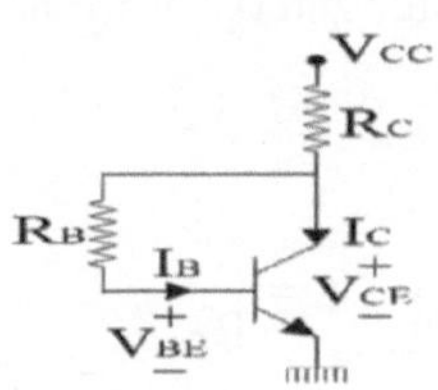

FIGURE 1.3 Transistor collector-to-base bias.

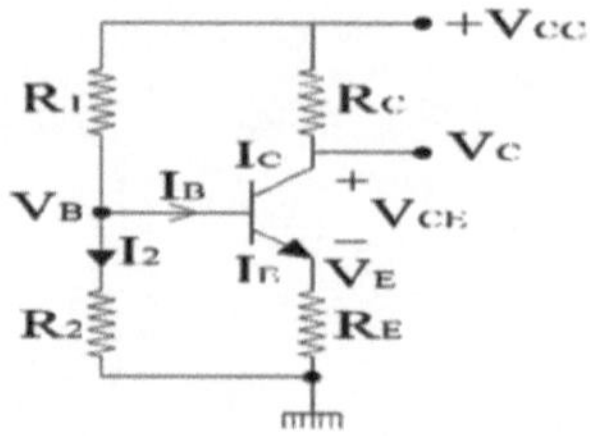

FIGURE 1.4 Transistor voltage divider bias.

The transistor voltage divider bias circuit is shown in Figure 1.4.

$$V_B = \frac{V_{CC}}{R_1 + R_2} R_2 \tag{1.11}$$

$$V_E = V_B - V_{BE} \tag{1.12}$$

$$I_E = \frac{V_E}{R_E} = \frac{V_B - V_{BE}}{R_E} \tag{1.13}$$

Choose I_2 such that

$$I_2 \sim \frac{I_C}{10} \tag{1.14}$$

$$V_C = V_{CC} - I_C R_C$$

$$V_{CE} = V_C - V_E$$

The collector current is approximately equal to emitter current

$$V_{CE} \approx V_{CC} - I_C \left(R_C + R_E\right) \tag{1.15}$$

1.3 TRANSISTOR SWITCHES

Figure 1.5 shows a simple transistor shunt switch. Let $V_C = \text{HIGH}(+V_{CC})$, the transistor base emitter terminals are forward biased and hence the transistor is ON.

$V_{BE} = 0.7$ V for silicon transistor and 0.3 V for germanium transistor

$$V_{CE} \approx 0\,\text{V}$$

$$I_C = \frac{V_1}{R_C} \tag{1.16}$$

$$I_C = \beta I_B$$

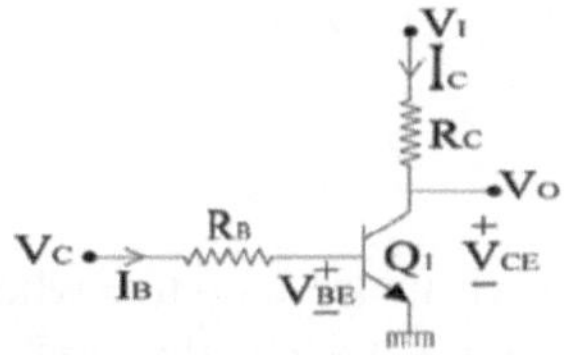

FIGURE 1.5 Transistor shunt switch.

$$V_O \sim O\,V$$

$$V_O \simeq O\,V \tag{1.17}$$

Let V_C = LOW ($-V_{CC}$), the transistor base emitter terminals are reverse biased and hence the transistor is OFF.

$$I_B = O$$

$$V_{CE} \approx V_I$$

$$I_C = O$$

$$V_O \sim V_{CE} = V_I$$

From the above, it is understood that when control input V_C is HIGH, the output voltage is zero volts, and when control input V_C is LOW, the output voltage is approximately equal to input voltage.

Figure 1.6 shows a transistor series switch. Let control input V_C be HIGH ($+V_{CC}$); the transistor base emitter junction is forward biased and hence it is ON

$$V_E \simeq V_I$$

$$V_O = V_I \tag{1.18}$$

Next, let control input V_C be LOW ($-V_{CC}$); the transistor base emitter junction is reverse biased and hence it is OFF

$$I_B = O$$

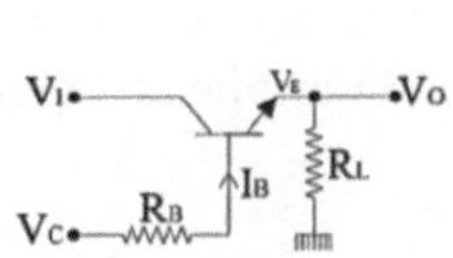

FIGURE 1.6 Transistor series switch.

$$V_E \neq V_I$$

$$V_O = O$$

From the above discussions, it is understood that when control input V_C is HIGH, the transistor is ON, and the output voltage is the input voltage. When control input V_C is LOW, the transistor is OFF, and the output voltage is zero volts.

1.4 TRANSISTOR MULTIPLEXERS

Figure 1.7 shows an analog multiplexer using transistors. If the control input CON is HIGH ($+V_{CC}$), transistor Q_1 is ON, Q_2 is OFF, and (V_1) will appear at V_N. If the control input CON is LOW ($-V_{CC}$), transistor Q_1 is OFF, Q_2 is ON, and (V_2) will appear at V_N. The operation is illustrated in Table I.

1.5 FIELD EFFECT TRANSISTORS

Like a bipolar transistor, field effect transistor (FET) is also a voltage-operated device to use in amplifiers and switching circuits. The advantages of an FET over a bipolar junction transistor (BJT) are

i. FET requires virtually no input current, and
ii. FET has high input resistance.

The two types of FETs are

i. Junction FET (JFET) and
ii. Metal-oxide-semiconductor field effect transistor (MOSFET).

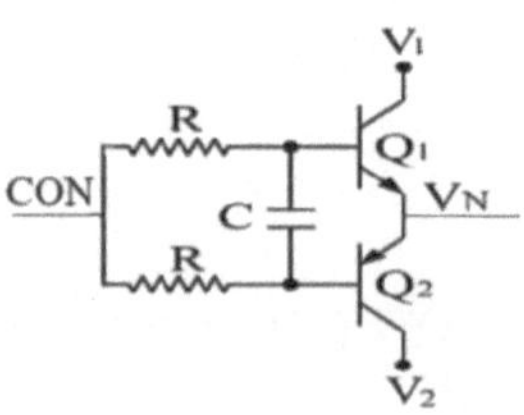

FIGURE 1.7 Transistor multiplexer.

TABLE I
Transistor Multiplexer Operation

CON	V_N
HIGH	V_1
LOW	V_2

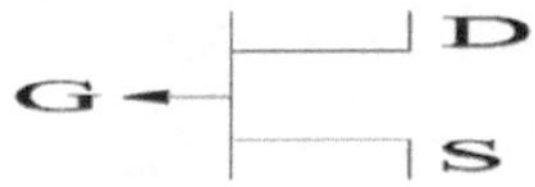

FIGURE 1.8(A) Symbol of n channel JFET.

FIGURE 1.8(B) Symbol of p channel JFET.

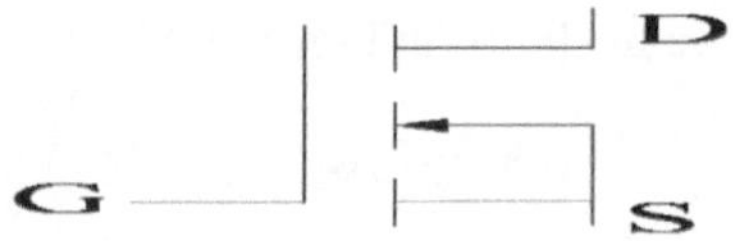

FIGURE 1.9(A) Symbol of n channel MOSFET.

FIGURE 1.9(B) Symbol of p channel MOSFET.

The two types of JFETs are

i. n channel JFET and
ii. p channel JFET.

Their symbols are shown in Figures 1.8(a) and 1.8(b).

The two types MOSFETs are

i. n channel MOSFET and
ii. p channel MOSFET.

Their symbols are shown in Figures 1.9(a) and 1.9(b).

1.6 JFET BIASING

Figure 1.10 shows JFET gate bias circuit.

$$V_{DS} = V_{DD} - I_D R_D \tag{1.19}$$

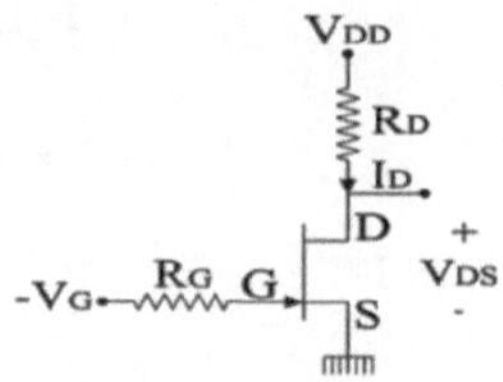

FIGURE 1.10 JFET gate bias circuit.

$$I_D = \frac{V_{DD} - V_{DS}}{R_D} \tag{1.20}$$

Consider the self-bias circuit shown in Figure 1.11.

$$V_S = I_D R_S \tag{1.21}$$

$$I_D = I_S$$

$$V_{GS} = -V_S = -I_D R_S$$

$$V_{DD} = I_D R_D + V_{DS} + I_D R_S \tag{1.22}$$

Consider the voltage divider bias circuit shown in Figure 1.12.

$$I_D = I_S$$

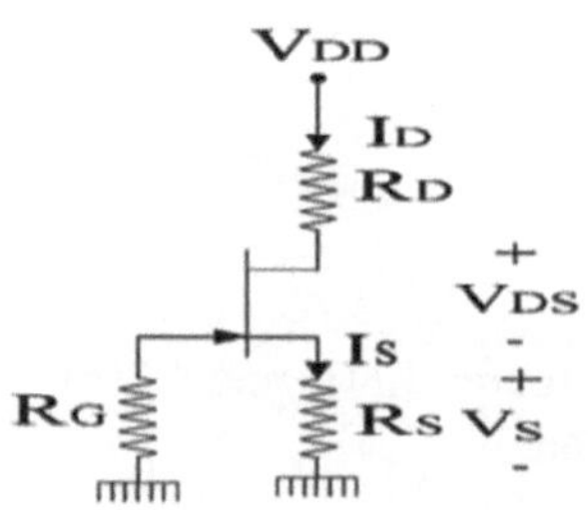

FIGURE 1.11 JFET self-bias circuit.

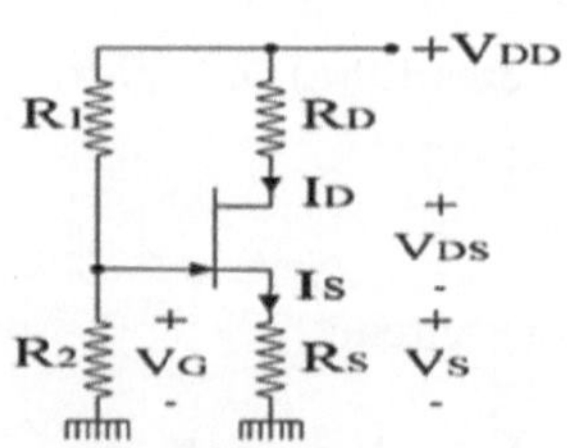

FIGURE 1.12 JFET voltage divider bias circuit.

$$V_{GS} = V_G - I_D R_S \tag{1.23}$$

$$V_G = \frac{V_{DD}}{R_1 + R_2} R_2 \tag{1.24}$$

1.7 JFET SWITCHES

Figure 1.13(a) shows junction field effect transistor (JFET) as a series switch. If the control input CON is HIGH ($+V_{CC}$), zero volts will exist on gate terminal, JFET is ON and acts as a closed switch. The output will be OUT ~ IN. If the control input CON is LOW ($-V_{CC}$), negative voltage will exist on gate terminal, JFET is OFF and acts as an open switch. The output will be OUT ~ 0.

Figure 1.13(b) shows a JFET shunt switching circuit. If the control input CON is HIGH ($+V_{CC}$), zero volts will exist on gate terminal, FET is ON and zero volts will be the output, OUT ~ 0. If the control input CON is LOW ($-V_{CC}$), negative voltage will exist on gate terminal, the FET operated on cut-off region and acts as an open circuit. The output will be OUT ~ IN.

1.8 JFET MULTIPLEXERS

Figure 1.14 shows analog multiplexer using FETs. If the control input CON is HIGH ($+V_{CC}$), FET Q_1 is ON, Q_2 is OFF, and (V_1) will appear at V_N. If the control input CON is LOW ($-V_{CC}$), FET Q_1 is OFF, Q_2 is ON, and (V_2) will appear at V_N. The operation is illustrated in Table II.

1.9 MOSFET BIASING

Consider the MOSFET circuit shown in Figure 1.15.

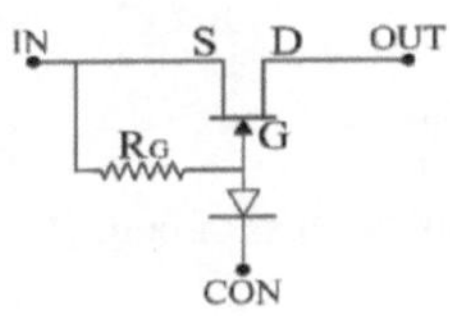

FIGURE 1.13(A) JFET series switch.

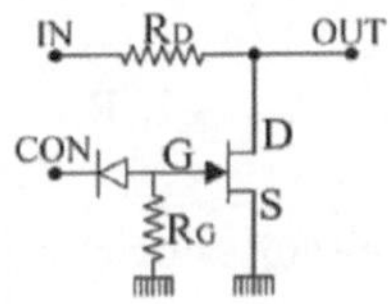

FIGURE 1.13(B) JFET shunt switch.

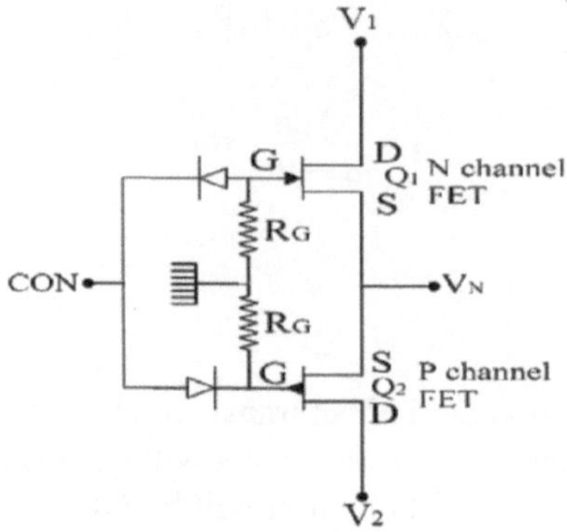

FIGURE 1.14 FET multiplexer.

TABLE II
JFET Multiplexer Operation

CON	V_N
HIGH	V_1
LOW	V_2

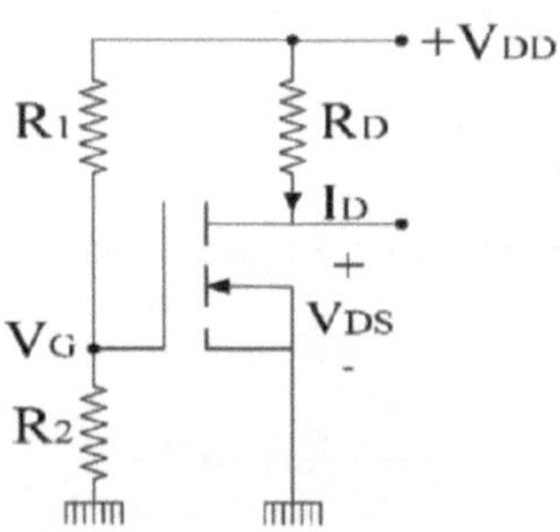

FIGURE 1.15 MOSFET voltage divider bias circuit.

$$V_G = \frac{V_{DD}}{R_1 + R_2} R_2 \tag{1.25}$$

$$V_{DS} = V_{DD} - I_D R_D \tag{1.26}$$

Consider the drain-to-gate bias circuit shown in Figure 1.16.

$$V_{GS} = V_{DS} \tag{1.27}$$

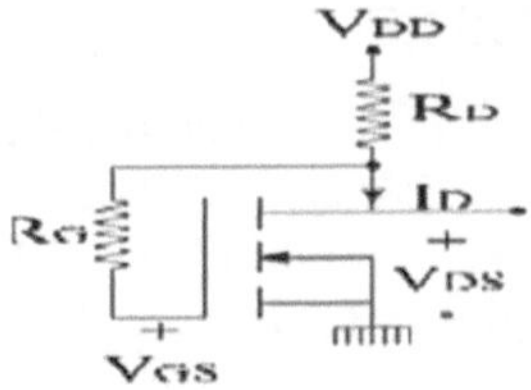

FIGURE 1.16 MOSFET drain-to-gate bias circuit.

$$V_{GS} = V_{DD} - I_D R_D \tag{1.28}$$

1.10 MOSFET SWITCHES

Figure 1.17(a) shows a MOSFET series switch. If the control input is HIGH (+V_{DD}), the channel resistance becomes so small and allows maximum drain current to flow. This is the saturation mode and the MOSFET is completely ON and acts as a closed circuit. The output will be OUT ~ IN. If the control input is LOW (V_{SS}), the channel resistance becomes HIGH, and no current flows from the drain. This is a cut-off region, and MOSFET is completely OFF and acts as an open switch. The output will be OUT ~ 0.

Figure 1.17(b) shows a MOSFET shunt switch. If the control input is HIGH (+V_{DD}), the channel resistance becomes so small and allows maximum drain current to flow. This is the saturation mode and the MOSFET is completely ON and acts as a closed circuit. The output will be OUT ~ 0. If the control input is LOW (V_{SS}), the channel resistance becomes HIGH, and no current flows from the drain. This is a cut-off region and MOSFET is completely OFF and acts as an open switch. The output will be OUT ~ IN.

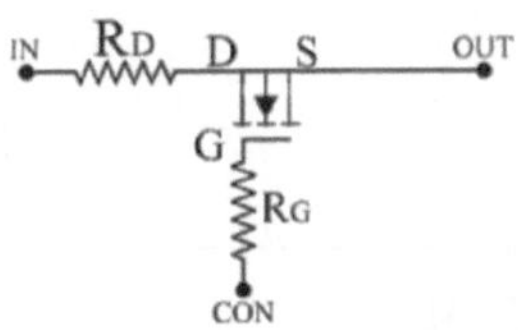

FIGURE 1.17(A) MOSFET series switch.

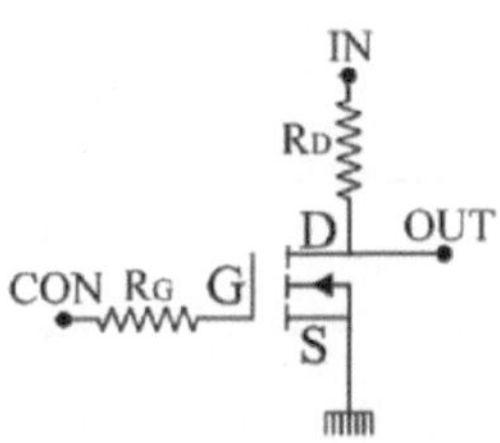

FIGURE 1.17(B) MOSFET shunt switch.

1.11 MOSFET MULTIPLEXERS

Figure 1.18 shows an analog multiplexer using MOSFETs. If the control input CON is HIGH ($+V_{CC}$), MOSFET Q_1 is ON, Q_2 is OFF, and (V_1) will appear at V_N. If the control input CON is LOW ($-V_{CC}$), MOSFET Q_1 is OFF, Q_2 is ON, and (V_2) will appear at V_N. The operation is illustrated in Table III.

1.12 ANALOG SWITCH INTEGRATED CIRCUITS

The symbol of an analog switch is shown in Figure 1.19. It has three terminals: CON, IN/OUT, and OUT/IN. If control (CON) pin is LOW, the switch S_1 is opened so that IN/OUT and OUT/IN terminals are disconnected. If the control (CON) pin is HIGH, the switch S_1 is closed so that IN/OUT and OUT/IN terminals are connected.

Analog switches are available in an IC PACKAGE of CMOS CD4066 IC. The pin details of this CD4066 IC are given in Figure 1.20.

The symbol of an inverted controlled analog switch is shown in Figure 1.21. It has three terminals: CON, IN/OUT, and OUT/IN. If control (CON) pin is HIGH, the

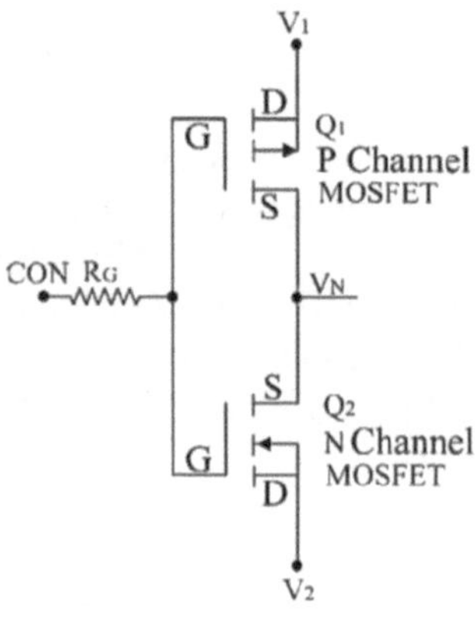

FIGURE 1.18 MOSFET multiplexer.

TABLE III
MOSFET Multiplexer Operation

CON	V_N
HIGH	V_1
LOW	V_2

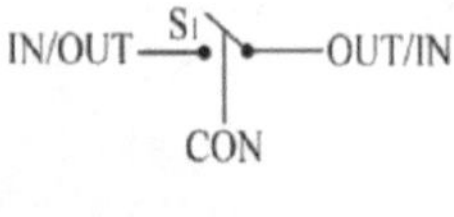

FIGURE 1.19 Non-inverted controlled switch symbol.

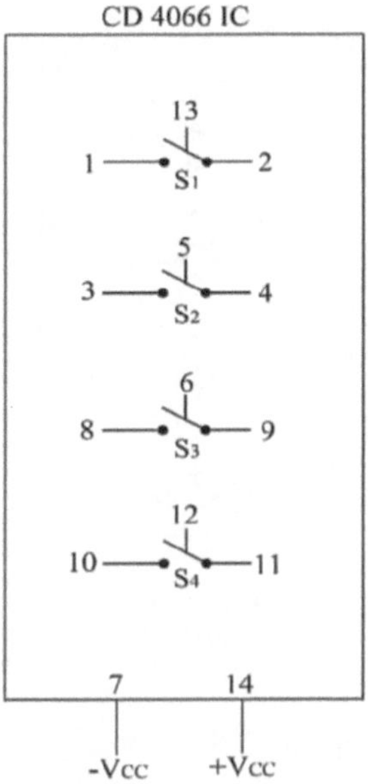

FIGURE 1.20 Pin details of CD4066 IC.

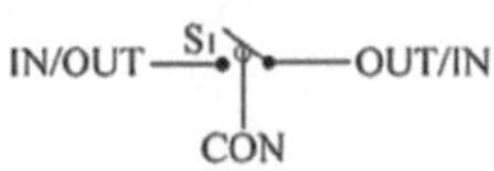

FIGURE 1.21 Inverted controlled switch symbol.

switch S_1 is opened so that IN/OUT and OUT/IN terminals are disconnected. If the control (CON) pin is LOW, the switch S_1 is closed so that IN/OUT and OUT/IN terminals are connected.

Inverted controlled analog switches are available in an IC PACKAGE of DG201 IC. The pin details of this DG201 IC are given in Figure 1.22.

1.13 ANALOG MULTIPLEXER IC CD4053

Figure 1.23 shows the symbol of analog triple 2 to 1 multiplexer. Each multiplexer has four terminals. In case of multiplexer M_1, it has 'ay', 'ax', 'a', and 'A' terminals. In case of multiplexer M_2, it has 'by', 'bx', 'b', and 'B' terminals. In case of multiplexer M_3, it has 'cy', 'cx', 'c', and 'C' terminals.

In multiplexer M_1, if the pin 'A' is HIGH, then 'ay' is connected to 'a', and if the pin 'A' is LOW, then 'ax' is connected to 'a'. In multiplexer M_2, if the pin 'B' is HIGH, then 'by' is connected to 'b', and if the pin 'B' is LOW, then 'bx' is connected to 'b'. In multiplexer M_3, if the pin 'C' is HIGH, then 'cy' is connected to 'c', and if the pin 'C' is LOW, then 'cx' is connected to 'c'.

All three multiplexers M_1, M_2 and M_3 are available in one IC PACKAGE of CMOS CD4053 IC. The pin details of this CD4053 IC are given in Figure 1.24.

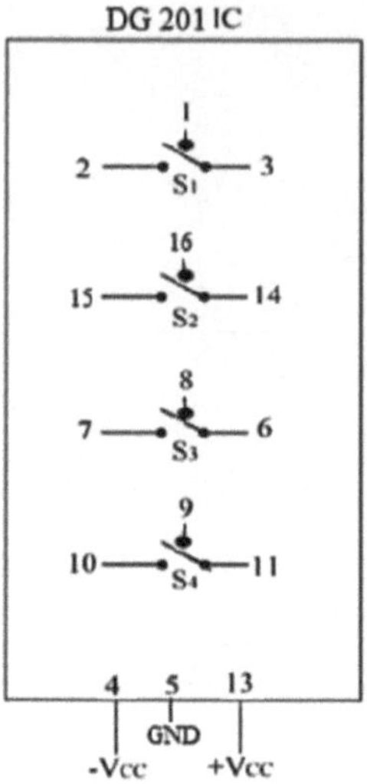

FIGURE 1.22 Pin details of DG201 IC.

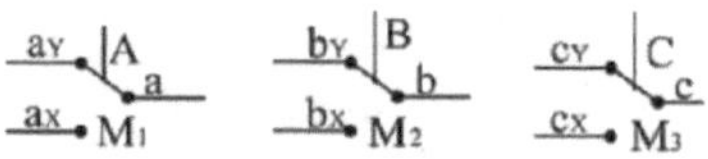

FIGURE 1.23 Triple 2 to 1 multiplexers.

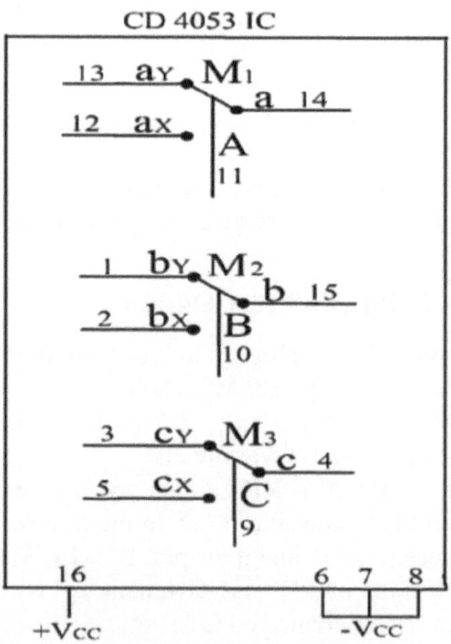

FIGURE 1.24 Pin details of CD4053 IC.

1.14 OPERATIONAL AMPLIFIERS

The operational amplifier (op-amp) is a powerful tool of electronic components and ICs. The op-amp was invented by John R. Ragazzini in 1947. LM 741 is the most popular op-amp IC.

The op-amp symbol is shown in Figure 1.25. Its 8 pin IC and its pin details are shown in Figure 1.26.

Pin 2: Inverting input terminal
Pin 3: Non-inverting input terminal
Pin 6: Output terminal
Pin 4: $-V_{CC}$ (negative power supply) terminal
Pin 7: $+V_{CC}$ (positive power supply) terminal

The other terminals are used for frequency compensation and offset adjustment.

The op-amp characteristics are described below:

i. Infinite input impedance: The signal current through the non-inverting terminal and the signal current through the inverting terminal are to be zero, as the op-amp has infinite input impedance.
ii. Zero output impedance: The output voltage should be $V_O = A(V_2 - V_1)$ where V_2 is the voltage at the non-inverting terminal, V_1 is the voltage at the inverting terminal, and A is the gain. This should be independent of the current drawn from output terminal to load impedance. That is, the op-amp has zero output impedance.
iii. Zero common-mode gain or infinite common-mode rejection: The op-amp output responds only to the difference $(V_2 - V_1)$ and ignores any signal common

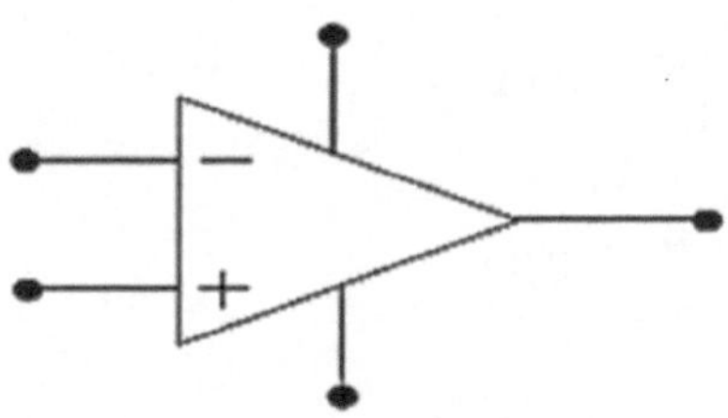

FIGURE 1.25 Op-amp symbol.

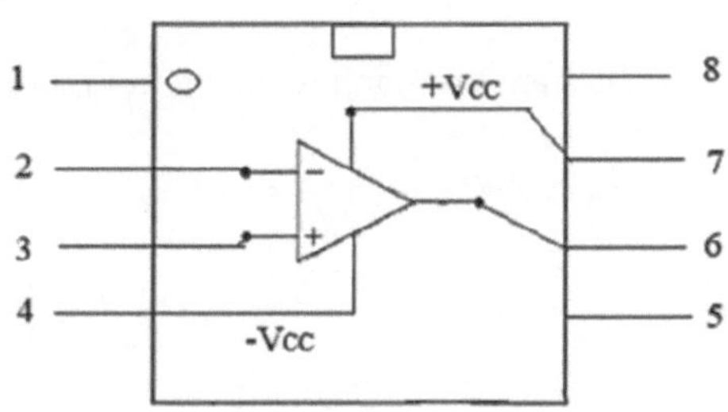

FIGURE 1.26 Op-amp IC.

to both input terminals, $V_1 = V_2$, then output should be ideally zero. This is called common-mode rejection, and an ideal op-amp should have zero common-mode gain.

iv. Infinite open-loop gain: The gain A is called differential gain or open-loop gain. This should be high or infinite.

v. Infinite bandwidth: The open-loop gain A of op-amp should remain constant from zero frequencies to infinite frequency. That is, ideal op-amp will amplify signals of any frequency with equal gain and thus has infinite bandwidth.

WORKED PROBLEMS

1. In the transistor circuit shown in Figure 1.27, find I_C, I_E and I_B.

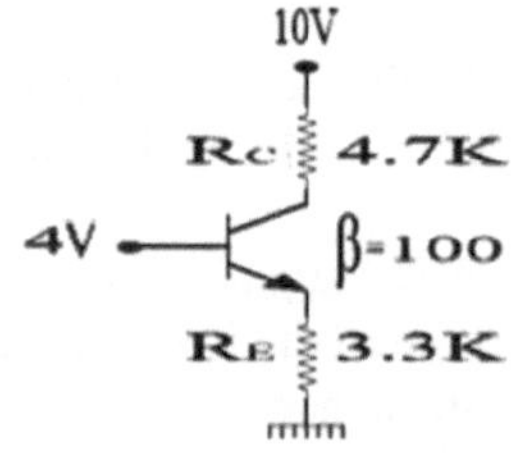

FIGURE 1.27

$$4V = V_{BE} + V_{RE}$$

$$V_{RE} = 4V - V_{BE} = 3.3\ V$$

$$I_E = \frac{V_{RE}}{R_E} = 1\ \mu A$$

$$\alpha = \frac{\beta}{1+\beta} = 0.99$$

$$I_C = \alpha I_E = 0.99\ mA$$

$$I_B = I_E - I_C = 1\ mA - 0.99\ mA = 0.01\ mA$$

2. Find I_C, I_B and I_E in the transistor circuit shown in Figure 1.28

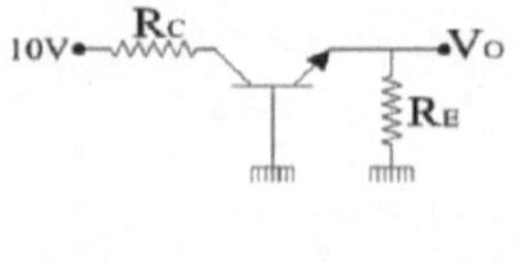

FIGURE 1.28

The transistor is at a cut-off region.

$$I_C = I_B = I_E = O\ mA$$

3. Find I_C, I_B and I_E in the transfer circuit shown in Figure 1.29.

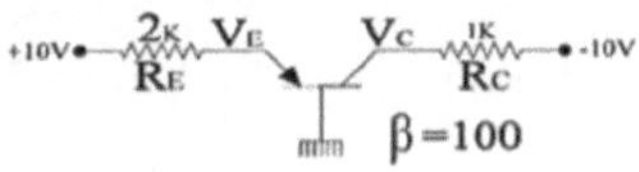

FIGURE 1.29

For silicon transistor

$$V_E = V_{BE} = 0.7 \text{ V}$$

$$I_E = \frac{10-0.7}{2} = 4.65 \text{ mA}$$

$$\alpha = \frac{\beta}{1+\beta} = \frac{100}{101} = 0.99$$

$$I_C = \alpha I_E = 4.6 \text{ mA}$$

$$I_B = I_E - I_C = 0.05 \text{ mA}$$

4. Find I_C, I_B and I_E in the transistor circuit shown in Figure 1.30.

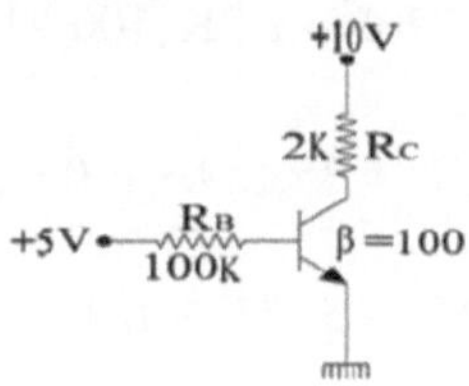

FIGURE 1.30

$$I_B = \frac{+5-0.7}{100K} = 43 \text{ µA}$$

$$I_C = \beta I_B = 4.3 \text{ mA}$$

$$\therefore I_E = I_B + I_C = 4.343 \text{ mA}$$

5. Find V_{CE} in the transistor shown in Figure 1.31.

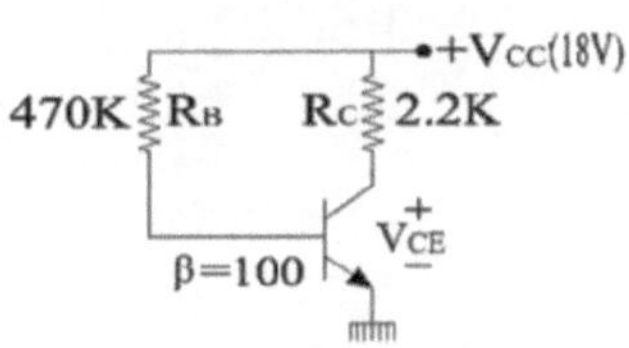

FIGURE 1.31

$$I_B = \frac{V_{CC} - V_{BE}}{R_B} = \frac{18-0.7}{470K} = 36.8 \mu A$$

$$I_C = \beta I_B = 100 \times 36.8\ \mu A = 3.68\ mA$$

$$V_{CE} = V_{CC} - (I_C R_C) = (18 - 3.68\ mA \times 2.2K) = 9.9V$$

6. Find I_B, I_C and V_{CE} in the transistor circuit shown in Figure 1.32.
From Eq. (1.10)

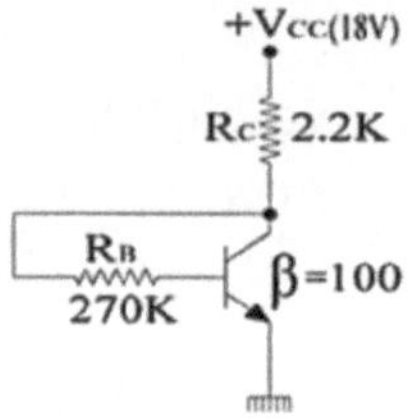

FIGURE 1.32

$$I_B = \frac{V_{CC} - V_{BE}}{R_B + R_C(\beta + 1)}$$

$$= \frac{18V - 0.7V}{270K + 2.2K(100 + 1)} = 35.1\ \mu A$$

$$I_C = \beta I_B = 35.1\mu A \times 100 = 3.51\ mA$$

$$V_{CE} = V_{CC} - R_C(I_C + I_B)$$

$$= 18 - 2.2K(3.51mA + 35.1\mu A)$$

$$= 10.2V$$

7. Find R_1, R_2, R_3 and R_4 in the transistor circuit shown in Figure 1.33.

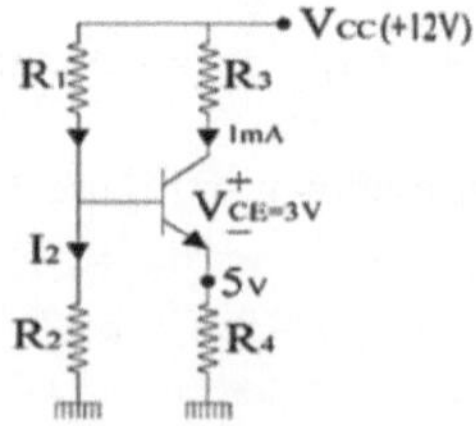

FIGURE 1.33

$$I_C = 1mA$$

$$I_E \approx 1mA$$

$$R_4 = \frac{V_E}{I_E} = \frac{5V}{1_{mA}} = 5K$$

$$V_C = V_{CE} + V_E = 3\,V + 5\,V = 8\,V$$

$$V_{R3} = V_{CC} - V_C = 12\,V - 8\,V = 4\,V$$

$$R_3 = \frac{V_{R3}}{I_C} = \frac{4V}{1_{mA}} = 4K$$

$$V_B = V_{BE} + V_E = 0.7V + 5V = 5.7\,V$$

From Eq. (1.14)

$$I_2 = \simeq \frac{I_C}{10} = \frac{1\text{ mA}}{10} = 100\ \mu A$$

$$R_2 = \frac{V_B}{I_2} = \frac{5.7V}{100\mu A} = 57\ K\Omega,\quad R_1 = \frac{V_{CC} - V_B}{I_2} = \frac{12V - 5.7V}{100\mu A} = 63\ K\Omega$$

8. Find I_2 in the circuit shown in Figure 1.34.

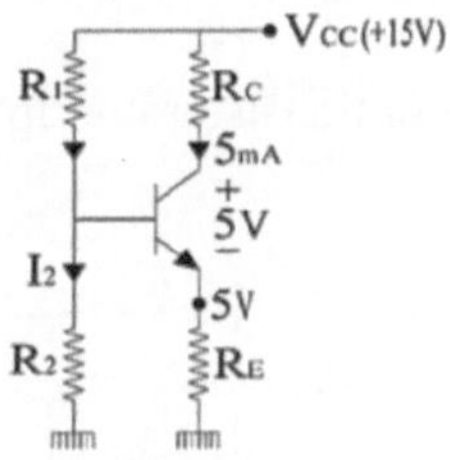

FIGURE 1.34

$$V_E = 5\ V,\ V_{CE} = 5\ V,\ I_C = 5\text{ mA}$$

$$I_E \simeq I_C = 5\text{ mA}$$

$$I_2 = \frac{I_C}{10} = \frac{5\text{ mA}}{10} = 500\ \mu A$$

TUTORIAL EXERCISES

1. Calculate β of a transistor whose collector current is measured to be 1mA and base current to be 25 μA.
   [Ans. β = 40]
2. A base bias circuit has $V_{CC} = 24$ V, $R_B = 390$ KΩ, $R_C = 3.3$ KΩ and $V_{CE} = 10$ V. Calculate the transistor h_{FE} value.
   [Ans. 71]

3. A collector-to-base bias V_{CC} = 24 V, R_B = 180 K, R_C = 3.3 K and V_{CE} is measured to be 10 V. Calculate the transistor h_{FE} value.
[Ans. 81]
4. A voltage divider bias circuit has V_{CC} = 20V, R_1 = 100 KΩ, R_2 = 33 KΩ, R_E = 3.9 KΩ and R_C = 6.8 KΩ. Calculate the appropriate values of V_E and $V_{C.}$
[Ans. 4.26 V, 12.50 V]
5. Find the output voltages at (a, b, c) of CD4053 multiplexers IC, as shown in Figure 1.35.

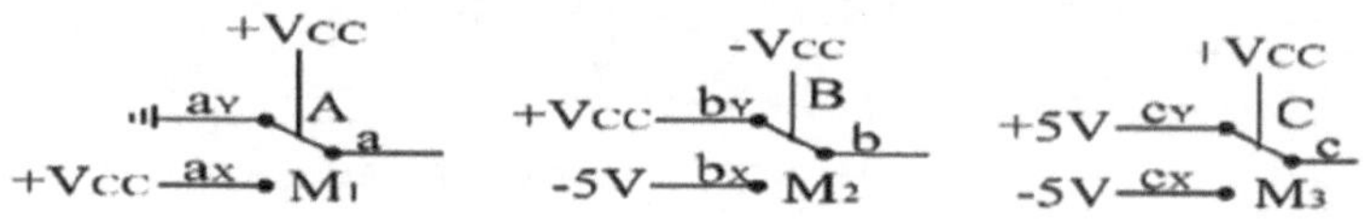

FIGURE 1.35

[Ans. 0 V, −5 V, +5 V]
6. What is the voltage across the resistor R_L in Figure 1.36?

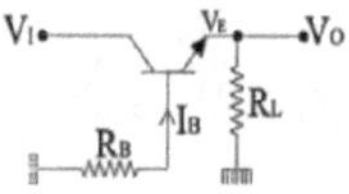

FIGURE 1.36

[Ans. 0 V]
7. Find the output voltage of transistor multiplexer shown in Figure 1.37.

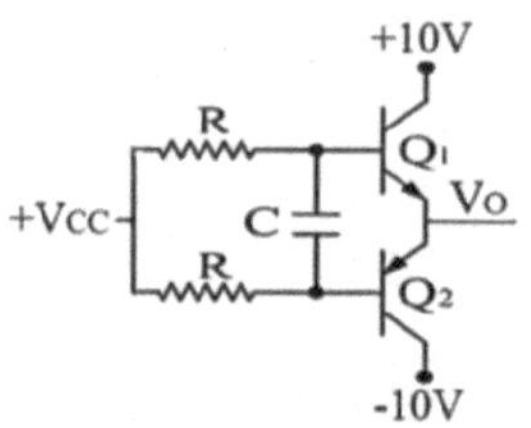

FIGURE 1.37

[Ans. 10 V]

2 Linear Circuits

Operational amplifier circuits are used in almost all modern communication, instrumentation, and computation systems. The output of an amplifier is proportional to input, and the proportional constant is called the "gain." An IC op-amp is made up of a large number of transistors, resistors, and the capacitor connected in a complex circuit. In this chapter we will discuss how operational amplifiers are used for buffers, inverting amplifiers, non-inverting amplifiers, inverting adders, non-inverting adders, differential amplifiers, instrumentation amplifiers, voltage-to-current (V/I) converters, current-to-voltage (I/V) converters, integrators, differential integrators, and differentiators. Many linear and non-linear analog systems are constructed with op-amp as a basic building block. Linear op-amp systems have negative feedback configuration and are described in this chapter.

2.1 BUFFER

Figure 2.1 shows op-amp buffer or voltage follower. The output terminal is connected directly to the inverting input terminal, the input voltage is applied to non-inverting terminal, and the load is directly coupled to the output. The buffer has high-input impedance, low-output impedance, and a voltage gain of 1. The buffer is in a negative closed loop feedback, and hence its inverting terminal voltage is equal to its non-inverting terminal voltage, i.e., $V_O = V_I$.

2.2 INVERTING AMPLIFIER

Figure 2.2 shows an inverting amplifier using op-amp. Since the non-inverting terminal (+) of op-amp is grounded through the resistor R_P, the voltage V_P at non-inverting terminal (+) of op-amp is zero volts. The op-amp is in a negative closed loop feedback, and hence its non-inverting terminal voltage will equal to its inverting terminal voltage, i.e., $V_N = V_P = 0$

The current through resistor R_1 will be

$$I = \frac{V_I - V_N}{R_1} = \frac{V_I}{R_1} \tag{2.1}$$

Since the op-amp has infinite input impedance, the current I does not enter into op-amp and flows through resistor R_2. The voltage across resistor R_2 will be

$$V_{R2} = IR_2 = \frac{V_I}{R_1} R_2 \tag{2.2}$$

DOI: 10.1201/9781003221449-3

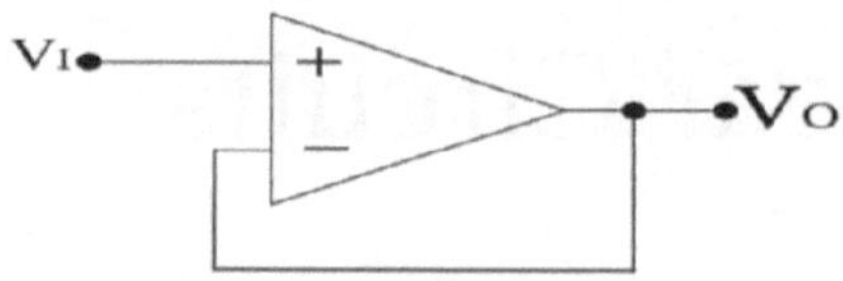

FIGURE 2.1 Op-amp buffer.

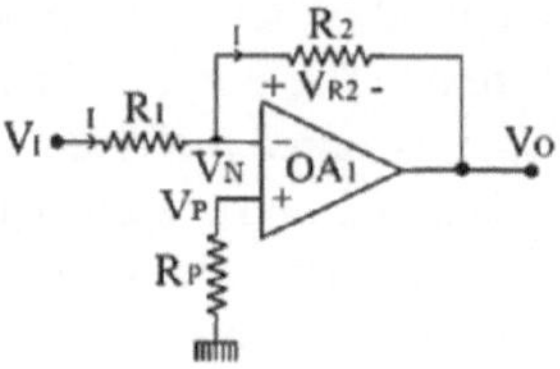

FIGURE 2.2 Inverting amplifier.

The negative feedback forces the op-amp to produce an output voltage that maintains a virtual ground at the op-amp inverting input. The output voltage is given as

$$V_O = V_N - V_{R2} = -V_{R2}$$

$$V_O = -\left(\frac{R_2}{R_1}\right)V_I \tag{2.3}$$

If $R_1 = R_2$ then

$$V_O = -V_I \tag{2.4}$$

2.3 NON-INVERTING AMPLIFIER

Figure 2.3 shows a non-inverting amplifier – type I. The op-amp is at a negative closed loop feedback, and its inverting terminal (–) voltage will equal its non-inverting terminal (+) voltage, i.e., $V_N = V_P = V_I$

The current through resistor R_1 will be

$$I = \frac{V_N}{R_1} = \frac{V_I}{R_1} \tag{2.5}$$

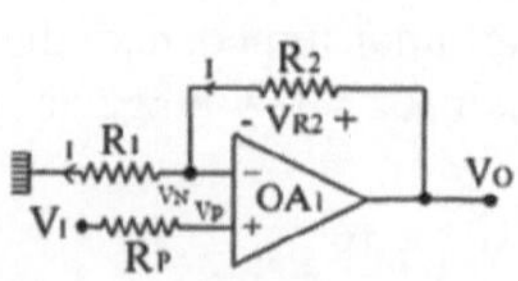

FIGURE 2.3 Non-inverting amplifier (type I).

The current I comes from resistor R_2 and it does not enter op-amp, as op-amp has infinite input impedance. The voltage across feedback resistor R_2 will be

$$V_{R2} = IR_2 = \frac{V_I}{R_1} R_2 \tag{2.6}$$

The output voltage is given as

$$V_O = V_N + V_{R2}$$

$$V_O = V_I + \left(\frac{R_2}{R_1} V_I \right)$$

$$V_O = V_I \left(1 + \frac{R_2}{R_1} \right) \tag{2.7}$$

If $R_1 = R_2$, then

$$V_O = 2V_I \tag{2.8}$$

Figure 2.4 shows unity gain non-inverting amplifier – type II. Let us analyze the circuit using the superposition principle.

First let the non-inverting terminal (+) be grounded through resistor R_P and the input voltage V_I be given to inverting terminal through resistor R_1. The circuit will work as an inverting amplifier as shown in Figure 2.2, and if $R_1 = R_2$, the output voltage will be

$$V_{O1} = -V_I \tag{2.9}$$

Next, the inverting terminal (−) is grounded through resistor R_1 and input voltage is given to non-inverting terminal (+) through resistor R_P. The circuit will work as non-inverting amplifier as shown in Figure 2.3, and if $R_1 = R_2$, the output voltage will be

$$V_{O2} = 2V_I \tag{2.10}$$

By the superposition principle, the actual output voltage will be addition of Eqs. (2.9) and (2.10)

$$V_O = V_{O1} + V_{O2} \tag{2.11}$$

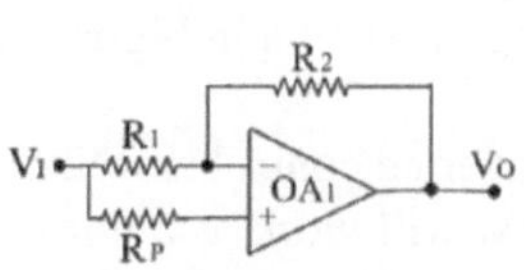

FIGURE 2.4 Unity gain non-inverting amplifier (type II).

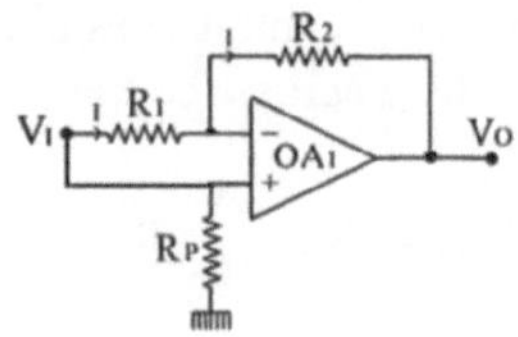

FIGURE 2.5 Unity gain non-inverting amplifier (type III).

$$V_O = V_I \tag{2.12}$$

Figure 2.5 shows a unity gain non-inverting amplifier – type III. Let us analyze the circuit using the superposition principle.

First let the non-inverting terminal (+) be grounded and the input voltage V_I be given to the inverting terminal through resistor R_1. The circuit will work as an inverting amplifier as shown in Figure 2.2 ('+' terminal may be directly connected to ground (GND) or through resistor R_P) and if $R_1 = R_2$, the output voltage will be

$$V_{O1} = -V_I \tag{2.13}$$

Next, the inverting terminal (−) is grounded through resistor R_1 and the input voltage is given to the non-inverting terminal (+). The circuit will work as a non-inverting amplifier as shown in Figure 2.3 ('+' terminal may be directly connected to V_2 or through resistor R_P) and if $R_1 = R_2$, the output voltage will be

$$V_{O2} = 2V_I \tag{2.14}$$

By the superposition principle, the actual output voltage will be addition of Eqs. (2.13) and (2.14)

$$V_O = V_{O1} + V_{O2} \tag{2.15}$$

$$V_O = V_I \tag{2.16}$$

2.4 CONTROL AMPLIFIER

Figure 2.6(a) shows an inverted control amplifier. When control input is HIGH, switch S_1 is closed, the non-inverting terminal is grounded, op-amp OA_1 will work as an inverting amplifier as shown in Figure 2.2, and $(-V_1)$ will appear at the output of op-amp OA_1. When control input is LOW, switch S_1 is opened, op-amp OA_1 will work as a non-inverting amplifier as shown in Figure 2.4 and $(+V_1)$ will appear at the output of op-amp OA_1.

Figure 2.6(b) shows a non-inverted control amplifier. When control input is HIGH, switch S_1 is closed, op-amp OA_1 will work as a non-inverting amplifier as shown in Figure 2.5, and $(+V_1)$ will appear at the output of op-amp OA_1. When control input

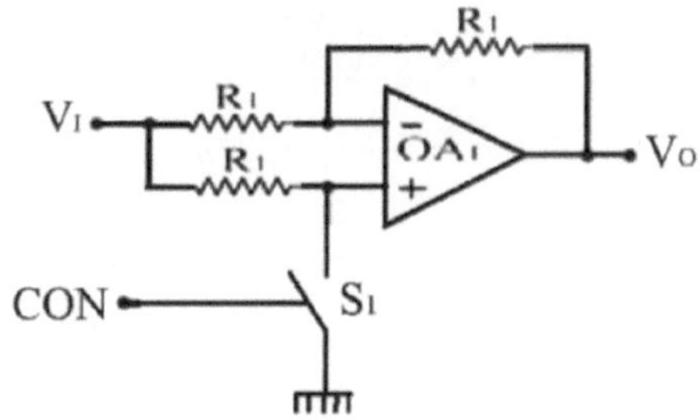

FIGURE 2.6(A) Inverted control amplifier.

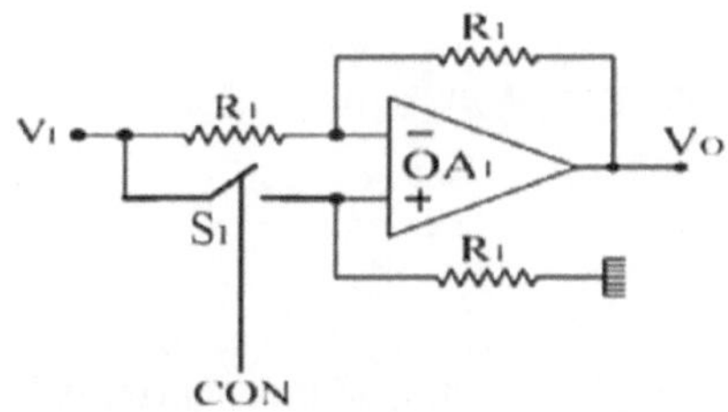

FIGURE 2.6(B) Non-inverted control amplifier.

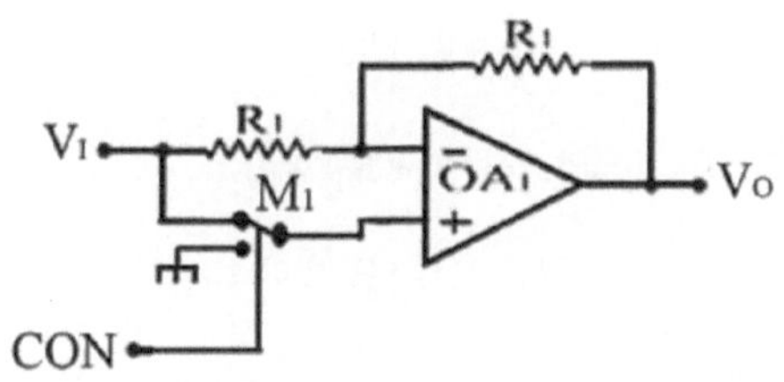

FIGURE 2.7 Control amplifier with multiplexer.

is LOW, switch S_1 is opened, op-amp OA_1 will work as an inverting amplifier as shown in Figure 2.2, and $(-V_1)$ will appear at the output of op-amp OA_1.

Figure 2.7 shows a control amplifier using a multiplexer. When control input CON is HIGH, the multiplexer M_1 connects V_1 to the non-inverting input of op-amp OA_1. The op-amp OA_1 will work as a non-inverting amplifier and $+V_1$ will be at its output. When control input CON is LOW, the multiplexer M_1 connects GND to the non-inverting input of op-amp OA_1. The op-amp OA_1 will work as an inverting amplifier and $-V_1$ will be at its output.

2.5 INVERTING ADDER

Figure 2.8 shows circuit diagram of an inverting adder. The op-amp OA_1 is at a negative closed loop feedback and its inverting terminal voltage will equal its non-inverting terminal voltage. Hence, $V_N = O$ V.

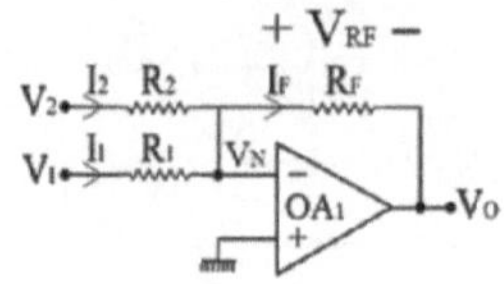

FIGURE 2.8 Inverting adder.

The current through resistor R_1 is given as

$$I_1 = \frac{V_1}{R_1} \tag{2.17}$$

The current through resistor R_2 is given as

$$I_2 = \frac{V_2}{R_2} \tag{2.18}$$

As the op-amp OA_1 has high input impedance, currents I_1 and I_2 are not entering into op-amp OA_1 and pass through the feedback resistor R_F. The current through feedback resistor R_F is given as

$$I_F = I_1 + I_2 \tag{2.19}$$

The voltage across the feedback resistor R_F will be

$$V_{RF} = I_F\left(R_F\right)$$

The output voltage is given as

$$V_O = V_N - V_{RF} = \ I_F\left(R_F\right)$$

$$= -\left(I_1 + I_2\right)R_F$$

$$= -\left(\frac{V_1}{R_1} + \frac{V_2}{R_2}\right)R_F$$

$$= -\left[\frac{R_F}{R_1}V_1 + \frac{R_F}{R_2}V_2\right]$$

Let us assume $R_F = R_1 = R_2$

$$V_O = -\left[V_1 + V_2\right] \tag{2.20}$$

Thus, the output voltage is the addition of all input voltages with sign inversion. Hence, it is called an "inverting adder."

2.6 NON-INVERTING ADDER

Figure 2.9 shows a non-inverting adder. Let us analyze the circuit using the superposition principle. V_2 is grounded and V_1 is acting. The voltage at a non-inverting terminal of op-amp OA_1 will be

$$V_P = V_1/2 \tag{2.21}$$

Next, V_1 is grounded and V_2 is acting. The voltage V_P will be

$$V_P = V_2/2 \tag{2.22}$$

Hence, the actual voltage will be the addition of Eqs. (2.21) and (2.22).

$$V_P = \frac{V_1}{2} + \frac{V_2}{2} = \frac{1}{2}(V_1 + V_2) \tag{2.23}$$

The op-amp OA_1 will work as a non-inverting amplifier with V_P as its input voltage. The output voltage is given as

$$V_O = V_P\left(1 + \frac{R_F}{R_1}\right)$$

Let us assume $R_1 = R_F = R$

$$V_O = V_P[2]$$

$$\therefore V_O = V_1 + V_2 \tag{2.24}$$

Thus, the output voltage is the addition of two input voltages without sign inversion, and hence it is called a "non-inverting adder."

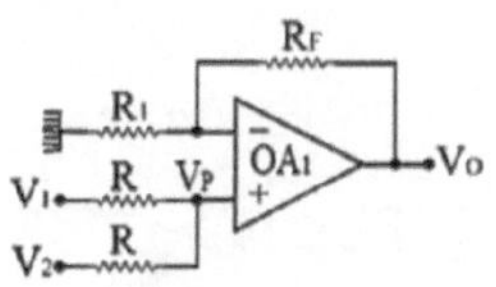

FIGURE 2.9 Non-inverting adder.

2.7 DIFFERENTIAL AMPLIFIER

Figure 2.10 shows a differential amplifier. Let us analyze the circuit using the superposition principle.

Case 1: V_1 is acting and V_2 is grounded. The op-amp OA_1 will work as an inverting amplifier. The output voltage will be

$$V_{O1} = -V_1\left(\frac{R_2}{R_1}\right) \tag{2.25}$$

Case 2: V_1 is grounded and V_2 is acting. The non-inverting terminal voltage of op-amp OA_1 will be

$$V_P = \frac{V_2}{R_1 + R_2} R_2 \tag{2.26}$$

The op-amp OA_1 will work as non-inverting amplifier with V_P as it input voltage. The output voltage is given as

$$V_{O2} = V_P\left(1 + \frac{R_2}{R_1}\right) = \frac{V_P}{R_1}(R_1 + R_2)$$

$$V_{O2} = \frac{V_2 R_2}{R_1(R_1 + R_2)}(R_1 + R_2)$$

$$V_{O2} = V_2\left(\frac{R_2}{R_1}\right) \tag{2.27}$$

The actual output voltage will be

$$V_O = V_{O1} + V_{O2}$$

$$= V_2\frac{R_2}{R_1} - V_1\frac{R_2}{R_1}$$

$$\therefore V_O = (V_2 - V_1)\frac{R_2}{R_1} \tag{2.28}$$

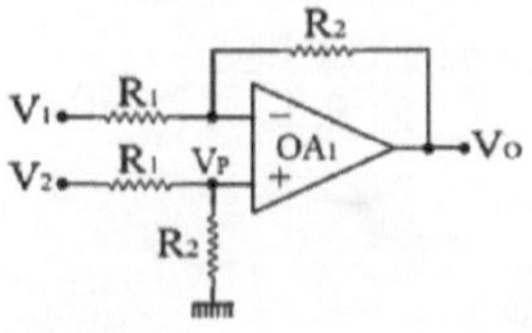

FIGURE 2.10 Differential amplifier.

Thus, the output voltage is the difference of the two input voltages with gain of R_2/R_1.

2.8 INSTRUMENTATION AMPLIFIER

Figure 2.11 shows an instrumentation amplifier.

The voltage across R_N is $V_{RN} = V_1 - V_2$.

The current I through the resistors R_M and R_N will be

$$I = \frac{V_{O1} - V_{O2}}{R_M + R_N + R_M}$$

$$V_{O1} - V_{O2} = I\left(2R_M + R_N\right)$$

The current is also given as

$$I = \frac{V_1 - V_2}{R_N}$$

$$V_{O1} - V_{O2} = \left(V_1 - V_2\right)\left(1 + \frac{2R_m}{R_N}\right)$$

$$V_{O2} - V_{O1} = \left(V_2 - V_1\right)\left(1 + \frac{2R_m}{R_N}\right) \quad (2.29)$$

The op-amp OA_3 functions as a differential amplifier as discussed in Section 2.7. The differential amplifier and hence the output of the instrumentation amplifier will be

$$V_O = \left(V_{O2} - V_{O1}\right)\frac{R_2}{R_1}$$

$$V_O = \left(V_2 - V_1\right)\left(1 + 2\frac{R_m}{R_N}\right)\left(\frac{R_2}{R_1}\right) \quad (2.30)$$

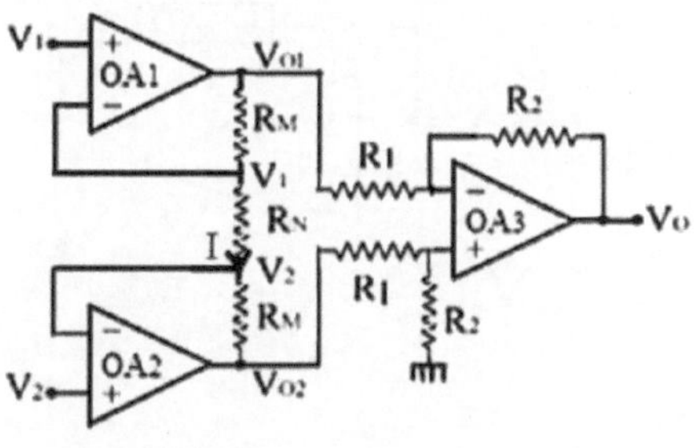

FIGURE 2.11 Instrumentation amplifier.

2.9 VOLTAGE-TO-CURRENT CONVERTER

A voltage-to-current (V/I) converter is called a "transconductance amplifier." It gives an output current proportional to an input voltage. Figure 2.12 shows a simple inverting V/I converter.

$$I = I_L = \frac{V_I}{R} \tag{2.31}$$

The load current is proportional to the input voltage V_I.

Figure 2.13 shows a simple non-inverting V/I converter. The op-amp OA_1 is kept in a negative closed loop configuration. The inverting terminal voltage must equal its non-inverting terminal voltage.

$$I = I_L = \frac{V_I}{R} \tag{2.32}$$

The load current is proportional to the input voltage V_I.

Figures 2.12 and 2.13 are floating load converters. It is better to have a V/I converter for grounded loads. Figure 2.14 shows such V/I converter for grounded loads.

$$I_1 + I_2 = I_L$$

$$I_1 = \frac{V_I - V_P}{R}$$

$$I_2 = \frac{V_O - V_P}{R}$$

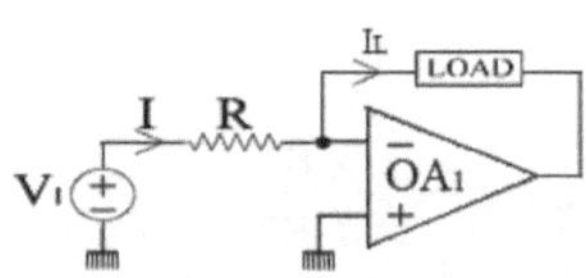

FIGURE 2.12 Inverting voltage-to-current (V/I) converter.

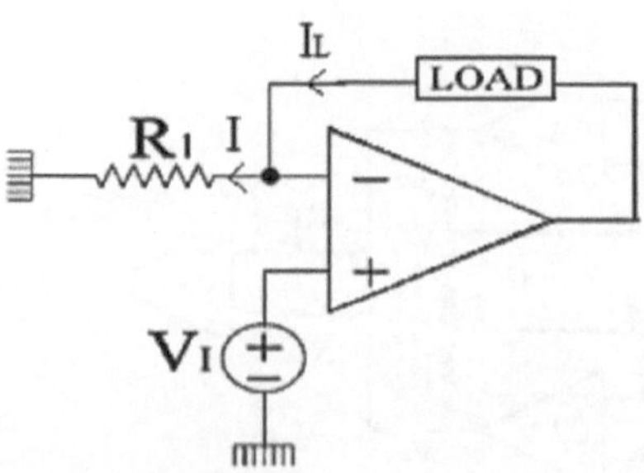

FIGURE 2.13 Non-inverting voltage-to-current (V/I) converter.

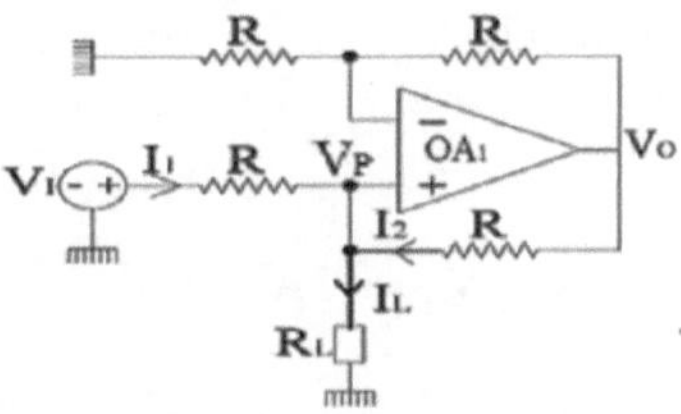

FIGURE 2.14 Grounded load voltage-to-current (V/I) converter.

$$\frac{V_I - V_P}{R} + \frac{V_O - V_P}{R} = I_L$$

$$V_P = \frac{V_I + V_O - I_L R}{2} \quad (2.33)$$

The op-amp OA_1 will work as non-inverting amplifiers for input V_1 with a gain 2 (both input and feedback resistors are equal)

$$\therefore \quad V_O = 2V_P$$

$$V_O = V_1 + V_O - I_L R$$

$$O = V_I - I_L R$$

$$\therefore \quad V_I = I_L R$$

$$\therefore \quad I_L = \frac{V_I}{R} \quad (2.34)$$

Thus the output load current is directly proportional to the input voltage V_I.

2.10 CURRENT-TO-VOLTAGE CONVERTER

The current-to-voltage (I/V) converter is also called a "transresistance amplifier." The output voltage is proportional to the input current. Figure 2.15 shows a simple I/V converter. The op-amp OA_1 is at a negative closed loop configuration. The inverting terminal voltage must equal its non-inverting terminal voltage, i.e., $V_N = 0$. The output voltage is

$$V_O = V_N - V_R = O - V_R \quad (2.35)$$

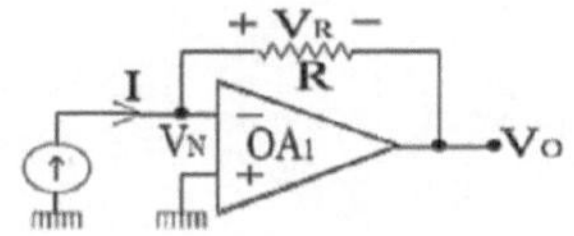

FIGURE 2.15 Current-to-voltage (I/V) converter.

The voltage across resistor will be

$$V_R = IR \tag{2.36}$$

$$\therefore \quad V_O = -IR$$

An output voltage V_O is developed at the output of op-amp OA_1 which is proportional to the input current I.

2.11 DIFFERENTIATOR

Figure 2.16(a) shows a differentiator using op-amp. The current through capacitor C will be

$$I = C\frac{dV_I}{dt} \tag{2.37}$$

The current I flows through feedback resistor R, as op-amp has high input impedance and no current flows into op-amp. The non-inverted terminal is grounded and hence the voltage at the inverting terminal is also grounded virtually. The output voltage is given as

$$V_O = 0 - V_R = 0 - IR = -IR$$

$$V_O = -RC\frac{dV_I}{dt} \tag{2.38}$$

The differentiator is more susceptible to noise than the integrator circuit. Input noise fluctuations of small amplitude may have large derivatives; when differentiated, these noise fluctuations may generate a large noise signal at the output, creating poor output signal-to-noise ratio. In order to avoid this, a practical differentiator is used as shown in Figure 2.16(b). The practical differentiator differentiates low frequency signals but a constant high frequency gain. The components values are chosen such that

$$RC_1 = R_1C, R_O = \frac{R_1R}{R_1 + R}$$

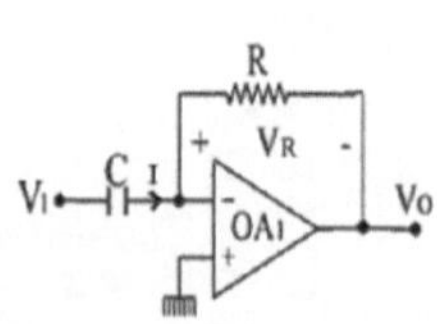

FIGURE 2.16(A) Differentiator.

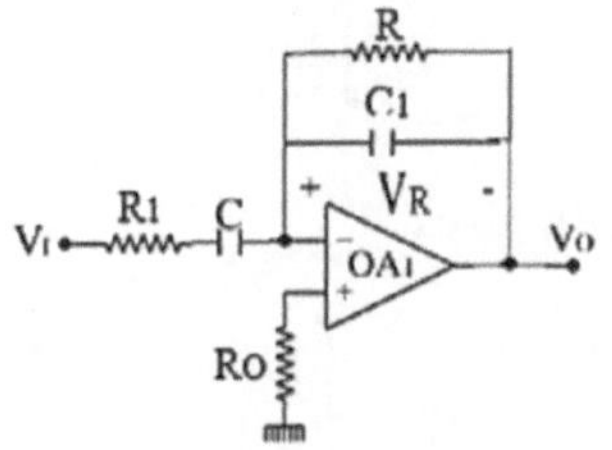

FIGURE 2.16(B) Practical differentiator.

2.12 INTEGRATOR

Figure 2.17(a) shows an integrator using op-amp. Since the non-inverting terminal (+) is at ground potential, the inverting terminal (–) will also be at ground potential by virtual ground. The current through the resistor R will be

$$I = \frac{V_I - 0}{R} = \frac{V_I}{R} \tag{2.39}$$

Due to op-amp's high input impedance, the current I will not enter into op-amp and flows through the capacitor C. The voltage across capacitor C will be

$$V_C = \frac{q}{C} \tag{2.40}$$

Where 'q' is charge to exist on plates of capacitor. The relation between current (i) and charge (q) is given as

$$i = \frac{dq}{dt}$$

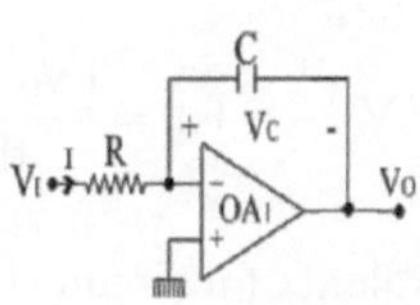

FIGURE 2.17(A) Integrator.

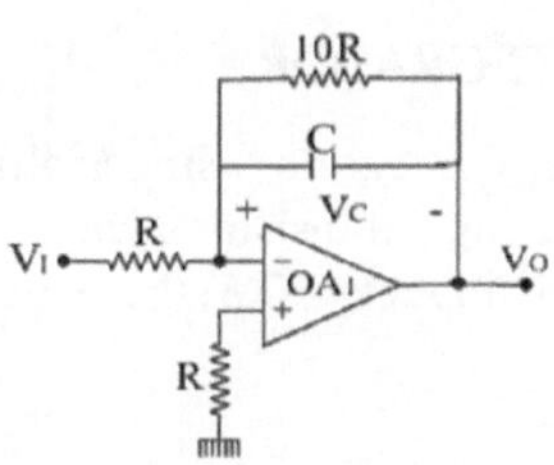

FIGURE 2.17(B) Practical integrator.

$$q = \int I dt \tag{2.41}$$

Equation (2.41) in Eq. (2.40) gives

$$V_C = \frac{\int I dt}{C} \tag{2.42}$$

The negative feedback forces the op-amp to produce an output voltage that maintains a virtual ground at the op-amp inverting input. The output voltage is given as

$$V_O = 0 - V_C = -V_C \tag{2.43}$$

Equation (2.42) in Eq. (2.43) gives

$$V_O = -\frac{1}{C}\int I dt \tag{2.44}$$

Equation (2.39) in Eq. (2.44) gives

$$V_O = -\frac{1}{RC}\int V_I dt = -\frac{V_I}{RC}t \tag{2.45}$$

Figure 2.16(b) shows a circuit of a practical integrator. If a square wave is given as the input to the integrator, then a triangular wave will be the output of the integrator.

2.13 DIFFERENTIAL INTEGRATOR

Figure 2.18(a) shows a differential integrator. Its output is given as

$$V_O = \frac{1}{RC}\int (V_D - V_I)\, dt = \frac{(V_D - V_I)}{RC}t \tag{2.46}$$

Figure 2.18(a) shows the equivalent of a circuit shown in Figure 2.18(a).

2.14 CONTROLLED INTEGRATOR

Figure 2.19(a) shows a controlled integrator using analog switch, and Figure 2.19(b) shows a controlled integrator using a transistor switch.

In Figure 2.19(a), when CON input is LOW, the switch S_1 opens and the integrator output is given as

$$V_O = -\frac{1}{R_1C_1}\int V_1\, dt = -\frac{V_1}{R_1C_1}t \tag{2.47}$$

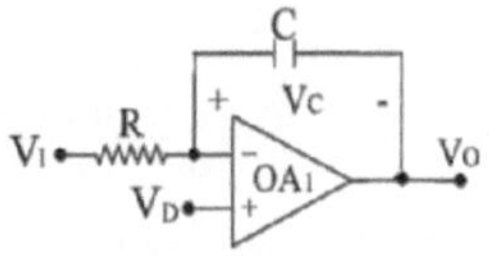

FIGURE 2.18(A) Differential integrator.

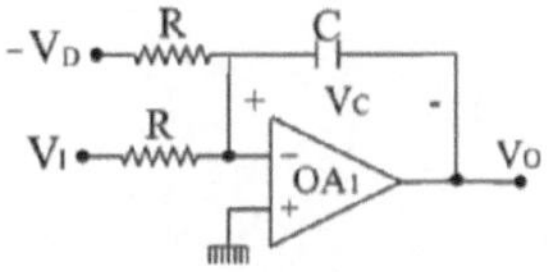

FIGURE 2.18(B) Equivalent circuit of Figure 2.18(a).

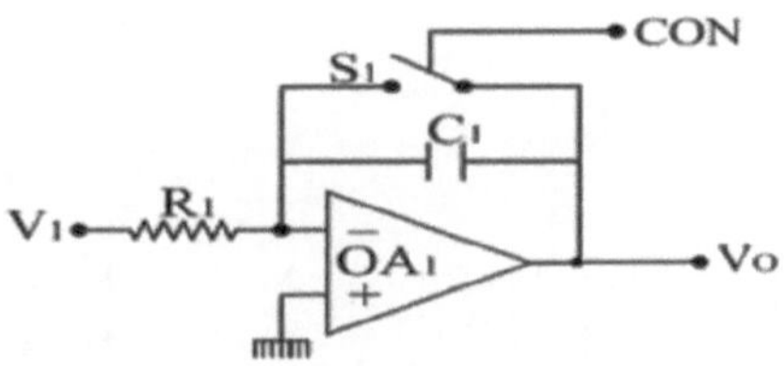

FIGURE 2.19(A) Controlled integrator with analog switch.

When CON input is HIGH, the switch S_1 is closed, the capacitor C_1 is short-circuited, and the integrator output is zero volts.

In Figure 2.19(b), when CON input is HIGH, the transistor Q_1 is OFF and the integrator output is given as

$$V_O = -\frac{1}{R_1C_1}\int V_I\, dt = -\frac{V_I}{R_1C_1}t \tag{2.48}$$

When CON input is LOW, the transistor Q_1 is ON, the capacitor C_1 is short-circuited, and the integrator output is zero volts.

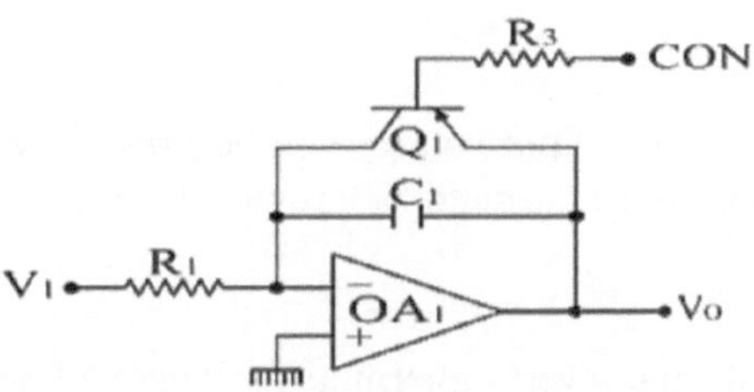

FIGURE 2.19(B) Controlled integrator with transistor switch.

WORKED PROBLEMS

1. Find output voltage in Figures 2.20(a)–2.20(d).

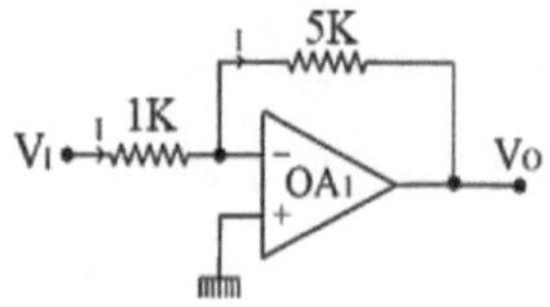

FIGURE 2.20(A)

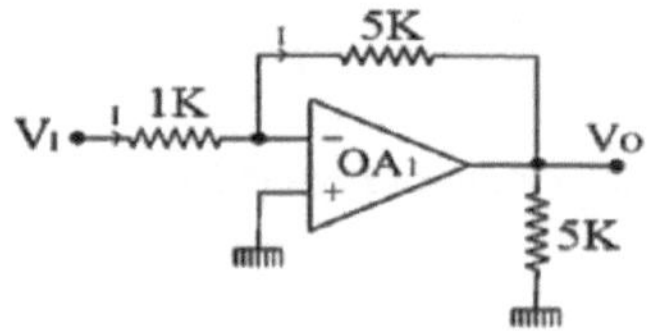

FIGURE 2.20(B)

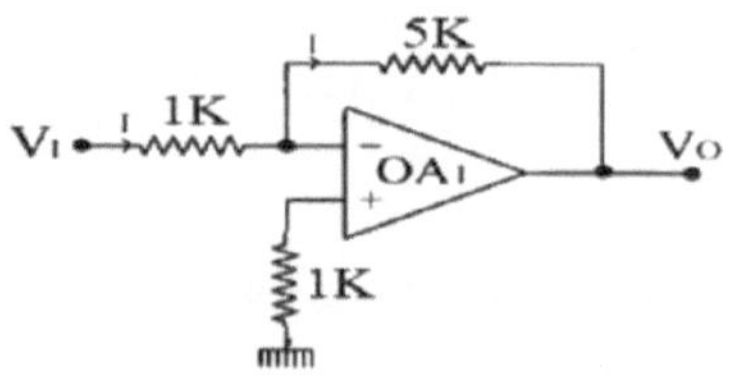

FIGURE 2.20(C)

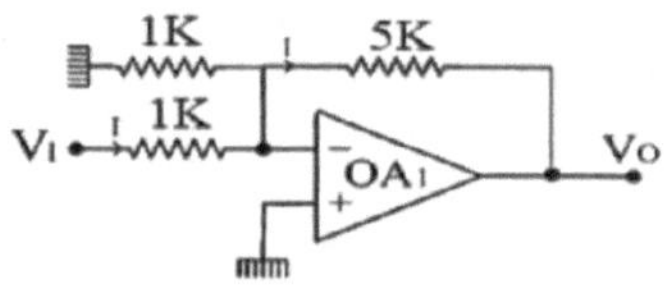

FIGURE 2.20(D)

In Figure 2.20(a), the output voltage is $-\left(\frac{5K}{1K}\right) = -5V$.

In Figure 2.20(b), the output voltage irrespective of load resistance will be $\left(\frac{5K}{1K}\right) = -5V$.

In Figure 2.20(c), the output voltage will be irrespective of whether its non-inverting terminal (+) is ground or grounded through some resistor value, $\left(\frac{5K}{1K}\right) = -5V$.

In Figure 2.20(d), the inverting terminal voltage will be zero volts virtually, as the non-inverting terminal is grounded and the op-amp in a negative closed loop feedback. Hence, a resistor connected between the inverting terminal and ground is of no meaning, and the circuit will work as usual as an inverting amplifier, the output voltage will be $\left(\frac{5K}{1K}\right) = -5V$.

2. Find output voltage in the circuit shown in Figure 2.21.

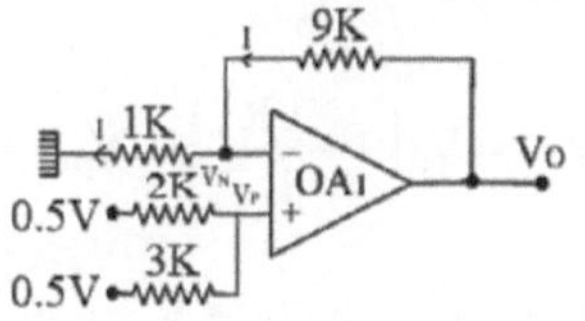

FIGURE 2.21

By the superposition principle, first let 3K resistor be at GND potential. Voltage at V_p will be $\frac{0.5V}{5K} \times 3K = 0.3V$. The output voltage will be $V_{O1} = 0.3\left(1+\frac{9K}{1K}\right) = 3V$. Next, let 2K resistor be at GND potential. Voltage at V_p will be $\frac{0.5V}{5K} \times 2K = 0.2V$. The output voltage will be $V_{O2} = 0.2\left(1+\frac{9K}{1K}\right) = 2V$. Then the actual output voltage $V_O = V_{O1} + V_{O2} = 5$ V.

3. Derive expression for output voltage V_O in Figure 2.22.

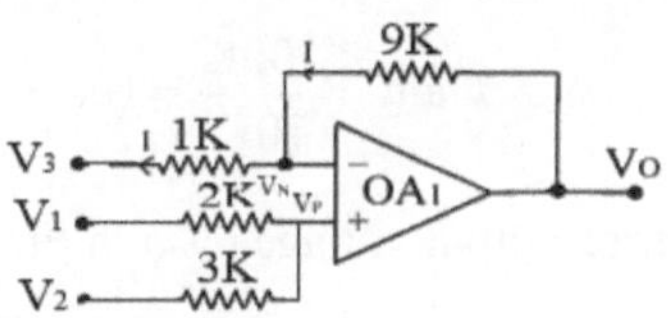

FIGURE 2.22

By the superposition principle, let V_3 be acting, and V_1, V_2 be grounded. The op-amp OA_1 will work an inverting amplifier and its output will be $V_{O1} = -V_3\left(\frac{9K}{1K}\right) = -9V_3$

Let V_1 be acting, and V_2, V_3 be grounded. Voltage at V_P will be

$$V_p = \frac{V_1}{5K} 3K = 0.6V_1$$

The output voltage will be

$$V_{O2} = V_P\left(1+\frac{9K}{1K}\right) = 10\,V_p = 6V_1$$

Let V_2 be acting, and V_1, V_3 be grounded. The voltage at V_P will be

$$\frac{V_2}{5K}2K = 0.4V_2$$

The output voltage will be

$$V_{O3} = V_P\left(1+\frac{9K}{1K}\right)10\,V_p = 4V_2$$

$$\text{The actual output voltage} = V_{O1} + V_{O2} + V_{O3} = 6V_1 + 4V_2 - 9V_3$$

4. Find input resistance and gain of the circuit shown in Figure 2.23.

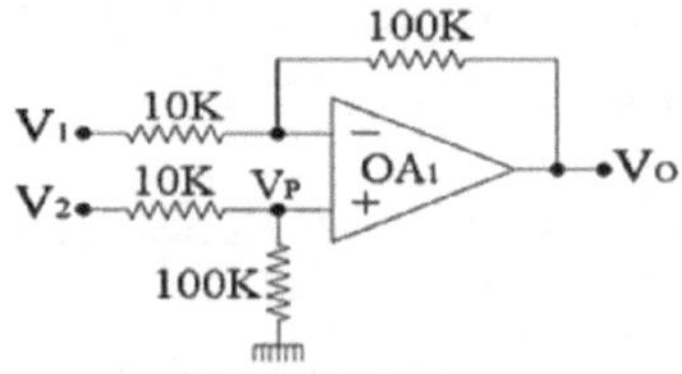

FIGURE 2.23

The input resistance is 10 K + 10 K = 20 K.

$$\text{Gain} = \frac{100K}{10K} = 10.$$

5. Derive expression for output voltage (V_O) in Figure 2.24.

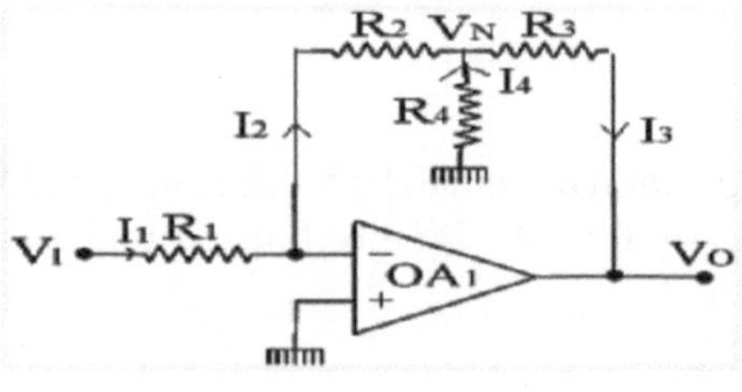

FIGURE 2.24

$$I_1 = \frac{V_1}{R_1}.$$

$$I_1 = I_2$$

$$V_N = -I_2R_2 = -V_1\frac{R_2}{R_1}$$

Applying KCL at node V_N

$$I_2 + I_4 = I_3$$

$$-\frac{V_N}{R_2} - \frac{V_N}{R_4} = \frac{V_N - V_0}{R_3}$$

$$-\frac{V_N}{R_2} - \frac{V_N}{R_4} - \frac{V_N}{R_3} = -\frac{V_0}{R_3}$$

$$V_N\left(\frac{1}{R_2} + \frac{1}{R_4} + \frac{1}{R_3}\right) = \frac{V_0}{R_3} \quad (2)$$

$$-V_1\left(\frac{R_2}{R_1}\right)\left(\frac{1}{R_2} + \frac{1}{R_4} + \frac{1}{R_3}\right) = \frac{V_0}{R_3}$$

$$\therefore V_O = -V_1\left(\frac{R_2}{R_1}\right)\left(\frac{R_3}{R_2} + \frac{R_3}{R_4} + 1\right)$$

TUTORIAL EXERCISES

1. An inverting amplifier with a gain of 10 is designed using op-amp. The output of this inverting amplifier is connected to a load resistance of $R_L = 1\ K\Omega$. If a + 1 V is applied to the circuit, find load current I_L passing through the load resistance R_L.
   [Ans. 10 mA]

2. Find the output voltage (V_O) in Figure 2.25.

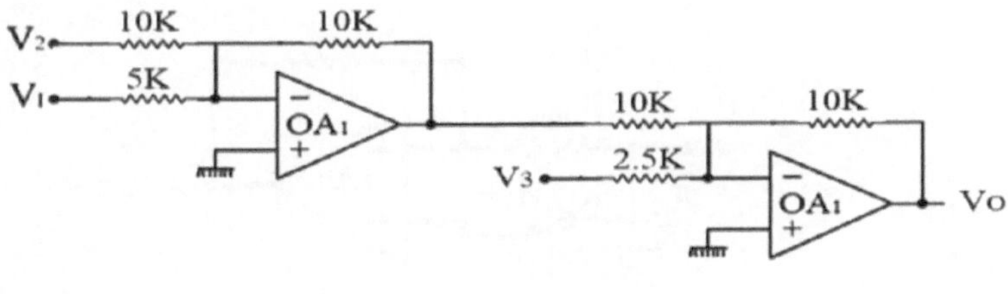

FIGURE 2.25

   [Ans. $2V_1 + V_2 - 4V_3$]

3. Find the resistance values in a differential amplifier circuit with an input resistance of 20 KΩ and a gain of 10.
   [Ans. $R_1 = 10$ K, $R_2 = 100$ K]

4. An integrator is designed with input resistance of 10 KΩ and integration time constant of 1 mS. Find the feedback capacitance value.
   [Ans. 0.1 μF]

5. Find the output voltage (V_O) in Figures 2.26–2.29.

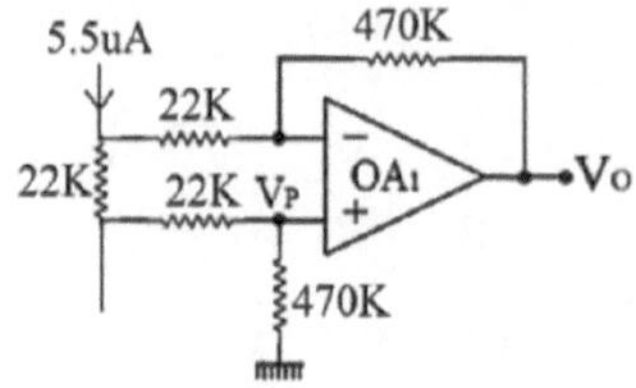

FIGURE 2.26

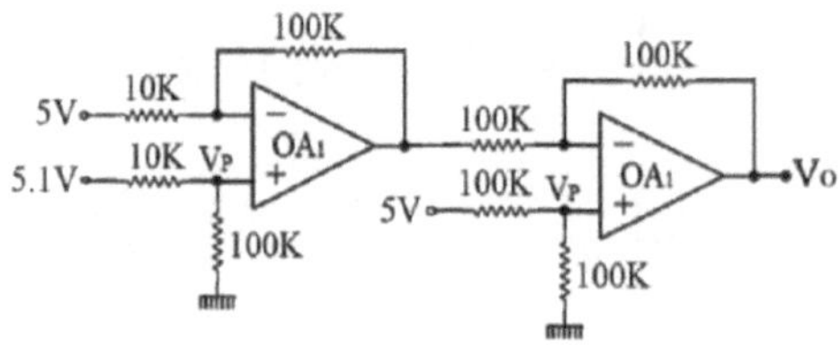

FIGURE 2.27

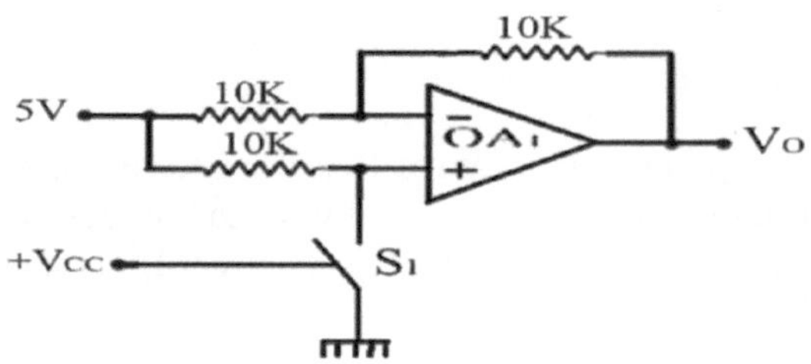

FIGURE 2.28

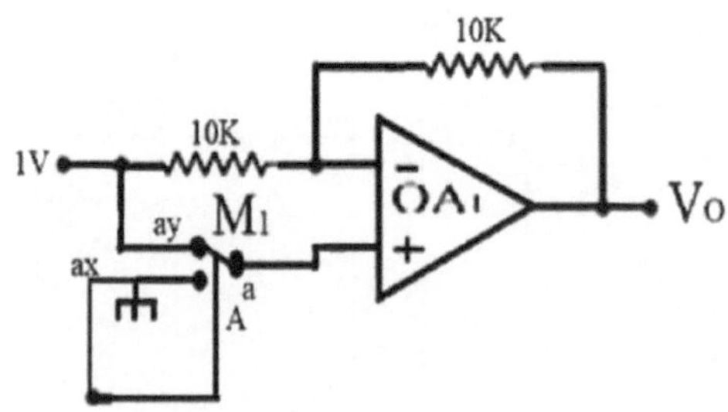

FIGURE 2.29

[Ans. −2.585 V, 4 V, −5 V, −1 V]

3 Non-Linear Circuits

Many linear and non-linear analog systems are constructed with op-amp as a basic building block. Linear circuits use negative feedback to force the op-amp to operate within its linear region and linear elements with feedback network. Non-linear circuits use high gain amplifiers with positive feedback or sometimes no feedback at all, which causes op-amp to operate in saturation made. Non-linear circuits use feedback network with non-linear elements like diodes and analog switches. Non-linear circuits like voltage comparator, Schmitt trigger, precision rectifier, peak detector, sample and hold circuits, log amplifier, and antilog amplifier are described in this chapter.

3.1 VOLTAGE COMPARATOR

Figures 3.1(a) and 3.1(b) show the voltage comparator. The figures compare its positive input voltage, V_P, with its negative input voltage, V_N, and produced output, V_O.

The output voltage, V_O, is

i. HIGH ($+V_{SAT}$) when $V_P > V_N$
ii. LOW ($-V_{SAT}$) when $V_P < V_N$

It is observed that V_P and V_N may be analog signals but output V_O is a binary signal.

When both the inputs V_P and V_N are shorted together and a single voltage V_D is applied, the output V_O will be

i. HIGH ($+V_{SAT}$) for $V_D >$ OV or positive
ii. LOW ($-V_{SAT}$) for $V_D <$ OV or negative

The voltage transfer curve is shown in Figure 3.1(b). Figure 3.2(a) shows a comparator as a zero crossing detector.

In Figure 3.2(a), the comparator compares input sine wave at its positive terminal and ground which is at its inverting terminal. The output is (i) HIGH during positive half-cycle of input sine wave, and (ii) LOW during negative half-cycle of input sine wave. A square waveform is generated at the output of comparator and is shown in Figure 3.2(b).

In Figure 3.3(a), the comparator compares input sine wave at its negative terminal and ground which is at its non-inverting terminal. The output is (i) HIGH during negative half-cycle of input sine wave, and (ii) LOW during positive half-cycle of input sine wave. A square waveform is generated at the output of comparator and is shown in Figure 3.3(b).

In Figure 3.4(a), the comparator compares a sawtooth waveform V_{SI} of peak value V_R and time period T at its inverting terminal with an input voltage V_1 at its non-inverting

DOI: 10.1201/9781003221449-4

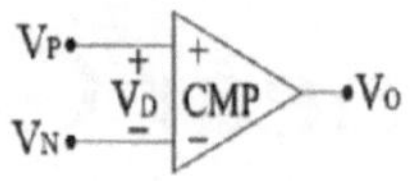

FIGURE 3.1(A) Comparator.

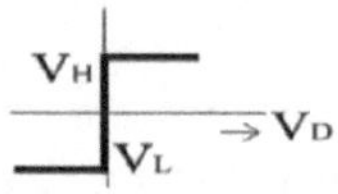

FIGURE 3.1(B) Voltage transfer curve.

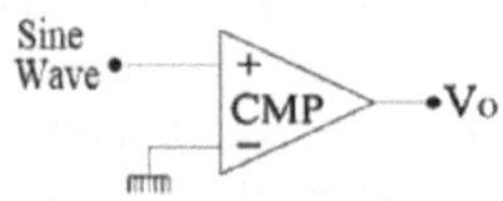

FIGURE 3.2(A) Zero crossing detector (case I).

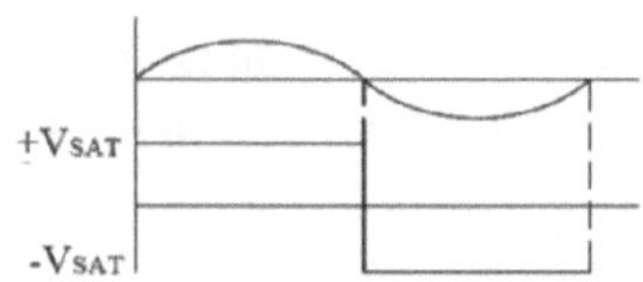

FIGURE 3.2(B) Associated waveforms of Figure 3.2(a).

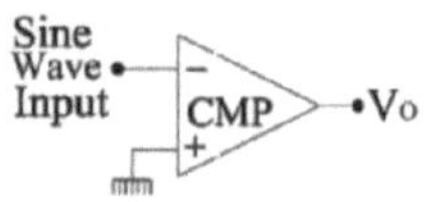

FIGURE 3.3(A) Zero crossing detector (case II).

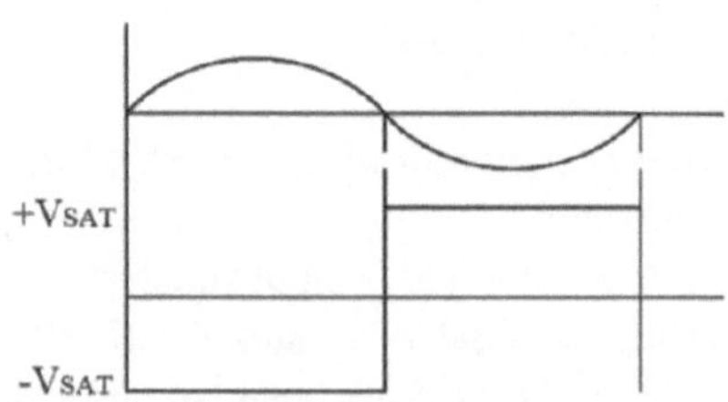

FIGURE 3.3(B) Associated waveforms of Figure 3.3(a).

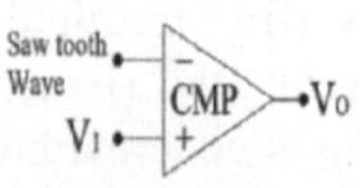

FIGURE 3.4(A) Comparator with sawtooth wave input (case I).

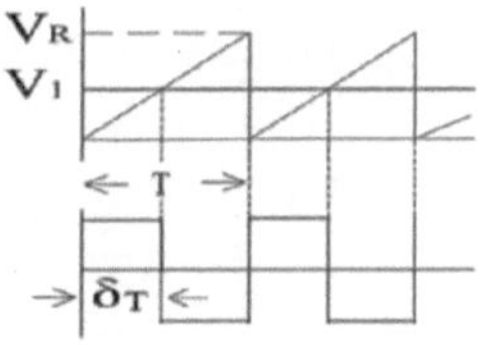

FIGURE 3.4(B) Associated waveforms of Figure 3.4(a).

terminal. A rectangular waveform is generated at the output of comparator and is shown in Figure 3.4(b). The HIGH or ON time of this rectangular waveform is given as

$$\delta_T = \frac{\text{Input voltage } V_1}{\text{Peak value of saw tooth}} \times \text{Time period} = \frac{V_1}{V_R} T \tag{3.1}$$

In Figure 3.5(a), the comparator compares a sawtooth waveform of peak value V_R and time period T at its non-inverting terminal with an input voltage V_1 at its inverting terminal. A rectangular waveform is generated at the output of comparator and is shown in Figure 3.5(b). The LOW or OFF time is given as0

$$\delta_T = \frac{\text{Input voltage } V_1}{\text{Peak value of saw tooth}} \times \text{Time period} = \frac{V_1}{V_R} T \tag{3.2}$$

In Figure 3.6(a), the comparator compares a triangular waveform of $\pm V_T$ peak value and time period T at its inverting terminal with an input voltage V_1 which is at its non-inverting terminal. A rectangular waveform is generated at the output of comparator and is shown in Figure 3.6(b). The OFF time T_1 is given as

$$T_1 = \frac{\text{Peak value of triangular wave} - \text{Input voltage}}{\text{Twice Peak value of triangular wave}} \times \text{Time period} = \frac{V_T - V_1}{2V_T} T \tag{3.3}$$

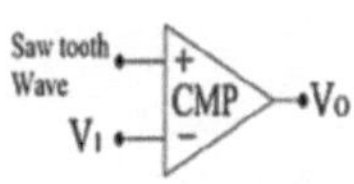

FIGURE 3.5(A) Comparator with sawtooth wave (case II).

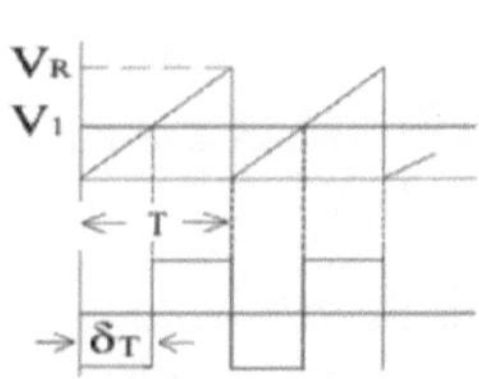

FIGURE 3.5(B) Associated waveform of Figure 3.5(a).

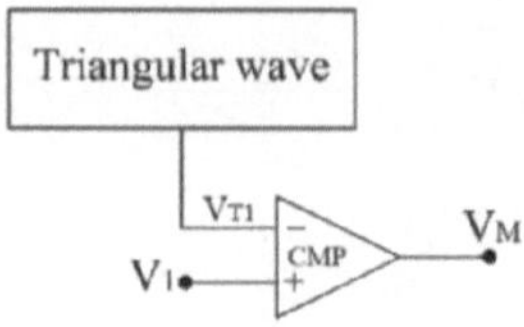

FIGURE 3.6(A) Comparison of triangular wave (case 1).

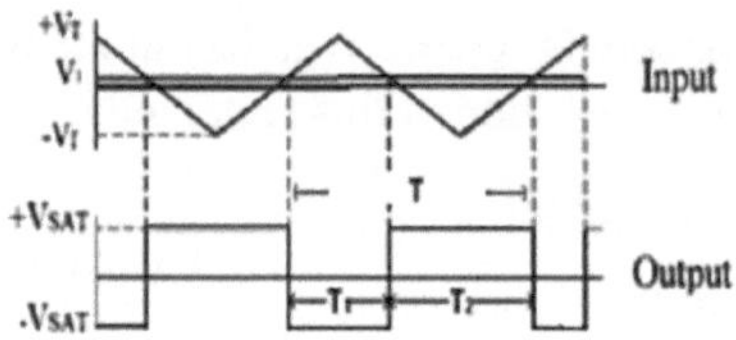

FIGURE 3.6(B) Associated waveforms of Figure 3.6(a).

The ON time T_2 is given as

$$T_2 = \frac{\text{Peak value of triangular wave} + \text{Input voltage}}{\text{Twice Peak value of triangular wave}} \times \text{Time period} = \frac{V_T + V_I}{2V_T} T \quad (3.4)$$

In Figure 3.7(a), the comparator compares a triangular wave form V_{T1} of $\pm V_T$ peak value and time period T at its non-inverting terminal with an input voltage V_1 which is at its inverting terminal. A rectangular waveform is generated at the output of comparator and is shown in Figure 3.7(b). The ON time T_1 is given as

$$T_1 = \frac{\text{Peak value of triangular wave} - \text{Input Voltage}}{\text{Twice Peak value of triangular wave}} \times \text{Time period} = \frac{V_T - V_I}{2V_T} T \quad (3.5)$$

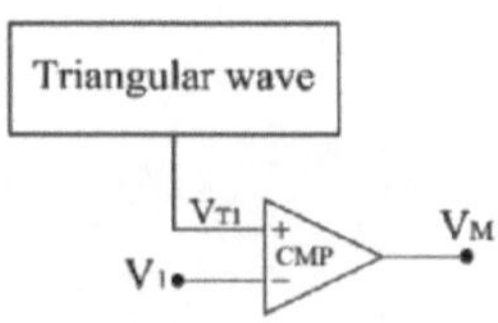

FIGURE 3.7(A) Comparison of triangular wave (case II).

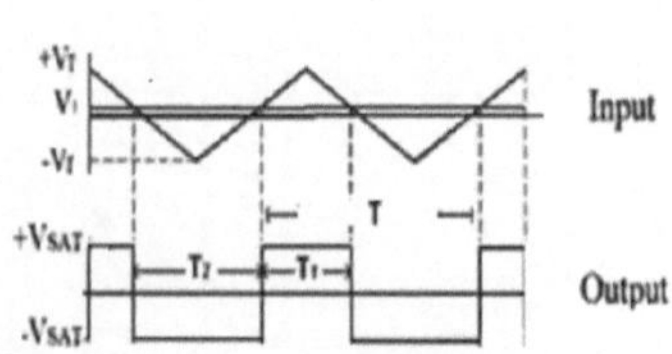

FIGURE 3.7(B) Associated waveforms of Figure 3.7(a).

The OFF time T_2 is given as

$$T_2 = \frac{\text{Peak value of triangular wave} + \text{Input Voltage}}{\text{Twice Peak value of triangular wave}} \times \text{Time period} = \frac{V_T + V_I}{2V_T} T \quad (3.6)$$

3.2 SCHMITT TRIGGER

A Schmitt trigger is an amplifier with positive feedback. The negative feedback of an op-amp is used for amplification, and the positive feedback of an op-amp is used for oscillation. The positive feedback makes the amplifier go to saturation.

Figure 3.8(a) shows a circuit diagram of an inverting Schmitt trigger. The output has only two states, namely V_{OH} and V_{OL}. Figure 3.8(b) shows a voltage transfer curve of Figure 3.8(a).

$$V_{TH} = \frac{R_1}{R_1 + R_2} V_{OH} \quad (3.7)$$

$$V_{TL} = \frac{R_1}{R_1 + R_2} V_{OL} \quad (3.8)$$

If the input voltage is less than zero (i.e., negative voltage), the op-amp goes to positive saturation ($+V_{SAT}$). When the input voltage is more than the V_{TH} value, the op-amp output goes to negative saturation ($-V_{SAT}$).

Figure 3.9(a) shows a non-inverting Schmitt trigger. For negative input voltage, the output will go to negative saturation. Figure 3.9(b) shows a voltage transfer curve of Figure 3.9(a).

$$V_{TH} = -\frac{R_1}{R_2} V_{OL} \quad (3.9)$$

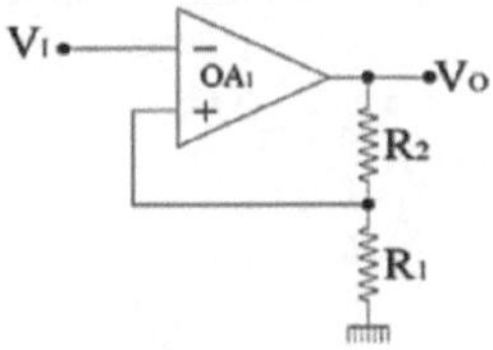

FIGURE 3.8(A) Inverting Schmitt trigger.

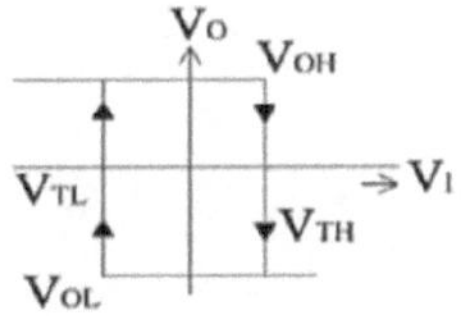

FIGURE 3.8(B) Voltage transfer curve of Figure 3.8(a).

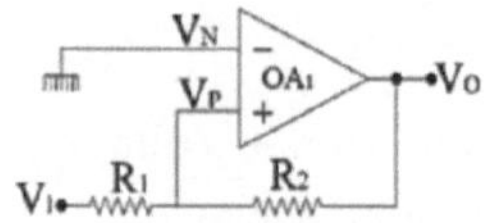

FIGURE 3.9(A) Non-inverting Schmitt trigger.

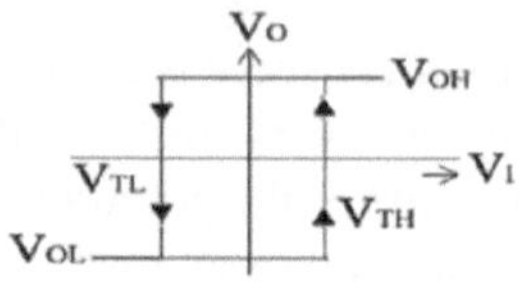

FIGURE 3.9(B) Voltage transfer curve of Figure 3.9(a).

$$V_{TL} = -\frac{R_1}{R_2} V_{OH} \tag{3.10}$$

3.3 HALF-WAVE RECTIFIER

A half-wave rectifier is a circuit which passes only one portion of an input wave (either positive or negative). The output of a positive half-wave rectifier is given as

$$V_O = V_I \text{ for positive input voltage}$$

$$V_O = O \text{ for negative input voltage}$$

Figure 3.10(a) shows a basic half-wave rectifier and its associated waveforms are shown in Figure 3.10(b). If the input voltage is positive, i.e., greater than zero volts, the op-amp output goes to positive saturation, the diode D_1 is forward-biased and conducting, the op-amp works as a buffer, and V_I appears across the load resister R_L

$$V_O = +V_I \tag{3.11}$$

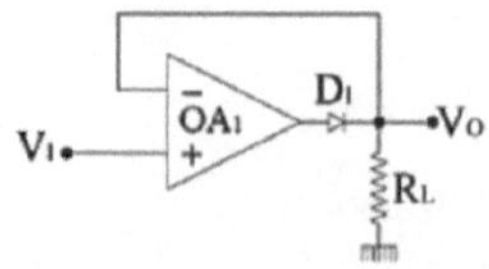

FIGURE 3.10(A) Basic half-wave rectifier.

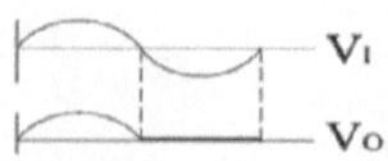

FIGURE 3.10(B) Associated waveforms of Figure 3.10(a).

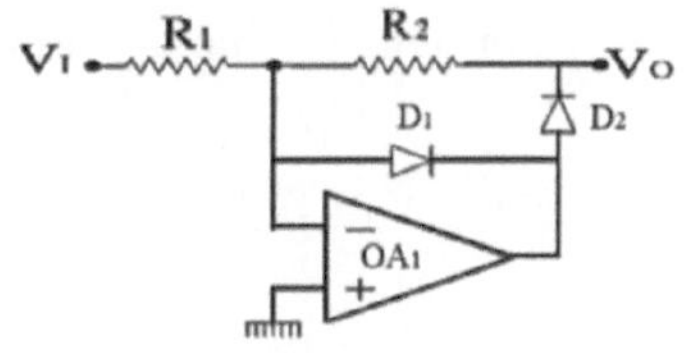

FIGURE 3.11(A) Half-wave rectifier.

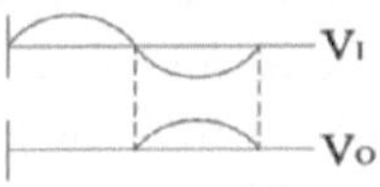

FIGURE 3.11(B) Associated waveforms of Figure 3.11(a).

If the input voltage is negative or is less than zero volts, the op-amp output goes to negative saturation, the diode D_1 is reverse-biased and not conducting, and zero volts appears across the load resister R_L

$$V_O = O \tag{3.12}$$

Figure 3.11(a) shows improved half-wave rectifier and its associated waveforms are shown in Figure 3.11(b).

If the input voltage is greater than zero or for positive input voltage, then diode D1 is ON which creates negative closed loop feedback, the inverting terminal voltage will be zero by virtual ground, and op-amp output will be $-V_{D1}$. Diode D_2 is OFF, no current flows through R_2, and hence output is zero volts.

$$V_O = O[(\text{for } V_I > O) \tag{3.13}$$

If the input voltage is less than zero or for negative input voltage, then op-amp output goes to positive saturation and diode D_2 is ON. Also, when diode D_1 is OFF, the circuit will work as an inverting amplifier with a gain of (R_2/R_1). The output voltage is given as

$$V_O = -V_I\left(\frac{R_2}{R_1}\right)[\text{for } V_I < O] \tag{3.14}$$

3.4 FULL WAVE RECTIFIER

A positive full wave rectifier is a circuit which passes a positive portion of input voltage, inverts, and then passes a negative portion of input voltage. The output of a positive full wave rectifier is given as

$$V_O = +V_I \text{ for positive input voltage}$$

$$V_O = -(-V_I) \text{for negative input voltage}$$

$$(\text{Or})\ V_O = |V_I|$$

A full wave rectifier is also known as an absolute value circuit.

Figure 3.12(a) shows a circuit diagram of a full wave rectifier.

The op-amp OA_1 constitutes a half-wave rectifier. Let $R_1 = R_2 = R$. During positive half-cycle of input V_I, diode D_1 is OFF, op-amp OA_1 output goes to negative saturation. When diode D_2 is ON, the op-amp OA_1 will work as inverting amplifiers, and $-V_I$ will appear at the junction V_X. During negative half-cycle of input V_I, diode D_1 is ON, op-amp OA_1 output goes to positive saturation. When diode D_2 is OFF, the op-amp OA_1 will be a negative feedback loop configuration. Its non-inverting voltage will be equal to its inverting terminal voltage. Zero volts will appear at junction V_X. The voltage at V_X is given as

$$V_X = -V_I \quad \text{when } V_I > O$$

$$V_X = O \quad \text{when } V_I < O$$

The op-amp OA_2 is an adder. It adds the voltages V_X and V_I and produces an output voltage V_O.

$$V_O = -\left[\frac{R_5}{R_3} V_x + \frac{R_5}{R_4} V_I\right]$$

During positive half-cycle of the input V_I, i.e., when $V_I > O$

$$V_O = -\left[-V_I\left(\frac{R_5}{R_3}\right) + \frac{R_5}{R_4} V_I\right]$$

Let $R_4 = R_5 = R$ and $R_3 = R/2$

$$V_O = [2V_I - V_I] = V_I \tag{3.15}$$

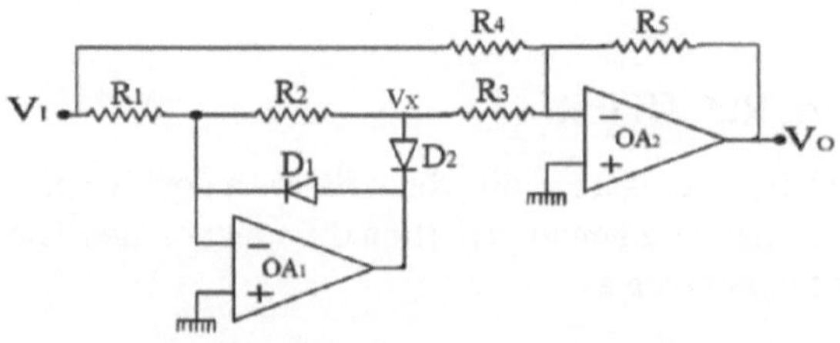

FIGURE 3.12(A) Full wave rectifier.

During negative half-cycle of the input V_I, i.e., when $V_I < O$

$$V_O = -\left[\frac{R_5}{R_3}(0) + \frac{R_5}{R_4} V_I\right]$$

$$V_O = -V_I \tag{3.16}$$

The input and output waveforms are shown in Figure 3.12(b).

Figure 3.13(a) shows another full wave rectifier with two matched resistors.

During positive half-cycle of the input V_I, i.e., $V_I > O$, diode D_1 is ON, the op-amp OA_1 will be at a negative closed loop configuration, its inverting terminal voltage will be equal to its non-inverting terminal voltage, and zero volts will be the output of op-amp OA_1. When diode D_2 is OFF, the op-amp OA_2 will work as a non-inverting amplifier. Its output will be

$$V_O = V_I\left(1 + \frac{R_3}{R_2}\right) \text{ when } V_I > 0 \tag{3.17}$$

During negative half-cycle of the input V_I, i.e., $V_I < 0$, diode D_1 is OFF. When diode D_2 is ON (forward biased by resistor R_4), op-amp OA_1 will still be in negative closed loop via D_2-OA_2 –R_3-R_2. Hence, its inverting terminal will be at zero volts virtually.

By Kirchoff Law,

$$I_1 = I_2$$

$$\frac{0 - V_I}{R_1} = \frac{V_O - 0}{R_2 + R_3}$$

$$-\frac{V_I}{R_1} = \frac{V_O}{R_2 + R_3}$$

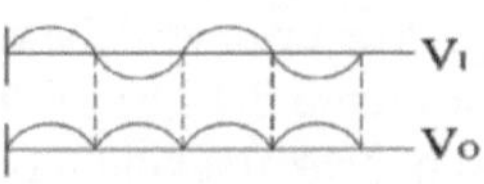

FIGURE 3.12(B) Input and output waveforms for Figure 3.12(a).

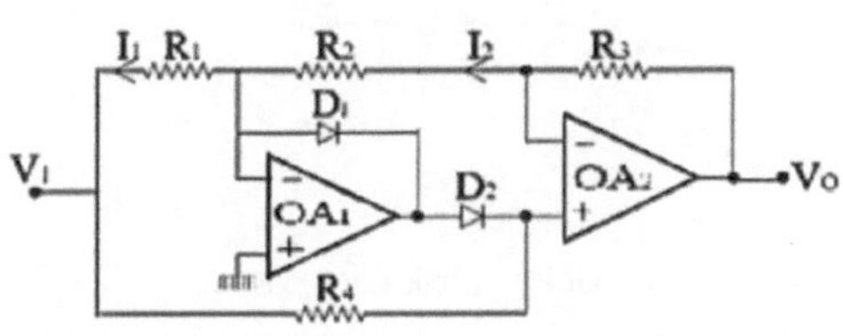

FIGURE 3.13(A) Full wave rectifier with two matched resistors.

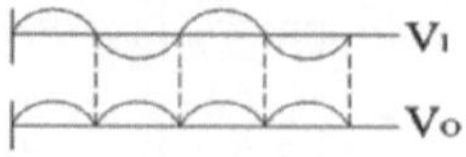

FIGURE 3.13(B) Associated waveforms of Figure 3.13(a).

$$\therefore V_O = -V_I \frac{(R_2 + R_3)}{R_1}$$

Let us assume

$$1 + \frac{R_3}{R_2} = \frac{R_2 + R_3}{R_1}$$

$$\frac{R_2 + R_3}{R_2} = \frac{R_2 + R_3}{R_1}$$

$$\text{Let} \quad R_1 = R_2 = R$$

$$V_O = -V_I \left(1 + \frac{R_3}{R_2}\right) \text{ when } V_I < 0 \tag{3.18}$$

The input and output waveforms are shown in Figure 3.13(b).

3.5 PEAK DETECTOR

The peak detector detects maximum value of a signal over a period of time. Figure 3.14(a) shows a simple diode-capacitor peak detector.

The capacitor C is charged by the input signal through the diode. When the input signal falls, the diode is reverse-biased and the capacitor voltage retains the peak value of the input signal. This simple circuit has errors because of the diode forward voltage drop. This sort of forward voltage drop errors can be removed by replacing the diode with a precise diode as shown in Figure 3.14(b). It operates in either peak tracking mode or peak storage mode. During peak tracking mode, the peak detector tracks the input toward a peak value. During peak storage mode, the peak detector held constant

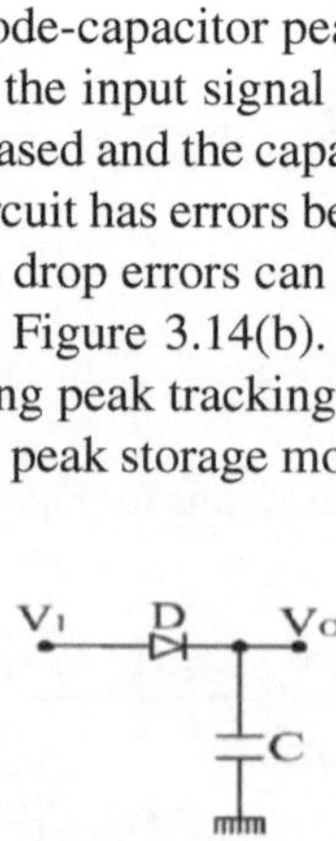

FIGURE 3.14(A) Simple diode-capacitor peak detector.

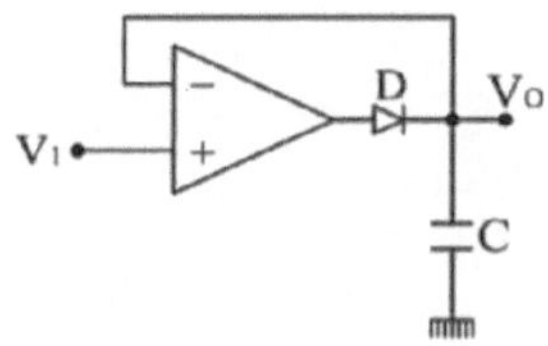

FIGURE 3.14(B) Op-amp peak detector.

of the peak value. Figures 3.14(a) and 3.14(b) are positive peak detectors and if we interchange the polarity of diode D, then they will become negative peak detectors.

If a sawtooth wave of peak value V_P is given to a peak detector, the peak detector output is a DC voltage of V_P. Similarly, if a triangular wave of $\pm V_P$ value is given to a peak detector, and if the peak detector is positive, then its output will be a DC voltage of $+V_P$; and if the peak detector is negative, then its output will be a DC voltage of $-V_P$.

3.6 SAMPLE AND HOLD CIRCUIT

The sample and hold circuit samples and holds the input signal at a particular instant of time determined by the sampling pulse. Figure 3.15(a) shows a simple sample and hold circuit.

Let a sawtooth wave of peak value V_P and time period T be given as input to the sample and hold circuit. As shown in Figure 3.15(a), a sampling pulse V_S is given to the sample and hold circuit. The switch S_1 is closed during the HIGH time of sampling pulse V_S, and at that particular time of the input signal is given to the capacitor C, and the capacitor C holds this signal even if the switch S_1 opens during LOW time of the sampling pulse V_S. Hence, the instant value of input signal is sampled and hold with the sampling pulse V_S. This is illustrated graphically in Figure 3.16. An op-amp buffer can be added at the output in order to avoid loading effect, as shown in Figure 3.15(b).

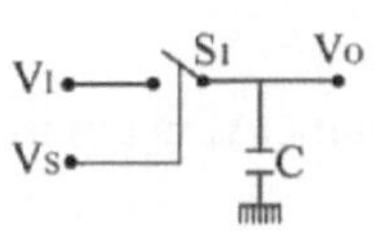

FIGURE 3.15(A) Simple sample and hold circuit.

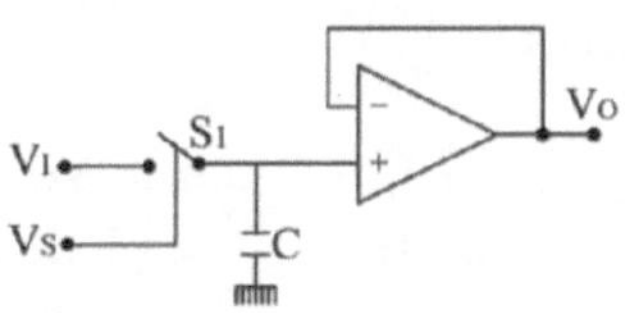

FIGURE 3.15(B) Op-amp sample and hold circuit.

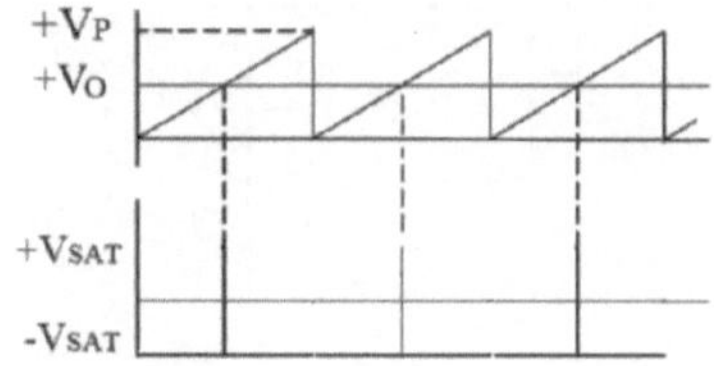

FIGURE 3.16 Waveforms of Figures 3.15(a) and 3.15(b).

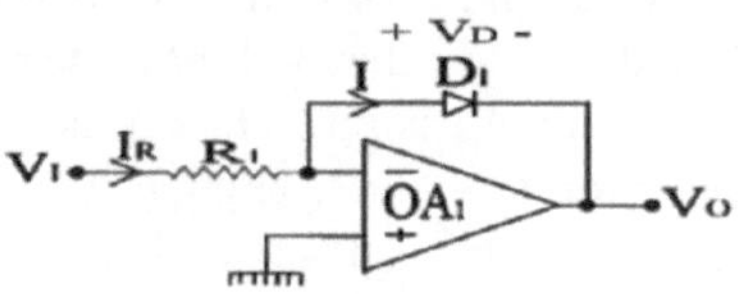

FIGURE 3.17 Log amplifier with diode.

3.7 LOG AMPLIFIER

The log amplifier using a diode is shown in Figure 3.17.

$$I_R = \frac{V_I}{R_1}$$

The voltage across diode V_D is given as

$$V_D = V_T \Big|_n \frac{I}{I_S} \tag{3.19}$$

Where

$V_T = KT/q = 25$ mV at room temperature
I = Current passing through diode
I_s = Reverse saturation current

The inverting terminal of op-amp OA_1 is at virtual ground, and since it has high input impedance

$$I_R = I$$

The output voltage is given as

$$-V_D = -V_T \Big|_n \left(\frac{I_R}{I_S}\right) = -V_T \Big|_n \left(\frac{V_I}{R_1 I_S}\right) \tag{3.20}$$

The diode log amplifier is simple but has many drawbacks. It has (i) very poor log conformity and (ii) drifting of output due to temperature variations. The drawbacks of diode log amplifiers are overcome by transistor log amplifiers. Figure 3.18 shows a transistor log amplifier.

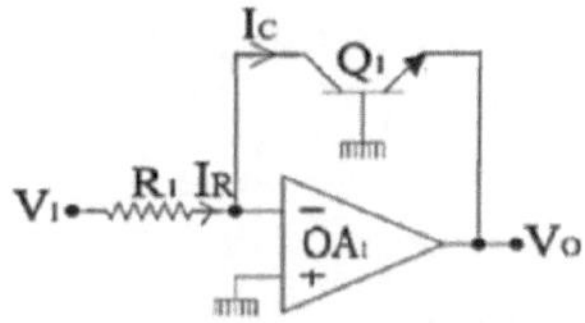

FIGURE 3.18 Log amplifier with transistor.

The logarithmic operation of bipolar transistor is given as

$$V_{BE} = V_T l_n \frac{I_C}{I_S} \tag{3.21}$$

Where
V_{BE} = Base emitter voltage
V_T = KT/q = 25 mV at room temperature
I_C = Collector current
I_S = Emitter saturation current

$$I_R = \frac{V_I}{R_1}$$

The inverting terminal of op-amp OA_1 is at virtual ground, and since it has high input impedance

$$I_R = I_C$$

Output voltage V_O is given as

$$V_O = -V_{BE}$$

$$= -V_T \ln\left(\frac{I_C}{I_S}\right)$$

$$V_O = -V_T \ln\left(\frac{V_I}{R_1 I_S}\right) \tag{3.22}$$

3.8 ANTILOG AMPLIFIER

The antilog amplifier using diode is shown in Figure 3.19. The inverting terminal of op-amp OA_1 will be at virtual ground, and hence the voltage at its inverting terminal is zero volts, as the op-amp has high input impedance. Hence

$$I_R = I_D$$

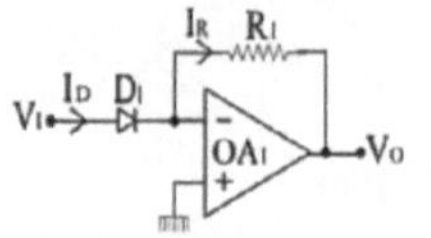

FIGURE 3.19 Antilog amplifier with diode.

The output voltage is given as

$$V_O = -I_R R_1 = -I_D R_1$$

The diode current I_D is given as

$$I_D = I_S e^{V_I/\eta V_T}$$

$$V_O = -\left[I_S e^{V_I/\eta V_T}\right] R_1 = -\left(I_S R_1\right) e^{V_I/\eta V_T} \tag{3.23}$$

Let us assume $I_S R_1 = V_R$

$$V_O = -V_R e^{V_I/\eta V_T} \tag{3.24}$$

Thus, the output voltage is proportional to the exponential function of V_I. Exponential function is nothing but an antilog function, and the circuit is working as an antilog amplifier. The antilog amplifier using a single transistor is shown in Figure 3.20.

The inverting terminal of op-amp OA_1 is at virtual ground, and the inverting terminal voltage of op-amp OA_1 is zero volts. Since op-amp OA_1 has high input impedance,

$$I_C = I_R$$

From Eq. (3.21), we can get

$$I_C = I_S e^{V_{BE}/V_T} \tag{3.25}$$

The output voltage is given as

$$V_O = -I_R R_1 = -I_C R_1$$

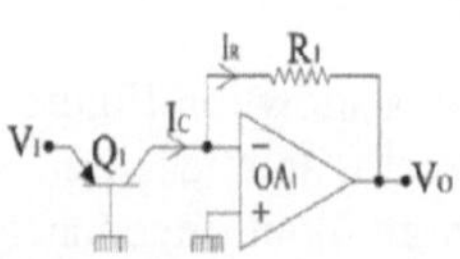

FIGURE 3.20 Antilog amplifier with transistor.

$$V_O = I_S R_1 e^{V_{BE}/V_T}$$

Let $I_S R_1 = V_R$, $V_{BE} = V_I$ and hence

$$V_O = V_R e^{V_I/V_T} \tag{3.26}$$

Thus, the output voltage is an exponential function of input or an antilog function of the input voltage.

WORKED PROBLEMS

1. Figure 3.21 shows a window detector using comparators. Explain its working operation.

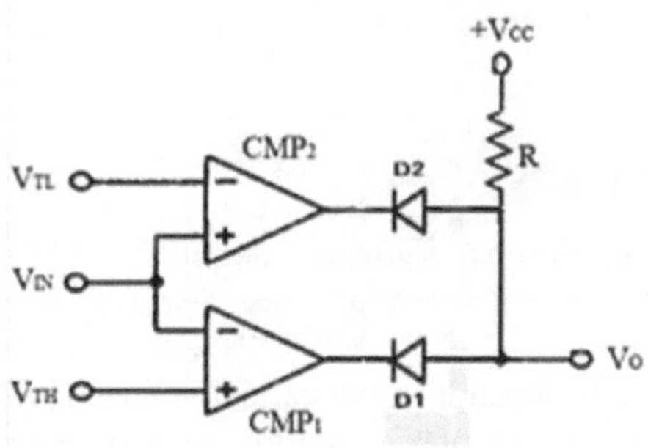

FIGURE 3.21

TABLE I
Operation of the Circuit

Input Range	CMP_1	CMP_2	Output
$V_{IN}<V_{TL}$	HIGH	LOW	LOW
$V_{TL}<V_{IN}<V_{TH}$	HIGH	HIGH	HIGH
$V_{IN}>V_{TH}$	LOW	HIGH	LOW

The operation of the circuit is shown in Table I. When V_{IN} is less than V_{TL}, comparator CMP_1 output is HIGH, comparator CMP_2 output is LOW, and diode D_1 is OFF. When diode D_2 is ON, output V_O will be LOW. When V_{IN} is greater than V_{TH}, comparator CMP_1 output is LOW, comparator CMP_2 output is HIGH, and diode D_1 is ON. When diode D_2 is OFF, output V_O will be LOW. When $V_{TL} < V_{IN} < V_{TH}$, both comparators CMP_1 and CMP_2 outputs are HIGH. Diodes D_1 and D_2 are OFF and the output voltage V_O will be HIGH.

2. A triangular wave with 10 Vpp and frequency of 1 KHz is compared with a DC voltage of 1 V. The triangular wave is given to inverted input of comparator. Draw output waveform of comparator and find ON and OFF times of this output waveform.

Given: $V_T = 5$ V, $V_1 = 1$ V, T = 1/f = 1 mS
The output waveforms are shown in Figure 3.22.

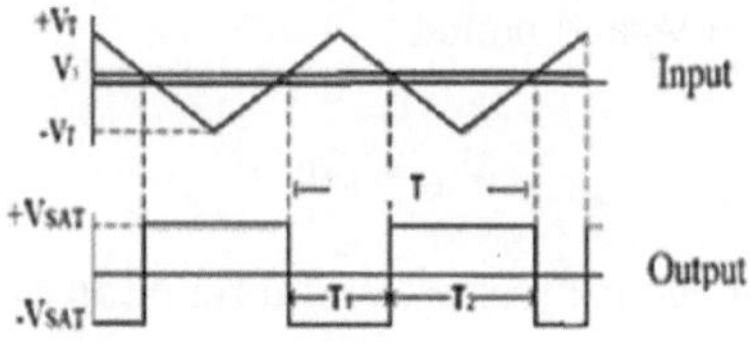

FIGURE 3.22 Waveforms for Problem 2.

$$T_1 = \frac{V_T - V_1}{2V_T} T = \frac{5-1}{10} 1 \text{ mS} = 0.4 \text{ mS}$$

$$T_2 = \frac{V_T + V_1}{2V_T} T = \frac{5+1}{10} 1 \text{ mS} = 0.6 \text{ mS}$$

TUTORIAL EXERCISES

1. Design a voltage comparator whose output V_O = O for input voltage is greater than 1 V, and V_O = 5 V for input is less than 1 V. The power supply to the circuit is ±15 V.
2. Design a window detector whose output voltage is HIGH between 2 to 5 V.
3. Design a half-wave rectifier circuit for negative input voltage.
4. Design a full wave rectifier circuit for negative input voltage.
5. In the full wave rectifier circuit shown in Figure 3.12(a), if all resistor values are equal, determine its output voltage and draw output waveforms.
6. Design a peak detector for negative-going input waveforms.
7. Design log and antilog amplifiers using diodes and op-amps.
8. Find the output voltage of a sample and hold circuit whose input is 1OV peak-to-peak triangular wave. A sampling pulse V_S is obtained during peak value of triangular wave.

4 Waveform Generators

The circuits described in Chapters 2 and 3 are called "processing circuits," and they operate on existing signals. In this chapter we will discuss circuits which generate sine, square, pulse, sawtooth, and triangular waveforms. These waveforms are used in timing and control, as signal carriers for information transmission and storage, as sweep signals for information display, as test signals for automatic test and measurement, and as audio signals for electronic music. Function generators produce a waveform of a particular frequency, amplitude, shape, and duty cycle. Sine wave oscillators are used to test the characteristics of lowpass, highpass, and bandpass filters. Pulse waveforms are used to test digital circuits. Sawtooth and triangular waves are required to develop function circuits either internally or externally. Operational amplifiers are used for generation of triangular, sinusoidal, sawtooth, and square waveforms and are described in this chapter.

4.1 WIEN BRIDGE OSCILLATOR

Figure 4.1 shows Wien bridge oscillator using op-amp. The frequency of oscillation is given as

$$f_0 = \frac{1}{2\pi\sqrt{R_1 R_2 C_1 C_2}}$$

$$_{\text{Gain}}\,A = 1 + \frac{R_4}{R_3}$$

$$\text{Feedback factor } \beta = \frac{R_2 C_1}{R_1 C_1 + R_2 C_2 + R_2 C_1}$$

Let us assure $R_1 = R_2 = R$ and $C_1 = C_2 = C$, then

$$f_0 = \frac{1}{2\pi RC} \tag{4.1}$$

$$\beta = \frac{1}{3} \tag{4.2}$$

For oscillations, the gain should be 3 and hence: $R_4 = 2\,R_3$.

DOI: 10.1201/9781003221449-5

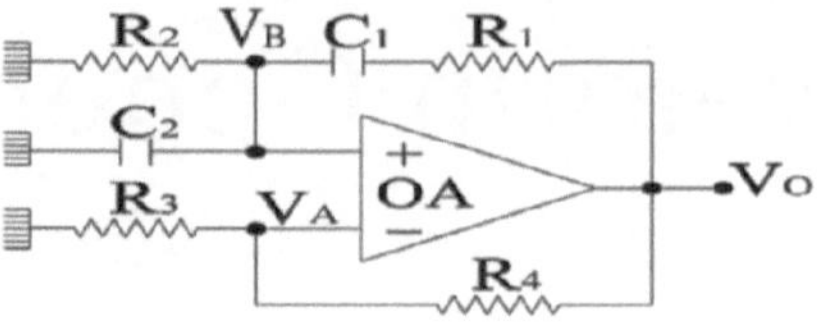

FIGURE 4.1 Wien bridge oscillator.

4.2 MONOSTABLE MULTIVIBRATOR

Figure 4.2 shows a monostable multivibrator using op-amp and Figure 4.3 shows its associated waveforms. It produces a single output pulse of adjustable time duration in response to its input triggering signal. The pulse width is proportional to the RC values connected to the op-amp.

Let us assume initially that the op-amp OA output is HIGH ($+V_{SAT}$), and diode D_1 clamps capacitor C_1 voltage to 0.7 V. The voltage at the non-inverting terminal of op-amp OA will be

$$V_A = \beta V_{SAT}$$

$$\beta = \frac{R_2}{R_1 + R_2}$$

Now, if a negative-going trigger pulse is applied at the input, then

$$V_A = \beta V_{SAT} - V_{IN}$$

The V_A is becoming less than 0.7 V (V_B), and the output goes to LOW ($-V_{SAT}$). The diode D_1 will get reverse-biased and the capacitor C_1 starts charging exponentially to $-V_{SAT}$ through resistor R_5. The voltage V_A will now be

$$V_A = -\beta V_{SAT}$$

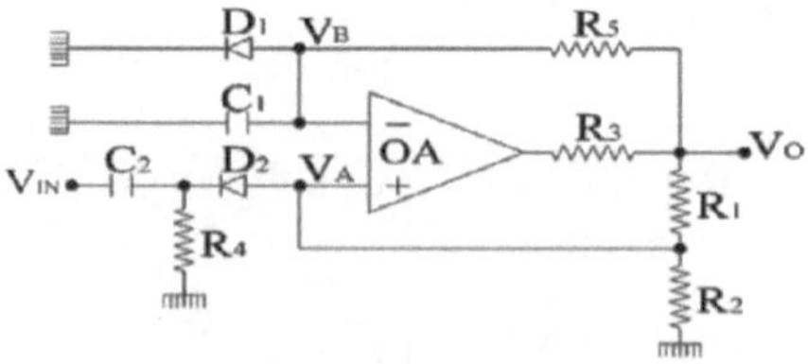

FIGURE 4.2 Monostable multivibrator using op-amp.

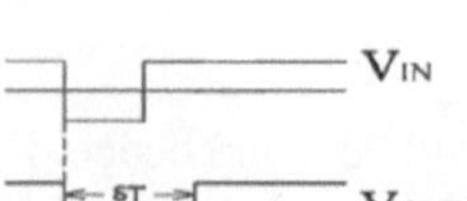

FIGURE 4.3 Associated waveforms of Figure 4.2.

When the capacitor voltage becomes slightly more than V_A, the output of op-amp is switched to HIGH ($+V_{SAT}$). The capacitor C_1 is now charging to $+V_{SAT}$ through resister R_5 until V_B is 0.7 V as capacitor C_1 gets clamped to the voltage.

The general solution for a single time constant lowpass RC circuit with V_1 and V_f is as initial and final values, respectively.

$$V_O = V_f\left(V_1 - V_f\right)e^{-t/RC} \tag{4.3}$$

For the circuit shown in Figure 4.2, $R = R_5$, $C = C_1$, final value $V_f = -V_{SAT}$ initial value $V_i = V_{D1}$ (diode D_1 forward voltage), V_C is capacitor voltage.

$$V_C = -V_{SAT} + \left(V_{D1} + V_{SAT}\right)e^{-t/R_5C_1}$$

at $t = T$, $V_C = -\beta V_{SAT}$

$$-\beta V_{SAT} = -V_{SAT} + \left(V_{D1} + V_{SAT}\right)e^{-t/R_5C_1}$$

The above equation can be simplified as

$$T = R_5C_1\Big|_n \frac{\left(1 + V_{D1}/V_{SAT}\right)}{1-\beta}$$

If $V_{SAT} \gg V_{D1}$ and $R_1 = R_2$, $\beta = 0.5$
then

$$T = 0.69\,R_5C_1 \tag{4.4}$$

The trigger pulse width should be less than T. The diode D_2 is used to avoid malfunctioning by blocking positive noise spikes that may present in the C_2 R_4 differentiator trigger input.

4.3 ASTABLE MULTIVIBRATOR

Figure 4.4 shows an astable multivibrator using op-amp. Let us assume initially op-amp output is LOW (i.e., negative saturation). The voltage at non-inverting terminal will be

$$V_A = \beta\left(-V_{SAT}\right)$$

$$\beta = \frac{R_1}{R_1 + R_2} \tag{4.5}$$

The voltage at inverting terminal V_B will be positive w.r.t. V_A and its potential is decreasing, i.e., C_1 charges down through R_3. When the potential difference between the two input terminals approaches zero, the op-amp comes out of saturation. The positive feedback from the output to terminal V_A causes regenerative switching which drives the op-amp to positive saturation. Capacitor C_1 charges up through R_3 and V_B potential rises

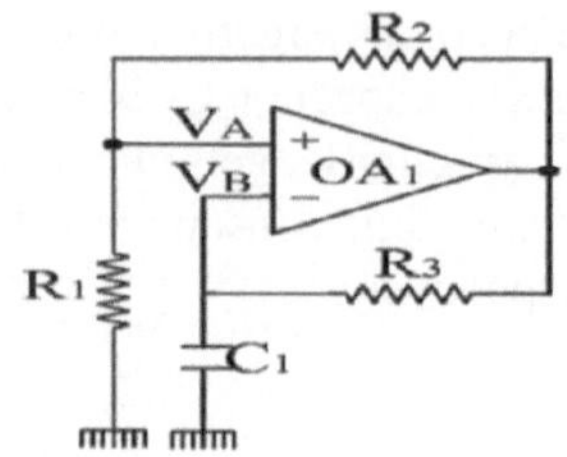

FIGURE 4.4 Astable multivibrator.

exponentially; when it reaches $V_B = \beta(+Vcc)$ the circuit switches back to the state in which op-amp is in negative saturation. The sequence therefore repeats to produce a square waveform of time period 'T' at its output. The time period 'T' is given as

$$T = 2R_3C_1 \ln\left(1 + 2\frac{R_1}{R_2}\right) \tag{4.6}$$

Voltage-to-Period Converter: If in the astable multivibrator shown in Figure 4.4, the R_2 terminal is removed from the output terminal and a controller is added between R_2 and output as shown in Figure 4.5, then the circuit will work as voltage-to-period converter. The time period 'T' is given as

$$T = V_C K_1 \tag{4.7}$$

where K_1 is a constant depending on Eq. (4.6) and op-amp saturation voltage or supply voltage Vcc.

Voltage-to-Frequency Converter: If in the astable multivibrator shown in Figure 4.4, the R_3 terminal is removed from the output terminal and a controller is added between R_3 and output as shown in Figure 4.6, then the circuit will work as a voltage-to-frequency converter. The frequency 'f' is given as

$$f = V_C K_1 \tag{4.8}$$

where K_1 is a constant depending on Eq. (4.6) and op-amp saturation voltage or supply voltage V_{CC}.

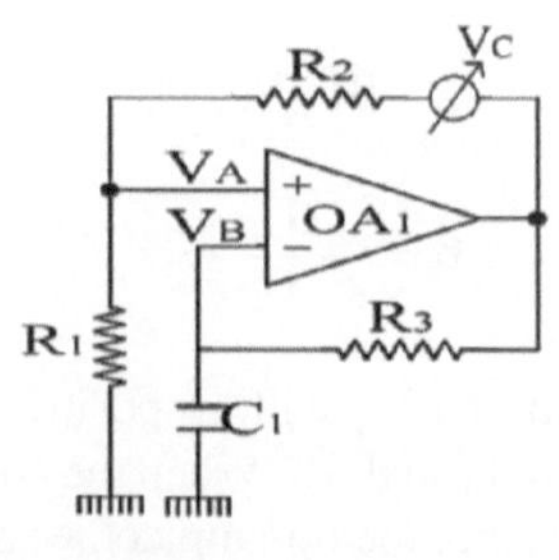

FIGURE 4.5 Voltage-to-period converter.

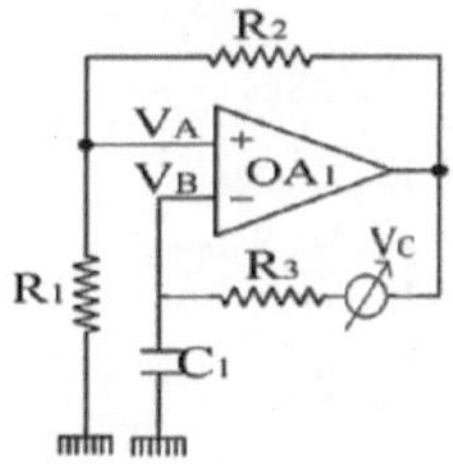

FIGURE 4.6 Voltage-to-frequency converter.

4.4 SAWTOOTH WAVE GENERATORS

Five circuits for generation of sawtooth wave are shown in Figures 4.7(a)–4.7(e), and their associated waveforms are shown in Figure 4.8. A sawtooth wave V_{S1} of peak value V_R and time period 'T' is generated by these circuits.

In Figure 4.7(a),

$$V_R = 2V_{BE} \tag{4.9}$$

$$T = 1.4R_1C_1 \tag{4.10}$$

In Figure 4.7(b), if initially op-amp OA_2 output is LOW, the multiplexer M_1 connects 'ax' to 'a' and the integrator is formed by resister R_1, capacitor C_1 and op-amp OA_1 integrates $(-V_R)$, and its output is given as

$$V_{S1} = -\frac{1}{R_1C_1}\int -V_R \, dt$$

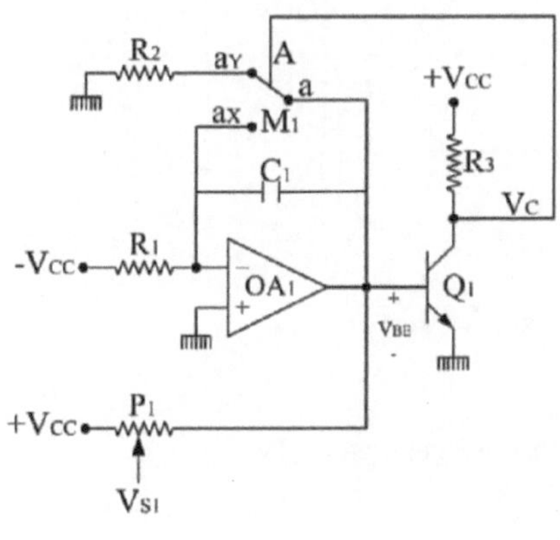

FIGURE 4.7(A) Sawtooth wave generator – I.

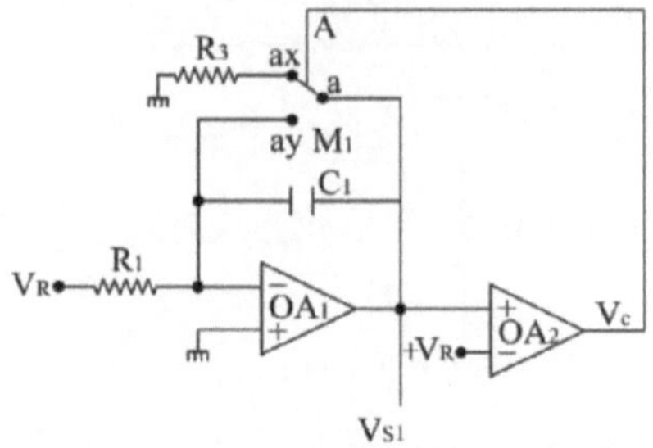

FIGURE 4.7(B) Sawtooth wave generator – II.

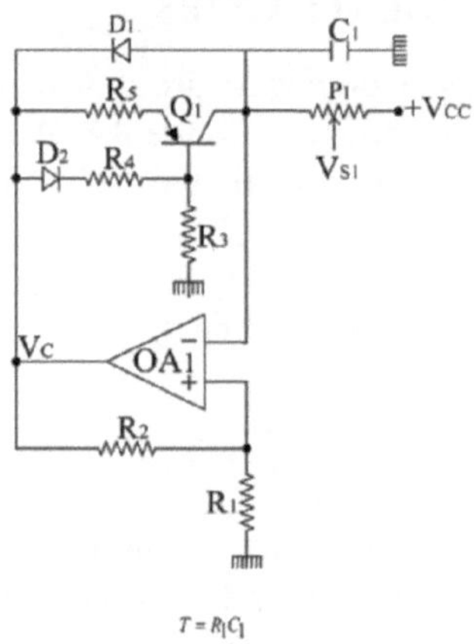

FIGURE 4.7(C) Sawtooth wave generator – III.

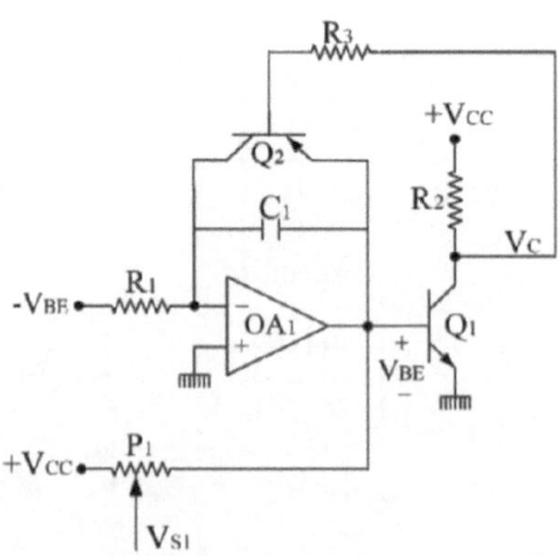

FIGURE 4.7(D) Sawtooth wave generator – IV.

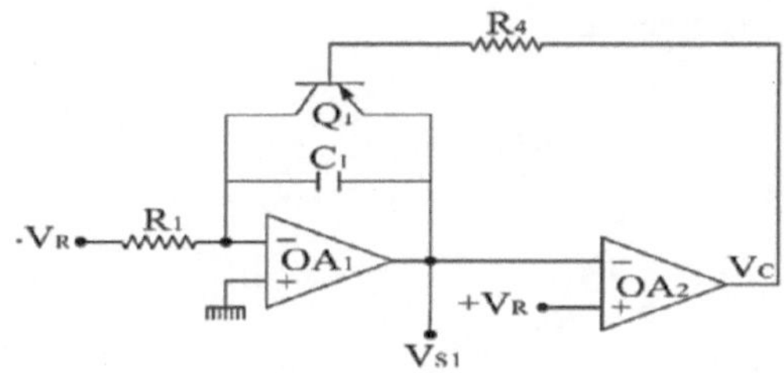

FIGURE 4.7(E) Sawtooth wave generator – V.

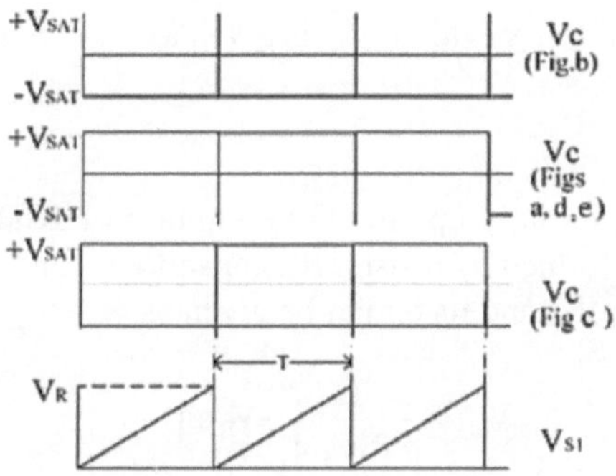

FIGURE 4.8 Associated waveforms of Figure 4.7.

$$V_{S1} = \frac{V_R}{R_1C_1}t \tag{4.11}$$

A positive-going ramp is generated at the output of op-amp OA_1, and when it reaches the value of reference voltage $+V_R$ the comparator OA_2 output becomes HIGH. The multiplexer M_1 now connects 'ay' to 'a' and shorts capacitor C_1, and hence integrator output becomes zero. Then comparator output is LOW and the sequence therefore repeats to give a perfect sawtooth wave V_{S1} of peak value V_R at the output of op-amp OA_1. From Eq. (4.11), Figure 4.8, and the fact that at t = T, $V_{S1} = V_R$.

$$V_R = \frac{V_R}{R_1C_1}T$$

$$T = R_1C_1 \tag{4.12}$$

In Figure 4.7(c), the time period 'T' is given as

$$T = 2R_5C_1 \ln\left(1 + 2\frac{R_1}{R_2}\right) \tag{4.13}$$

The peak value V_R is given as

$$V_R = \beta(V_{SAT}) + \frac{\beta(V_{SAT})}{1.5} \tag{4.14}$$

Where β is given as $\beta = \frac{R_1}{R_1 + R_2}$

In Figure 4.7(d), the time period 'T' is given as

$$T = 2R_1C_1$$

The peak value V_R of this sawtooth V_{S1} is given as

$$V_R = 2V_{BE}$$

In Figure 4.7(e), let initially op-amp OA_2 output be HIGH, the transistor Q_1 be OFF and the integrator formed by resister R_1, capacitor C_1 and op-amp OA_1 integrate the reference voltage ($-V_R$) and its output be given as

$$V_{S1} = -\frac{1}{R_1C_1}\int -V_R \, dt$$

$$V_{S1} = \frac{V_R}{R_1C_1} t \tag{4.15}$$

A positive-going ramp is generated at the output of op-amp OA_1 and when it reaches the value of reference voltage $+V_R$ the comparator OA_2 output becomes LOW. The transistor Q_1 is ON and shorts capacitor C_1, and hence integrator output becomes zero. Then comparator output is HIGH and the sequence therefore repeats to give a perfect sawtooth wave V_{S1} of peak value V_R at op-amp OA_1 output, and a short pulse waveform V_C at op-amp OA_2 output as shown in Figure 4.8. From Eq. (4.15), Figure 4.8, and the fact that at $t = T$, $V_{S1} = +V_R$.

$$V_R = \frac{V_R}{R_1C_1} T$$

$$T = R_1C_1 \tag{4.16}$$

4.5 TRIANGULAR WAVE GENERATOR

A triangular wave V_{T1} with $\pm V_T$ peak-to-peak value and time period 'T' is generated by the triangular wave generators shown in Figures 4.9(a) and 4.9(b), and their associated waveforms are shown in Figure 4.10. The value of V_T and time period 'T' are given in the following figures.

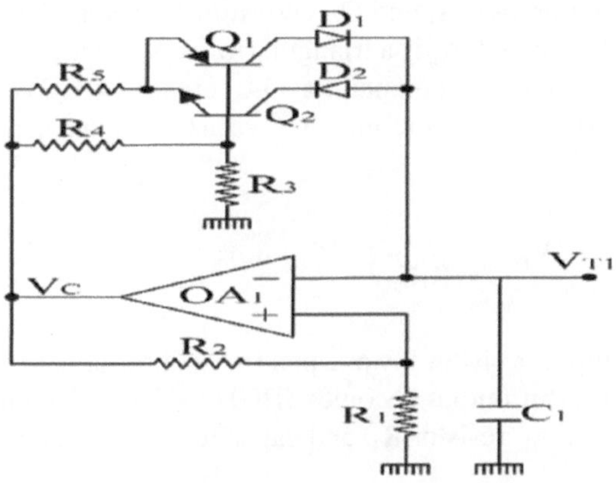

FIGURE 4.9(A) Triangular wave generator – I.

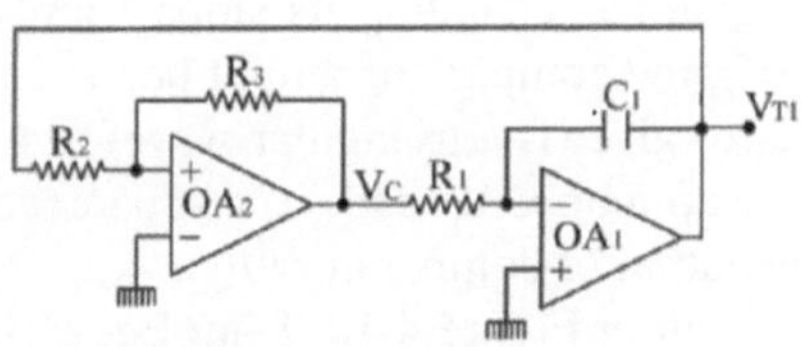

FIGURE 4.9(B) Triangular wave generator – II.

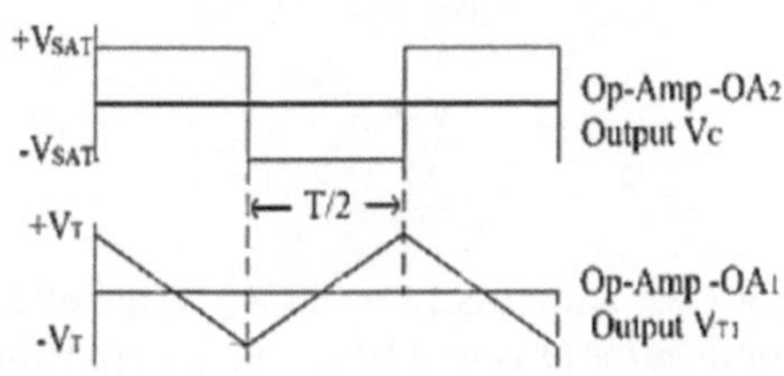

FIGURE 4.10 Associated waveforms of Figure 4.9(b).

In Figure 4.9(a)

$$V_T = \beta\left(V_{SAT}\right) \tag{4.17}$$

$$T = 4R_5C_1\frac{R_1}{R_2} \tag{4.18}$$

Where β is given as $\beta = \dfrac{R_1}{R_1 + R_2}$

In Figure 4.9(b), op-amps OA_1 and OA_2 constitute a triangular/square wave generator. The output of op-amp OA_1 is a triangular wave V_{T1} with $\pm V_T$ peak values and time period T. Let initially the comparator OA_2 output be LOW $(-V_{SAT})$, and the output of integrator composed by op-amp OA_1, resistor R_1, and capacitor C_1 be given as

$$V_{T1} = -\frac{1}{R_1C_1}\int -V_{SAT}\,dt = \frac{V_{SAT}}{R_1C_1}t \tag{4.19}$$

The integrator output is rising toward positive saturation and when it reaches a value $+V_T$, the comparator output becomes HIGH $(+V_{SAT})$. The output of integrator composed by op-amp OA_1, resistor R_1, and capacitor C_1 is given as

$$V_{T1} = -\frac{1}{R_1C_1}\int +V_{SAT}\,dt = -\frac{V_{SAT}}{R_1C_1}t$$

Now the output of integrator is changing its slope from $+V_T$ toward $(-V_T)$ and when it reaches a value $-V_T$, the comparator output becomes LOW $(-V_{SAT})$ and the sequence therefore repeats to give (i) a triangular waveform V_{T1} with $\pm V_T$ peak-to-peak values at the output of op-amp OA_1, and (ii) a square waveform V_C with $\pm V_{SAT}$ peak-to-peak values at the output of comparator OA_2.

From the waveforms shown in Figure 4.10, from Eq. (4.19), and the fact that at $t = T/2$, $V_{T1} = 2V_T$

$$2V_T = \frac{V_{SAT}}{R_1C_1}\frac{T}{2}$$

$$T = \frac{4V_TR_1C_1}{V_{SAT}} \tag{4.20}$$

When the comparator OA_2 output is LOW $(-V_{SAT})$, the effective voltage at non-inverting terminal of comparator OA_2 will be by the superposition principle

$$\frac{(-V_{SAT})}{(R_2+R_3)}R_2 + \frac{(+V_T)}{(R_2+R_3)}R_3$$

When this effective voltage at non-inverting terminal of comparator OA_2 becomes zero

$$\frac{(-V_{SAT})R_2 + (+V_T)R_3}{(R_2+R_3)} = 0$$

$$(+V_T) = (+V_{SAT})\frac{R_2}{R_3}$$

When the comparator OA_2 output is HIGH ($+V_{SAT}$), the effective voltage at non-inverting terminal of comparator OA_2 will be by the superposition principle

$$\frac{(+V_{SAT})}{(R_2+R_3)}R_2+\frac{(-V_T)}{(R_2+R_3)}R_3$$

When this effective voltage at non-inverting terminal of comparator OA_2 becomes zero

$$\frac{(+V_{SAT})R_2+(-V_T)R_3}{(R_2+R_3)}=0$$

$$(-V_T)=(-V_{SAT})\frac{R_2}{R_3}$$

$$\pm V_T=\pm V_{SAT}\frac{R_2}{R_3};\ 0.76(\pm V_{CC})\frac{R_2}{R_3} \qquad (4.21)$$

From Eqs. (4.20) and (4.21), time period 'T' of the generated triangular/square waveforms is given as

$$T=4R_1C_1\frac{R_2}{R_3} \qquad (4.22)$$

4.6 VOLTAGE CONTROLLED FUNCTION GENERATOR

The circuit diagram of the dual slope voltage-to-frequency converter is shown in Figure 4.11 and its associated waveforms are shown in Figure 4.12. In Figure 4.11 the triangular waveform is generated by the comparator OA_3 and the integrator formed by OA_2, resistor R, and capacitor C. The output of comparator OA_3 is a square wave of amplitude equal to $\pm V_{SAT}$ and is applied to the transistor switch Q_1.

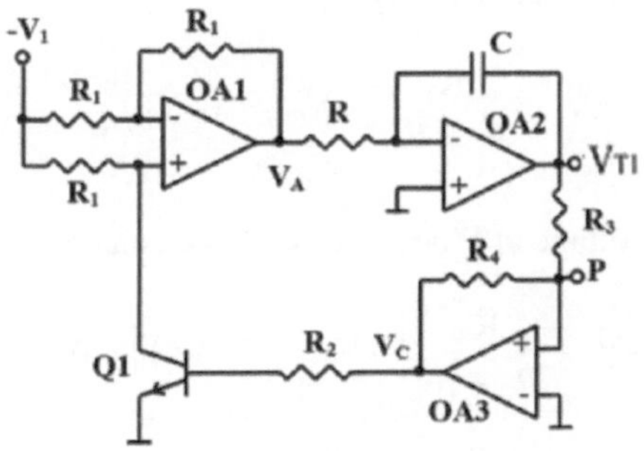

FIGURE 4.11 Circuit diagram of dual slope voltage-to-frequency converter.

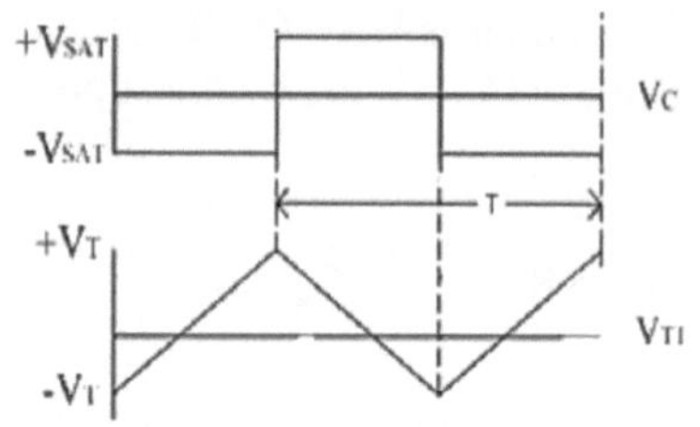

FIGURE 4.12 Associated waveforms of Figure 4.11.

The output of OA_2 is a triangular wave of peak value $\pm V_T$ and is fed back as input to the comparator OA_3 through resistors R_3 and R_4. Let us assume initially that output of comparator OA_3 is at $+V_{SAT}$. This forces the transistor Q_1 to ON and the non-inverting terminal of the op-amp OA_1 to GROUND. Hence, OA_1 acts as an inverting amplifier and $+V_1$ is applied to the integrator formed by OA_2, resistor R, and capacitor C. A constant current $+V_1/R$ flows through capacitor C to give a negative-going ramp at the output of the integrator formed by OA_2, resistor R, and capacitor C. Therefore, one end of resistor R_4 (comparator OA_3 output) is at a voltage $+V_{SAT}$, and the other end of R_3 (integrator formed by OA_2, resistor R, and capacitor C output) is at the negative-going ramp. When the negative-going ramp reaches a certain value $-V_T$, the effective voltage at point 'P' becomes slightly below 0 V.

As a result, the output of comparator OA_3 switches from $+V_{SAT}$ to $-V_{SAT}$. The transistor Q_1 becomes OFF and the amplifier OA_1 will work as non-inverting amplifier. $-V_1$ is given to the integrator formed by OA_2, resistor R, and capacitor C. This forces a reverse constant current through the capacitor C to give a positive-going ramp at the output of the integrator formed by OA_2, resistor R, and capacitor C. When the positive-going ramp reaches $+V_T$, the effective voltage at the point 'P' becomes slightly above 0 V. As a result, the output of comparator OA_3 switches from $-V_{SAT}$ to $+V_{SAT}$. This sequence therefore repeats to give a triangular wave with $\pm V_T$ peak-to-peak value at the output of integrator formed by OA_2, resistor R, and capacitor C, a square wave with $\pm V_{SAT}$ peak-to-peak value at the output of OA_3, and another square wave of $\pm V_I$ peak-to-peak value at the output of control amplifier OA_1. When the comparator output is $+V_{SAT}$, the input to the integrator formed by OA_2, resistor R and capacitor C is $+V_1$, negative ramp $-V_T$ is generated at the output of OA_2, and the effective voltage at point 'P' is given by

$$-V_T + \frac{R_3}{R_3 + R_4}\left[+V_{SAT} - (-V_T)\right]$$

When the effective voltage at P becomes 0 V, we can write the above equation as

$$-V_T + \frac{R_3}{R_3 + R_4}\left[+V_{SAT} - (-V_T)\right] = 0 \quad (4.23)$$

The above Eq. (4.23) can be simplified as

$$-V_T = -\frac{R_3}{R_4}\left(+V_{SAT}\right) \tag{4.24}$$

When the comparator output is $-V_{SAT}$, the input to the integrator is $-V_1$, positive-going ramp $+V_T$ is generated at the output of OA_2 and the effective voltage at point 'P' is given by

$$+V_T + \frac{R_3}{R_3 + R_4}\left[-V_{SAT} - \left(+V_T\right)\right]$$

When the effective voltage at P becomes 0 V, we can write the above equation as

$$+V_T + \frac{R_3}{R_3 + R_4}\left[-V_{SAT} - \left(+V_T\right)\right] = 0 \tag{4.25}$$

The Eq. (4.25) can be simplified as

$$+V_T = \frac{R_3}{R_4}\left(-V_{SAT}\right) \tag{4.26}$$

The integrator formed by OA_2, resistor R, and capacitor C output V_S for one transition or during one-half period, i.e., when $+V_1$ is at its input will be

$$V_S\left(t\right) = -\frac{1}{RC}\int_0^{t_1}\left(+V_1\right)dt = -\frac{V_1}{RC}t_1 \tag{4.27}$$

From the waveform shown in Figure 4.12, and from Eq. (4.27) and the fact that at $t_1 = T/2$, $V_S(t) = [+V_T\text{-}(-V_T)] = 2V_T$, we get

$$2V_T = \frac{-V_1}{RC}\frac{T}{2}$$

$$T = \frac{V_T}{-V_1}4RC;\ f = \frac{-V_I}{4RCV_T} \tag{4.28}$$

The improved voltage-to-frequency converter or the analog dividing function generator circuit is shown in Figure 4.13 and its associated waveforms are shown in Figure 4.14. Let initially the SR flip flop output Q be LOW. $\bar{Q}$ will be HIGH which makes transistor Q_1 to conduct, thereby enabling the control amplifier OA_1 to work as an inverting amplifier. $-V_1$ will be given to the integrator formed by OA_2, resistor R, and capacitor C. Its output will be

$$V_{T1} = -\frac{1}{RC}\int -V_1\,dt = \frac{V_1}{RC}t \tag{4.29}$$

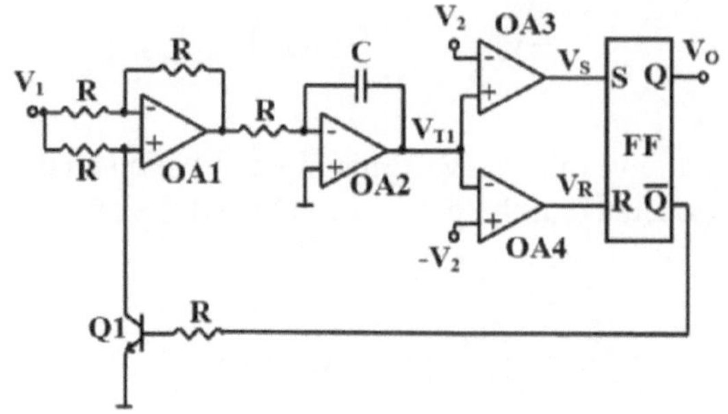

FIGURE 4.13 The improved voltage-to-frequency converter circuit.

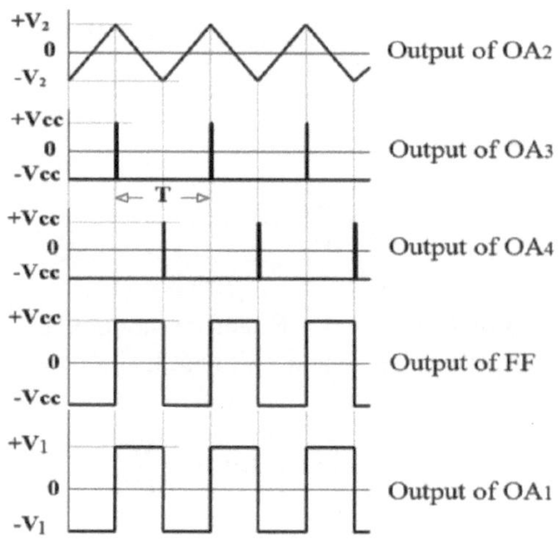

FIGURE 4.14 Associated waveforms of Figure 4.13.

The output of integrator OA_2 is a positive-going ramp. When the output of the integrator exceeds the other input voltage V_2, SR flip flop output Q is set to HIGH by the comparator OA_3. $\bar{Q}$ will become LOW, thereby enabling the control amplifier OA_1 to work as a non-inverting amplifier, and $+V_1$ is given to the integrator OA_2. Now the integrator output will be

$$V_{T1} = -\frac{1}{RC}\int V_1\, dt = -\frac{V_1}{RC}t \tag{4.30}$$

The output of integrator is changing its slope from positive to negative. When the output of the integrator exceeds the input voltage $-V_2$, SR flip flop output Q will be reset to LOW by the comparator OA_4. $\bar{Q}$ will become HIGH, and the cycle therefore repeats. From the Eq. (4.29) and from the waveforms shown in Figure 4.14 at $t = T/2$, $V_T(t) = 2V_2$

$$2V_2 = \frac{V_1}{2RC}T \tag{4.31}$$

$$T = \frac{V_2}{V_1} 4RC, \quad f = \frac{V_1}{V_2} \frac{1}{4RC} \tag{4.32}$$

WORKED PROBLEMS

1. Design a square wave generator for frequency of 1 KHz. Power supply is ±15 V.

   The square wave generator or astable multivibrator is given in Figure 4.4. From Eq. (4.6)

$$T = 2R_3C_1 \ln\left(1 + 2\frac{R_1}{R_2}\right)$$

   Given: frequency = 1 kHz
   $T = 1/f = 1$ mS
   Let $R_2 = 1.16\ R_1$
   Let $R_1 = 10$ K, then $R_2 = 11.6$ K

$$T = 2R_3C_1 \ln 2.7241 = 2R_3C_1$$

$$R_3 = \frac{T}{2C_1}$$

   Let $C_1 = 0.05\ \mu F$,

$$R_3 = \frac{1 \times 10^{-3}}{2 \times 0.05 \times 10^{-6}} = 10K$$

2. Design a Wien bridge oscillator for frequency of 1 KHz. The Wien bridge oscillator circuit is shown in Figure 4.1.

$$f_0 = \frac{1}{2\pi RC}, \; G = 3 = \left(1 + \frac{R_4}{R_3}\right)$$

   Given: $f_O = 1$ KHz. Let $C_1 = C_2 = C = 0.05\ \mu F$.
   Let $R_1 = R_2 = R$

$$R = \frac{1}{2\pi f_O C} = \frac{1}{2\pi \times 1 \times 10^3 \times 0.05 \times 10^{-6}} = 3.1K$$

   Let $R_3 = 10$ K, then $R_4 = 2R_3 = 20$ K.

3. Design a sawtooth wave generator of 5 V peak value and time period of 1 mS. The sawtooth wave generator is shown in Figure 4.7(b). Choose $V_R = 5$ V with LM3365V reference diode.

   Given: T = 1 mS. Let $R_1 = 1M$.

From Eq. (4.12)

$$T = R_1C_1, C_1 = \frac{T}{R_1} = \frac{1\times10^{-3}}{1\times10^{6}} = 1\text{n F}$$

4. Design a triangular wave generator of time period of 20 mS with 10 V peak value. Power supply is ±15 V.
   The triangular wave generator circuit is shown in Figure 4.9(b).
   Given: T = 20 mS, V_T = 10 V, V_{CC} = 15 V
   From Eq. (4.21)

$$\pm V_T = \pm V_{SAT}\frac{R_2}{R_3}; 0.76(\pm V_{CC})\frac{R_2}{R_3}$$

Let R_2 = 10 K.

$$R_3 = \frac{R_2 0.76 V_{CC}}{V_T} = 11.4\text{K}$$

From Eq. (4.22)

$$T = 4R_1C_1\frac{R_2}{R_3} = 3.5R_1C_1$$

Let C_1 = 0.1 μF.

$$R_1 = \frac{T}{3.5C_1} = \frac{20\times10^{-3}}{3.5\times0.1\times10^{-6}} = 57.14\text{K}$$

5. Design a voltage-controlled triangular wave generator of 1 KHz frequency and 10 V peak value. The given control voltage is (−1 V). Power supply is ±15 V.
   Given: f = 1 KHz; T = 1/f = 1 mS, V_I = −1 V, V_T = 10 V, V_{CC} = 15 V
   The voltage-controlled triangular wave generator is shown in Figure 4.11.

$$\pm V_T = \pm V_{SAT}\frac{R_3}{R_4}; 0.76(\pm V_{CC})\frac{R_3}{R_4}$$

$$R_4 = 0.76(V_{CC})\frac{R_3}{V_T}$$

Let R_3 = 10 K.

$$R_4 = 0.76\times15\times\frac{10\times10^{3}}{10} = 11.4\text{K}$$

From Eq. (4.28)

$$f = \frac{-V_I}{4RCV_T}$$

Let C = 0.1 μF.

$$R = \frac{-V_I}{4fCV_T} = \frac{1}{4 \times 1 \times 10^3 \times 0.1 \times 10^{-6} \times 10} = 250\Omega$$

5 Active Filters

A filter is a frequency-selective electric circuit that passes electrical signals of specified frequency and attenuates other frequencies. Filters are classified into passive and active filters. Passive filters are realized with inductors, capacitors, and resistors. Active filters are realized with op-amp in addition to capacitors and resistors. The problem with passive filters is the size of inductors used for low-frequency applications. Active filters do not require inductors at all. For frequencies of more than 1MHz where the op-amp performances are limited to beyond this 1 MHz, passive filters are recommended. However, in many high-frequency applications such as radio transmitters and receiver passive filters have been replaced by a crystal surface acoustic wave (SAW) filter.

A lowpass filter passes low frequencies below the cut-off frequency f_c and attenuates the input signal for frequencies above the cut-off frequency f_c, as shown in Figure 5.1(a). A highpass filter attenuates frequencies above the cut-off frequency f_c and passes the input signal for frequencies above the cut-off frequency f_c, as shown in Figure 5.1(b). A bandpass filter attenuates signal frequencies up to f_L and passes signal frequencies from f_L to f_H, as shown in Figure 5.1(c). A bandstop filter passes signal frequencies up to f_L and attenuates signals from f_L to f_H, as shown in Figure 5.1(d).

5.1 LOWPASS FILTER CIRCUITS

Figure 5.2(a) shows a first-order inverting lowpass filter, and Figure 5.2(b) shows a first-order non-inverting lowpass filter.

The transfer function of both Figures 5.2(a) and 5.2(b) is given as

$$T(s) = \frac{G\omega_0}{s + \omega_0}$$

The passband gain

$$G = -\frac{R_1}{R_2} \text{for Fig}(a) \tag{5.1}$$

$$G = \left(1 + \frac{R_3}{R_2}\right) \text{for Fig}(b) \tag{5.2}$$

The cut-off frequency w_o is given as

$$w_0 = \frac{1}{R_1C_1}, \quad f_0 = \frac{1}{2\pi R_1C_1} \tag{5.3}$$

DOI: 10.1201/9781003221449-6

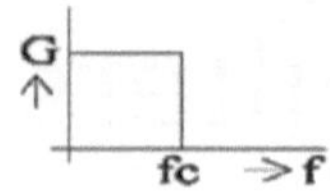

FIGURE 5.1(A) Lowpass filter (LPF).

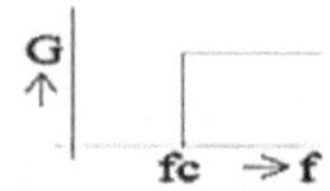

FIGURE 5.1(B) Highpass filter (HPF).

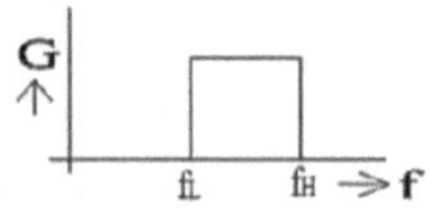

FIGURE 5.1(C) Bandpass filter (BPF).

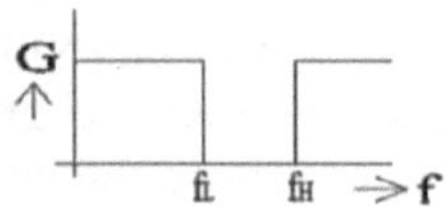

FIGURE 5.1(D) Bandstop filter (BSF).

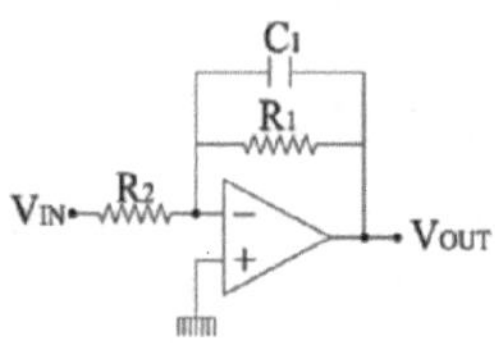

FIGURE 5.2(A) First-order inverting lowpass filter.

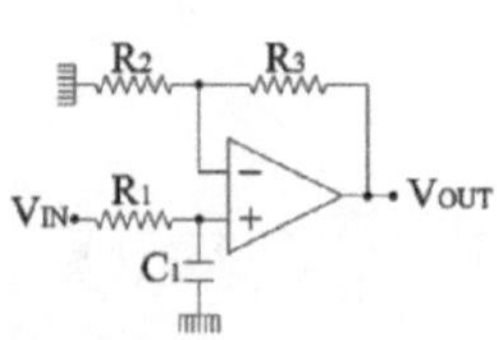

FIGURE 5.2(B) First-order non-inverting lowpass filter.

For both circuits (a) and (b).

These circuits are single op-amp amplifiers with capacitor added to work for first-order lowpass filter. The non-inverting lowpass filter is better for high input impedance to avoid loading effect. Figure 5.3 shows second-order lowpass filter.

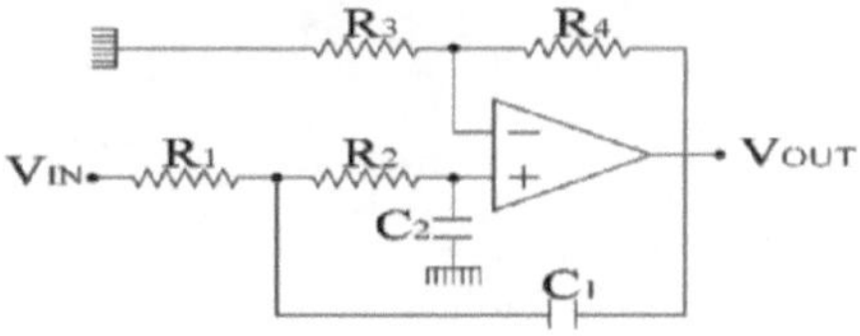

FIGURE 5.3 Second-order lowpass filter.

$$\text{Transfer function} = \frac{\dfrac{1}{R_1R_2C_1C_2}\left(1+\dfrac{R_4}{R_3}\right)}{s^2+\left(\dfrac{1}{R_2C_1}+\dfrac{1}{R_1C_1}-\dfrac{R_4}{R_2R_3C_2}\right)s+\dfrac{1}{R_1R_2C_1C_2}}$$

$$\text{Gain } G = 1+\frac{R_4}{R_3} \tag{5.4}$$

$$\omega_0 = \frac{1}{\sqrt{R_1R_2C_1C_2}}$$

Let $R_1 = R_2$, $C_1 = C_2$

$$f_0 = \frac{1}{2\pi RC} \tag{5.5}$$

5.2 HIGHPASS FILTER CIRCUITS

Figure 5.4 shows first-order highpass filter circuits. Figure 5.4(a) shows an inverting highpass filter, and Figure 5.4(b) shows a non-inverting highpass filter.

The circuits are simple inverting and non-inverting amplifiers with capacitor added to perform highpass filter operation.

$$\text{Transfer function } T(s) = \frac{G_s}{s+\omega_0}$$

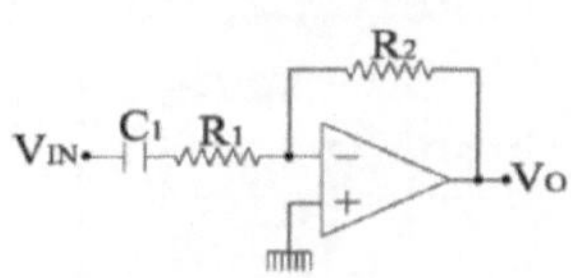

FIGURE 5.4(A) First-order inverting highpass filter.

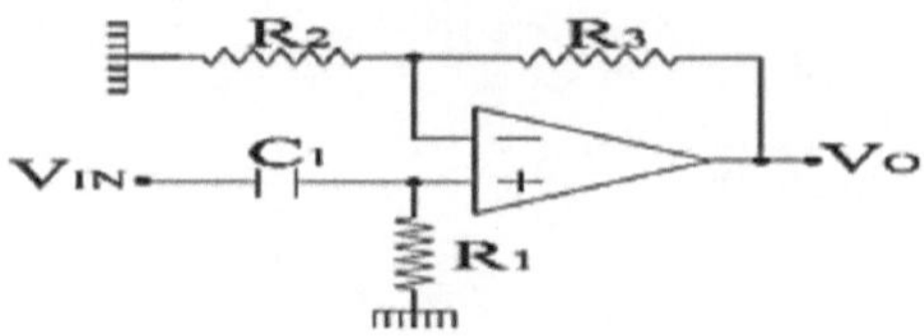

FIGURE 5.4(B) First-order non-inverting highpass filter.

$$\text{Gain } G = -\frac{R_2}{R_1} \text{ for Fig (a)} \tag{5.6}$$

$$G = \left(1 + \frac{R_3}{R_2}\right) \text{ for Fig (b)} \tag{5.7}$$

$$\text{pole frequency } \omega_0 = \frac{1}{R_1 C_1} \tag{5.8}$$

Figure 5.5 shows a second-order highpass filter.

$$\text{Transfer function } T(s) = \frac{\left(1 + \frac{R_4}{R_3}\right) s^2}{s^2 + \left(\frac{1}{R_2 C_1} + \frac{1}{R_2 C_2} - \frac{R_4}{R_1 R_3 C_1}\right) s + \frac{1}{R_1 R_2 C_1 C_2}}$$

$$\text{Gain } G = 1 + \frac{R_4}{R_3} \tag{5.9}$$

$$\text{Pole frequency} = \omega_0 = \frac{1}{\sqrt{R_1 R_2 C_1 C_2}}$$

Let $R_1 = R_2$, $C_1 = C_2$

$$f_0 = \frac{1}{2\pi RC} \tag{5.10}$$

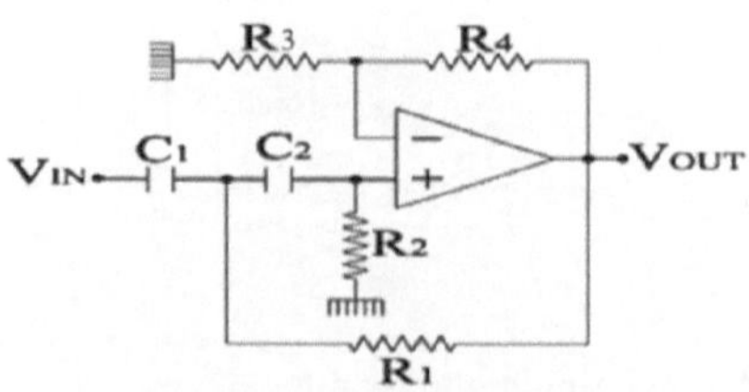

FIGURE 5.5 Second-order highpass filter.

5.3 BANDPASS FILTER CIRCUITS

Figure 5.6(a) shows a first-order bandpass filter.

$$\text{Transfer function } T(s) = \frac{-\dfrac{1}{R_1C_2}s}{\left(s+\dfrac{1}{R_1C_1}\right)\left(s+\dfrac{1}{R_2C_2}\right)}$$

$$\text{Pole frequency } \omega_1 = \frac{1}{R_1C_1} \text{ for low frequency}$$

$$f_L = \frac{1}{2\pi R_1C_1} \tag{5.11}$$

$$\omega_2 = \frac{1}{R_2C_2} \text{ for high frequency}$$

$$f_H = \frac{1}{2\pi R_2C_2} \tag{5.12}$$

$$\text{Mid frequency gain} = G_O = -\frac{R_2}{R_1} \tag{5.13}$$

The circuit is a simple inverting amplifier with two capacitors added to perform bandpass filtering.

Figure 5.7 shows a multiple feedback bandpass filter.

$$\text{Transfer function} = \frac{-\dfrac{1}{R_1C_2}s}{s^2+\left(\dfrac{1}{C_1}+\dfrac{1}{C_2}\right)\dfrac{1}{R_3}s+\dfrac{1}{R_3C_1C_2}\left(\dfrac{1}{R_1}+\dfrac{1}{R_2}\right)}$$

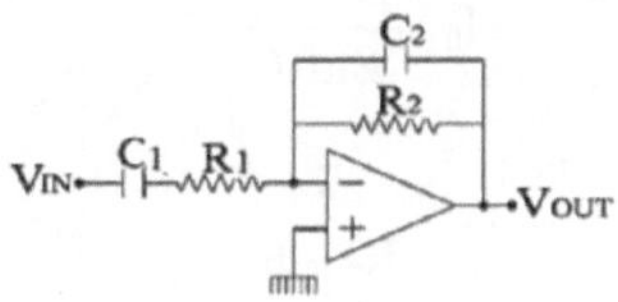

FIGURE 5.6(A) First-order bandpass filter.

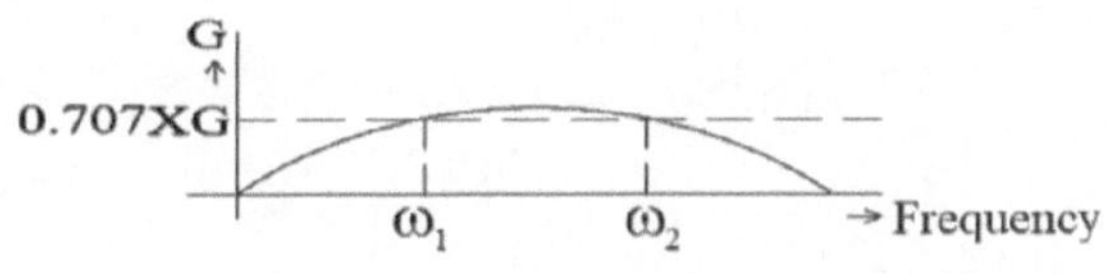

FIGURE 5.6(B) Characteristics of a bandpass filter.

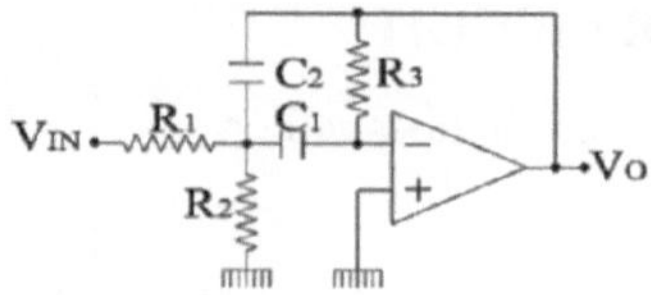

FIGURE 5.7 Multiple feedback bandpass filter.

$$\text{Gain}\, G = \frac{R_3}{R_1}\left(\frac{C_1}{C_1 + C_2}\right) \tag{5.14}$$

$$\omega_0 = \sqrt{\frac{1}{R_3 C_1 C_2}\left(\frac{1}{R_1} + \frac{1}{R_2}\right)} \tag{5.15}$$

5.4 BANDSTOP FILTER CIRCUITS

Figure 5.8 shows a multiple loop feedback band reject filter.

The transfer function is given as

$$\text{Transfer function} = \frac{K\left[s^2 + s\left(\frac{1}{R_2C_1} + \frac{1}{R_2C_2} - \frac{R_3}{R_4}\frac{1}{R_1C_1}\right) + \frac{1}{R_1R_2C_1C_2}\right]}{\left[s^2 + s\left(\frac{1}{R_2C_1} + \frac{1}{R_2C_2}\right) + \frac{1}{R_1R_2C_1C_2}\right]}$$

$$\text{Notch frequency} = \omega_0 = \frac{1}{\sqrt{R_1R_2C_1C_2}}$$

$$\text{Let } R_1 = R_2, C_1 = C_2$$

$$f_0 = \frac{1}{2\pi RC} \tag{5.16}$$

$$G = \frac{R_4}{R_3 + R_4} \tag{5.17}$$

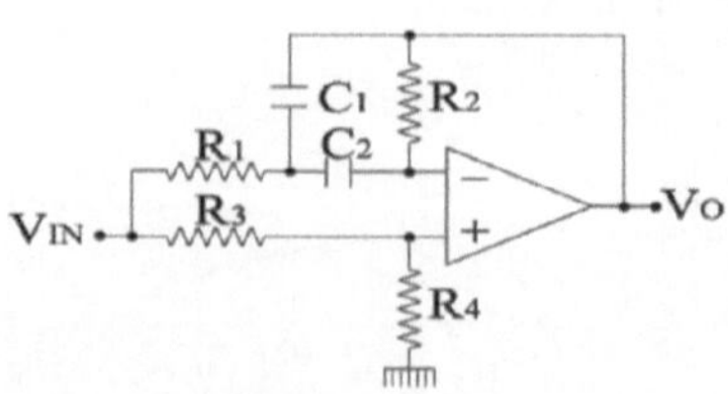

FIGURE 5.8 Multiple loop feedback band reject filter.

$$Q_F = \frac{\left(\frac{R_2}{R_1}\right)^{1/2}}{\left(\frac{C_2}{C_1}\right)^{1/2} + \left(\frac{C_1}{C_2}\right)^{1/2}} \tag{5.18}$$

5.5 ALLPASS FILTER CIRCUITS

An allpass filter is nothing but a phase adjusting circuit with gain magnitude constant.

Figure 5.9(a) shows an allpass filter where the phase difference can vary from 0 to −180°, and Figure 5.9(b) shows an allpass filter where the phase difference can vary from −180° to −360°.

In Figure 5.9(a), if R = 0, the input signal is directly connected to non-inverting amplifier input and produces zero phase shift. If R = HIGH value compared to impedance of C value, the circuit functions like an inverter with a phase difference of −180°.

The gain response of Figure 5.9(a) is given as

$$A_a = \frac{1 - j\omega RC}{1 + j\omega RC} \tag{5.19}$$

The gain response of Figure 5.9(b) is given as

$$A_b = -A_a \tag{5.20}$$

The phase shifts developed by the two circuits are

$$\varnothing_a = -2\tan^{-1}\omega RC \tag{5.21}$$

$$\varnothing_b = -180° + \varnothing_a \tag{5.22}$$

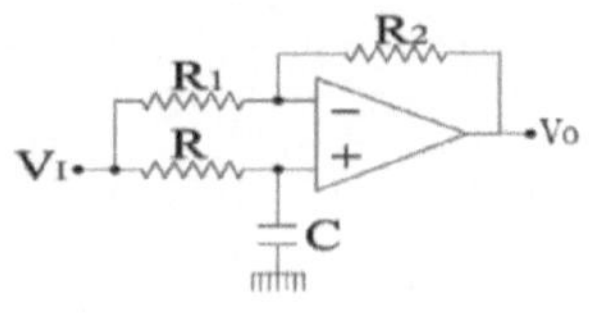

FIGURE 5.9(A) Phase angle between 0 and −180°.

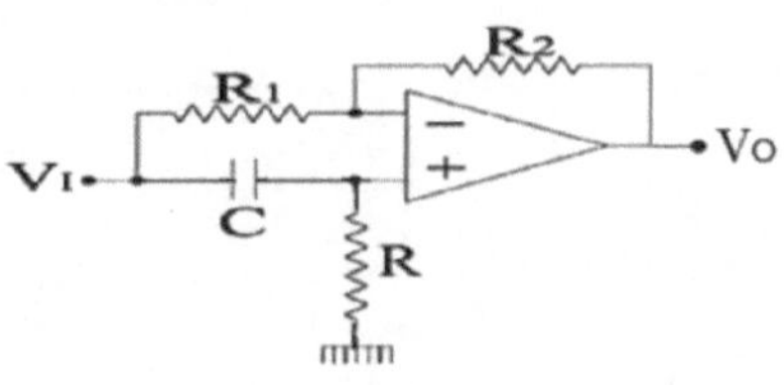

FIGURE 5.9(B) Phase angle between −180° and −360°.

5.6 UNIVERSAL ACTIVE FILTER CIRCUITS

A filter circuit which includes a lowpass filter, highpass filter, and bandpass filter is called a "universal filter" or a "state variable filter." Figure 5.10(a) shows an inverting universal filter, and Figure 5.10(b) shows a non-inverting universal filter. The bandpass response is generated by integrating the highpass response, and lowpass is generated by integrating the bandpass response.

Let us consider the circuit shown in Figure 5.10(a). Using the superposition principle

$$V_{HP} = -\frac{R_5}{R_3}V_I - \frac{R_5}{R_4}V_{LP} + \left(1 + \frac{R_5}{R_3//R_4}\right)\frac{R_1}{R_1 + R_2}V_{BP}$$

$$V_{HP} = -\frac{R_5}{R_3}V_I - \frac{R_5}{R_4}V_{LP} + \frac{1 + R_5/R_3 + R_5/R_4}{1 + R_2/R_1}V_{BP} \tag{5.23}$$

The output of integrator OA_2 is given as

$$V_{BP} = -\frac{1}{R_6C_1s}V_{HP} \tag{5.24}$$

The output of integrator OA_3 is given as

$$V_{LP} = -\frac{1}{R_7C_2s}V_{BP} \tag{5.25}$$

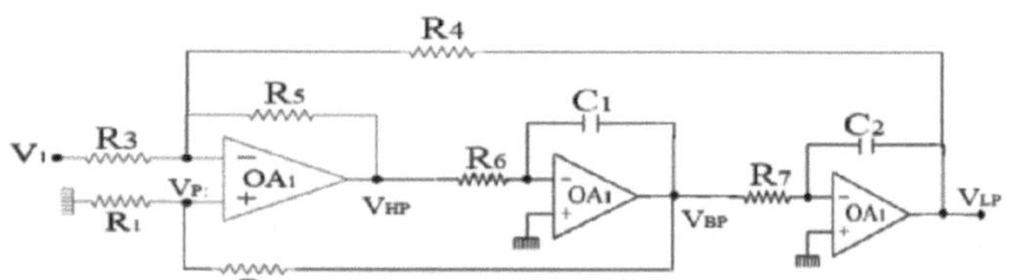

FIGURE 5.10(A) Inverting universal filter.

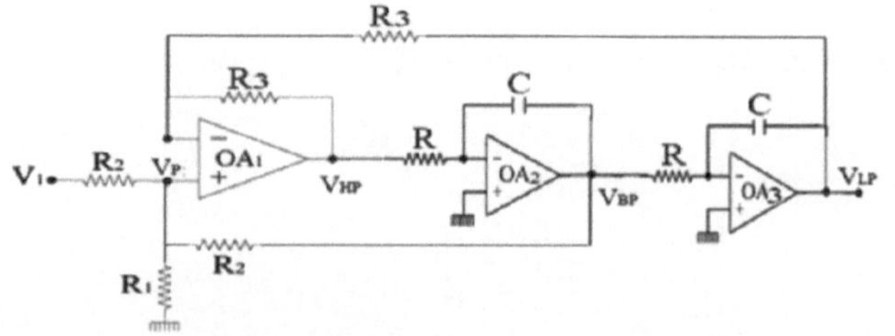

FIGURE 5.10(B) Non-inverting universal filter.

$$\frac{V_{HP}}{V_I} = -\frac{R_5}{R_3}\frac{R_4R_6C_1R_7C_2s^2/R_5}{R_4R_6C_1R_7C_2s^2/R_5 + R_4\left(1+R_5/R_3+R_5/R_4\right)s/\left(1+R_2/R_1\right)R_5+1}$$

The standard form for a highpass filter is given as

$$\frac{V_{HP}}{V_I} = H_{OHP}H_{HP}$$

$$H_{OHP} = -\frac{R_5}{R_3} \tag{5.26}$$

$$\omega_O = \frac{\sqrt{R_5/R_4}}{\sqrt{R_6C_1R_7C_2}}$$

$$Q = \frac{\left(1+R_2/R_1\right)\sqrt{R_5R_6C_1/R_4R_7C_2}}{1+R_5/R_3+R_5/R_4}$$

$$\frac{V_{BP}}{V_I} = \frac{\left(-1/R_6C_1s\right)V_{HP}}{V_I}$$

The standard form for a bandpass filter is given as

$$\frac{V_{BP}}{V_I} = H_{OBP}H_{BP}$$

Comparing equations

$$H_{OBP} = \frac{1+R_2/R_1}{1+R_3/R_4+R_3/R_5} \tag{5.27}$$

$$\frac{V_{LP}}{V_I} = \frac{\left(-1/R_7C_2s\right)V_{BP}}{V_I}$$

The standard form for a lowpass filter is given as

$$\frac{V_{LP}}{V_I} = H_{OLP}H_{LP}$$

Comparing equations

$$H_{OLP} = -\frac{R_4}{R_3} \tag{5.28}$$

Let $R_3 = R_4 = R_5$, $R_6 = R_7 = R$, $C_1 = C_2 = C$.

$$\omega_O = \frac{1}{RC}, \; f_O = \frac{1}{2\pi RC} \tag{5.29}$$

$$Q = \frac{1}{3}\left(1 + \frac{R_2}{R_1}\right)$$

$$H_{OHP} = -1, H_{OBP} = Q, H_{OLP} = -1 \tag{5.30}$$

Let us consider the non-inverting state variable filter shown in Figure 5.10(b). In the way we obtained Figure 5.10(a), Figure 5.10(b) can also be obtained as follows:

$$\omega_O = \frac{1}{RC} \tag{5.31}$$

$$Q = 1 + \frac{R_2}{2R_1}$$

$$H_{OHP} = 1/Q, H_{OBP} = -1, H_{OLP} = 1/Q \tag{5.32}$$

Figure 5.10(c) shows a biquad filter. Let us analyze the circuit. At negative input terminal (–) of op-amp OA_1

$$\frac{V_I}{R_1} + \frac{-V_{LP}}{R_5} + \frac{V_{BP}}{R_2} + \frac{V_{BP}}{1/C_1 s} = 0$$

In the way we obtained Figure 5.10(a), Figure 5.10(c) can also be obtained as follows:

Let $R_4 = R_5 = R$ and $C_1 = C_2 = C$.

$$\omega_O = \frac{1}{RC} \tag{5.33}$$

$$Q = \frac{R_2}{R} \tag{5.34}$$

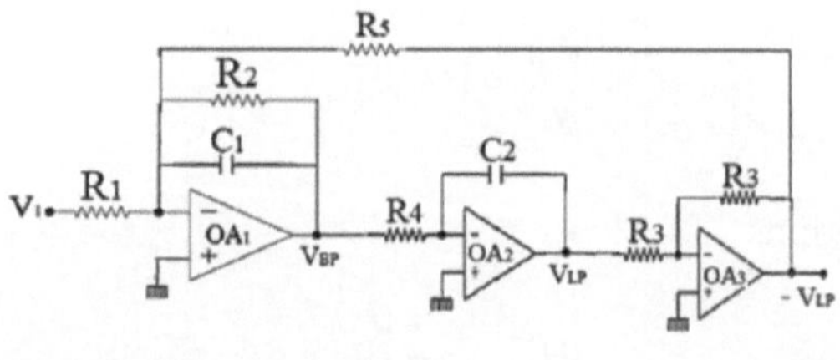

FIGURE 5.10(C) Biquad filter.

$$H_{OBP} = -\frac{R_2}{R_1} \tag{5.35}$$

$$H_{OLP} = \frac{R}{R_1} \tag{5.36}$$

DESIGN EXAMPLES

1. Design a second-order non-inverting lowpass filter having an upper cut-off frequency of 1 KHz and passband gain of 1.586. The second-order non-inverting lowpass filter is shown in Figure 5.3.
   From Eq. (5.5)

$$f_0 = \frac{1}{2\pi R_1 C_1}$$

Let C = 0.1 μF.

$$R_1 = \frac{1}{2\pi f_O C_1} = \frac{1}{2\pi 1 \times 10^3 \times 0.1 \times 10^{-6}} = 1.6K$$

From Eq. (5.4)

$$G = \left(1 + \frac{R_4}{R_3}\right)$$

Let $R_4 = 5.86$ K

$$R_3 = \frac{R_4}{G-1} = \frac{5.86 \times 10^3}{1.586 - 1} = 10K$$

2. Design a second-order non-inverting highpass filter having an upper cut-off frequency of 1 KHz and passband gain of 1.586. The second-order non-inverting highpass filter is shown in Figure 5.5.
   From Eq. (5.5)

$$f_0 = \frac{1}{2\pi R_1 C_1}$$

Let C = 0.1 μF.

$$R_1 = \frac{1}{2\pi f_O C_1} = \frac{1}{2\pi 1 \times 10^3 \times 0.1 \times 10^{-6}} = 1.6K$$

From Eq. (5.9)

$$G = \left(1 + \frac{R_4}{R_3}\right)$$

Let $R_4 = 5.86$ K.

$$R_3 = \frac{R_4}{G-1} = \frac{5.86 \times 10^3}{1.586 - 1} = 10\text{K}$$

3. Design a wide bandpass filter having $f_L = 400$ Hz and $f_H = 2$ KHz and mid-frequency gain of −4. The wide bandpass filter is shown in Figure 5.6(a).
   From Eq. (5.11)

$$f_L = \frac{1}{2\pi R_1 C_1}$$

Let $C_1 = 0.01\mu\text{F}$, $R_1 = \dfrac{1}{2\pi f_L C_1} = \dfrac{1}{2\pi 400 \times 0.01 \times 10^{-6}} = 39.77\text{K}$

From Eq. (5.12)

$$f_H = \frac{1}{2\pi R_2 C_2}$$

Let $C_2 = 0.01$ μF.

$$R_2 = \frac{1}{2\pi f_H C_2} = \frac{1}{2\pi \times 2 \times 10^3 \times 0.01 \times 10^{-6}} = 7.95\text{K}$$

From Eq. (5.13)

Mid-frequency gain $= G = -\dfrac{R_2}{R_1} = 4$, Let $R_2 = 4\,\text{K}$, $R_1 = \dfrac{R_2}{-G_1} = \dfrac{4 \times 10^3}{4} = 1\text{K}$

4. Design a notch filter with a frequency of 1 KHz and a gain of 0.5. The notch filter is shown in Figure 5.8.
   From Eq. (5.16)

$$f_O = \frac{1}{2\pi RC}$$

Let C = 0.01 μF.

$$R = \frac{1}{2\pi f_O C} = \frac{1}{2\pi \times 1 \times 10^3 \times 0.01 \times 10^{-6}} = 15.92\text{K}$$

From Eq. (5.17)

$$G = \frac{R_4}{R_3 + R_4}$$

Let $R_4 = 1$ K.

$$R_3 = \frac{R_4(1-G)}{G} = \frac{1\times10^3(1-0.5)}{0.5} = 1K$$

5. Design a 90° phase shifter at a frequency of 1 KHz. The phase shifter is shown in Figure 5.9(a).
   Let $R_2 = R_1 = 10$ K, C = 0.01 μF.
   From Eq. (5.21)

$$\varnothing_a = -2\tan^{-1}\omega RC$$

$$R = \frac{\tan\varnothing/2}{\omega C} = \frac{1}{2\pi\times1\times10^3\times0.01\times10^{-6}} = 15.9K$$

6. Design a biquad filter with *f*o = 8 KHz, BW = 200 Hz, $H_{OLP} = 0.254$ V/V. The biquad filter circuit is shown in Figure 5.10(c).
   Let $C_1 = C_2 = 1$ nF, $R_4 = R_5 = R$.
   From Eq. (5.33)

$$R = \frac{1}{\omega_O C} = \frac{1}{2\pi 8\times10^3\times10^{-9}} = 19.89K\Omega$$

$$Q = \frac{f_O}{BW} = \frac{8\times10^3}{200} = 40$$

From Eq. (5.34)

$$R_2 = QR = 40\times19.89K = 795.6K\Omega$$

From Eq. (5.36)

$$R_1 = \frac{R}{H_{OLP}} = \frac{19.89\times10^3}{0.254} = 78.3K\Omega$$

Part B

Principles of Function Circuits

6 Principles of Multipliers – Multiplexing

Multipliers are classified into: (i) time division multipliers, (ii) peak responding multipliers, and (iii) pulse position responding multipliers. Peak responding multipliers are further classified into (i) peak detecting multipliers and (ii) peak sampling multipliers. Types of peak responding multipliers are (i) double single slope, (ii) double dual slope, and (iii) pulse width integrated multipliers.

If the width of a pulse train is made proportional to one voltage and the amplitude of the same pulse train to a second voltage, then the average value of this pulse train is proportional to the product of two voltages and is called "time division multiplier," "pulse averaging multiplier," or "sigma delta multiplier." The time division multiplier can be implemented using a triangular wave, a sawtooth wave, and without using any reference waves.

A short pulse/sawtooth waveform whose time period 'T' is proportional to one voltage is generated. Another input voltage is integrated during the time period 'T'. The peak value of the integrated voltage is proportional to the product of the two input voltages. This is called a "double single slope peak responding multiplier." A square/triangular waveform whose time period 'T' is proportional to one voltage is generated. Another input voltage is integrated during the time period 'T'. The peak value of the integrated voltage is proportional to the product of the two input voltages. This is called a "double dual slope peak responding multiplier." A rectangular pulse waveform whose OFF time is proportional to one voltage is generated. Another voltage is integrated during this OFF time. The peak value of integrated output is proportional to the product of the two input voltages. This is called a "pulse width integrated peak responding multiplier." At the output stage of a peak responding multiplier, if peak detector is used, it is called a "peak detecting multiplier" and if sample and hold is used, it is called a "peak sampling multiplier."

In general, multipliers can be implemented with (i) analog multiplexers and (ii) analog switches. The principles of multipliers designed with analog multiplexers are described in this chapter, and the principles of multipliers designed with analog switches are described in Chapter 7.

6.1 SAWTOOTH WAVE–BASED TIME DIVISION MULTIPLIER

The sawtooth wave–based time division multiplier is shown in Figure 6.1, and its associated waveforms are shown in Figure 6.2. A sawtooth wave V_{S1} of peak value V_R and time period 'T' is generated by a sawtooth wave generator.

DOI: 10.1201/9781003221449-8

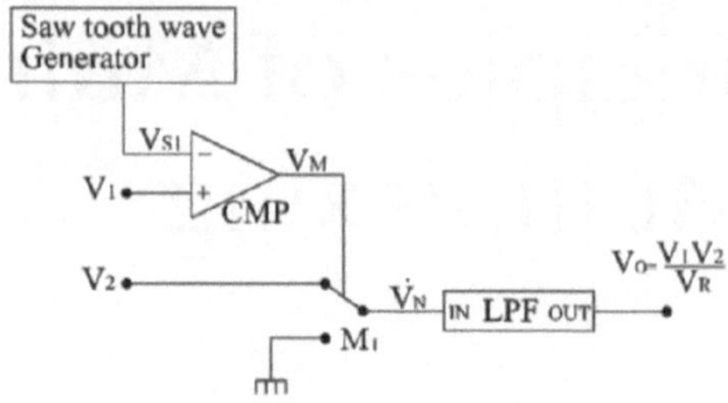

FIGURE 6.1 Sawtooth wave–based time division multiplier.

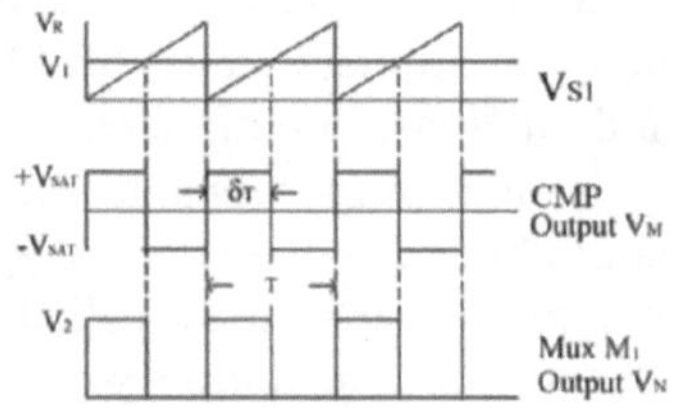

FIGURE 6.2 Associated waveforms of Figure 6.1.

The comparator (CMP) compares the sawtooth wave V_{S1} of peak value V_R with the input voltage V_1 and produces a rectangular waveform V_M at its output. The ON time δ_T of this rectangular waveform V_M is given as

$$\delta_T = \frac{V_1}{V_R} T \tag{6.1}$$

The rectangular pulse V_M controls the multiplexer M_1. When V_M is HIGH, another input voltage V_2 is connected to a lowpass filter. When V_M is LOW, zero volts is connected to a lowpass filter. Another rectangular pulse V_N with maximum value of V_2 is generated at the multiplexer M_1 output. The lowpass filter gives an average value of this pulse train V_N and is given as

$$V_O = \frac{1}{T}\int_0^{\delta_T} V_2 dt \tag{6.2}$$

$$V_O = \frac{V_2}{T}\delta_T \tag{6.3}$$

Equation (6.1) in Equation (6.3) gives

$$V_O = \frac{V_1 V_2}{V_R} \tag{6.4}$$

6.2 TRIANGULAR WAVE–BASED TIME DIVISION MULTIPLIER

The triangular wave–based multiplier is shown in Figure 6.3 and its associated waveforms are shown in Figure 6.4. A triangular wave V_{T1} with $\pm V_T$ peak-to-peak value and time period 'T' is generated by a triangular wave generator.

One input voltage V_1 is compared with the generated triangular wave V_{T1} by the CMP. An asymmetrical rectangular waveform V_M is generated at the CMP output. From the waveforms shown in Figure 6.4, it is observed that

$$T_1 = \frac{V_T - V_1}{2V_T} T$$

$$T_2 = \frac{V_T + V_1}{2V_T} T$$

$$T = T_1 + T_2 \tag{6.5}$$

This rectangular wave V_M is given as control input to the multiplexer M_1. The multiplexer M_1 connects the other input voltage $(+V_2)$ during T_2 and $(-V_2)$ during T_1.

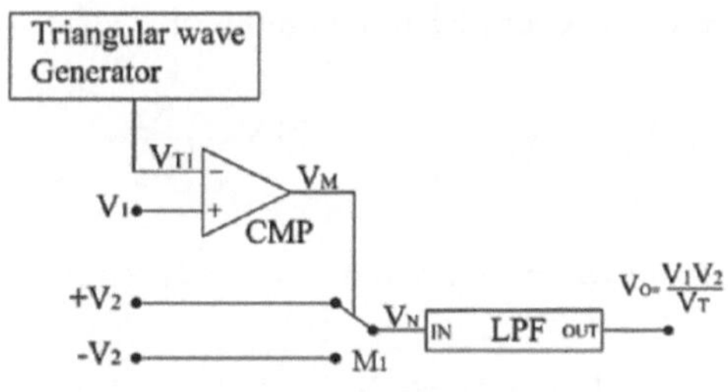

FIGURE 6.3 Triangular wave–based time division multiplier.

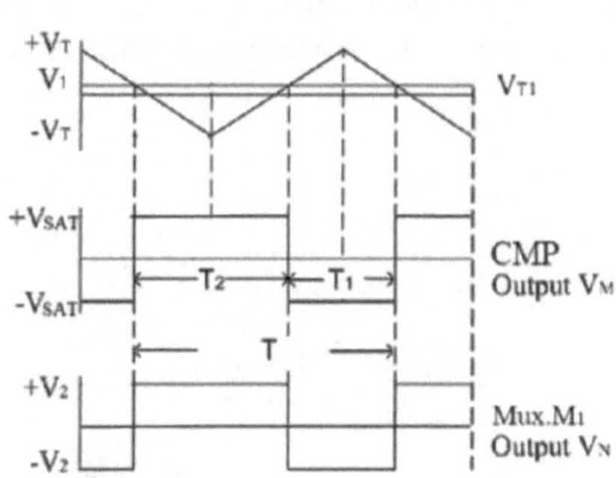

FIGURE 6.4 Associated waveforms of Figure 6.3.

Another rectangular asymmetrical wave V_N with peak-to-peak value of $\pm V_2$ is generated at the multiplexer M_1 output. The lowpass filter gives average value of the asymmetrical wave V_N which is given as

$$V_O = \frac{1}{T}\left[\int_O^{T_2} V_2\,dt + \int_{T_2}^{T_1+T_2} (-V_2)\,dt\right] \tag{6.6}$$

$$V_O = \frac{V_2}{T}\left(T_2 - T_1\right) \tag{6.7}$$

Equation (6.5) in Equation (6.7) gives

$$V_O = \frac{V_1 V_2}{V_T} \tag{6.8}$$

6.3 TIME DIVISION MULTIPLIER WITH NO REFERENCE WAVES

The time division multiplier using no triangular or sawtooth waves as reference is shown in Figure 6.5, and its associated waveforms are shown in Figure 6.6. The differential integrator and CMP constitute an asymmetrical rectangular wave generator, and a rectangular wave V_C is generated at the output of CMP.

$$\pm V_T = \pm V_{SAT}\frac{R_2}{R_3};\ 0.76\left(\pm V_{CC}\right)\frac{R_2}{R_3} \tag{6.9}$$

From the waveforms shown in Figure 6.6. It is observed that

$$T_1 = \frac{V_{SAT} - V_1}{2V_{SAT}}T \quad , T_2 = \frac{V_{SAT} + V_1}{2V_{SAT}}T \quad , T = T_1 + T_2 \tag{6.10}$$

The rectangular wave V_C controls multiplexer M_1, and this multiplexer selects $(+V_2)$ during T_2 and $(-V_2)$ during T_1 to its output. Another asymmetrical rectangular

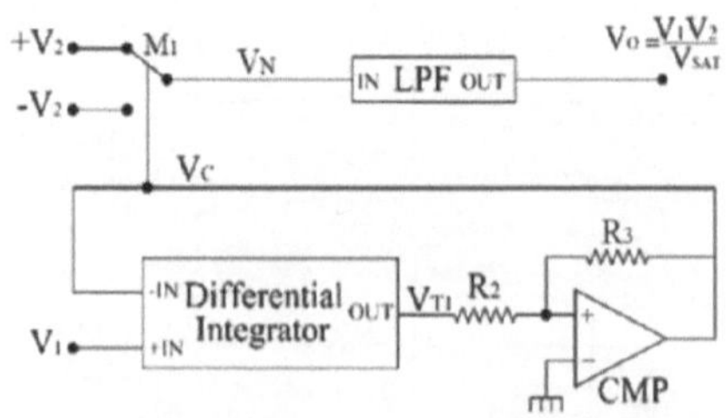

FIGURE 6.5 Time division multiplier with no reference waves.

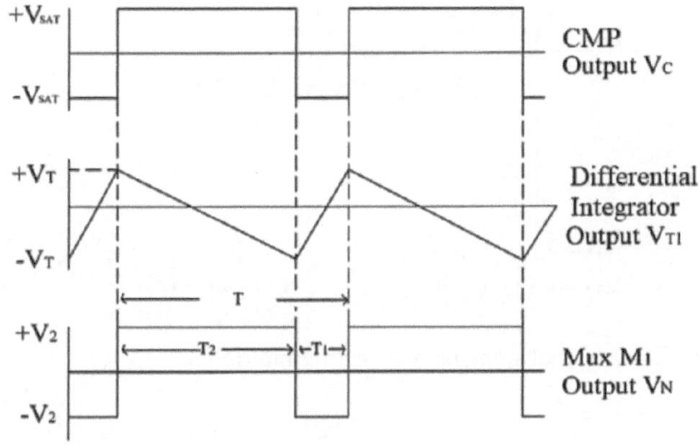

FIGURE 6.6 Associated waveforms of Figure 6.5.

wave V_N is generated at the multiplexer M_1 output with $\pm V_2$ as peak-to peak-values. The lowpass filter gives an average value of this pulse train V_N and is given as

$$V_O = \frac{1}{T}\left[\int_O^{T_2} V_2\,dt + \int_{T_2}^{T_1+T_2} (-V_2)\,dt\right] = \frac{V_2}{T}\left[T_2 - T_1\right] \tag{6.11}$$

Equations (6.10) in (6.11) give

$$V_O = \frac{V_1 V_2}{V_{SAT}} \tag{6.12}$$

6.4 DOUBLE DUAL SLOPE PEAK RESPONDING MULTIPLIERS WITH FEEDBACK COMPARATOR

The double dual slope peak responding multipliers are shown in Figure 6.7, and their associated waveforms are shown in Figure 6.8. Figure 6.7(a) shows a double dual slope peak detecting multiplier, and Figure 6.7(b) shows a double dual slope peak sampling multiplier. Let initially the comparator CMP_1 output be LOW. The multiplexer M_1 selects $(-V_1)$ to the one end of resistor R_3, $(-V_{SAT})$ is given to integrator-I. The multiplexer M_2 selects $(-V_2)$ to integrator-II. The integrator-I output is given as

$$V_{T1} = -\frac{1}{R_1 C_1}\int -V_{SAT}\,dt = \frac{V_{SAT}}{R_1 C_1}t \tag{6.13}$$

where R_1C_1 are resistor capacitor values in integrator-I.

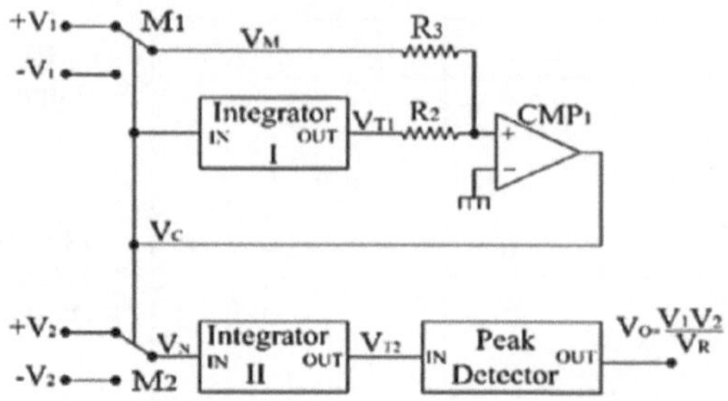

FIGURE 6.7(A) Double dual slope peak detecting multiplier.

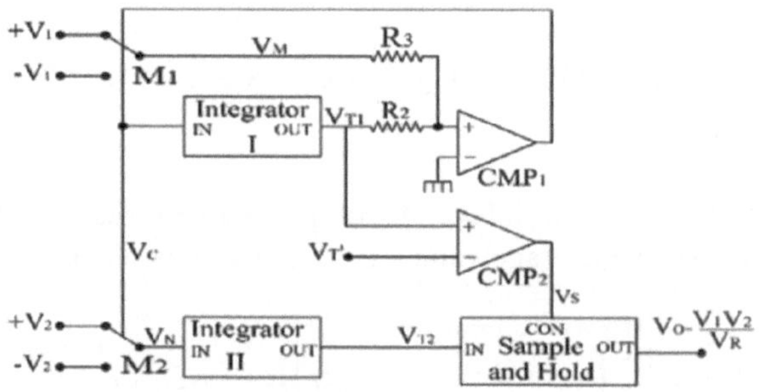

FIGURE 6.7(B) Double dual slope peak sampling multiplier.

The integrator-II output is given as

$$V_{T2} = -\frac{1}{R_4C_2}\int -V_2\,dt = \frac{V_2}{R_4C_2}t \tag{6.14}$$

where R_4C_2 are resistor capacitor values in integrator-II.

The output of integrator-I is rising toward positive saturation and when it reaches a value of $+V_T$, the comparator CMP_1 output becomes HIGH. The multiplexer M_1 selects $(+V_1)$ to the one end of resistor R_3, $(+V_{SAT})$ is given to integrator-I. The multiplexer M_2 selects $(+V_2)$ to integrator-II.

The output of integrator-I will now be

$$V_{T1} = -\frac{1}{R_1C_1}\int V_{SAT}\,dt = -\frac{V_{SAT}}{R_1C_1}t \tag{6.15}$$

and the output of integrator-II will be

$$V_{T2} = -\frac{1}{R_4C_2}\int V_2\,dt = -\frac{V_2}{R_4C_2}t \tag{6.16}$$

The output of integrator-I changes its slope and is going toward negative saturation. When the output of integrator-I comes down to a value $(-V_T)$, the comparator

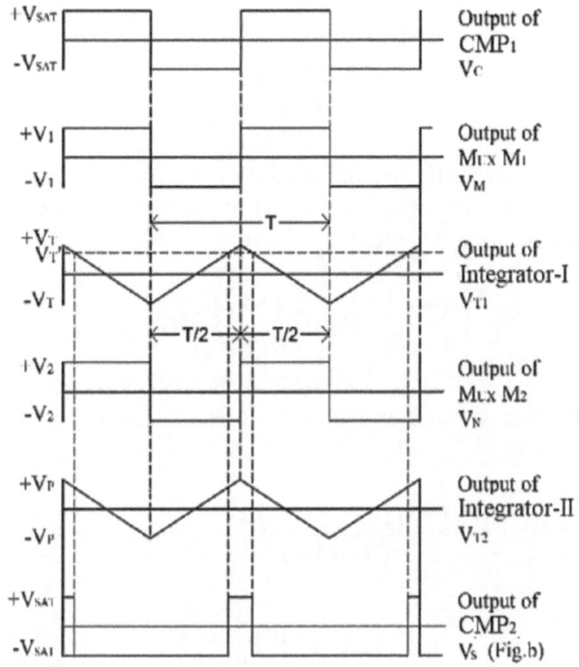

FIGURE 6.8 Associated waveforms of Figure 6.7.

CMP_1 output will become LOW; therefore, the cycle repeats itself to give (1) a triangular wave V_{T1} at the output of integrator-I with $\pm V_T$ peak-to- peak value, (2) another triangular wave V_{T2} at the output of integrator-II with $\pm V_P$ as peak-to-peak value, (3) first square waveform V_C with $\pm V_{SAT}$ as peak-to-peak values at the output of comparator CMP_1, (4) second square waveform V_M with $\pm V_1$ as peak-to-peak values at the output of multiplexer M_1, and (5) third square waveform V_N with $\pm V_2$ as peak-to-peak values at the output of multiplexer M_2. From the waveforms shown in Figure 6.8, from Eqs. (6.13) and (6.14), and the fact that at t = T/2, $V_{T1} = 2V_T$, $V_{T2} = 2V_P$.

$$2V_T = \frac{V_{SAT}}{R_1C_1}\frac{T}{2} \tag{6.17}$$

$$2V_P = \frac{V_2}{R_4C_2}\frac{T}{2} \tag{6.18}$$

From Eqs. (6.17) and (6.18)

$$2V_P = \frac{V_2}{R_4C_2}\frac{2V_TR_1C_1}{V_{SAT}}$$

Let us assume $R_1 = R_4$, $C_1 = C_2$.

$$V_P = \frac{V_2}{V_{SAT}} V_T \tag{6.19}$$

When the comparator CMP_1, output is LOW ($-V_{SAT}$), ($-V_1$) will be at output of multiplexer M_1, the effective voltage at the non-inverting terminal of comparator OA_2 will be by the superposition principle

$$\frac{(-V_1)}{(R_2 + R_3)} R_2 + \frac{(+V_T)}{(R_2 + R_3)} R_3$$

When this effective voltage at the non-inverting terminal of comparator OA_2 becomes zero

$$\frac{(-V_1)R_2 + (+V_T)R_3}{(R_2 + R_3)} = 0$$

$$(+V_T) = (+V_1)\frac{R_2}{R_3}$$

When the comparator OA_2 output is HIGH ($+V_{SAT}$), the effective voltage at the non-inverting terminal of comparator OA_2 will be by the superposition principle.

$$\frac{(+V_1)}{(R_2 + R_3)} R_2 + \frac{(-V_T)}{(R_2 + R_3)} R_3$$

When this effective voltage at the non-inverting terminal of comparator OA_2 becomes zero

$$\frac{(+V_1)R_2 + (-V_T)R_3}{(R_2 + R_3)} = 0$$

$$(-V_T) = (-V_1)\frac{R_2}{R_3}$$

$$\pm V_T = \pm V_1 \frac{R_2}{R_3} \tag{6.20}$$

Equation (6.20) in (6.19) gives

$$V_P = \frac{V_1 V_2}{V_{SAT}} \frac{R_2}{R_3}$$

Let

$$V_R = \frac{V_{SAT}}{R_2} R_3$$

$$V_P = \frac{V_1 V_2}{V_R} \tag{6.21}$$

In the circuit shown in Figure 6.7(a), the peak detector gives this peak value V_P at the output: $V_O = V_P$.

In the circuit shown in Figure 6.7(b), the peak value V_P is obtained by the sample and hold circuit. The sampling pulse V_S is generated by comparator CMP_2 by comparing a slightly less than voltage of V_T called V_T' with the triangular wave V_{T1}. The sample and hold operation is illustrated in Figure 6.8. The sampled output is given as $V_O = V_P$.

From Eq. (6.21), the output voltage is given as $V_O = V_P$.

$$V_O = \frac{V_1 V_2}{V_R} \tag{6.22}$$

6.5 DOUBLE DUAL SLOPE PEAK RESPONDING MULTIPLIER WITH FLIP FLOP

The multiplier using double dual slope peak responding principle with flip flop is shown in Figure 6.9 and its associated waveforms are shown in Figure 6.10. Figure 6.9(a) shows peak detecting multiplier and Figure 6.9(b) shows peak sampling multiplier. Let initially flip flop output be LOW. ($-V_{SAT}$) is given to integrator-I. The integrator-I output is given as

$$V_{T1} = -\frac{1}{R_1 C_1} \int (-V_{SAT})\, dt = V_{SAT} t \tag{6.23}$$

where $R_1 C_1$ are resistor capacitor values in integrator-I.

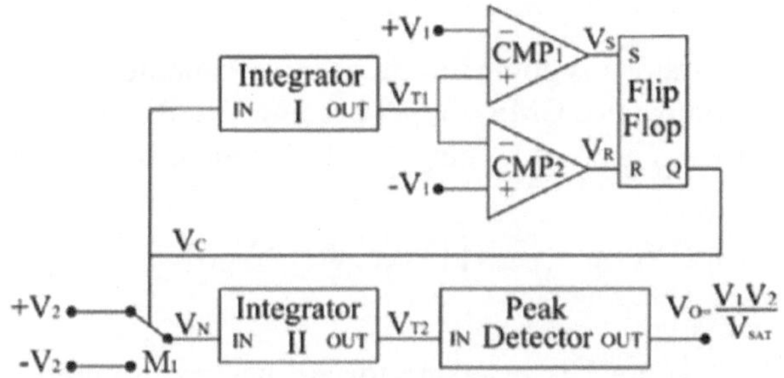

FIGURE 6.9(A) Double dual slope peak detecting multiplier with flip flop.

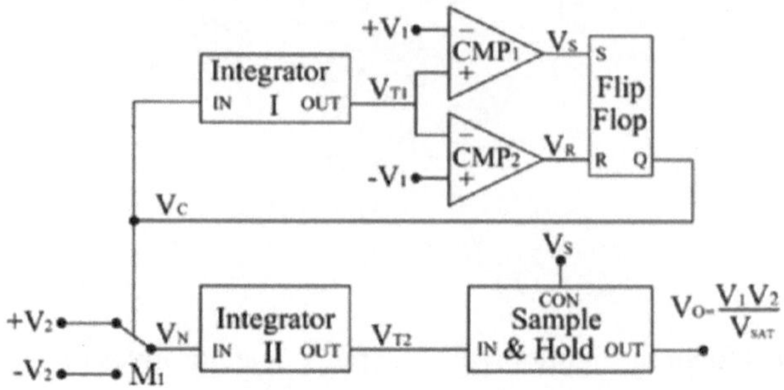

FIGURE 6.9(B) Double dual slope peak sampling multiplier with flip flop.

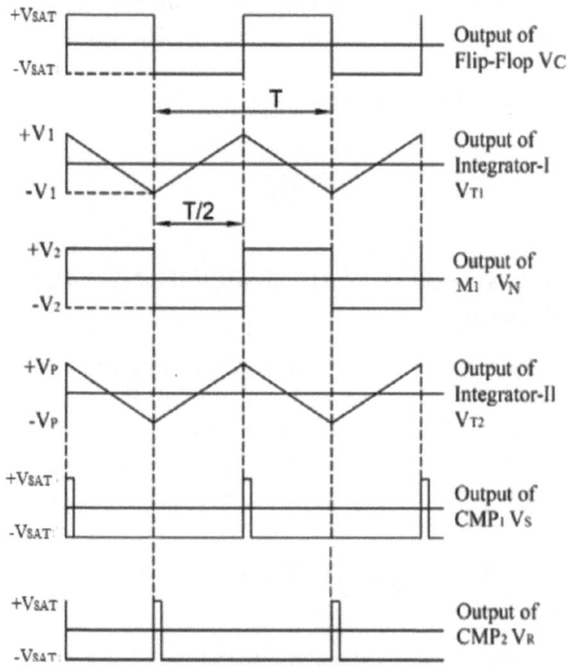

FIGURE 6.10 Associated waveforms of Figure 6.9.

The output of integrator-I is going toward positive saturation, and when it reaches the value $+V_1$, the comparator CMP_1 output becomes HIGH, and it sets the flip flop output to HIGH. ($+V_{SAT}$) is given to integrator-I. The integrator-I output is given as

$$V_{T1} = -\frac{1}{R_1C_1}\int\left(V_{SAT}\right)dt = -V_{SAT}t \tag{6.24}$$

The output of integrator-I is reversing toward negative saturation, and when it reaches the value $(-V_1)$, the comparator CMP_2 output becomes HIGH and resets the flip flop so that its output becomes LOW. ($-V_{SAT}$) is given to integrator-I and the sequence repeats to give (i) a triangular waveform V_{T1} of $\pm V_1$ peak-to-peak values with a time period 'T' at the output of integrator-I, and (ii) a square waveform V_C at the output of flip flop. From the waveforms shown in Figure 6.10, Eq. (6.23), and the fact that at $t = T/2$, $V_{T1} = 2V_1$.

$$2V_1 = V_{SAT}\frac{T}{2}$$

$$T = \frac{4V_1}{V_{SAT}} \tag{6.25}$$

The multiplexer M_1 connects $+V_2$ during ON time of V_C and $-V_2$ during OFF time of V_C. Another square waveform V_N with $\pm V_2$ peak-to-peak value is generated at the output of multiplexer M_1. This square waveform V_N is converted into a triangular wave V_{T2} by integrator-II with $\pm V_P$ as peak-to-peak values of same time period 'T'. For one transition, integrator-II output is given as

$$V_{T2} = -\frac{1}{R_2C_2}\int(-V_2)\,dt = V_2t \tag{6.26}$$

where R_2C_2 are resistor capacitor values in integrator-II.

From the waveforms shown in Figure 6.10, Eq. (6.26), and the fact that at $t = T/2$, $V_{T2} = 2V_p$.

$$2V_P = V_2\frac{T}{2}$$

$$V_P = \frac{V_1V_2}{V_{SAT}} \tag{6.27}$$

In Figure 6.9(a), the peak detector at output stage gives peak value V_P of triangular wave V_{T2}. Hence $V_O = V_P$.

In Figure 6.9(b), the peak value V_P is obtained by the sample and hold circuit. The comparator CMP_1 output V_S is acting as a sampling pulse. The sampled output is given as $V_O = V_P$.

$$V_O = \frac{V_1V_2}{V_{SAT}} \tag{6.28}$$

6.6 PEAK RESPONDING MULTIPLIERS WITH V/T CONVERTER

The multiplexing type peak responding multipliers using voltage-to-period (V/T) converter are shown in Figure 6.11, and their associated waveforms are shown in

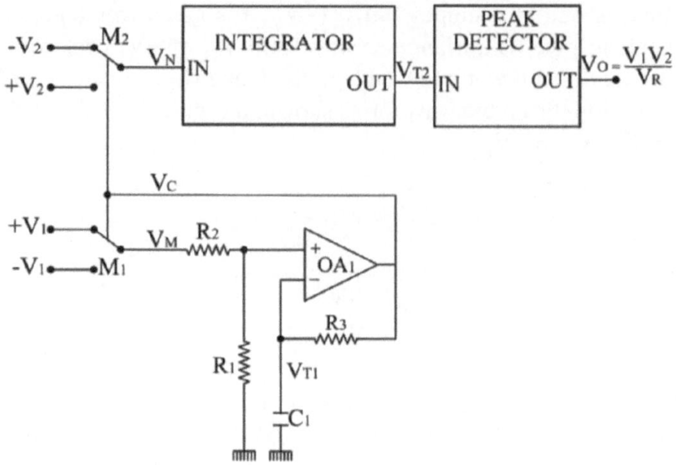

FIGURE 6.11(A) Peak detecting multiplier using voltage-to-period converter.

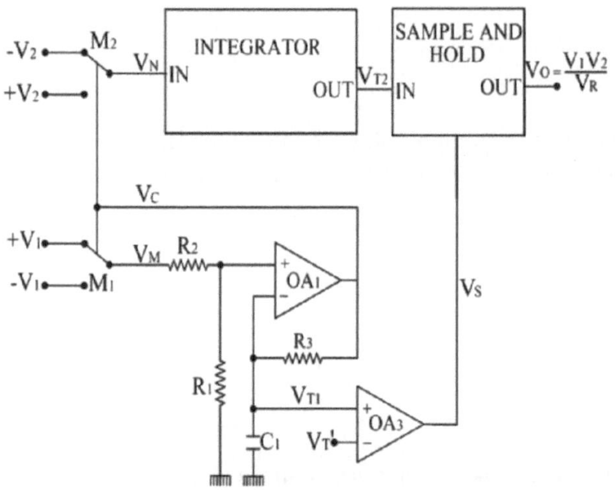

FIGURE 6.11(B) Peak sampling multiplier using voltage-to-period converter.

Figure 6.12. Figure 6.11(a) shows a peak detecting multiplier, and Figure 6.11(b) shows a peak sampling multiplier. A square wave V_C is generated by op-amp OA_1, resistors R_1, R_2, R_3, capacitor C_1, and multiplexer M_1. The square wave V_C controls multiplexer M_2. The multiplexer M_2 connects $(-V_2)$ during HIGH time of V_C and $(+V_2)$ during LOW time of V_C. Another square wave V_N with $\pm V_2$ peak-to-peak values is generated at the multiplexer M_2 output. The integrator converts the square wave V_N into a triangular wave V_{T2} of $\pm V_P$ peak-to-peak values. V_P is proportional to V_2.

$$V_P = K_1 V_2 \tag{6.29}$$

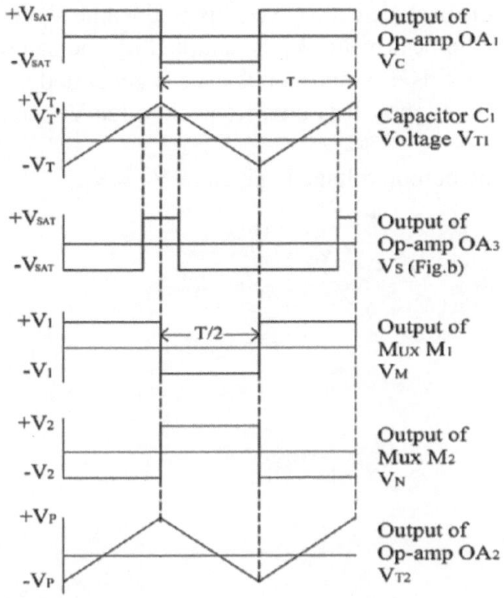

FIGURE 6.12 Associated waveforms of Figure 6.11.

The time period 'T' of square wave V_C is proportional to V_1

$$T = K_2 V_1 \tag{6.30}$$

The integrator output is given as

$$V_{T2} = \frac{1}{R_4C_2}\int V_2\,dt = \frac{V_2}{R_4C_2}t \tag{6.31}$$

where R_4C_2 are the resistor capacitor values of the integrator.

For a half time period t = T/2. From the waveforms shown in Figure 6.12 and from Eq. (6.31) at $t = T/2$, $V_{T2} = 2\,V_P$.

$$2V_P = \frac{V_2}{R_4C_2}\frac{T}{2};\; V_P = \frac{V_2}{4R_4C_2}V_1K_2 \tag{6.32}$$

K_1 and K_2 are constant values

Let

$$V_R = \frac{4R_4C_2}{K_2}$$

$$V_P = \frac{V_1V_2}{V_R} \tag{6.33}$$

In Figure 6.11(a), the peak detector gives this peak value V_P at the output: $V_O = V_P$.

In Figure 6.11(b), the peak value V_P is sampled by the sample and hold circuit with a sampling pulse V_S. The sampling pulse V_S is generated by comparing capacitor C_1 voltage V_{T1} with slightly less than its peak value V_T, i.e., V_T'. The sampled output $V_O = V_P$.

From Eq. (6.33), the output voltage is given as $V_O = V_P$.

$$V_O = \frac{V_1 V_2}{V_R} \tag{6.34}$$

7 Principles of Multipliers – Switching

As discussed next in Chapter 8, multipliers are classified into (i) time division multipliers (TDMs), (ii) peak responding multipliers, and (iii) pulse position responding multipliers. Peak responding multipliers are further classified into (i) peak detecting multipliers and (ii) peak sampling multipliers. Types of peak responding multipliers are (i) double single slope, (ii) double dual slope, and (iii) pulse width integrated multipliers.

If the width of a pulse train is made proportional to one voltage, and the amplitude of the same pulse train to a second voltage, then the average value of this pulse train is proportional to the product of two voltages, and it is called "time division multiplier," "pulse averaging multiplier," or "sigma delta multiplier." The TDM can be implemented using (1) triangular wave (2) saw tooth wave and (3) without using any reference wave.

A short pulse/sawtooth waveform whose time period 'T' is proportional to one voltage is generated. Another input voltage is integrated during the time period 'T'. The peak value of the integrated voltage is proportional to the product of the two input voltages. This is called "double single slope peak responding multiplier." A square/triangular waveform whose time period 'T' is proportional to one voltage is generated. Another input voltage is integrated during the time period 'T'. The peak value of the integrated voltage is proportional to the product of the two input voltages. This is called "double dual slope peak responding multiplier." A rectangular pulse waveform whose OFF time is proportional to one voltage is generated. Another voltage is integrated during this OFF time. The peak value of integrated output is proportional to the product of the two input voltages. This is called "pulse width integrated peak responding multiplier." At the output stage of a peak responding multiplier, if peak detector is used, it is called "peak detecting multiplier" and if sample and hold is used, it is called "peak sampling multiplier."

In general, multipliers can be implemented with (i) analog multiplexers and (ii) analog switches. The principles of analog multipliers designed with analog multiplexers are described in Chapter 6, and the principles of multipliers with analog switches are described in this chapter.

7.1 SAWTOOTH WAVE–BASED TIME DIVISION MULTIPLIERS

The sawtooth wave–based TDMs are shown in Figure 7.1, and their associated waveforms are shown in Figure 7.2. Figure 7.1(a) shows series switching TDM, and Figure 7.1(b) shows shunt or parallel switching TDM.

The comparator (CMP) compares the sawtooth wave V_{S1} of peak value V_R with an input voltage V_1 and produces a rectangular waveform V_M.

DOI: 10.1201/9781003221449-9

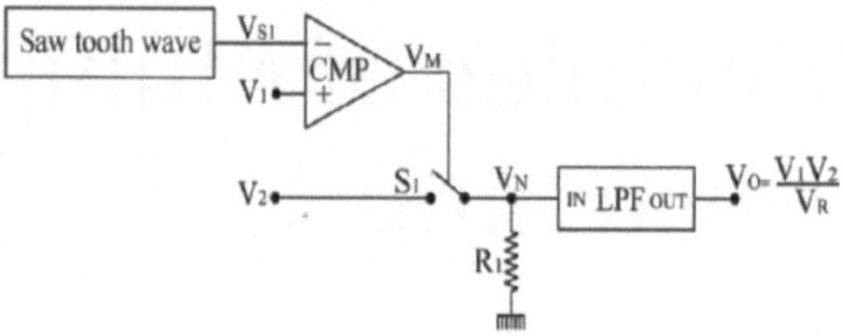

FIGURE 7.1(A) Series switching sawtooth wave–based time division multiplier.

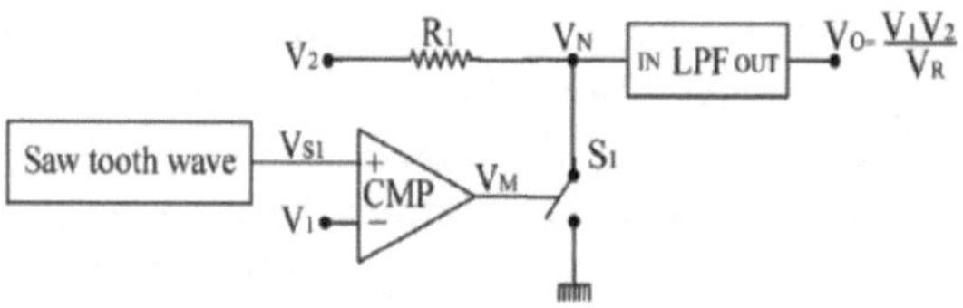

FIGURE 7.1(B) Shunt switching sawtooth wave–based time division multiplier.

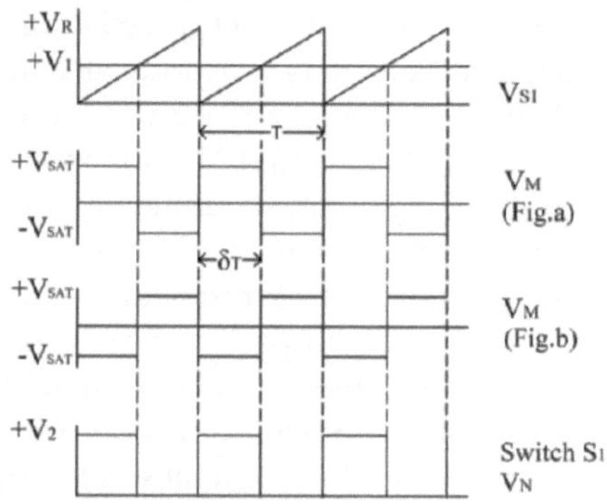

FIGURE 7.2 Associated waveforms of Figure 7.1.

The ON time, Figure 7.1(a), or the OFF time, Figure 7.1(b), of V_M is given as

$$\delta_T = \frac{V_1}{V_R} T \tag{7.1}$$

The rectangular pulse V_M controls the switch S_1. In Figure 7.1(a), when V_M is HIGH, the switch S_1 is closed and another input voltage V_2 is connected to a lowpass filter. When V_M is LOW, the switch S_1 is opened and zero volts exists on the lowpass filter. In Figure 7.1(b), when V_M is HIGH, the switch S_1 is closed and zero volts exists on the lowpass filter. When V_M is LOW, the switch S_1 is opened and another input voltage V_2 is connected to the lowpass filter.

Another rectangular pulse V_N with maximum value V_2 is generated at the switch S_1 output. The lowpass filter gives average value of this pulse train V_N and is given as

$$V_O = \frac{1}{T}\int_0^{\delta_T} V_2 dt \tag{7.2}$$

$$V_O = \frac{V_2}{T}\delta_T \tag{7.3}$$

Equation (7.1) in (7.3) gives

$$V_O = \frac{V_1 V_2}{V_R} \tag{7.4}$$

7.2 TRIANGULAR WAVE–BASED TIME DIVISION MULTIPLIER

The triangular wave–based multiplier is shown in Figure 7.3, and its associated waveforms are shown in Figure 7.4.

One input voltage V_1 is compared with the triangular wave V_{T1} by the CMP. An asymmetrical rectangular waveform V_M is generated at the CMP output. From the waveforms shown in Figure 7.4, it is observed that

$$T_1 = \frac{V_T - V_1}{2V_T}T, \qquad T_2 = \frac{V_T + V_1}{2V_T}T, \qquad T = T_1 + T_2 \tag{7.5}$$

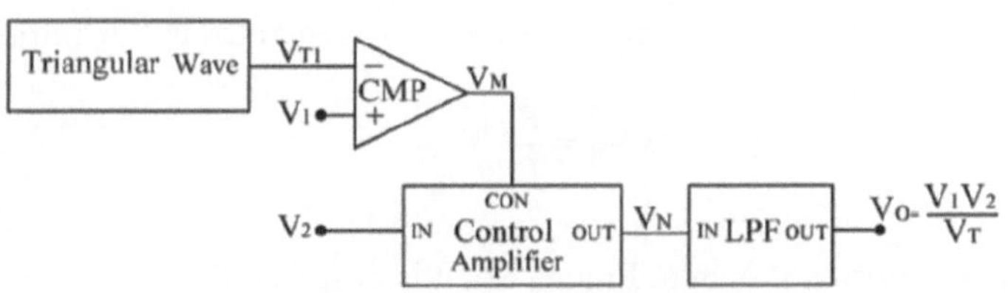

FIGURE 7.3 Triangular wave–based time division multiplier.

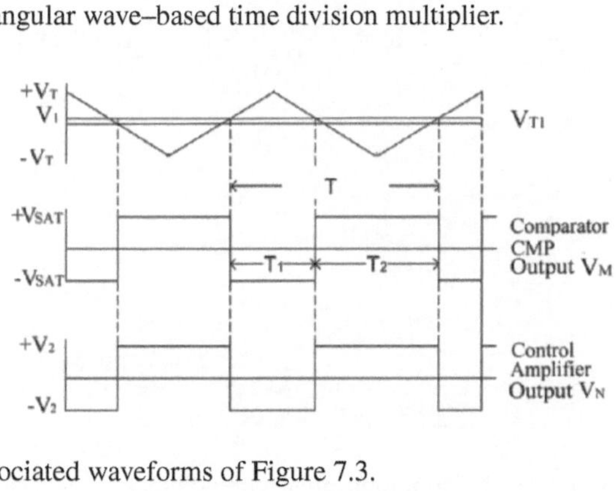

FIGURE 7.4 Associated waveforms of Figure 7.3.

The rectangular wave V_M controls the control amplifier. During HIGH(T_2) of the rectangular waveform V_M, the control amplifier will work as a non-inverting amplifier and $+V_2$ will appear at its output ($V_N = +V_2$). During LOW(T_1) of the rectangular waveform V_M, the control amplifier will work as an inverting amplifier and $-V_2$ will appear at its output ($V_N = -V_2$).

Another rectangular wave V_N with peak-to-peak values of $\pm V_2$ is generated at the output of control amplifier. The lowpass filter gives average value of this pulse train V_N and is given as

$$V_O = \frac{1}{T}\left[\int_{O}^{T_2} V_2\,dt + \int_{T_2}^{T_1+T_2} (-V_2)\,dt\right] \tag{7.6}$$

$$V_O = \frac{V_2}{T}(T_2 - T_1) \tag{7.7}$$

Equations (7.5) in (7.7) give

$$V_O = \frac{V_1 V_2}{V_T} \tag{7.8}$$

7.3 TIME DIVISION MULTIPLIER WITH NO REFERENCE

The time division multiplier without using either sawtooth wave or triangular wave as reference is shown in Figure 7.5, and its associated waveforms are shown in Figure 7.6. The differential integrator and CMP constitutes an asymmetrical rectangular wave generator, and a rectangular wave V_M is generated at the output of CMP.

$$\pm V_T = \frac{R_2}{R_3} V_{SAT} \tag{7.9}$$

From the waveforms shown in Figure 7.6, it is observed that

$$T_1 = \frac{V_{SAT} - V_1}{2V_{SAT}} T \qquad T_2 = \frac{V_{SAT} + V_1}{2V_{SAT}} T \qquad T = T_1 + T_2 \tag{7.10}$$

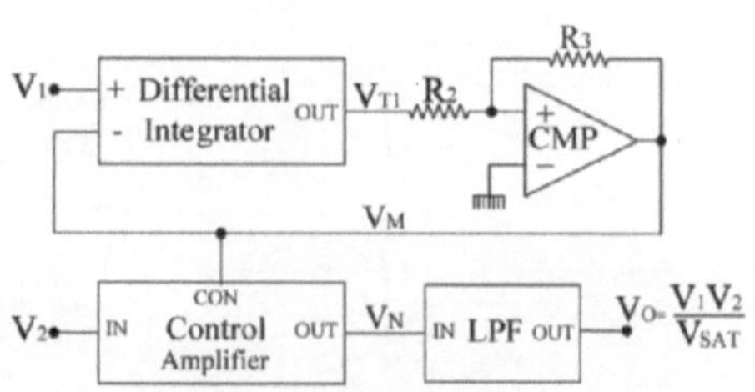

FIGURE 7.5 Switching time division multiplier with no clock.

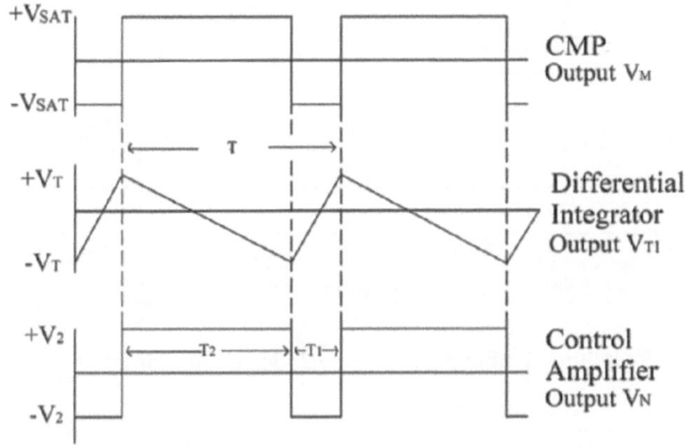

FIGURE 7.6 Associated waveforms of Figure 7.5.

The rectangular wave V_M controls the control amplifier. During HIGH(T_2) of the rectangular waveform V_M, the control amplifier will work as a non-inverting amplifier and $+V_2$ will appear at its output ($V_N = +V_2$). During LOW(T_1) of the rectangular waveform V_M, the control amplifier will work as an inverting amplifier and $-V_2$ will appear at its output ($V_N = -V_2$).

Another asymmetrical rectangular wave V_N is generated at the output of control amplifier with $\pm V_2$ as maximum values. The lowpass filter gives an average value of this pulse train V_N and is given as

$$V_O = \frac{1}{T}\left[\int_{O}^{T_2} V_2\,dt + \int_{T_2}^{T_1+T_2} (-V_2)\,dt\right] = \frac{V_2}{T}\left[T_2 - T_1\right] \tag{7.11}$$

Equations (7.10) in (7.11) give

$$V_O = \frac{V_1 V_2}{V_{SAT}} \tag{7.12}$$

7.4 DOUBLE DUAL SLOPE PEAK RESPONDING MULTIPLIERS

The double dual slope peak responding multipliers are shown in Figure 7.7, and their associated waveforms are shown in Figure 7.8. Figure 7.7(a) shows switching peak detecting multiplier, and Figure 7.7(b) shows switching peak sampling multiplier.

Let initially CMP_1 output be LOW ($-V_{SAT}$).

i. Control amplifier-I will work as an inverting amplifier, and $-V_1$ will appear at its output ($V_M = -V_1$).
ii. Control amplifier-II will work as an inverting amplifier, and $-V_2$ will appear at its output ($V_N = -V_2$).
iii. ($-V_{SAT}$) is given to integrator-I.

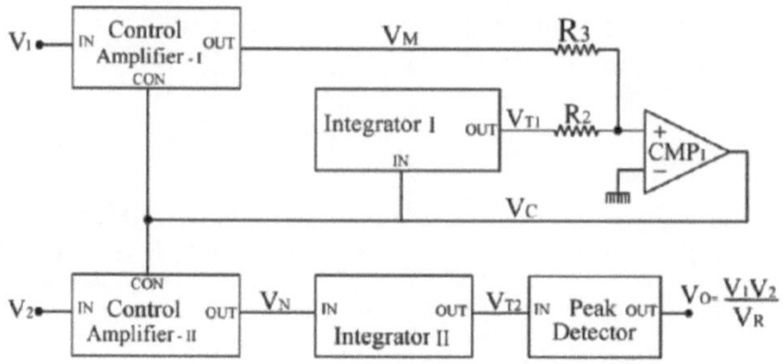

FIGURE 7.7(A) Switching double dual slope peak detecting multiplier.

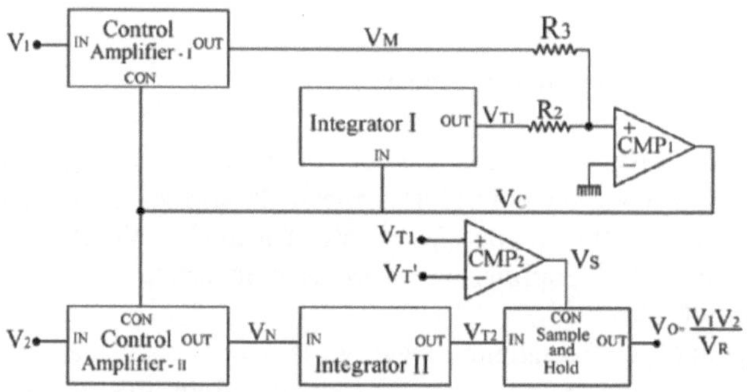

FIGURE 7.7(B) Switching double dual slope sampling multiplier.

Integrator-I output is given as

$$V_{T1} = -\frac{1}{R_1C_1}\int -V_{SAT}\,dt = \frac{V_{SAT}}{R_1C_1}t \tag{7.13}$$

where R_1, C_1 are resistor and capacitor values used in integrator-I.

Integrator-II output is given as

$$V_{T2} = -\frac{1}{R_2C_2}\int -V_2\,dt = \frac{V_2}{R_2C_2}t \tag{7.14}$$

where R_2, C_2 are resistor capacitor values used in integrator-II.

The output of integrator-I is rising toward positive saturation, and when it reaches a value of $+V_T$, CMP_1 output becomes HIGH ($+V_{SAT}$).

i. Control amplifier-I will work as a non-inverting amplifier, and $+V_1$ will appear at its output ($V_M = +V_1$).
ii. Control amplifier-II will work as a non-inverting amplifier, and $+V_2$ will appear at its output ($V_N = +V_2$).
iii. ($+V_{SAT}$) is given to integrator-I.

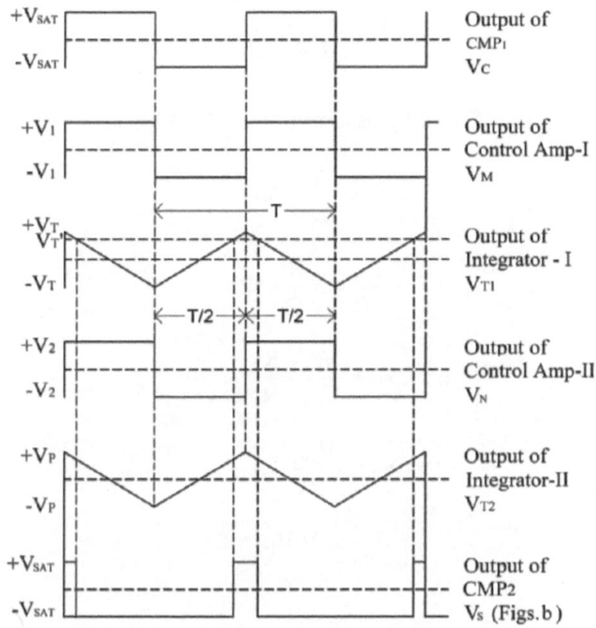

FIGURE 7.8 Associated waveforms of Figure 7.7.

The output of integrator-I will now be

$$V_{T1} = -\frac{1}{R_1C_1}\int V_{SAT}\,dt = -\frac{V_{SAT}}{R_1C_1}t \tag{7.15}$$

And the output of integrator-II will be

$$V_{T2} = -\frac{1}{R_2C_2}\int V_2\,dt = -\frac{V_2}{R_2C_2}t \tag{7.16}$$

The output of integrator-I changes its slope and is going toward negative saturation. When the output of integrator-I comes down to a value ($-V_T$), CMP_1 output will become LOW; therefore, the cycle repeats itself to give (1) a triangular wave V_{T1} at the output of integrator-I with $\pm V_T$ peak-to-peak value, (2) another triangular wave V_{T2} at the output of integrator-II with $\pm V_P$ peak-to-peak value, (3) first square waveform V_C with $\pm V_{SAT}$ peak-to-peak values at the output of comparator CMP_1, (4) second square waveform V_M with $\pm V_1$ as peak-to-peak values at the output of control amplifier-I, and (5) third square waveform V_N with $\pm V_2$ as peak-to-peak values at the output of control amplifier-II. From the waveforms shown in Figure 7.8, from Eqs. (7.13) and (7.14), and the fact that at t = T/2, $V_{T1} = 2V_T$, $V_{T2} = 2V_P$.

$$2V_T = \frac{V_{SAT}}{R_1C_1}\frac{T}{2} \tag{7.17}$$

$$2V_P = \frac{V_2}{R_2C_2}\frac{T}{2} \tag{7.18}$$

From Eqs. (7.17) and (7.18), by assuming $R_1 = R_2$, $C_1 = C_2$

$$V_P = \frac{V_2}{V_{SAT}} V_T$$

$$\pm V_T = \pm \frac{R_2}{R_3} V_1$$

$$V_P = \frac{V_1V_2}{V_{SAT}}\frac{R_2}{R_3}$$

$$\text{Let } V_R = \frac{V_{SAT}}{R_2} R_3$$

$$V_P = \frac{V_1V_2}{V_R} \tag{7.19}$$

In Figure 7.7(a), the peak detector gives this peak value V_p at its output: $V_O = V_P$. In Figure 7.7(b), the peak value V_P is obtained by the sample and hold circuit. The sampling pulse V_S is generated by CMP_2 by comparing a slightly less than voltage V_T called V_T' with the triangular wave V_{T1}. The sampled output is given as $V_O = V_P$.

From Eq. (7.19), the output voltage will be $V_O = V_P$

$$V_O = \frac{V_1V_2}{V_R} \tag{7.20}$$

7.5 DOUBLE DUAL SLOPE PEAK RESPONDING MULTIPLIER WITH FLIP FLOP

The double dual slope peak responding multiplier with flip flop is shown in Figure 7.9, and its associated waveforms are shown in Figure 7.10. Figure 7.9(a) shows peak detecting multiplier, and Figure 7.9(b) shows peak sampling multiplier.

Let initially flip flop output be LOW. $(-V_{SAT})$ is given to integrator-I. The integrator-I output is given as

$$V_{T1} = -\frac{1}{R_1C_1}\int(-V_{SAT})\,dt = \frac{V_{SAT}}{R_1C_1}t \tag{7.21}$$

The output of integrator-I is going toward positive saturation, and when it reaches the value $+V_1$, CMP_1 output becomes HIGH and it sets the flip flop output to HIGH. $(+V_{SAT})$ is given to integrator-I. The integrator-I output is given as

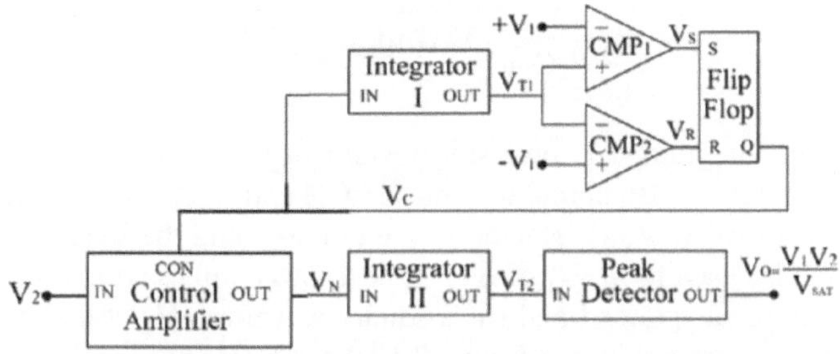

FIGURE 7.9(A) Double dual slope peak detecting multiplier with flip flop.

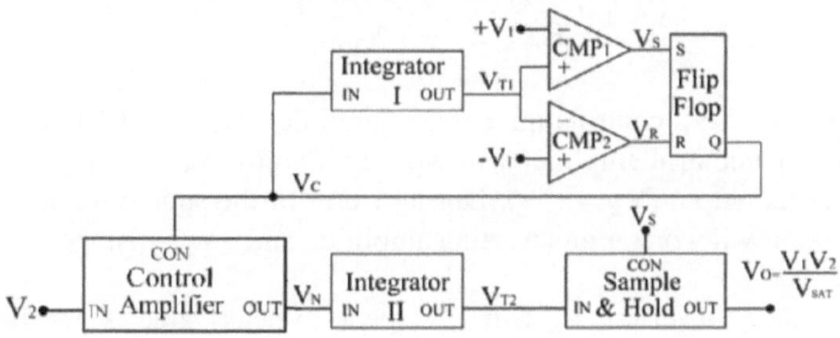

FIGURE 7.9(B) Double dual slope peak sampling multiplier with flip flop.

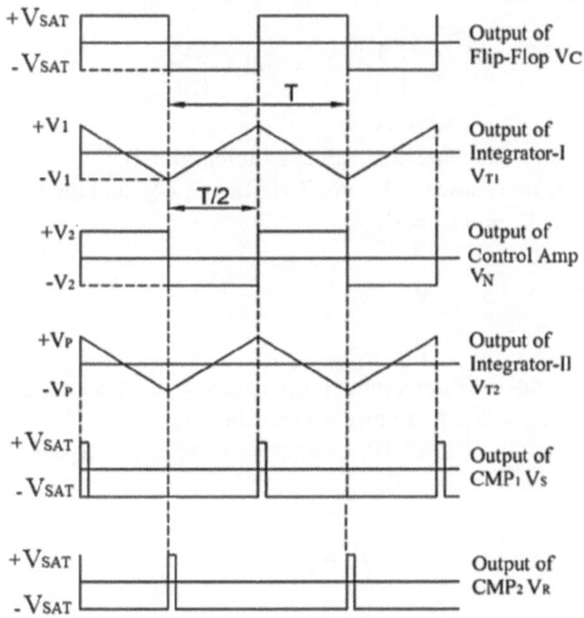

FIGURE 7.10 Associated waveforms of Figure 7.9.

$$V_{T1} = -\frac{1}{R_1C_1}\int (V_{SAT})\,dt = -\frac{V_{SAT}}{R_1C_1}t \tag{7.22}$$

The output of integrator-I is reversing toward negative saturation, and when it reaches the value $-V_1$, CMP_2 output becomes HIGH and resets the flip flop so that its output becomes LOW. ($-V_{SAT}$) is given to integrator-I, and the sequence repeats to give (i) a triangular waveform V_{T1} of $\pm V_1$ peak-to-peak values with a time period of 'T' at the output of integrator-I and (ii) a square waveform V_C at the output of flip flop. From the waveforms shown in Figure 7.10, Eq. (7.21) and the fact that at $t = T/2$, $V_{T1} = 2V_1$

$$2V_1 = \frac{V_{SAT}}{R_1C_1}\frac{T}{2}, \; T = \frac{4V_1}{V_{SAT}}R_1C_1 \tag{7.23}$$

The square wave V_C controls the control amplifier. During HIGH of the square waveform V_C, the control amplifier will work as a non-inverting amplifier, and $+V_2$ will appear at its output ($V_N = +V_2$). During LOW of the square waveform V_C, the control amplifier will work as an inverting amplifier, and $-V_2$ will appear at its output ($V_N = -V_2$).

Another square waveform V_N with $\pm V_2$ peak-to-peak value is generated at the output of the control amplifier. This square waveform V_N is converted to a triangular wave V_{T2} by integrator-II with $\pm V_P$ as peak-to-peak values of same time period 'T'. For one transition, integrator-II output is given as

$$V_{T2} = -\frac{1}{R_2C_2}\int (-V_2)\,dt = \frac{V_2}{R_2C_2}t \tag{7.24}$$

where R_2C_2 is resistor and capacitor values of integrator-II.

From the waveforms shown in Figure 7.10, Eq. (7.24), and the fact that at $t = T/2$, $V_{T2} = 2V_p$. Let $R_1 = R_2$ and $C_1 = C_2$.

$$2V_P = V_2\frac{T}{2}, \; V_P = \frac{V_1V_2}{V_{SAT}} \tag{7.25}$$

In Figure 7.9(a), the peak detector at output stage gives peak value V_P of triangular wave V_{T2}. Hence, $V_O = V_P$. In Figure 7.9(b), the peak value V_P is obtained by the sample and hold circuit. The CMP_1 output V_S is acting as a sampling pulse. The sampled output is given as $V_O = V_P$.

$$V_O = \frac{V_1V_2}{V_{SAT}} \tag{7.26}$$

7.6 PEAK RESPONDING MULTIPLIERS WITH V/T CONVERTER

The switching type peak responding multipliers using a voltage-to-period (V/T) converter are shown in Figure 7.11, and their associated waveforms are shown in Figure 7.12. Figure 7.11(a) shows switching peak detecting multiplier, and Figure 7.11(b) shows switching peak sampling multiplier. A square wave V_C is generated by op-amp OA, resistors R_1, R_2, R_3, and capacitor C_1.

During HIGH of V_C,

i. Control amplifier-I will work as a non-inverting amplifier, and $+V_1$ will appear at its output ($V_M = +V_1$).
ii. Control amplifier-II will work as a non-inverting amplifier, and $-V_2$ will appear at its output ($V_N = -V_2$).

During LOW of V_C,

i. Control amplifier-I will work as an inverting amplifier, and $-V_1$ will appear at its output ($V_M = -V_1$).

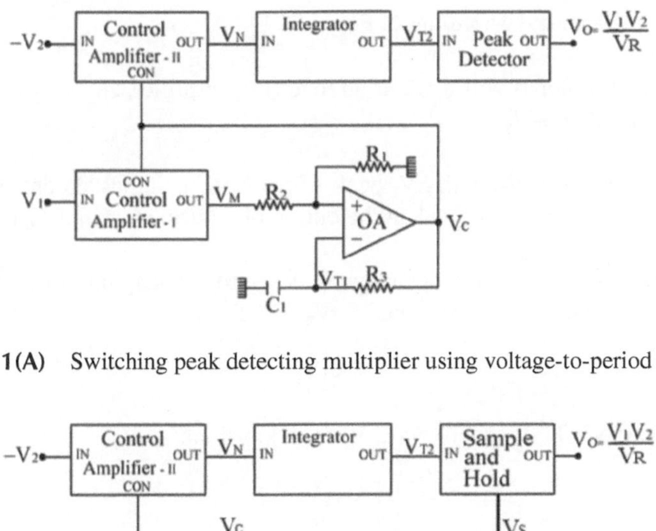

FIGURE 7.11(A) Switching peak detecting multiplier using voltage-to-period converter.

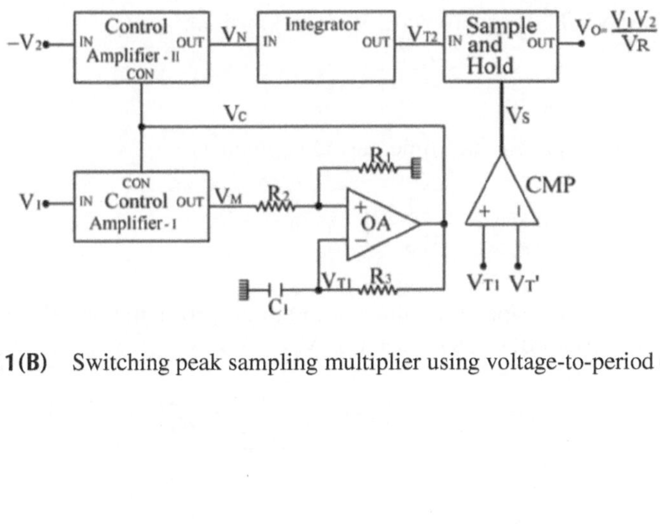

FIGURE 7.11(B) Switching peak sampling multiplier using voltage-to-period converter.

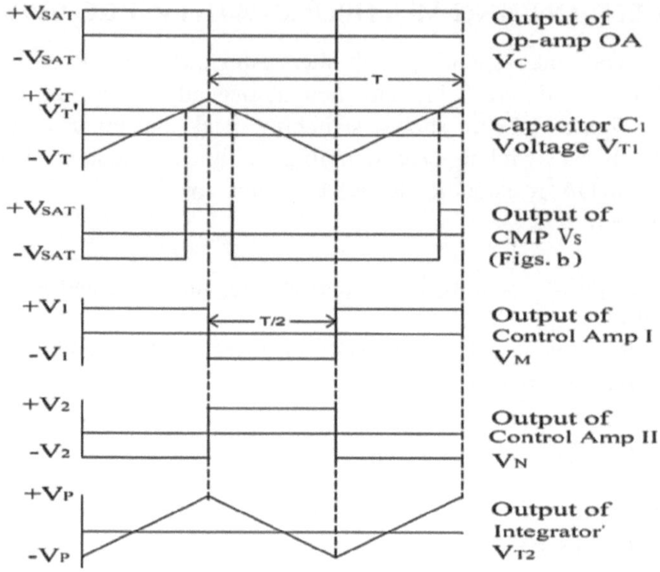

FIGURE 7.12 Associated waveforms of Figure 7.11.

ii. Control amplifier-II will work as an inverting amplifier, and $+V_2$ will appear at its output ($V_N = +V_2$).

Two square waves (i) V_M with $\pm V_1$ peak-to-peak at the output of control amplifier-I and (ii) V_N with $\pm V_2$ peak-to-peak at the output of control amplifier-II are generated from the astable clock V_C.

The integrator converts the square wave V_N into a triangular wave V_{T2} of $\pm V_P$ peak-to-peak values. V_P is proportional to V_2.

$$V_P = K_1 V_2 \tag{7.27}$$

The time period 'T' of square wave V_S is proportional to V_1

$$T = K_2 V_1 \tag{7.28}$$

The integrator output for half time period is given as

$$V_{T2} = -\frac{1}{R_4C_2}\int -V_2\,dt = \frac{V_2}{R_4C_2}t \tag{7.29}$$

where R_4C_2 are resistor capacitor values of integrator, from the waveforms shown in Figure 7.12, and from Eq. (7.29) at $t = T/2$, $V_{T2} = 2\,V_P$.

$$2V_P = \frac{V_2}{R_4C_2}\frac{T}{2};\ V_P = \frac{V_2}{4R_4C_2}V_1K_2 \tag{7.30}$$

K_1 and K_2 are constant values

Let
$$V_R = \frac{4R_4C_2}{K_2}$$

$$V_P = \frac{V_1V_2}{V_R} \tag{7.31}$$

In Figure 7.11(a), the peak detector gives this peak value V_P at the output: $V_O = V_P$. In Figure 7.11(b), the peak value V_P is sampled by the sample and hold circuit with a sampling pulse V_S. The sampling pulse V_S is generated by comparing capacitor C_1 voltage V_{T1} with slightly less than its peak value V_T, i.e., V_T' by CMP. The sampled output V_O is the peak value V_P.

From Eq. (7.31), the output voltage will be $V_O = V_P$

$$V_O = \frac{V_1V_2}{V_R} \tag{7.32}$$

8 Principles of Analog Dividers

Analog dividers are classified into (i) time division dividers and (ii) peak responding dividers. Peak responding dividers (PRDs) are further classified into (i) peak detecting dividers and (ii) peak sampling dividers. A rectangular waveform is generated whose duty cycle is (i) proportional to one input voltage V_2 and (ii) inversely proportional to another input voltage V_1. The amplitude of this rectangular waveform is constant. The average value of this rectangular waveform is proportional to the division V_2/V_1. This is called "time division divider."

A square waveform is generated whose time period 'T' is inversely proportional to one input voltage (V_1). Another input voltage (V_2) is integrated during the time period 'T'. The peak value of the integrated output is proportional to the division V_2/V_1. This is called a "double dual slope peak responding divider." If peak detector is used at the output stage, then it is called a "double dual slope peak detecting divider," and if sample and hold is used at the output stage, then it is called a "double dual slope peak sampling divider." The principles of analog dividers are described in this chapter.

8.1 TIME DIVISION DIVIDER – MULTIPLEXING

The time division divider without using either a sawtooth wave or a triangular wave as reference is shown in Figure 8.1, and its associated waveforms are shown in Figure 8.2. The differential integrator and comparator (CMP) constitute an asymmetrical rectangular wave generator, and a rectangular wave V_C is generated at the output of CMP.

$$\pm V_T = \frac{R_2}{R_3} V_{SAT} \tag{8.1}$$

From waveforms shown in Figure 8.2, it is observed that

$$T_1 = \frac{V_1 - V_2}{2V_1} T, \qquad T_2 = \frac{V_1 + V_2}{2V_1} T \qquad T = T_1 + T_2 \tag{8.2}$$

The lowpass filter gives an average value of the rectangular wave V_C and is given as

$$V_O = \frac{1}{T}\left[\int_O^{T_2} V_{SAT}\,dt + \int_{T_2}^{T_1+T_2} (-V_{SAT})\,dt\right] \tag{8.3}$$

$$= \frac{V_{SAT}}{T}\left[T_2 - T_1\right],$$

DOI: 10.1201/9781003221449-10

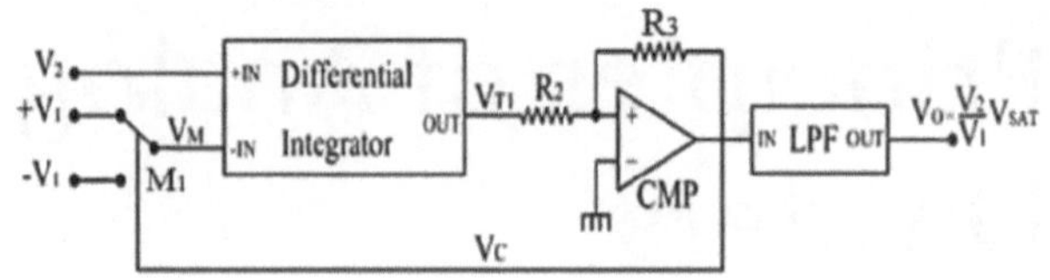

FIGURE 8.1 Time division divider without reference.

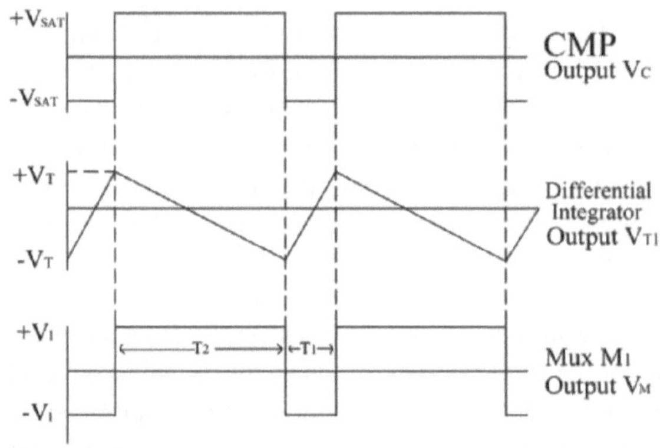

FIGURE 8.2 Associated waveforms of Figure 8.1.

$$V_O = \frac{V_2}{V_1} V_{SAT} \tag{8.4}$$

8.2 TIME DIVISION DIVIDER – SWITCHING

The circuit diagram of a time division divider without using triangular or sawtooth wave as reference is shown in Figure 8.3, and its associated waveforms are shown in Figure 8.4. A rectangular waveform V_C is generated at the output of CMP. During HIGH of V_C, control amplifier-I will work as a non-inverting amplifier, and $+V_1$ will exist at its output ($V_M = +V_1$). During LOW of V_C, control amplifier-I will work as an inverting amplifier, and $-V_1$ will exist at its output ($V_M = -V_1$). Another rectangular

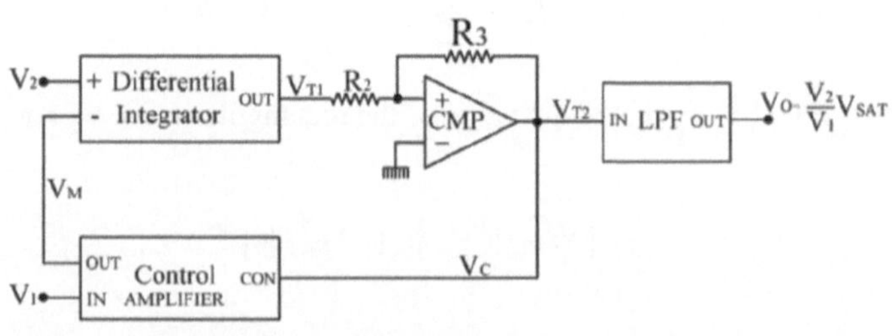

FIGURE 8.3 Switching time division divider with no reference.

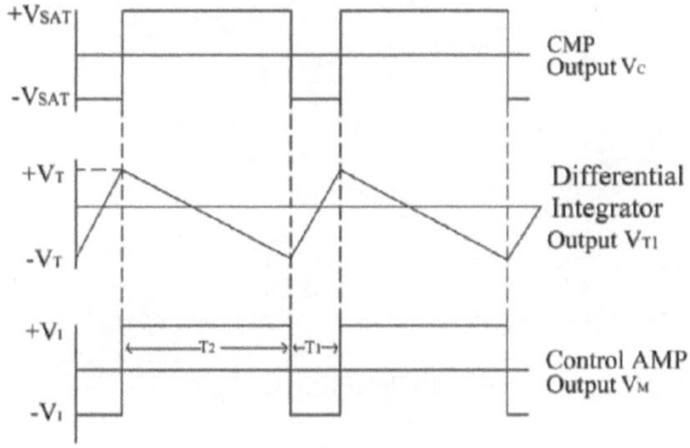

FIGURE 8.4 Associated waveforms of Figure 8.3.

waveform V_M with $\pm V_1$ peak-to-peak value is generated at the output of control amplifier.

$$\pm V_T = \frac{R_2}{R_3} V_{SAT} \tag{8.5}$$

From waveforms shown in Figure 8.5, it is observed that

$$T_1 = \frac{V_1 - V_2}{2V_1} T \qquad T_2 = \frac{V_1 + V_2}{2V_1} T \qquad T = T_1 + T_2 \tag{8.6}$$

The lowpass filter gives an average value of the rectangular wave V_C and is given as

$$V_O = \frac{1}{T}\left[\int_O^{T_2} V_{SAT}\,dt + \int_{T_2}^{T_1+T_2} \left(-V_{SAT}\right)dt\right]$$

$$= \frac{V_{SAT}}{T}\left[T_2 - T_1\right] \tag{8.7}$$

Equation (8.6) in (8.7) gives

$$V_O = \frac{V_2}{V_1} V_{SAT} \tag{8.8}$$

8.3 DOUBLE DUAL SLOPE PEAK RESPONDING DIVIDER – MULTIPLEXING

The double dual slope PRDs are shown in Figure 8.5, and their associated waveforms in Figure 8.6. Figure 8.5(a) shows a peak detecting divider, and Figure 8.5(b) shows

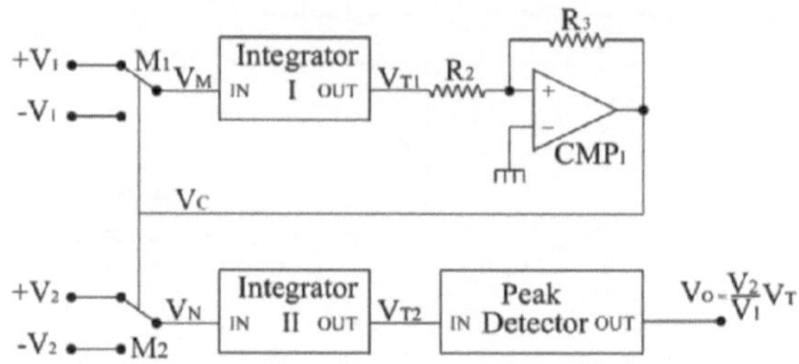

FIGURE 8.5(A) Double dual slope peak detecting divider.

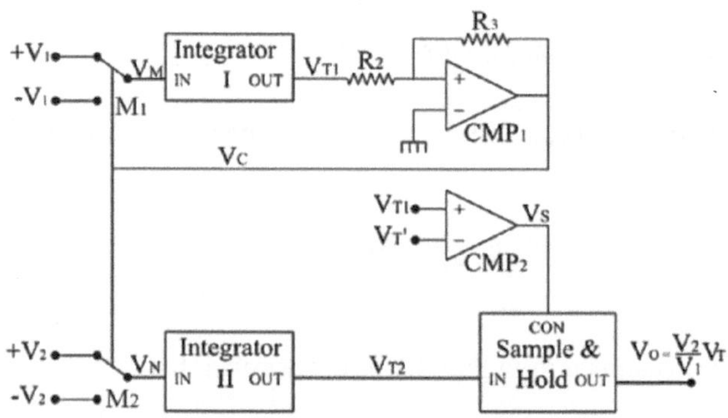

FIGURE 8.5(B) Double dual slope peak sampling divider.

a peak sampling divider. Let initially CMP_1 output be LOW, $(-V_1)$ be given to integrator-I by multiplexer M_1, and $(-V_2)$ be given to integrator-II by multiplexer M_2. The output of integrator-I will be

$$V_{T1} = -\frac{1}{R_1C_1}\int -V_1\,dt = \frac{V_1}{R_1C_1}t \tag{8.9}$$

where R_1C_1 are resistor capacitor values of integrator-I.

The output of integrator-II will be

$$V_{T2} = -\frac{1}{R_4C_2}\int -V_2\,dt = \frac{V_2}{R_4C_2}t \tag{8.10}$$

where R_4C_2 are resistor capacitor values of integrator-II.

The output of integrator-I is a positive-going ramp, and when it reaches a value $+V_T$, the output of CMP_1 becomes HIGH. The multiplexer M_1 selects $(+V_1)$ to integrator-I, and the multiplexer M_2 selects $(+V_2)$ to integrator-II. The output of integrator-I will now be

$$V_{T1} = -\frac{1}{R_1C_1}\int V_1\,dt = -\frac{V_1}{R_1C_1}t \tag{8.11}$$

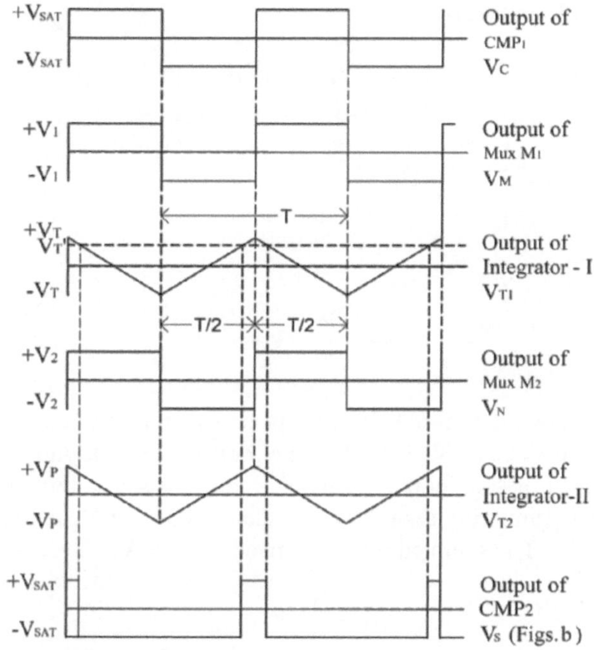

FIGURE 8.6 Associated waveforms of Figure 8.5.

The output of integrator-II will now be

$$V_{T2} = -\frac{1}{R_4C_2}\int V_2\,dt = -\frac{V_2}{R_4C_2}t \tag{8.12}$$

The output of integrator-I changes slope from $(+V_T)$ toward $(-V_T)$, and when it reaches a value of $(-V_T)$, the output of CMP_1 becomes LOW, and the cycle therefore repeats to give (1) a triangular wave V_{T1} at the output of integrator-I with $\pm V_T$ peak-to-peak value, (2) another triangular wave V_{T2} at the output of integrator-II with $\pm V_P$ peak-to-peak value, (3) first square waveform V_C with $\pm V_{SAT}$ peak-to-peak values at the output of CMP_1, (4) second square waveform V_M with $\pm V_1$ peak-to-peak values at the output of multiplexer M_1, and (5) third square waveform V_N with $\pm V_2$ as peak-to-peak values at the output of multiplexer M_2.

From Eqs. (8.9) and (8.10) and the fact that at $t = T/2$, $V_{T1} = 2V_T$, $V_{T2} = 2V_P$

$$2V_T = \frac{V_1}{R_1C_1}\frac{T}{2} \tag{8.13}$$

$$2V_P = \frac{V_2}{R_4C_2}\frac{T}{2} \tag{8.14}$$

From Eqs. (8.13) and (8.14)

$$2V_P = \frac{V_2}{R_4C_2}\frac{2V_TR_1C_1}{V_1}$$

Let us assume $R_1 = R_4$, $C_1 = C_2$

$$V_P = \frac{V_2}{V_1}V_T \tag{8.15}$$

$$V_T = \frac{R_2}{R_3}V_{SAT}$$

In the circuit shown in Figure 8.5(a), the peak detector gives the peak value V_P of the triangular wave V_{T2}, i.e., $V_O = V_P$. In the circuit shown in Figure 8.5(b), the peak value V_P is obtained by the sample and hold circuit. The sampling pulse V_S is generated by CMP_2 by comparing a slightly less than voltage of V_T called V_T' with the triangular wave V_{T1}. The sampled output is given as $V_O = V_P$

$$V_O = \frac{V_2}{V_1}V_T \tag{8.16}$$

8.4 DOUBLE DUAL SLOPE PEAK RESPONDING DIVIDER – SWITCHING

The circuit diagrams of double dual slope PRDs are shown in Figure 8.7, and their associated waveforms are shown in Figure 8.8. Figure 8.7(a) shows a switching double dual slope peak detecting divider, and Figure 8.7(b) shows a switching double dual slope peak sampling divider.

Let initially CMP_1 output be LOW, control amplifier-I will work as an inverting amplifier, and $-V_1$ will appear at its output ($V_M = -V_1$). Control amplifier-II will work as an inverting amplifier, and $-V_2$ will appear at its output ($V_N = -V_2$).

The output of integrator-I will be

$$V_{T1} = -\frac{1}{R_1C_1}\int -V_1\,dt = \frac{V_1}{R_1C_1}t \tag{8.17}$$

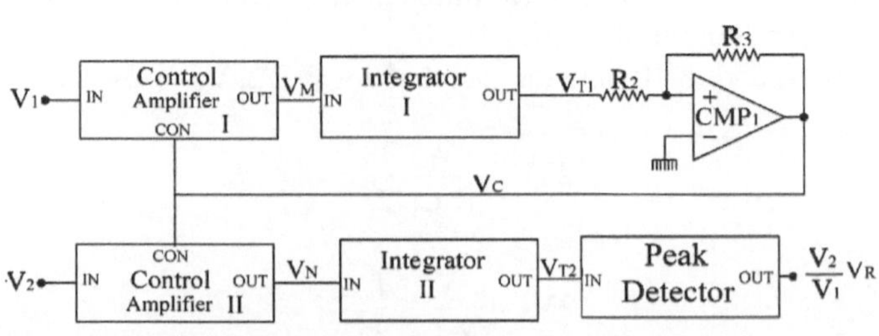

FIGURE 8.7(A) Switching double dual slope peak detecting divider.

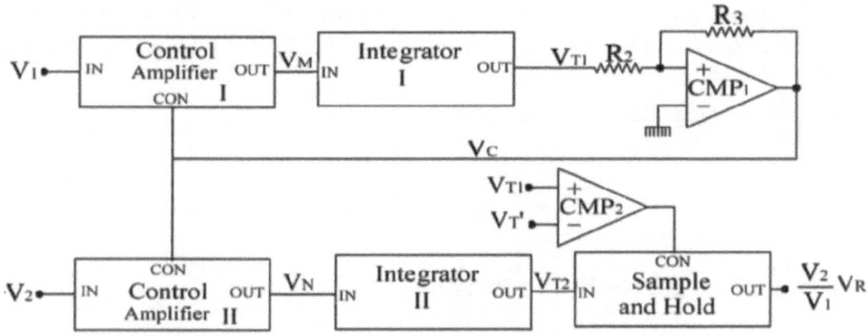

FIGURE 8.7(B) Switching double dual slope peak sampling divider.

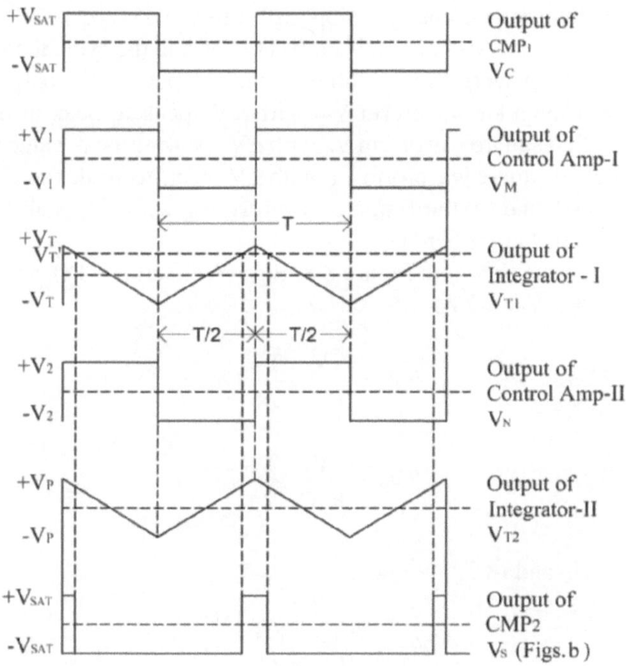

FIGURE 8.8 Associated waveforms of Figure 8.7.

where R_1C_1 are resistor capacitor values of integrator-I.

The output of integrator-II will be

$$V_{T2} = -\frac{1}{R_4C_2}\int -V_2\,dt = \frac{V_2}{R_4C_2}t \qquad (8.18)$$

where R_4C_2 are resistor capacitor values of integrator-I.

The output of integrator-I is a positive-going ramp, and when it reaches a value $+V_T$, CMP_1 output becomes HIGH. The control amplifier-I will work as a non-inverting amplifier, and $+V_1$ will appear at its output ($V_M = +V_1$). Control amplifier-II will work as a non-inverting amplifier, and $+V_2$ will appear at its output ($V_N = +V_2$).

The output of integrator-I will now be

$$V_{T1} = -\frac{1}{R_1C_1}\int V_1\,dt = -\frac{V_1}{R_1C_1}t \tag{8.19}$$

The output of integrator-II will now be

$$V_{T2} = -\frac{1}{R_4C_2}\int V_2\,dt = -\frac{V_2}{R_4C_2}t \tag{8.20}$$

The output of integrator-I changes slope from ($+V_T$) toward ($-V_T$) and when it reaches a value of ($-V_T$), CMP_1 output becomes LOW, and the cycle therefore repeats to give (i) a triangular waveform V_{T1} with $\pm V_T$ peak-to-peak at the output of integrator-I, (ii) another triangular waveform V_{T2} with $\pm V_P$ peak-to-peak at the output of integrator-II, (iii) first square waveform V_C with $\pm V_T$ peak-to-peak value at the output of CMP_1, (iv) second square waveform V_M with $\pm V_1$ peak-to-peak value at the output of control amplifier-I, and (v) third square waveform V_N with $\pm V_2$ peak-to-peak value at the output of control amplifier-II.

From the waveforms shown in Figure 8.8, Eqs. (8.17) and (8.18), and the fact that at $t = T/2$, $V_{T1} = 2V_T$, $V_{T2} = 2V_P$

$$2V_T = \frac{V_1}{R_1C_1}\frac{T}{2} \tag{8.21}$$

$$2V_P = \frac{V_2}{R_4C_2}\frac{T}{2} \tag{8.22}$$

From Eqs. (8.21) and (8.22)

$$V_P = \frac{V_2V_T}{V_1}\frac{R_1C_1}{R_4C_2}$$

Let us assume $R_1 = R_4$, $C_1 = C_2$

$$V_P = \frac{V_2}{V_1}V_T \tag{8.23}$$

$$V_T = \frac{R_2}{R_3}V_{SAT}$$

In Figure 8.7(a), the peak detector gives the peak value V_P of the triangular wave V_{T2}. Hence $V_O = V_P$. In Figure 8.7(b), the peak value V_P is obtained by the sample and hold circuit. The sampling pulse is generated by CMP2 by comparing a slightly less than voltage of VT called VT' with the triangular wave VT1. The sample and hold output is $V_O = V_P$.

From Eq. (8.23)

$$V_O = \frac{V_2}{V_1} V_T \tag{8.24}$$

8.5 DIVIDER USING VOLTAGE-TO-FREQUENCY CONVERTER – MULTIPLEXING

The PRDs using a voltage-to-frequency (V/F) converter are shown in Figure 8.9, and their associated waveforms are shown in Figure 8.10. Figure 8.9(a) shows peak detecting divider, and Figure 8.9(b) shows peak sampling divider. A square wave V_C is generated by op-amp OA_1, resistors R_1, R_2, R_3, capacitor C_1, and multiplexer M_1. The square wave V_C controls multiplexer M_2. Multiplexer M_2 connects $(-V_2)$ during HIGH time of V_C and $(+V_2)$ during LOW time of V_C. Another square wave V_N with $\pm V_2$ peak-to-peak values is generated at the multiplexer M_2 output. The integrator converts the square wave V_N into a triangular wave V_{T2} of $\pm V_P$ peak-to-peak values. V_P is proportional to V_2

$$V_P = K_1 V_2 \tag{8.25}$$

The time period 'T' of square wave V_C is inversely proportional to V_1

$$T = \frac{K_2}{V_1} \tag{8.26}$$

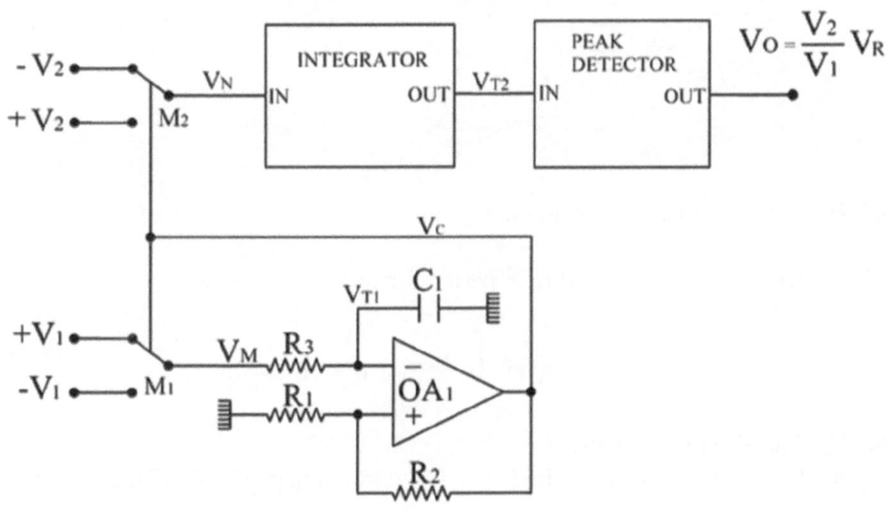

FIGURE 8.9(A) Peak detecting divider using voltage-to-frequency converter.

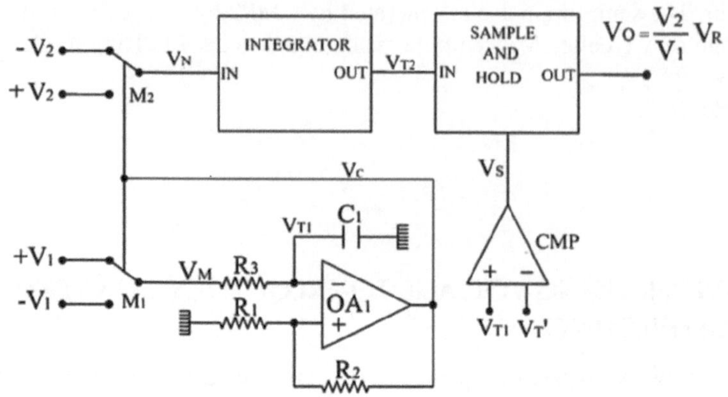

FIGURE 8.9(B) Peak sampling divider using voltage-to-frequency converter.

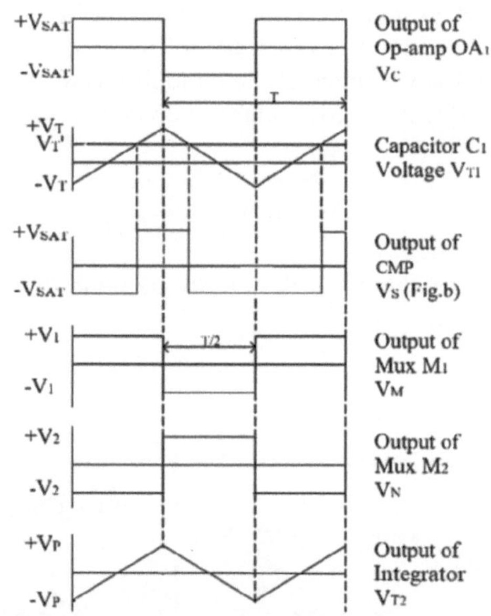

FIGURE 8.10 Associated waveforms of Figure 8.9.

The integrator output for half time period for t = T/2 is given as

$$V_{T2} = -\frac{1}{R_4C_2}\int -V_2\,dt = \frac{V_2}{R_4C_2}t \tag{8.27}$$

where K_1 and K_2 are constants.

From the waveforms shown in Figure 8.10 and from Eq. (8.27) at $t = T/2$, $V_{T2} = 2\,V_P$.

$$2V_P = \frac{V_2}{R_4C_2}\frac{T}{2};\; V_P = \frac{V_2}{4R_4C_2}\frac{K_2}{V_1} \tag{8.28}$$

Let

$$V_R = \frac{K_2}{4R_4C_2}$$

$$V_P = \frac{V_2V_R}{V_1} \tag{8.29}$$

In the circuit shown in Figure 8.9(a), the peak detector gives this peak value V_P at the output: $V_O = V_P$. In the circuit shown in Figure 8.9(b), the sample and hold circuit gives the peak value V_P by the sampling pulse V_S generated by CMP1. The sampling pulse V_S is generated by comparing the capacitor C1 voltage VT1 of ±VT peak-to-peak value with a voltage of VT' which is slightly less than the value of VT by CMP1. The sampled output V_O is the peak value V_P: VO = VP.

From Eq. (8.29)

$$V_O = \frac{V_2}{V_1}V_R \tag{8.30}$$

8.6 PEAK RESPONDING DIVIDERS USING VOLTAGE-TO-FREQUENCY CONVERTER – SWITCHING TYPE

PRDs using a V/F converter are shown in Figure 8.11, and their associated waveforms are shown in Figure 8.12. Figure 8.11(a) shows a switching peak detecting divider, and Figure 8.11(b) shows a switching peak sampling divider. A square wave V_C is generated by op-amp OA_1, resistors R_1, R_2, R_3, capacitor C_1, and control amplifier-I.

During HIGH of the square waveform V_C, (i) control amplifier-I will work as a non-inverting amplifier, and $+V_1$ will appear at its output ($V_M = +V_1$), and (ii) control amplifier-II will work as a non-inverting amplifier and $-V_2$ will appear at its output ($V_N = -V_2$). During LOW of the square waveform V_C, (i) control amplifier-I will work as an inverting amplifier, and $-V_1$ will appear at its output ($V_M = -V_1$), and (ii) control amplifier-II will work as an inverting amplifier and $+V_2$ will appear at its output ($V_N = +V_2$).

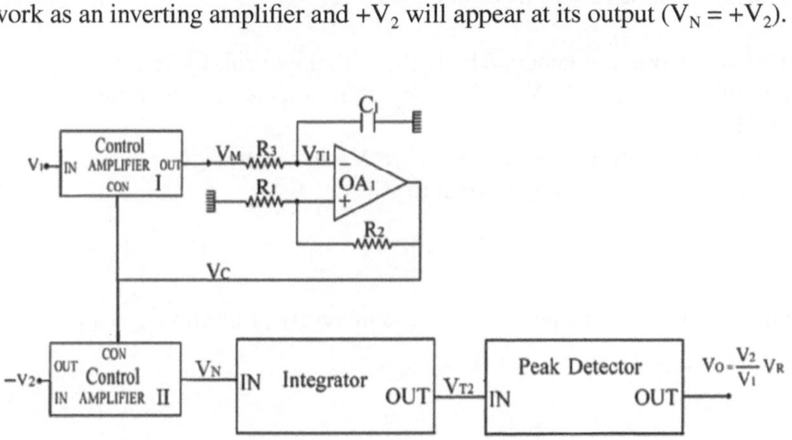

FIGURE 8.11(A) Switching peak detecting divider using voltage-to-frequency converter.

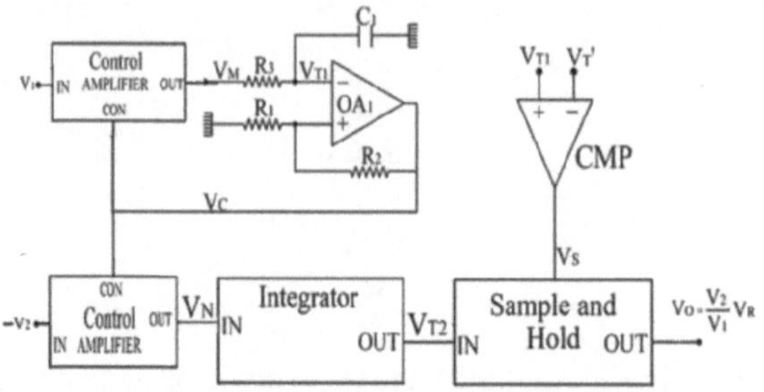

FIGURE 8.11(B) Switching peak sampling divider using voltage-to-frequency converter.

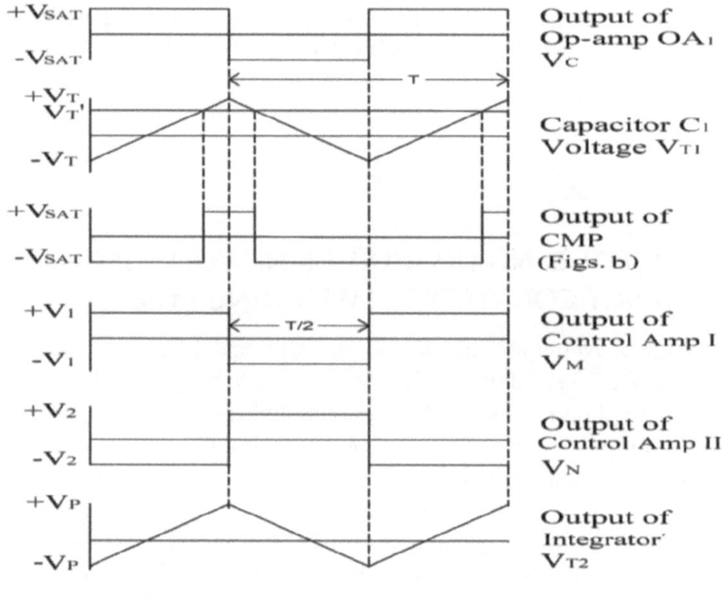

FIGURE 8.12 Associated waveforms of Figure 8.11.

Two square waves are generated: (i) V_M with $\pm V_1$ peak-to-peak at the output of control amplifier-I and (ii) V_N with $\pm V_2$ peak-to-peak at the output of control amplifier-II.

The integrator converts the square wave V_N into a triangular wave V_{T2} of $\pm V_P$ peak-to-peak values. V_P is proportional to V_2

$$V_P = K_1 V_2 \tag{8.31}$$

The time period 'T' of square wave V_S is inversely proportional to V_1

$$T = \frac{K_2}{V_1} \tag{8.32}$$

where K_1 and K_2 are constants.

The integrator output for half time period for t = T/2 is given as

$$V_{T2} = -\frac{1}{R_4C_2}\int -V_2\,dt = \frac{V_2}{R_4C_2}t \tag{8.33}$$

where R_4C_2 is the resistor–capacitor values of integrator.

From the waveforms shown in Figure 8.12 and from Eq. (8.33) at t = T/2, $V_{T2} = 2V_P$.

$$2V_P = \frac{V_2}{R_4C_2}\frac{T}{2};\ V_P = \frac{V_2}{4R_4C_2}\frac{K_2}{V_1}$$

Let

$$V_R = \frac{K_2}{4R_4C_2}$$

$$V_P = \frac{V_2}{V_1}V_R \tag{8.34}$$

In Figure 8.11(a), the peak detector gives this peak value V_P at the output: $V_O = V_P$.

In Figure 8.11(b), this peak value V_P is sampled by the sample and hold circuit with a sampling pulse V_S. The sampling pulse V_S is generated with capacitor C_1 voltage V_{T1} with slightly less than its peak value V_T, i.e., V_T'. The sampled output V_O is the peak value V_P.

$$V_O = V_P \tag{8.35}$$

From Eqs. (8.34) and (8.35)

$$V_O = \frac{V_2}{V_1}V_R \tag{8.36}$$

9 Principles of Time Division Multipliers-Cum-Dividers

Like multipliers, multipliers-cum-dividers (MCDs) also classified into (i) time division MCDs, (ii) peak responding MCDs, and (iii) pulse position responding MCDs. The peak responding MCDs are further classified into (i) peak detecting MCDs and (ii) peak sampling MCDs.

If width of a pulse train is made proportional to one voltage and its amplitude is made proportional to another voltage, then the average value of this pulse train is proportional to the product of the two voltages. This is called a "time division principle." MCD obtained using this principle is called a "time division MCD." Time division MCDs can be implemented using (i) analog multiplexers and (ii) analog switches and are described in this chapter.

9.1 SAWTOOTH WAVE–BASED MULTIPLIER-CUM-DIVIDER – MULTIPLEXING

The circuit diagram of a double multiplexing–averaging time division MCD is shown in Figure 9.1, and its associated waveforms are shown in Figure 9.2. A sawtooth wave V_{S1} of peak value V_R and time period 'T' is generated by the sawtooth wave generator as discussed in Figure 4.7(a). The time period 'T' is given as

$$T = R_1C_1 \tag{9.1}$$

The comparator (CMP) compares the sawtooth wave V_{S1} with the voltage V_Y and produces a rectangular waveform V_K. The ON time δ_T of $V_{K,}$ is given as

$$\delta_T = \frac{V_Y}{V_R}T \tag{9.2}$$

The rectangular pulse V_K controls the first multiplexer M_1. When V_K is HIGH, input voltage V_1 is connected to lowpass filter I. When V_K is LOW, zero volts is connected to lowpass filter I. Another rectangular pulse V_M with maximum value of V_1 is generated at the multiplexer M_1 output. The lowpass filter I gives average value of this pulse train V_M and is given as

$$V_X = \frac{1}{T}\int_0^{\delta_T} V_1\,dt = \frac{V_1}{T}\delta_T \tag{9.3}$$

DOI: 10.1201/9781003221449-11

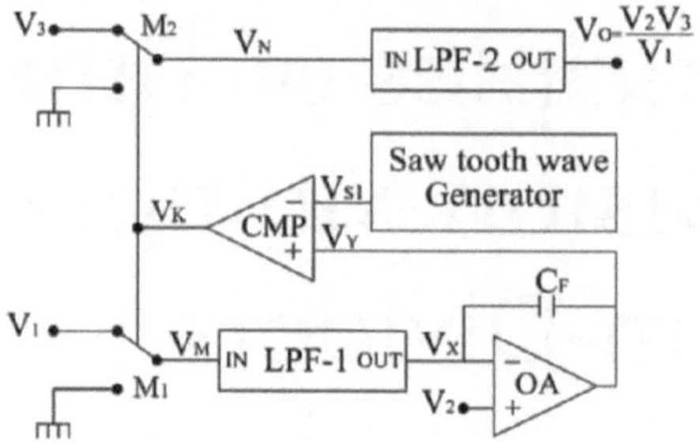

FIGURE 9.1 Double multiplexing–averaging time division multiplier-cum-divider.

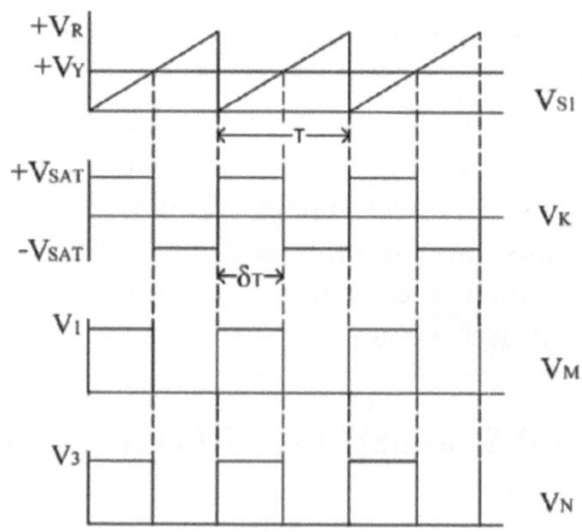

FIGURE 9.2 Associated waveforms of Figure 9.1.

$$V_X = \frac{V_1 V_Y}{V_R} \tag{9.4}$$

The op-amp, OA, is configured in a negative closed loop feedback, and a positive DC voltage is ensured in the feedback loop. Hence, its inverting terminal voltage must equal to its non-inverting terminal voltage.

$$V_2 = V_X \tag{9.5}$$

From Eqs. (9.4) and (9.5)

$$V_Y = \frac{V_2 V_R}{V_1} \tag{9.6}$$

The rectangular pulse V_K also controls the second multiplexer M_2. When V_K is HIGH, another input voltage V_3 is connected to lowpass filter II. When V_K is LOW, zero volts is connected to lowpass filter II. Another rectangular pulse V_N with maximum value of V_3 is generated at the multiplexer M_2 output. The lowpass filter gives an average value of this pulse train V_N and is given as

$$V_O = \frac{1}{T}\int_0^{\delta_T} V_3\,dt = \frac{V_3}{T}\delta_T \tag{9.7}$$

Eqs. (9.2) and (9.6) in (9.7) gives

$$V_O = \frac{V_2 V_3}{V_1} \tag{9.8}$$

9.2 TRIANGULAR WAVE–BASED MULTIPLIER-CUM-DIVIDER – MULTIPLEXING

The circuit diagram of triangular wave–based time division MCD is shown in Figure 9.3, and its associated waveforms are shown in Figure 9.4. A triangular wave V_{T1} of $\pm V_T$ peak-to-peak values and time period 'T' is generated by the triangular wave generator. The CMP compares the triangular wave V_{T1} with the voltage V_Y and produces the asymmetrical rectangular wave V_K.

From Figure 9.4, it is observed that

$$T_1 = \frac{V_T - V_Y}{2V_T}T \qquad T_2 = \frac{V_T + V_Y}{2V_T}T \qquad T = T_1 + T_2 \tag{9.9}$$

The rectangular wave V_K controls multiplexer M_1 which connects $(+V_1)$ to its output during T_2 and $(-V_1)$ to its output during T_1. Another asymmetrical rectangular waveform V_M is generated at the multiplexer M_1 output with $\pm V_1$ peak-to-peak values. The lowpass filter I gives the average value of V_M and is given as

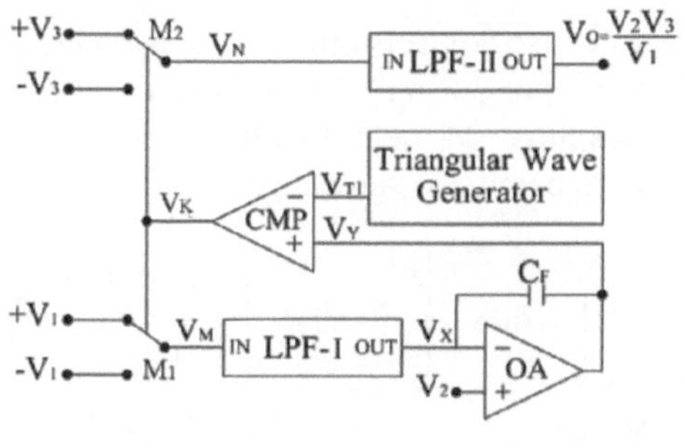

FIGURE 9.3 Triangular wave–based time division multiplier-cum-divider.

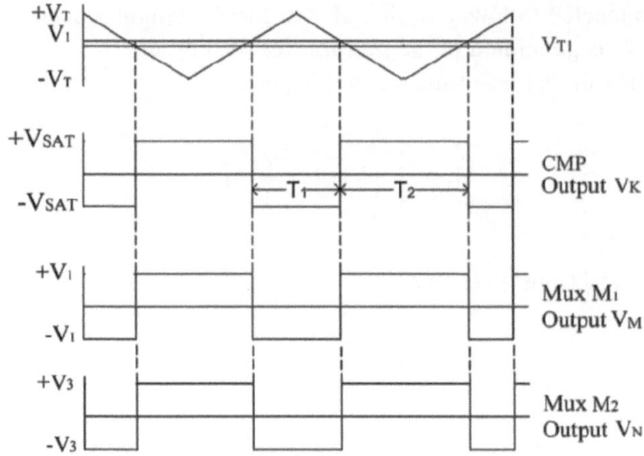

FIGURE 9.4 Associated waveforms of Figure 9.3.

$$V_X = \frac{1}{T}\left[\int_{0}^{T_2} V_1\,dt + \int_{T_2}^{T_1+T_2} (-V_1)\,dt\right] = \frac{V_1}{T}\left[T_2 - T_1\right]$$

$$V_X = \frac{V_1 V_Y}{V_T} \tag{9.10}$$

The op-amp, OA, is configured in a negative closed loop feedback, and a positive DC voltage is ensured in the feedback loop. Hence, its inverting terminal voltage must equal to its non-inverting terminal voltage, i.e.,

$$V_X = V_2 \tag{9.11}$$

From Eqs. (9.10) and (9.11)

$$V_Y = \frac{V_2 V_T}{V_1} \tag{9.12}$$

The rectangular wave V_K also controls multiplexer M_2 which connects $(+V_3)$ to its output during T_2 $(-V_3)$ to output during T_1. Another asymmetrical rectangular wave V_N is generated at the multiplexer M_2 output with $\pm V_3$ peak-to-peak values. The low-pass filter II gives the average value V_O and is given as

$$V_O = \frac{1}{T}\left[\int_{0}^{T_2} V_3\,dt + \int_{T_2}^{T_1+T_2} (-V_3)\,dt\right] = \frac{V_3}{T}\left[T_2 - T_1\right]$$

$$V_O = \frac{V_3 V_Y}{V_T} \tag{9.13}$$

Equation (9.12) in (9.13) gives

$$V_O = \frac{V_2 V_3}{V_1} \quad (9.14)$$

9.3 TIME DIVISION MULTIPLIER-CUM-DIVIDER WITH NO REFERENCE; TYPE I – MULTIPLEXING

An MCD using the time division principle without any reference clock is shown in Figure 9.5, and its associated waveforms are shown in Figure 9.6. Let initially the CMP output be LOW. The multiplexer M_1 connects $(-V_1)$ to the differential integrator. The output of the differential integrator will be

$$V_{T1} = \frac{1}{R_1C_1}\int (V_2 + V_1)dt - V_T$$

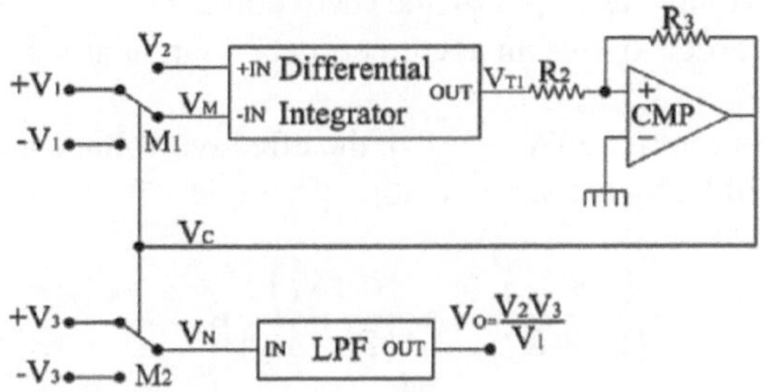

FIGURE 9.5 Time division multiplier-cum-divider without reference clock.

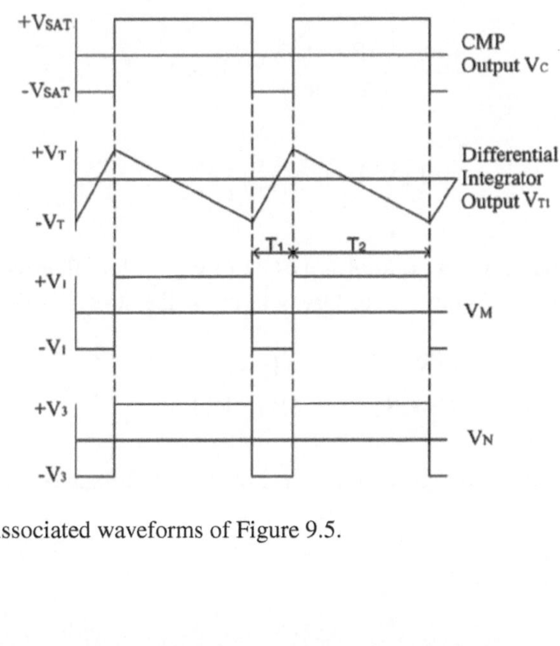

FIGURE 9.6 Associated waveforms of Figure 9.5.

$$V_{T1} = \frac{(V_1 + V_2)}{R_1C_1} t - V_T \quad (9.15)$$

where R_1C_1 are resistor–capacitor values used in the differential integrator.

The output of the differential integrator rises toward positive saturation, and when it reaches the voltage level of $(+V_T)$, the CMP output becomes HIGH. The multiplexer M_1 connects $(+V_1)$ to the differential integrator. Now, the output of the differential integrator will be

$$V_{T1} = \frac{1}{R_1C_1} \int (V_2 - V_1) dt + V_T$$

$$V_{T1} = -\frac{(V_1 - V_2)}{R_1C_1} t + V_T \quad (9.16)$$

The output of the differential integrator reverses toward negative saturation, and when it reaches the voltage level $(-V_T)$, the CMP output becomes LOW and the cycle therefore repeats in order to give an asymmetrical rectangular wave V_C at the output of the CMP.

When the CMP output is LOW $(-V_{SAT})$, the effective voltage at the non-inverting terminal of CMP will be by the superposition principle

$$\frac{(-V_{SAT})}{(R_2 + R_3)} R_2 + \frac{(+V_T)}{(R_2 + R_3)} R_3$$

When this effective voltage at the non-inverting terminal of comparator OA_2 becomes zero

$$\frac{(-V_{SAT})R_2 + (+V_T)R_3}{(R_2 + R_3)} = 0$$

$$(+V_T) = (+V_{SAT}) \frac{R_2}{R_3}$$

When the comparator OA_2 output is HIGH $(+V_{SAT})$, the effective voltage at the non-inverting terminal of comparator OA_2 will be by the superposition principle

$$\frac{(+V_{SAT})}{(R_2 + R_3)} R_2 + \frac{(-V_T)}{(R_2 + R_3)} R_3$$

When this effective voltage at the non-inverting terminal of comparator OA_2 becomes zero

$$\frac{(+V_{SAT})R_2 + (-V_T)R_3}{(R_2 + R_3)} = 0$$

$$(-V_T) = (-V_{SAT})\frac{R_2}{R_3}$$

$$\pm V_T = \pm V_{SAT}\frac{R_2}{R_3}; 0.76(\pm V_{CC})\frac{R_2}{R_3} \tag{9.17}$$

From the waveforms shown in Figure 9.6, it is observed that

$$T_1 = \frac{V_1 - V_2}{2V_1}T, \quad T_2 = \frac{V_1 + V_2}{2V_1}T, \quad T = T_1 + T_2 \tag{9.18}$$

The asymmetrical rectangular wave V_C controls another multiplexer M_2. The multiplexer M_2 selects $(+V_3)$ during ON time T_2 and $(-V_3)$ during OFF time T_1 of the rectangular wave V_C. Another rectangular wave V_N is generated at the multiplexer M_2 output. The lowpass filter gives an average value of this pulse train V_N and is given as

$$V_O = \frac{1}{T}\left[\int_O^{T_2} V_3\,dt + \int_{T_2}^{T_1+T_2} (-V_3)\,dt\right]$$

$$V_O = \frac{V_3(T_2 - T_1)}{T} \tag{9.19}$$

Equations (9.18) in (9.19) give

$$V_O = \frac{V_2 V_3}{V_1} \tag{9.20}$$

9.4 TIME DIVISION MULTIPLIER-CUM-DIVIDER WITH NO REFERENCE; TYPE II – MULTIPLEXING

An MCD using the time division principle without any reference clock is shown in Figure 9.7, and its associated waveforms are shown in Figure 9.8. Let initially the CMP output be LOW. The multiplexer M_1 connects $(-V_3)$ to the differential integrator. The output of the differential integrator will be

$$V_{T1} = \frac{1}{R_1 C_1}\int (V_O + V_3)\,dt - V_T$$

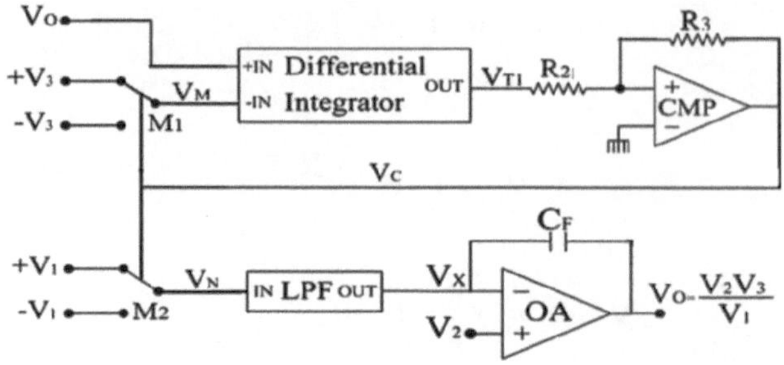

FIGURE 9.7 Multiplier-cum-divider without reference clock – type II.

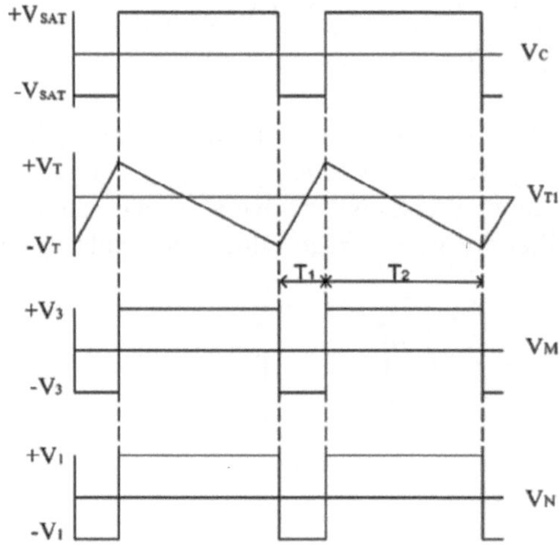

FIGURE 9.8. Associated waveforms of Figure 9.7.

$$V_{T1} = \frac{(V_3 + V_O)}{R_1C_1}t - V_T \tag{9.21}$$

The output of the differential integrator OA_1 rises toward positive saturation, and when it reaches the voltage level of $(+V_T)$, the CMP output becomes HIGH. The multiplexer M_1 connects $(+V_3)$ to the differential integrator. Now the output of the differential integrator OA_1 will be

$$V_{T1} = \frac{1}{R_1C_1}\int (V_O - V_3)\,dt + V_T$$

$$V_{T1} = -\frac{(V_3 - V_O)}{R_1C_1}t + V_T \tag{9.22}$$

The output of the differential integrator reverses toward negative saturation, and when it reaches the voltage level ($-V_T$), the CMP output becomes LOW and the cycle therefore repeats in order to give an asymmetrical rectangular wave V_C at the output of the CMP.

When the CMP output is LOW ($-V_{SAT}$), the effective voltage at the non-inverting terminal of CMP will be by the superposition principle

$$\frac{(-V_{SAT})}{(R_2+R_3)}R_2+\frac{(+V_T)}{(R_2+R_3)}R_3$$

When this effective voltage becomes zero

$$\frac{(-V_{SAT})R_2+(+V_T)R_3}{(R_2+R_3)}=0$$

$$(+V_T)=(+V_{SAT})\frac{R_2}{R_3}$$

When the comparator OA_2 output is HIGH ($+V_{SAT}$), the effective voltage at the non-inverting terminal of comparator OA_2 will be by the superposition principle

$$\frac{(+V_{SAT})}{(R_2+R_3)}R_2+\frac{(-V_T)}{(R_2+R_3)}R_3$$

When this effective voltage becomes zero

$$\frac{(+V_{SAT})R_2+(-V_T)R_3}{(R_2+R_3)}=0$$

$$(-V_T)=(-V_{SAT})\frac{R_2}{R_3}$$

$$\pm V_T=\pm\frac{R_2}{R_3}V_{SAT} \tag{9.23}$$

From the waveforms shown in Figure 9.8, it is observed that

$$T_1=\frac{V_3-V_O}{2V_3}T,\quad T_2=\frac{V_3+V_O}{2V_3}T,\quad T=T_1+T_2 \tag{9.24}$$

The asymmetrical rectangular wave V_C controls another multiplexer M_2. The multiplexer M_2 selects ($+V_1$) during ON time T_2 and ($-V_1$) during OFF time T_1 of the rectangular wave V_C. Another rectangular wave V_N is generated at the multiplexer M_2 output. The lowpass filter gives average value of this pulse train V_N and is given as

$$V_x = \frac{1}{T}\left[\int_{O}^{T_2} V_1\,dt + \int_{T_2}^{T_1+T_2} (-V_1)\,dt\right]$$

$$V_x = \frac{V_1(T_2 - T_1)}{T} \tag{9.25}$$

Equations (9.24) in (9.25) give

$$V_x = \frac{V_o}{V_3} V_1$$

The op-amp, OA, is at a negative closed loop feedback configuration and a positive DC voltage is ensured in the feedback loop. Hence, its inverting terminal voltage will equal to its non-inverting terminal voltage.

$$V_X = V_2$$

$$V_O = \frac{V_2 V_3}{V_1} \tag{9.26}$$

9.5 SAWTOOTH WAVE–BASED MULTIPLIER-CUM-DIVIDER – SWITCHING

The double switching–averaging time division MCDs are shown in Figure 9.9, and their associated waveforms are shown in Figure 9.10. Figure 9.9(a) shows a series switching MCD, and Figure 9.9(b) shows a shunt switching MCD.

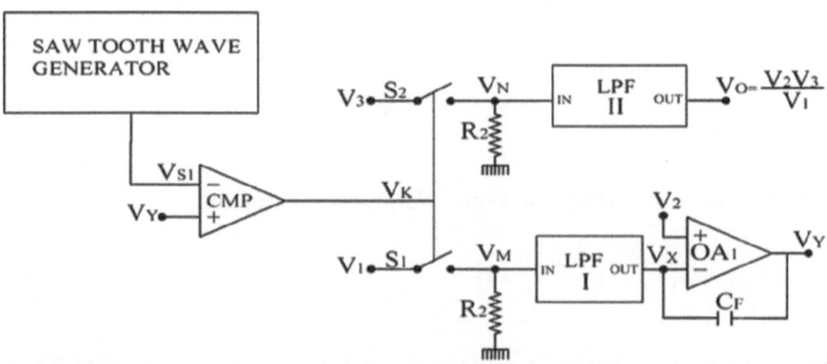

FIGURE 9.9(A) Double series switching – averaging time division multiplier-cum-divider.

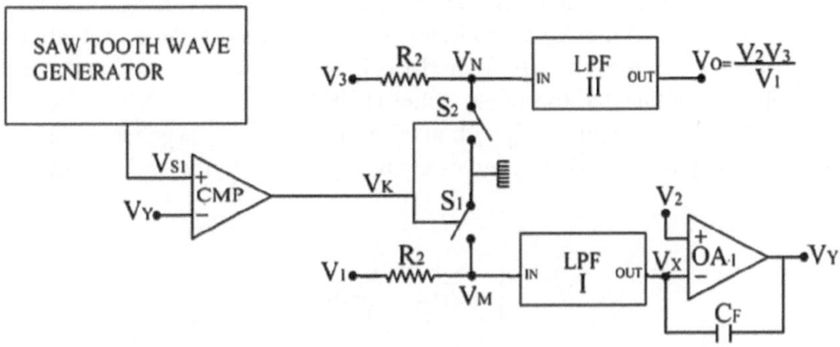

FIGURE 9.9(B) Double shunt switching – averaging time division multiplier-cum-divider.

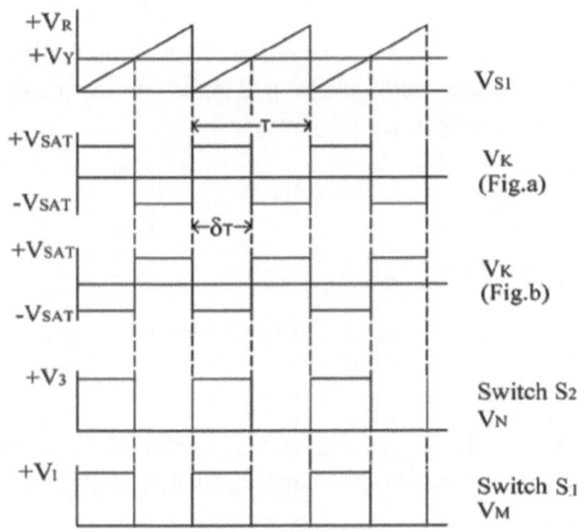

FIGURE 9.10 Associated waveforms of Figure 9.9(b).

The CMP compares the sawtooth wave with the voltage V_Y and produces a first rectangular waveform V_K. The ON time, Figure 9.9(a), or OFF time, Figure 9.9(b), δ_T of V_K is given as

$$\delta_T = \frac{V_Y}{V_R} T \tag{9.27}$$

The rectangular pulse V_K controls switches S_1 and S_2.

In Figure 9.9(a), when V_K is HIGH, switch S_1 is closed and the first input voltage V_1 is connected to lowpass filter (LPF I), switch S_2 is closed and third input voltage V_3 is connected to lowpass filter (LPF II). When V_K is LOW, switch S_1 is opened and zero volts exists on lowpass filter (LPF I), switch S_2 is opened and zero volts exists on lowpass filter (LPF II). In Figure 9.9(b), when V_K is HIGH, switch S_1 is closed and zero volts exists on lowpass filter (LPF I), switch S_2 is closed and zero volts exists on lowpass filter (LPF II). When V_K is LOW, switch S_1 is opened and the first input voltage V_1 is connected to the lowpass filter (LPF I), switch S_2 is opened and the third input voltage V_3 is connected to lowpass filter (LPF II).

The second rectangular pulse V_M with maximum value of V_1 is generated at switch S_1 output. The lowpass filter (LPF I) gives an average value of this pulse train V_M and is given as

$$V_X = \frac{1}{T}\int_0^{\delta_T} V_1\,dt = \frac{V_1}{T}\delta_T$$

$$V_X = \frac{V_1 V_Y}{V_R} \tag{9.28}$$

The op-amp, OA_1, is configured in a negative closed loop feedback, and a positive DC voltage is ensured in the feedback loop. Hence, its inverting terminal voltage must equal to its non-inverting terminal voltage

$$V_2 = V_X \tag{9.29}$$

From Eqs. (9.28) and (9.29)

$$V_Y = \frac{V_2 V_R}{V_1} \tag{9.30}$$

The third rectangular pulse V_N with maximum value V_3 is generated at switch S_2 output. The lowpass filter (LPF-II) gives average value of this pulse train V_N and is given as

$$V_O = \frac{1}{T}\int_0^{\delta_T} V_3\,dt = \frac{V_3}{T}\delta_T \tag{9.31}$$

Equations (9.27) and (9.30) in (9.31) give

$$V_O = \frac{V_2 V_3}{V_1} \tag{9.32}$$

9.6 TRIANGULAR WAVE–BASED TIME DIVISION MULTIPLIER-CUM-DIVIDER – SWITCHING

The circuit diagram of triangular wave–based time division MCD is shown in Figure 9.11, and its associated waveforms are shown in Figure 9.12.

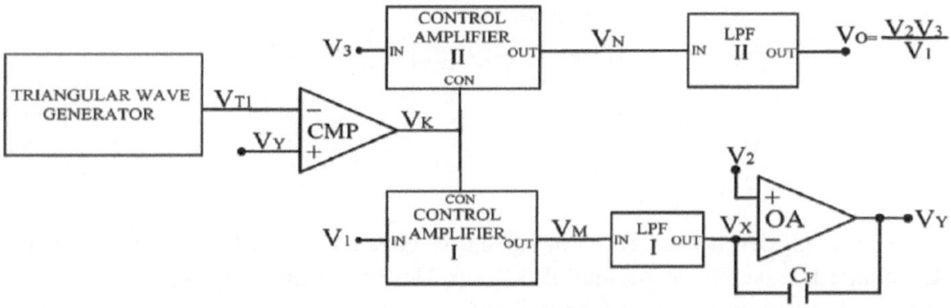

FIGURE 9.11 Switching time division multiplier-cum-divider using triangular wave generator.

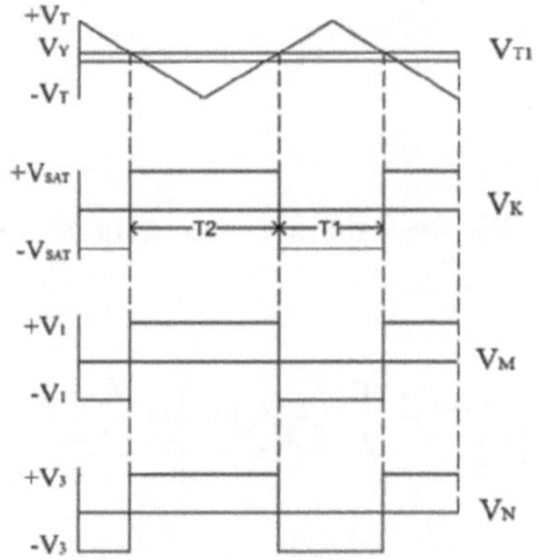

FIGURE 9.12 Associated waveforms of Figure 9.11.

The CMP compares the triangular wave V_{T1} with the voltage V_Y and produces an asymmetrical rectangular wave V_K. From Figure 9.12, it is observed that

$$T_1 = \frac{V_T - V_Y}{2V_T}T \qquad T_2 = \frac{V_T + V_Y}{2V_T}T \qquad T = T_1 + T_2 \tag{9.33}$$

The rectangular wave V_K controls amplifiers I and II. During ON time T_2 of this rectangular wave V_K, the control amplifier-I will work as a non-inverting amplifier, and $+V_1$ will exist on its output ($V_M = +V_1$). The control amplifier-II will work as a non-inverting amplifier, and $+V_3$ will exist on its output ($V_N = +V_3$). During OFF time T_1 of this rectangular wave V_K, the control amplifier-I will work as an inverting amplifier, and $-V_1$ will exist on its output ($V_M = -V_1$). The control amplifier-II will work as an inverting amplifier and $-V_3$ will exist on its output ($V_N = -V_3$).

Two asymmetrical rectangular waveforms are generated: (i) V_M at the output of control amplifier-I and (ii) V_N at the output of control amplifier-II. The lowpass filter (LPF-I) gives the average value of V_M and is given as

$$V_X = \frac{1}{T}\left[\int_{O}^{T_2} V_1\,dt + \int_{T_2}^{T_1+T_2} (-V_1)dt\right] = \frac{V_1}{T}\left[T_2 - T_1\right]$$
$$V_X = \frac{V_1 V_Y}{V_T} \tag{9.34}$$

The op-amp, OA, is configured in a negative closed loop feedback, and a positive DC voltage is ensured in the feedback loop. Hence, its inverting terminal voltage must equal to its non-inverting terminal voltage, i.e.,

$$V_X = V_2 \tag{9.35}$$

From Eqs. (9.34) and (9.35)

$$V_Y = \frac{V_2 V_T}{V_1} \tag{9.36}$$

The lowpass filter (LPF-II) gives the average value of the rectangular waveform V_N and is given as

$$V_O = \frac{1}{T}\left[\int_{O}^{T_2} V_3\,dt + \int_{T_2}^{T_1+T_2} (-V_3)dt\right] = \frac{V_3}{T}\left[T_2 - T_1\right]$$
$$V_O = \frac{V_3 V_Y}{V_T} \tag{9.37}$$

Equation (9.36) in (9.37) gives

$$V_O = \frac{V_2 V_3}{V_1} \tag{9.38}$$

9.7 TIME DIVISION MULTIPLIER-CUM-DIVIDER WITH NO REFERENCE; TYPE I – SWITCHING

The MCD using time division principle without any reference clock type I is shown in Figure 9.13, and its associated waveforms are shown in Figure 9.14.

Let initially the CMP output be LOW. The control amplifier-I will work as an inverting amplifier, and $(-V_1)$ will be at its output $(V_M = -V_1)$. The control amplifier-II will work as an inverting amplifier, and $(-V_3)$ will be at its output $((V_N = -V_3)$.

The output of the differential integrator will be

$$V_{T1} = \frac{1}{R_1 C_1}\int (V_2 + V_1)dt - V_T$$

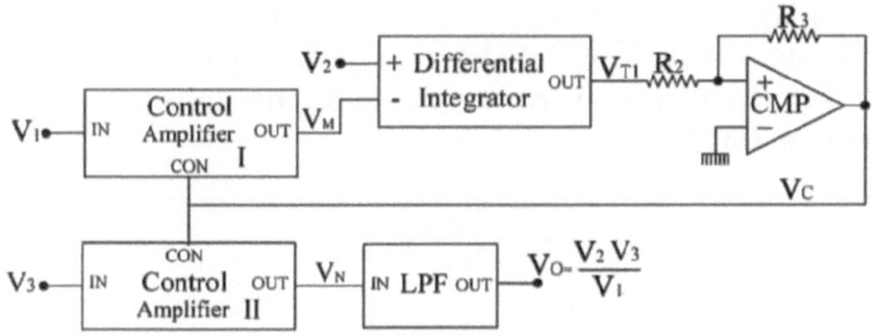

FIGURE 9.13 Switching time division multiplier-cum-divider without reference clock.

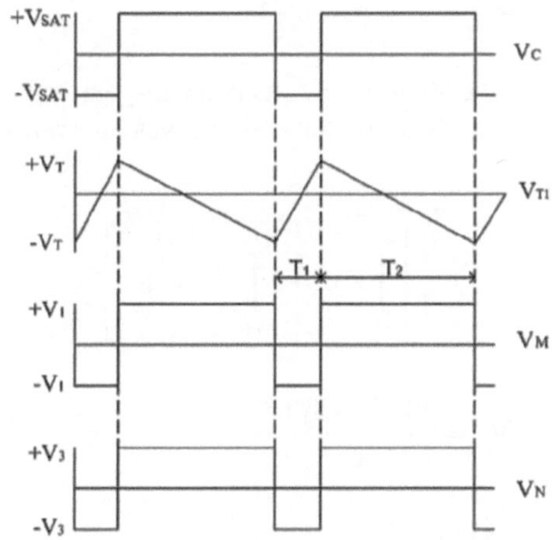

FIGURE 9.14 Associated waveforms of Figure 9.13.

$$V_{T1} = \frac{(V_1 + V_2)}{R_1C_1} t - V_T \tag{9.39}$$

The output of the differential integrator rises toward positive saturation, and when it reaches the voltage level of $(+V_T)$, the comparator OA_2 output becomes HIGH. The control amplifier-I will work as a non-inverting amplifier, and $(+V_1)$ will be at its output $(V_M = +V_1)$. The control amplifier-II will work as a non-inverting amplifier, and $(+V_3)$ will be at its output $(V_N = +V_3)$.

Now, the output of the differential integrator will be

$$V_{T1} = \frac{1}{R_1C_1}\int (V_2 - V_1)dt + V_T$$

$$V_{T1} = -\frac{(V_1 - V_2)}{R_1C_1} t + V_T \tag{9.40}$$

The output of the differential integrator reverses toward negative saturation, and when it reaches the voltage level ($-V_T$), the comparator OA_2 output becomes LOW and the cycle therefore repeats in order to give an asymmetrical rectangular wave V_C at the output of the CMP.

$$V_T = \frac{R_2}{R_3} V_{SAT} \tag{9.41}$$

From the waveforms shown in Figure 9.14, it is observed that

$$T_1 = \frac{V_1 - V_2}{2V_1} T, \quad T_2 = \frac{V_1 + V_2}{2V_1} T, \quad T = T_1 + T_2 \tag{9.42}$$

Another rectangular wave V_N with $\pm V_3$ as peak-to-peak values is generated at the output of control amplifier-II. The lowpass filter gives an average value of this pulse train V_N and is given as

$$V_O = \frac{1}{T}\left[\int_O^{T_2} V_3\,dt + \int_{T_2}^{T_1+T_2} (-V_3)\,dt\right]$$

$$V_O = \frac{V_3\left(T_2 - T_1\right)}{T} \tag{9.43}$$

Equation (9.42) in (9.43) gives

$$V_O = \frac{V_2 V_3}{V_1} \tag{9.44}$$

9.8 TIME DIVISION MULTIPLIER-CUM-DIVIDER WITH NO REFERENCE; TYPE II – SWITCHING

The MCD using time division principle without any reference clock type II is shown in Figure 9.15, and its associated waveforms are shown in Figure 9.16. Let initially the CMP output be LOW. The control amplifier-I will work as an inverting amplifier, and ($-V_3$) will be at its output ($V_M = -V_3$). The control amplifier-II will work as an inverting amplifier and ($-V_1$) will be at its output ($V_N = -V_1$).

The output of the differential integrator will be

$$V_{T1} = \frac{1}{R_1 C_1} \int \left(V_O + V_3\right) dt - V_T$$

$$V_{T1} = \frac{\left(V_3 + V_O\right)}{R_1 C_1} t - V_T \tag{9.45}$$

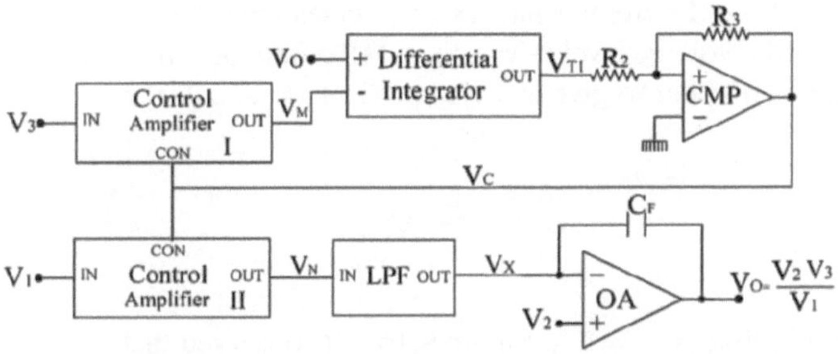

FIGURE 9.15 Switching multiplier-cum-divider without reference clock – type II.

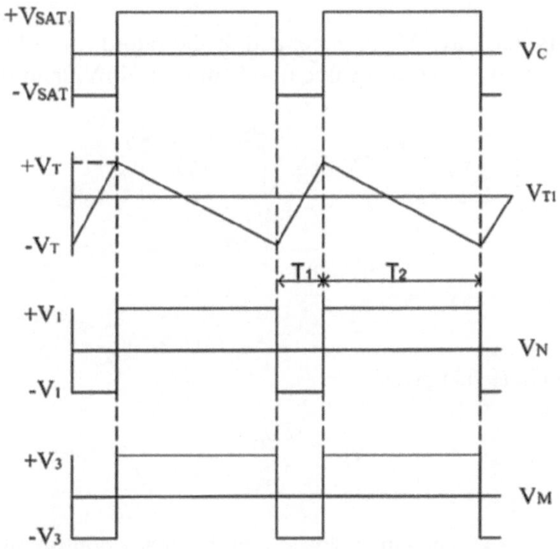

FIGURE 9.16 Associated waveforms of Figure 9.15.

The output of the differential integrator rises toward positive saturation, and when it reaches the voltage level of ($+V_T$), the CMP output becomes HIGH. The control amplifier-I will work as a non-inverting amplifier, and ($+V_3$) will be at its output ($V_M = +V_3$).

The control amplifier-II will work as a non-inverting amplifier, and ($+V_1$) will be at its output ($V_N = +V_1$).

Now, the output of the differential integrator will be

$$V_S = \frac{1}{R_1C_1}\int (V_O - V_3)\,dt + V_T$$

$$V_S = -\frac{(V_3 - V_O)}{R_1C_1}t + V_T \tag{9.46}$$

The output of the differential integrator reverses toward negative saturation, and when it reaches the voltage level ($-V_T$), the CMP output becomes LOW and the cycle therefore repeats in order to give an asymmetrical rectangular wave V_C at the output of the CMP.

$$V_T = \frac{R_2}{R_3} V_{SAT} \tag{9.47}$$

From the waveforms shown in Figure 9.16, it is observed that

$$T_1 = \frac{V_3 - V_O}{2V_3} T, \quad T_2 = \frac{V_3 + V_O}{2V_3} T, \quad T = T_1 + T_2 \tag{9.48}$$

Another rectangular wave V_N is generated at the output of control amplifier-II. The lowpass filter gives an average value of this pulse train V_N and is given as

$$V_X = \frac{1}{T}\left[\int_O^{T_2} V_1\,dt + \int_{T_2}^{T_1+T_2} (-V_1)\,dt\right]$$

$$V_X = \frac{V_1(T_2 - T_1)}{T} \tag{9.49}$$

Equation (9.48) in (9.49) gives

$$V_X = \frac{V_1 V_O}{V_3} \tag{9.50}$$

The op-amp, OA, is at a negative closed loop feedback configuration, and a positive DC voltage is ensured in the feedback loop. Its non-inverting terminal voltage will equal to its inverting terminal voltage. Hence

$$V_X = V_2 \tag{9.51}$$

From Eqs. (9.50) and (9.51)

$$V_O = \frac{V_2 V_3}{V_1} \tag{9.52}$$

10 Principles of Peak Responding Multipliers-Cum-Dividers

Peak responding multipliers-cum-dividers (MCDs) are classified into (i) peak detecting MCDs and (ii) peak sampling MCDs. A short pulse/sawtooth waveform is generated whose time period 'T' is (i) proportional to one voltage (V_2) and (ii) inversely proportional to another voltage (V_1). The third input voltage (V_3) is integrated during this time period 'T'. The peak value of the integrated output is proportional to $\frac{V_2V_3}{V_1}$. This is called "double single slope peak responding MCD." A square wave/triangular waveform is generated whose time period 'T' is (i) proportional to one voltage (V_2) and (ii) inversely proportional to one voltage (V_1). The third input voltage (V_3) is integrated during this time period 'T'. The peak value of the integrated output is proportional to $\frac{V_2V_3}{V_1}$. This is called "double dual slope peak responding MCD." A rectangular waveform is generated whose (i) time period 'T' is inversely proportional to first input voltage (V_1) (ii) OFF time proportional to second input voltage (V_2). The third input voltage (V_3) is integrated during this OFF time. The peak value of the integrated output is proportional to $\frac{V_2V_3}{V_1}$. This is called "pulse width integrated peak responding MCD." At the output stage of a peak responding MCD, if peak detector is used, it is called "peak detecting MCD," and if sample and hold is used, it is called "peak sampling MCD." The principles of peak responding MCDs are described in this chapter.

10.1 DOUBLE SINGLE SLOPE PEAK RESPONDING MULTIPLIERS-CUM-DIVIDERS

MCDs using the double single slope peak responding principle are shown in Figure 10.1, and their associated waveforms are shown in Figure 10.2. Figure 10.1(a) shows a peak detecting MCD, and Figure 10.1(b) shows a peak sampling MCD. Let initially the comparator (CMP_1) output be LOW, the controlled integrator-I integrate the input voltage ($-V_1$), and its output V_{S1} be given as

$$V_{S1} = -\frac{1}{R_1C_1}\int(-V_1)dt = \frac{V_1}{R_1C_1}t \tag{10.1}$$

DOI: 10.1201/9781003221449-12

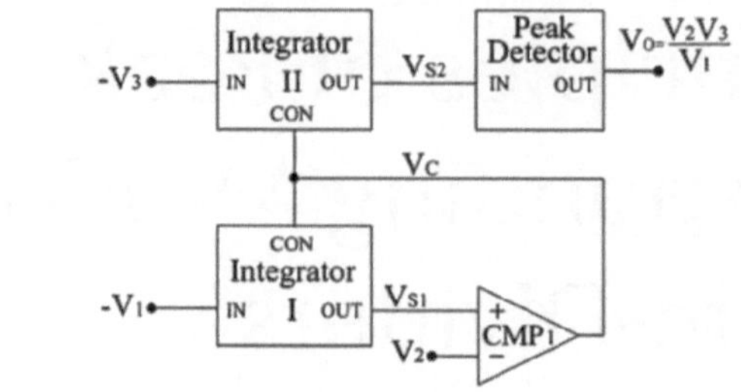

FIGURE 10.1(A) Double single slope peak detecting multiplier-cum-divider.

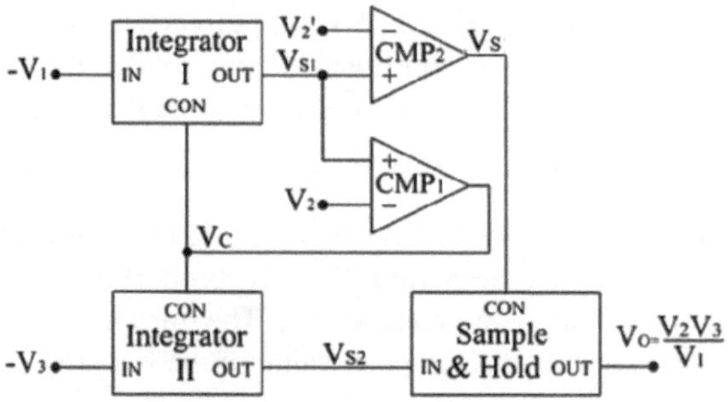

FIGURE 10.1(B) Double single slope peak sampling multiplier-cum-divider.

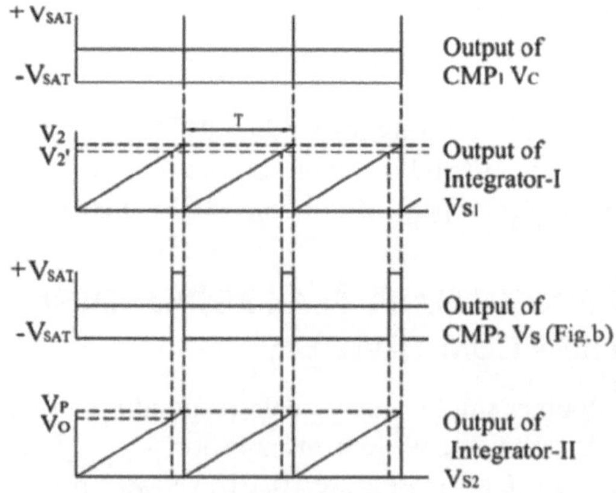

FIGURE 10.2 Associated waveforms of Figure 10.1.

where R_1C_1 are resistor-capacitor values of integrator-I.

$$V_2 = \frac{V_1}{R_1C_1}T, \quad T = \frac{V_2}{V_1}R_1C_1 \tag{10.2}$$

The output of integrator-I is a positive-going ramp, and when it reaches the value V_2, the CMP_1 output becomes HIGH which makes the controlled integrator-I output fall to zero volts. Then output of CMP_1 becomes LOW, and the cycle therefore repeats to give a perfect sawtooth wave V_{S1} at the output of controlled integrator-I, with peak value V_2 and time period 'T'. From the waveforms shown in Figure 10.2, Eq. (10.1), and the fact that at $t = T$, $V_{S1} = V_2$

The controlled integrator-II is switched to zero during ON time of short pulse V_C and integrates the input voltage ($-V_3$) during OFF time of V_C. During OFF time of V_C, the controlled integrator-II output is given as

$$V_{S2} = -\frac{1}{R_2C_2}\int(-V_3)dt = \frac{V_3}{R_2C_2}t \tag{10.3}$$

where R_2C_2 are resistor-capacitor values of controlled integrator-II.

Another sawtooth wave V_{S2} is generated at the output of controlled integrator-II with a peak value V_P and the same time period 'T'. From the waveforms shown in Figure 10.2, Eq. (10.3), and the fact that at $t = T$, $V_{S2} = V_P$

$$V_P = \frac{V_3}{R_2C_2}T$$

$$T = \frac{V_P}{V_3}R_2C_2 \tag{10.4}$$

Comparing Eqs. (10.2) and (10.4), let $R_1C_1 = R_2C_2$

$$V_P = \frac{V_2V_3}{V_1} \tag{10.5}$$

In Figure 10.1(a), the peak detector at the output stage gives peak value V_P of the sawtooth wave V_{S2} generated at the output of controlled integrator-II. Hence $V_O = V_P$.

In Figure 10.1(b), the sawtooth waveform V_{S1} is compared with slightly less than voltage V_2, i.e., V_2' by the comparator CMP_2. A short pulse V_S is generated at the CMP_2 output and is acting as a sampling pulse to the sample and hold circuit. The sampled output is $V_O = V_P$.

$$V_O = \frac{V_2V_3}{V_1} \tag{10.6}$$

10.2 DOUBLE DUAL SLOPE PEAK RESPONDING MULTIPLIERS-CUM-DIVIDERS WITH FEEDBACK COMPARATOR

MCDs using the double dual slope principle with feedback comparator (FBC) are shown in Figure 10.3, and their associated waveforms are shown in Figure 10.4. Figure 10.3(a) shows a peak detecting MCD, and Figure 10.3(b) shows peak

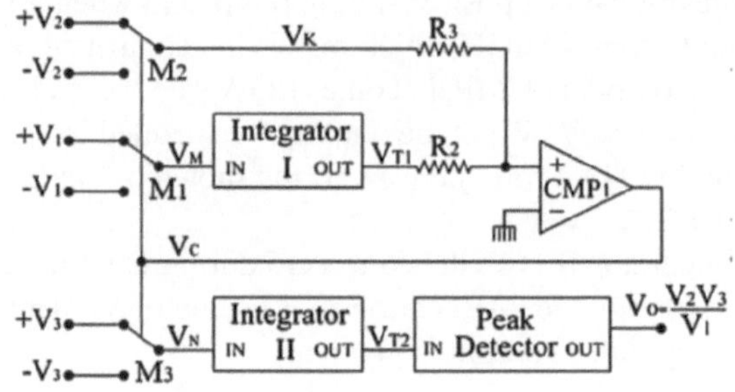

FIGURE 10.3(A) Double dual slope peak detecting multiplier-cum-divider.

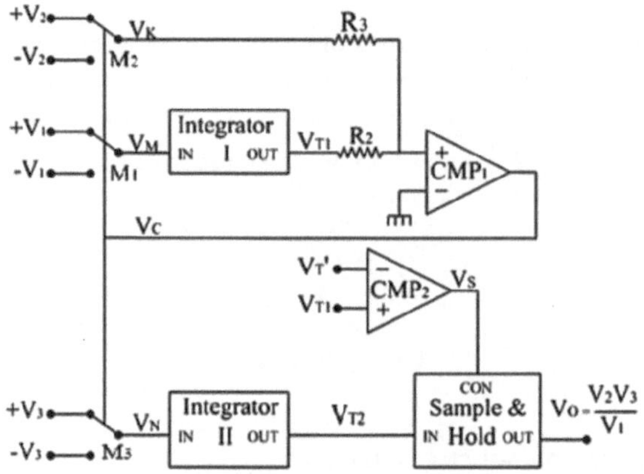

FIGURE 10.3(B) Double dual slope peak sampling multiplier-cum-divider.

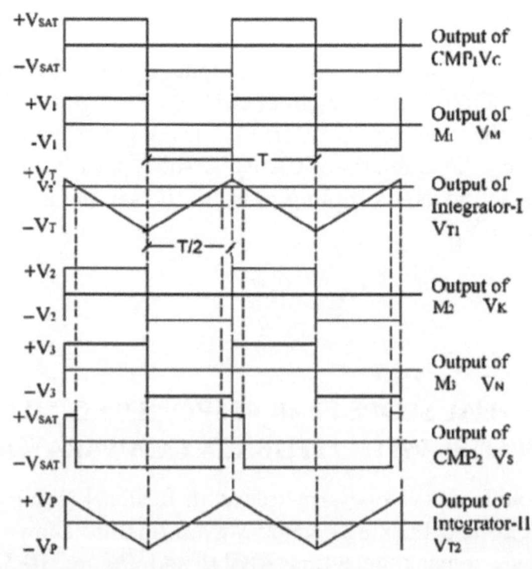

FIGURE 10.4 Associated waveforms of Figure 10.3.

sampling MCD. Let initially CMP_1 output be LOW. Multiplexer M_1 selects $(-V_1)$ to integrator-I. The integrator-I output is given as

$$V_{T1} = -\frac{1}{R_1C_1}\int(-V_1)dt = \frac{V_1}{R_1C_1}t \tag{10.7}$$

where R_1C_1 are resistor-capacitor values of integrator-I.

The output of integrator-I is going toward positive saturation and when it reaches a value $+V_T$, the CMP_1 output becomes HIGH. Multiplexer M_1 selects $(+V_1)$ to integrator-I. The integrator-I output is given as

$$V_{T1} = -\frac{1}{R_1C_1}\int(+V_1)dt = -\frac{V_1}{R_1C_1}t \tag{10.8}$$

The output of integrator-I is reversing toward negative saturation and when it reaches a value $(-V_T)$, the CMP_1 output becomes LOW. Multiplexer M_1 selects $(-V_1)$, and the sequence repeats to give a triangular waveform V_{T1} of $\pm V_T$ peak-to-peak values with a time period 'T' at the output of integrator-I. From the waveforms shown in Figure 10.4, Eq. (10.7), and the fact that at $t = T/2$, $V_{T1} = 2V_T$

$$2V_T = \frac{V_1}{R_1C_1}\frac{T}{2}, \quad T = \frac{4V_T}{V_1}R_1C_1 \tag{10.9}$$

V_T is given as

$$V_T = \frac{R_2}{R_3}V_2 \tag{10.10}$$

Multiplexer M_3 connects $(+V_3)$ during ON time of V_C and $(-V_3)$ during OFF time of V_C. Another square waveform V_N with $\pm V_3$ peak-to-peak value is generated at the output of multiplexer M_3. This square wave V_N is converted into triangular wave V_{T2} by integrator-II with $\pm V_P$ as peak-to-peak values of the same time period 'T'. For one transition the integrator-II output is given as

$$V_{T2} = -\frac{1}{R_4C_2}\int(-V_3)dt = \frac{V_3}{R_4C_2}t \tag{10.11}$$

where R_4C_2 are resistor-capacitor values of integrator-II.

From the waveforms shown in Figure 10.4, Eq. (10.11), and the fact that at $t = T/2$, $V_{T2} = 2V_p$.

$$2V_P = \frac{V_3}{R_4C_2}\frac{T}{2}$$

Let $R_1C_1 = R_4C_2$

$$V_P = \frac{V_2V_3}{V_1}\frac{R_2}{R_3}$$

In Figure 10.3(a), the peak detector at output stage gives peak value V_P of triangular wave V_{T2} and hence $V_O = V_P$.

In Figure 10.3(b), the peak value of triangular waveform V_{T2} is sampled by a sample and hold circuit with a sampling pulse V_S. The sampling pulse V_S is generated by comparing first triangular wave V_{T1} with slightly less than its peak value, i.e., V_T'. The sample and hold output is $V_O = V_P$.

Let $R_2/R_3 = 1$(But in practical $R_2/R_3 < 1$)

$$V_O = \frac{V_2 V_3}{V_1} \tag{10.12}$$

10.3 DOUBLE DUAL SLOPE MULTIPLIERS-CUM-DIVIDERS USING FLIP FLOP

MCDs using the double dual slope principle with flip flop are shown in Figure 10.5, and their associated waveforms are shown in Figure 10.6. Figure 10.5(a) shows a peak detecting MCD, and Figure 10.5(b) shows a peak sampling MCD. Let initially flip flop output be LOW. The multiplexer M_1 selects $(-V_1)$ to integrator-I. The integrator-I output is given as

$$V_{T1} = -\frac{1}{R_1C_1}\int(-V_1)\,dt = \frac{V_1}{R_1C_1}t \tag{10.13}$$

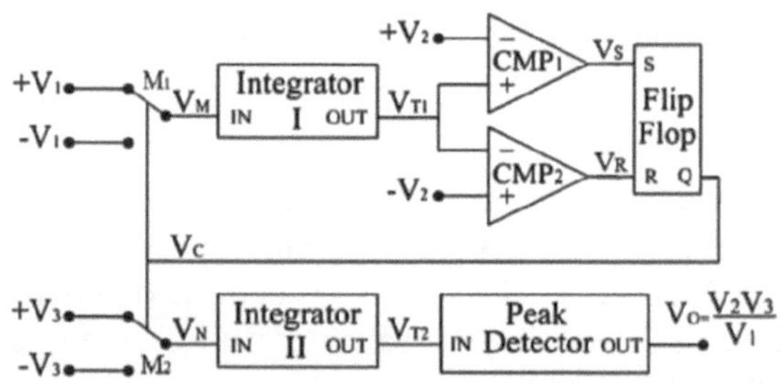

FIGURE 10.5(A) Double dual slope peak responding multiplier-cum-divider.

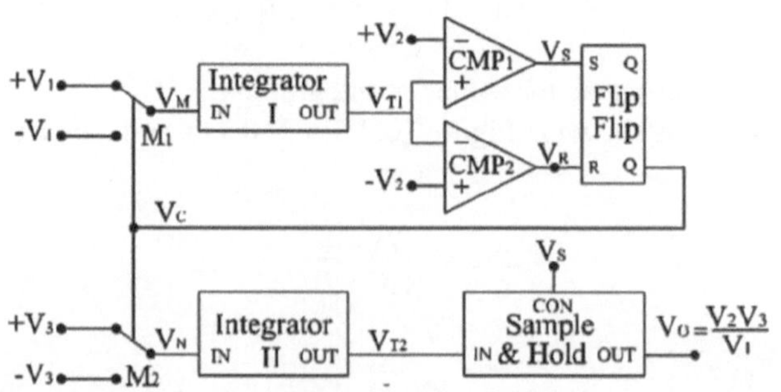

FIGURE 10.5(B) Double dual slope peak detecting multiplier-cum-divider.

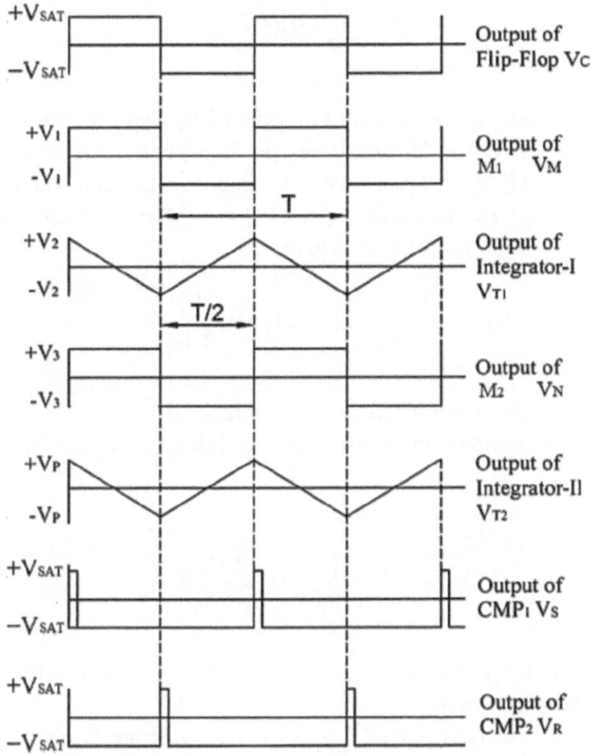

FIGURE 10.6 Associated waveforms of Figure 10.5.

where R_1C_1 are resistor-capacitor values of integrator-I.

The output of integrator-I is going toward positive saturation, and when it reaches the value $+V_2$, the CMP_1 output becomes HIGH, and it sets the flip flop output to HIGH. Multiplexer M_1 selects $(+V_1)$ to integrator-I. The integrator-I output is given as

$$V_{T1} = -\frac{1}{R_1C_1}\int(+V_1)\,dt = -\frac{V_1}{R_1C_1}t \tag{10.14}$$

The output of integrator-I is reversing toward negative saturation, and when it reaches the value $(-V_2)$, the CMP_2 output becomes HIGH and resets the flip flop so that its output becomes LOW. Multiplexer M_1 selects $(-V_1)$, and the sequence repeats to give a triangular waveform V_{T1} of $\pm V_2$ peak-to-peak values with a time period 'T' at the output of integrator-I. From the waveforms shown in Figure 10.6, Eq. (10.13), and the fact that at $t = T/2$, $V_{T1} = 2V_2$

$$2V_2 = \frac{V_1}{R_1C_1}\frac{T}{2}$$

$$T = \frac{4V_2}{V_1} R_1C_1$$

Multiplexer M_2 connects $+V_3$ during ON time of V_C and $-V_3$ during OFF time of V_C. Another square waveform V_N with $\pm V_3$ peak-to-peak value is generated at the output of multiplexer M_2. This square wave V_N is converted into triangular wave V_{T2} by the integrator-II with $\pm V_P$ as peak-to-peak values of the same time period 'T'. For one transition, the integrator-II output is given as

$$V_{T2} = -\frac{1}{R_2C_2}\int(-V_3)dt = \frac{V_3}{R_2C_2}t \tag{10.15}$$

where R_2C_2 are resistor-capacitor values of integrator-II.

From the waveforms shown in Figure 10.6, Eq. (10.15), and the fact that at $t = T/2$, $V_{T2} = 2V_p$.

Let $R_1C_1 = R_2C_2$.

$$2V_P = \frac{V_3}{R_2C_2}\frac{T}{2}, \quad V_P = \frac{V_2V_3}{V_1}$$

In Figure 10.5(a), the peak detector at output stage gives peak value V_P of triangular wave V_{T2} and hence $V_O = V_P$.

In Figure 10.5(b), the short pulse V_S generated at the output of CMP_1 is acting as a sampling pulse to the sample and hold circuit. The integrator-II output is given to the sample and hold circuit. The sampled output V_O is given as $V_O = V_P$

$$V_O = \frac{V_2V_3}{V_1} \tag{10.16}$$

10.4 PULSE WIDTH INTEGRATED PEAK DETECTING MULTIPLIER-CUM-DIVIDER

The MCD using the pulse width integrated peak detecting principle is shown in Figure 10.7, and its associated waveforms are shown in Figure 10.8. Let initially CMP_1 output be LOW, the controlled integrator-I integrate the input voltage $(-V_1)$, and its output V_{S1} be given as

$$V_{S1} = -\frac{1}{R_1C_1}\int(-V_1)dt = \frac{V_1}{R_1C_1}t \tag{10.17}$$

where R_1C_1 are resistor-capacitor values of integrator-I.

The output of integrator-I is a positive-going ramp, and when it reaches the value of $+V_R$, the CMP_1 output becomes HIGH which makes the controlled integrator-I output fall to zero volts. Then output of CMP_1 becomes LOW, and the cycle therefore

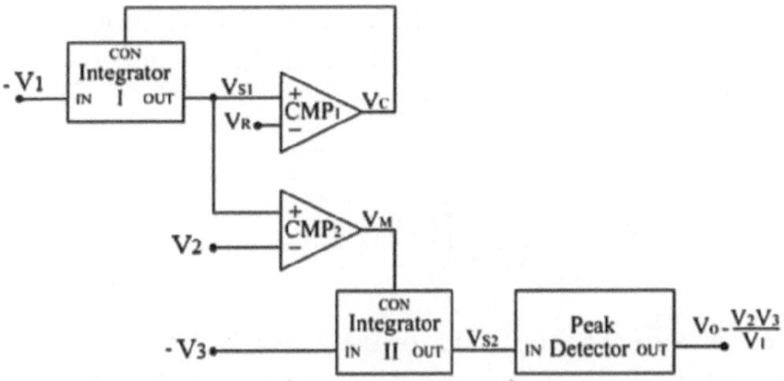

FIGURE 10.7 Pulse width integrated peak detecting multiplier-cum-divider.

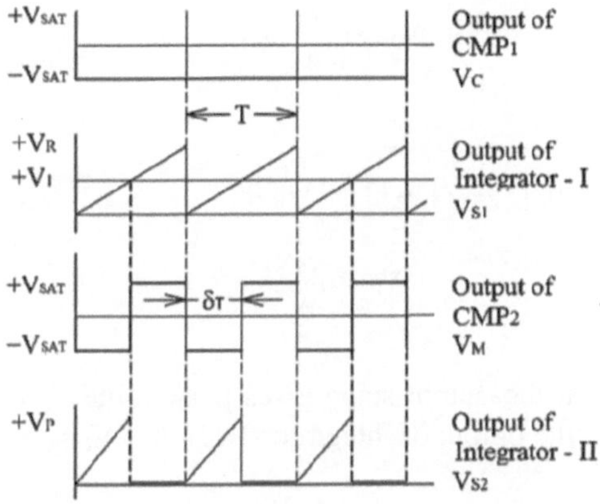

FIGURE 10.8 Associated waveforms of Figure 10.7.

repeats to give a perfect sawtooth wave V_{S1} at the output of integrator-I, with peak value of V_R and time period 'T'. From the waveforms shown in Figure 10.8, from Eq. (10.17), and fact that at t = T, $V_{S1} = V_R$

$$V_R = \frac{V_1}{R_1C_1}T$$

$$T = \frac{V_R}{V_1}R_1C_1 \tag{10.18}$$

The sawtooth wave V_{S1} is compared with another input voltage V_2 by CMP_2, and a rectangular pulse V_M is generated at the output of CMP_2.

The OFF time δ_T of this rectangular wave V_M is given as

$$\delta_T = \frac{V_2}{V_R}T \tag{10.19}$$

The rectangular wave V_M controls the controlled integrator-II. The controlled integrator-II is switched to zero during ON time of V_M and integrates the input voltage $(-V_3)$ during OFF time of V_M. During OFF time of V_M, the integrator-II output is given as

$$V_{S2} = -\frac{1}{R_2C_2}\int(-V_3)\,dt = \frac{V_3}{R_2C_2}t \tag{10.20}$$

where R_2C_2 are resistor-capacitor values of integrator-II.

A semi-sawtooth wave V_{S2} is generated at the output of integrator-II with a peak value V_P and same time period 'T'. From the waveforms shown in Figure 10.8, Eq. (10.20), and fact that at $t = \delta_T$, $V_{S2} = V_P$.

$$V_P = \frac{V_3}{R_2C_2}\delta_T \tag{10.21}$$

Let $R_1C_1 = R_2C_2$.

Equations (10.18) and (10.19) in (10.21) give

$$V_P = \frac{V_2V_3}{V_1}$$

The peak detector at the output stage gives peak value V_P of the semi-sawtooth wave V_{S2} generated at the output of integrator-II. Hence $V_O = V_P$.

$$V_O = \frac{V_2V_3}{V_1} \tag{10.22}$$

10.5 PULSE POSITION PEAK RESPONDING MULTIPLIER-CUM-DIVIDER

MCDs using the pulse position peak responding principle are shown in Figure 10.9, and their associated waveforms are shown in Figure 10.10. Figure 10.9(a) shows a peak detecting MCD, and Figure 10.9(b) shows a peak sampling MCD. Let initially

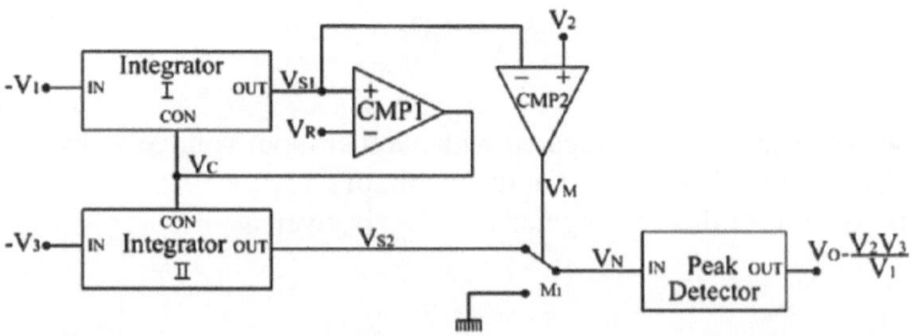

FIGURE 10.9(A) Pulse position peak detecting multiplier-cum-divider.

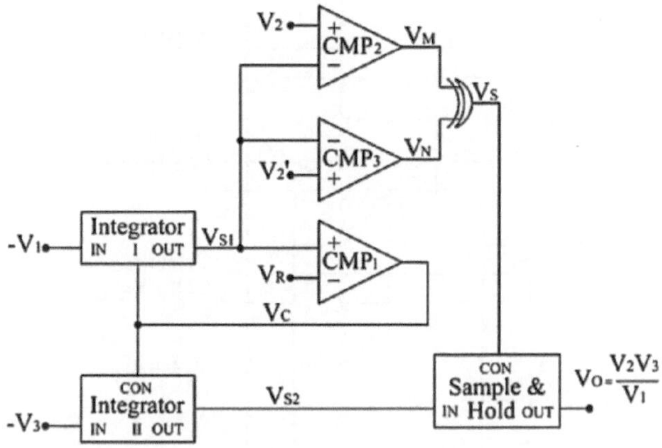

FIGURE 10.9(B) Pulse position peak sampling multiplier-cum-divider.

CMP_1 output be LOW, integrator-I integrate the input voltage ($-V_1$), and its output V_{S1} be given as

$$V_{S1} = -\frac{1}{R_1C_1}\int(-V_1)dt = \frac{V_1}{R_1C_1}t \tag{10.23}$$

where R_1C_1 are resistor-capacitor values of integrator-I.

The output of integrator-I is a positive-going ramp, and when it reaches the value of $+V_R$, CMP_1 output becomes HIGH which makes the integrator-I output fall to zero volts. Then output of CMP_1 becomes LOW, and the cycle therefore repeats to give a perfect sawtooth wave V_{S1} at the output of integrator-I, with peak value V_R and time period 'T'. From the waveforms shown in Figure 10.10(a), Eq. (10.23), and the fact that at $t = T$, $V_{S1} = V_R$

$$V_R = \frac{V_1}{R_1C_1}T, \quad T = \frac{V_R}{V_1}R_1C_1 \tag{10.24}$$

Integrator II is switched to zero during ON time of short pulse V_C and integrates the input voltage ($-V_3$) during OFF time of V_C. During OFF time of V_C, the integrator-II output is given as

$$V_{S2} = -\frac{1}{R_2C_2}\int(-V_3)dt = \frac{V_3}{R_2C_2}t \tag{10.25}$$

where R_2C_2 are resistor-capacitor values of integrator-II.

Another sawtooth wave V_{S2} is generated at the output of integrator-II with a peak value V_P and the same time period 'T'. From the waveforms shown in Figure 10.10(b), Eq. (10.25), and the fact that at $t = T$, $V_{S2} = V_p$

$$V_P = \frac{V_3}{R_2C_2}T, \quad T = \frac{V_P}{V_3}R_2C_2 \tag{10.26}$$

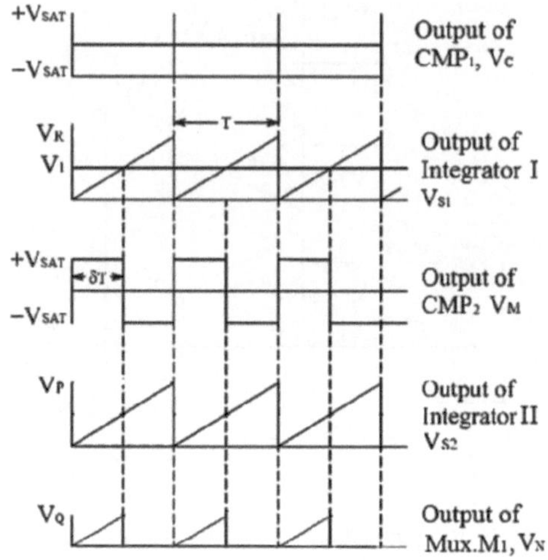

FIGURE 10.10(A) Associated waveforms of Figure 10.9(a).

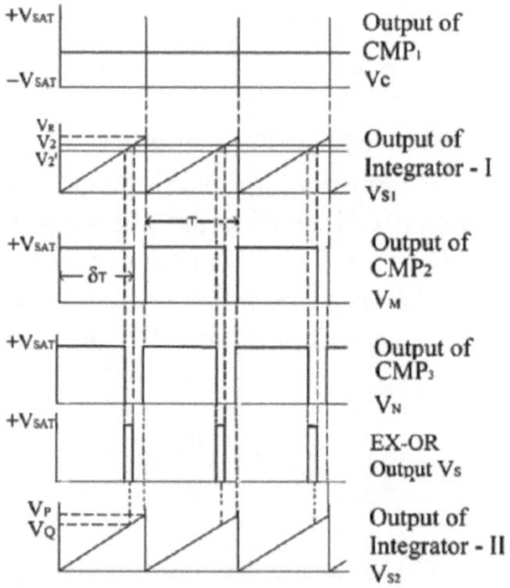

FIGURE 10.10(B) Associated waveforms of Figure 10.9(a).

The input voltage V_2 is given to CMP_2 which compares V_2 with the sawtooth wave V_{S1}. A rectangular wave V_M is generated at the output of CMP_2. The ON time of this rectangular wave V_M is given as

$$\delta_T = \frac{V_2}{V_R} T \tag{10.27}$$

In Figure 10.9(a), this rectangular wave V_M controls multiplexer M_1. Multiplexer M_1 connects the sawtooth wave V_{S2} during ON time of V_M and connects zero volt during OFF time of V_M. A semi-sawtooth wave V_N with peak value of V_Q is generated at the multiplexer M_1 output. The peak detector at the output stage gives this peak value V_Q at the output. Hence $V_O = V_Q$.

In Figure 10.9(b), the sawtooth waveform V_{S1} is compared with slightly less than V_2 voltage, i.e., V_2' by CMP_3. Another rectangular pulse V_N is generated at the output of CMP_3. The rectangular pulses V_M and V_N are given to an EX-OR gate to get a short pulse V_S. This short pulse V_S is acting as a sampling pulse to the sample and hold circuit. The sampled output V_O is given as $V_O = V_Q$.

$$V_Q = \frac{V_P}{T}\delta_T, \quad V_Q = \frac{V_2 V_3}{V_1}, \quad V_O = \frac{V_2 V_3}{V_1} \tag{10.28}$$

10.6 PEAK RESPONDING MULTIPLIER-CUM-DIVIDER USING VOLTAGE TUNABLE ASTABLE MULTIVIBRATOR

The peak responding MCDs using a voltage tunable astable multivibrator are shown in Figure 10.11, and their associated waveforms are shown in Figure 10.12. Figure 10.11(a) shows a peak detecting MCD, and Figure 10.11(b) shows a peak sampling MCD. The op-amp OA_1 in association with multiplexers M_1 and M_2 along with components R_1,R_2,R_3, and C_1 produces a square waveform V_C. The time period of this square waveform V_C is proportional to the input voltage V_2 and inversely proportional to another input voltage V_1.

$$T = K\frac{V_2}{V_1} \tag{10.29}$$

During HIGH value of V_C, multiplexer M_3 connects $-V_3$ to the integrator, and during LOW value of V_C, $(+V_3)$ is connected to the integrator by multiplexer M_3. Another square wave V_K with $\pm V_3$ peak-to-peak value is generated at the multiplexer M_3 output. This square wave V_K is converted into triangular wave V_{T2} with $\pm V_P$ peak-to-peak value. For one transition, the integrator output is given as

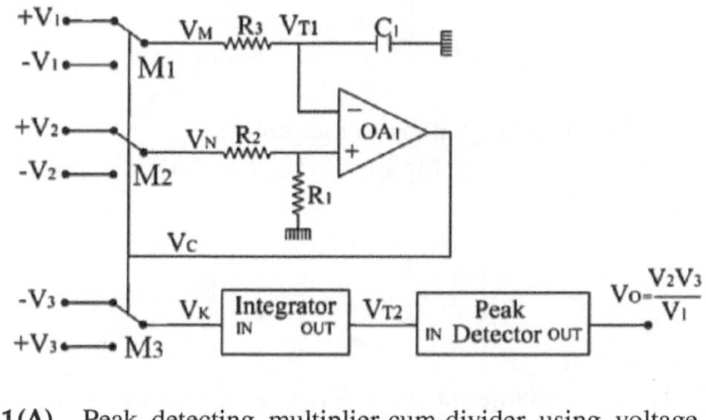

FIGURE 10.11(A) Peak detecting multiplier-cum-divider using voltage tunable astable multivibrator.

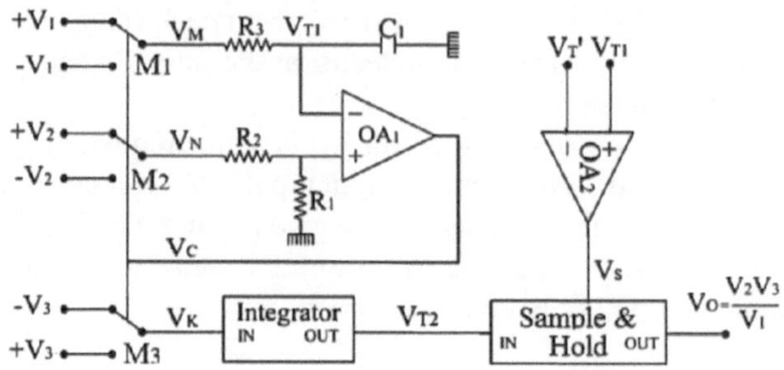

FIGURE 10.11(B) Peak sampling multiplier-cum-divider using voltage tunable astable multivibrator.

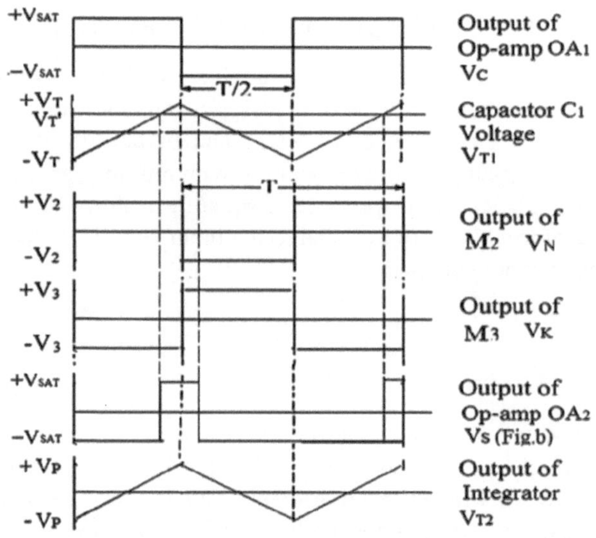

FIGURE 10.12 Associated waveforms of Figure 10.12.

$$V_{T2} = -\frac{1}{R_4C_4}\int(-V_3)dt = \frac{V_3}{R_4C_4}t \tag{10.30}$$

where R_4C_4 are resistor-capacitor values of integrator.

From the waveforms shown in Figure 10.12, Eq. (10.30), and the fact that at $t = T/2$, $V_{T2} = 2V_p$

$$2V_P = \frac{V_3}{R_4C_4}\frac{T}{2}, \quad V_P = \frac{V_2V_3}{V_1}K_1 \tag{10.31}$$

In Figure 10.11(a), the triangular wave V_{T2} is given to a peak detector to get its peak value V_P. Hence, the output of peak detector will be $V_O = V_P$.

In Figure 10.11(b), the triangular wave V_{T2} is given to the sample and hold circuit to get its peak value V_P. The sampling pulse V_S is generated by comparing the capacitance C_1 voltage V_{T1} with slightly less than its peak value V_T, i.e., V_T'. The output of the sample and hold circuit will be $V_O = V_P$.

$$V_O = \frac{V_2 V_3}{V_1} K_1 \tag{10.32}$$

Part C

Design of Function Circuits

11 Design of Analog Multipliers – Multiplexing

If the width of a pulse train is made proportional to one voltage and the amplitude of the same pulse train to a second voltage, then the average value of this pulse train is proportional to the product of two voltages, and it is called "time division multiplier," "pulse averaging multiplier," or "sigma delta multiplier." The time division multiplier (TDM) can be implemented using a (1) triangular wave, (2) sawtooth wave, and (3) without using any reference wave.

Peak responding multipliers are classified into (i) peak detecting multipliers and (ii) peak sampling multipliers. A short pulse/sawtooth waveform is generated whose time period 'T' is proportional to one voltage. Another input voltage is integrated during the time period 'T'. The peak value of the integrated voltage is proportional to the product of the two input voltages. This is called "double single slope peak responding multiplier." A square/triangular waveform whose time period 'T' is proportional to one voltage is generated. Another input voltage is integrated during the time period 'T'. The peak value of the integrated voltage is proportional to the product of the two input voltages. This is called "double dual slope peak responding multiplier." A rectangular pulse waveform is generated whose OFF time is proportional to one voltage. Another voltage is integrated during this OFF time. The peak value of integrated output is proportional to the product of the two input voltages. This is called "pulse width integrated peak responding multiplier." At the output stage of a peak responding multiplier, if peak detector is used, it is called "peak detecting multiplier," and if sample and hold is used, it is called "peak sampling multiplier."

A multiplier uses either analog switches or analog multiplexers for its operation. If analog switches are used, they are called "switching multipliers," and if analog multiplexers are used, they are called "multiplexing multipliers." Multiplexing multipliers are discussed in this chapter, and switching multipliers are described next in Chapter 12.

11.1 TRIANGULAR WAVE–BASED TIME DIVISION MULTIPLIERS

The circuit diagrams of triangular wave–based TDMs are shown in Figure 11.1, and their associated waveforms are shown in Figure 11.2. In Figure 11.1(a), a triangular wave V_{T1} with $\pm V_T$ peak-to-peak value and time period 'T' is generated by the op-amp OA_1 and transistors Q_1 and Q_2. The value of V_T is given as

$$V_T = \beta(V_{SAT}) \approx \beta(0.76)(V_{CC}) \tag{11.1}$$

DOI: 10.1201/9781003221449-14

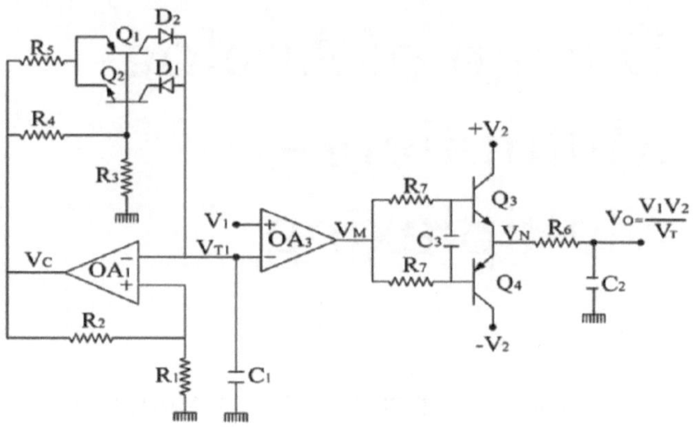

FIGURE 11.1(A) Triangular wave–based time division multiplier – type I.

where β is given as $\beta = \dfrac{R_1}{R_1 + R_2}$ and time period 'T' is given as

$$T = 4R_5C_1\frac{R_1}{R_2} \tag{11.2}$$

In Figure 11.1(b), op-amps OA_1 and OA_2 constitute a triangular/square wave generator. The output of op-amp OA_1 is a triangular wave V_{T1} with $\pm V_T$ peak values and time period 'T'. Let initially the comparator OA_2 output be LOW($-V_{SAT}$), and the integrator ouput composed by op-amp OA_1, resistor R_1, and capacitor C_1 be given as

$$V_{T1} = -\frac{1}{R_1C_1}\int -V_{SAT}\,dt = \frac{V_{SAT}}{R_1C_1}t \tag{11.3}$$

The integrator output is rising toward positive saturation, and when it reaches a value $+V_T$, the comparator output becomes HIGH($+V_{SAT}$). The integrator output composed by op-amp OA_1, resistor R_1, and capacitor C_1 is given as

$$V_{T1} = -\frac{1}{R_1C_1}\int +V_{SAT}\,dt = -\frac{V_{SAT}}{R_1C_1}t \tag{11.4}$$

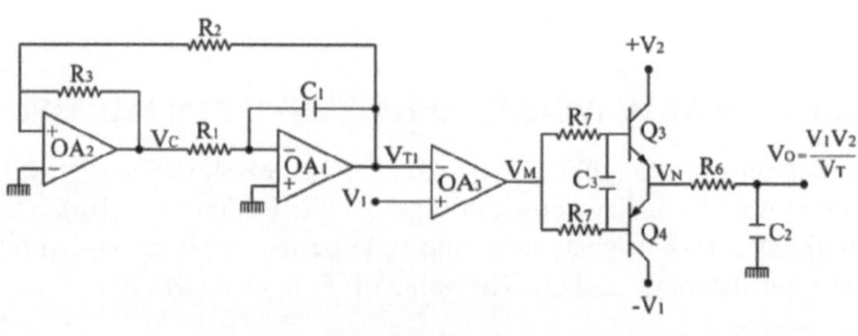

FIGURE 11.1(B) Triangular wave–based time division multiplier – type II.

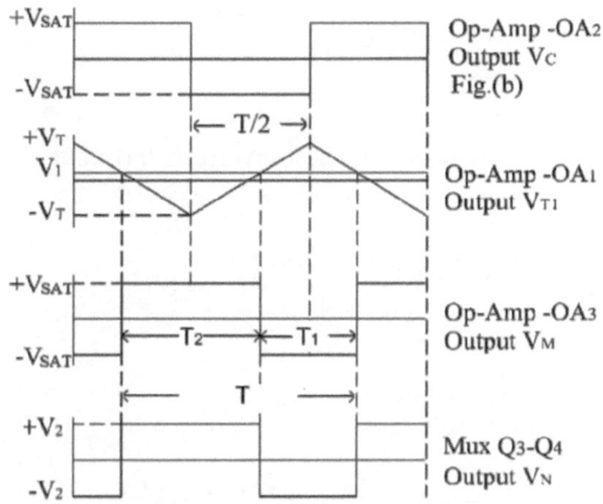

FIGURE 11.2 Associated waveforms of Figure 11.1.

Now the integrator output is changing its slope from $+V_T$ toward $(-V_T)$ and when it reaches a value $-V_T$, the comparator output becomes LOW$(-V_{SAT})$, and the sequence therefore repeats to give (i) a triangular waveform V_{T1} with $\pm V_T$ peak-to-peak values at the output of op-amp OA_1, and (ii) a square waveform V_C with $\pm V_{SAT}$ peak-to-peak values at the output of comparator OA_2. From the waveforms shown in Figure 11.2, Eq. (11.3), and the fact that at $t = T/2$, $V_{T1} = 2V_T$

$$2V_T = \frac{V_{SAT}}{R_1C_1}\frac{T}{2}, \quad T = \frac{4V_TR_1C_1}{V_{SAT}} \tag{11.5}$$

When the comparator OA_2 output is LOW $(-V_{SAT})$, the effective voltage at the non-inverting terminal of comparator OA_2 will be by the superposition principle

$$\frac{(-V_{SAT})}{(R_2+R_3)}R_2 + \frac{(+V_T)}{(R_2+R_3)}R_3$$

When this effective voltage at the non-inverting terminal of comparator OA_2 becomes zero

$$\frac{(-V_{SAT})R_2 + (+V_T)R_3}{(R_2+R_3)} = 0$$

$$(+V_T) = (+V_{SAT})\frac{R_2}{R_3}$$

When the comparator OA_2 output is HIGH $(+V_{SAT})$, the effective voltage at the non-inverting terminal of comparator OA_2 will be by the superposition principle

$$\frac{(+V_{SAT})}{(R_2+R_3)}R_2+\frac{(-V_T)}{(R_2+R_3)}R_3$$

When this effective voltage at the non-inverting terminal of comparator OA_2 becomes zero

$$\frac{(+V_{SAT})R_2+(-V_T)R_3}{(R_2+R_3)}=0$$

$$(-V_T)=(-V_{SAT})\frac{R_2}{R_3}$$

$$\pm V_T=\pm V_{SAT}\frac{R_2}{R_3}\simeq 0.76(\pm V_{CC})\frac{R_2}{R_3} \tag{11.6}$$

From Equations (11.5) and (11.6), time period 'T' of the generated triangular/square waveforms is given by

$$T=4R_1C_1\frac{R_2}{R_3} \tag{11.7}$$

In both circuits of Figures 11.1(a) and 11.1(b), one input voltage V_1 is compared with the generated triangular wave V_{T1} by the comparator OA_3. An asymmetrical rectangular waveform V_M is generated at the comparator OA_3 output. From the waveforms shown in Figure 11.2, it is observed that

$$T_1=\frac{V_T-V_1}{2V_T}T \qquad T_2=\frac{V_T+V_1}{2V_T}T \qquad T=T_1+T_2 \tag{11.8}$$

This rectangular wave V_M is given as control input to the transistor multiplexer Q_3–Q_4. The transistor multiplexer Q_3–Q_4 connects the other input voltage ($+V_2$) during T_2 (transistor Q_3 is ON and Q_4 is OFF) and ($-V_2$) during T_1 (transistor Q_3 is OFF and Q_4 is ON). Another rectangular asymmetrical wave V_N with peak-to-peak value of $\pm V_2$ is generated at the transistor multiplexer Q_3–Q_4 output. The R_6C_2 lowpass filter gives the average value of the pulse train V_N which is given as

$$V_O=\frac{1}{T}\left[\int_{O}^{T_2}V_2\,dt+\int_{T_2}^{T_1+T_2}(-V_2)\,dt\right]=\frac{V_2}{T}(T_2-T_1) \tag{11.9}$$

Equation (11.8) in (11.9) gives

$$V_O=\frac{V_1V_2}{V_T} \tag{11.10}$$

The output voltage depends on sharpness and linearity of triangular waveform. Since the output voltage depends on V_T, a highly stable precision voltage V_T must be obtained. The output voltage is also proportional to attenuation constant α of the lowpass filter. The attenuation constant α will be in the range of 0.80 to 0.90. The polarity of both input voltages V_1 and V_2 are of any polarity. Hence, this multiplier is of four-quadrant type.

DESIGN EXERCISES

1. The transistor multiplexers Q_3–Q_4 in Figures 11.1(a) and 11.1(b) are to be replaced with JFET multiplexers and MOSFET multiplexers. In each, (i) draw the circuit diagrams, (ii) explain their working operation, (iii) draw waveforms at appropriate places, and (iv) deduce the expression for the output voltages.

PRACTICAL EXERCISES

1. Verify the multiplier circuits shown in Figure 11.1 by experimenting with the AFC trainer kit.
2. As discussed later in Chapter 18 and Section 18.2, convert the analog multipliers of Figure 11.1 into analog dividers by both type I and II methods. In each, (i) draw the circuit diagrams, (ii) explain their working operation, (iii) draw waveforms at appropriate places, and (iv) deduce expression for the output voltage. Verify this divider by experimenting with the AFC trainer kit.
3. As discussed later in Chapter 18 and Section 18.3, convert the analog multipliers of Figure 11.1 into analog square rooters by both type I and II methods. In each, (i) draw the circuit diagrams, (ii) explain their working operation (iii) draw waveforms at appropriate places, and (iv) deduce expression for the output voltage. Verify this square rooter by experimenting with the AFC trainer kit.
4. Replace the transistor multiplexers in Figure 11.1 with CD4053 analog multiplexers IC. (i) Draw the circuit diagrams, (ii) explain their working operation, (iii) draw waveforms at appropriate places, and (iv) deduce expression for the output voltage. Verify this analog multiplier by experimenting with the AFC trainer kit.
5. Convert the multiplier obtained in Problem 4 into a divider (as discussed later in Chapter 18 and Section 18.2). (i) Draw the circuit diagrams, (ii) explain their working operation, (iii) draw waveforms at appropriate places, and (iv) deduce expression for the output voltage. Verify this analog divider by experimenting with the AFC trainer kit.
6. Convert the multiplier obtained in Problem 4 into a square rooter (as discussed later in Chapter 18 and Section 18.3). (i) Draw the circuit diagrams, (ii) explain their working operation, (iii) draw waveforms at appropriate places, and (iv) deduce expression for the output voltage. Verify this square rooter by experimenting with the AFC trainer kit.

11.2 TIME DIVISION MULTIPLIER WITH NO REFERENCE

The circuit diagrams of TDMs without using either triangular or sawtooth waves as reference are shown in Figure 11.3, and their associated waveforms are shown in Figure 11.4. The op-amps OA_1 and OA_2 along with R_1, C_1, R_2 and R_3 constitute an asymmetrical rectangular wave generator, and a rectangular wave V_C is generated at the output of op-amp OA_2.

$$\pm V_T = \pm V_{SAT}\frac{R_2}{R_3};\ 0.76\left(\pm V_{CC}\right)\frac{R_2}{R_3} \tag{11.11}$$

From waveforms shown in Figure 11.4, it is observed that

$$T_1 = \frac{V_{SAT} - V_1}{2V_{SAT}}T \qquad T_2 = \frac{V_{SAT} + V_1}{2V_{SAT}}T \qquad T = T_1 + T_2 \tag{11.12}$$

The rectangular wave V_C controls transistor multiplexers Q_1–Q_2, and the transistor multiplexer selects ($+V_2$) during T_2 (transistor Q_1 is ON and Q_2 is OFF) and ($-V_2$) during T_1 (transistor Q_1 is OFF and Q_2 is ON) to its output. Another asymmetrical rectangular wave V_N is generated at the transistor multiplexers Q_1–Q_2 output with

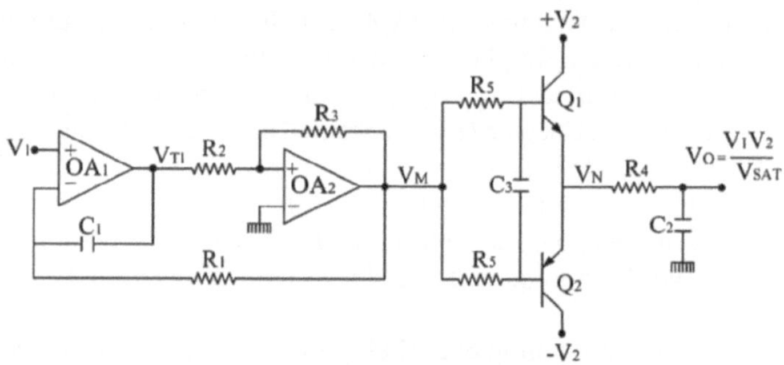

FIGURE 11.3(A) Time division multiplier with no reference.

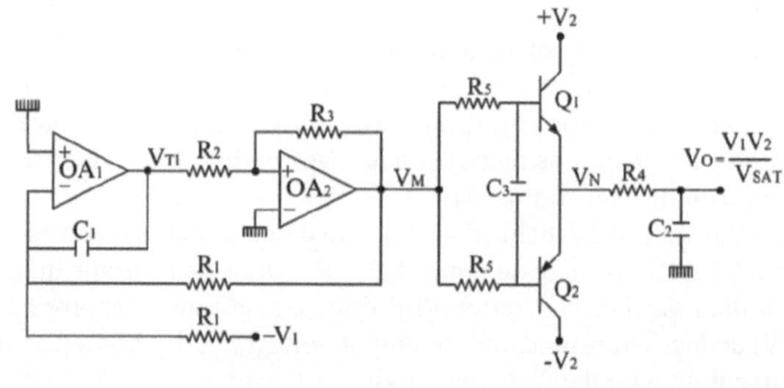

FIGURE 11.3(B) Equivalent circuit of Figure 11.3(a).

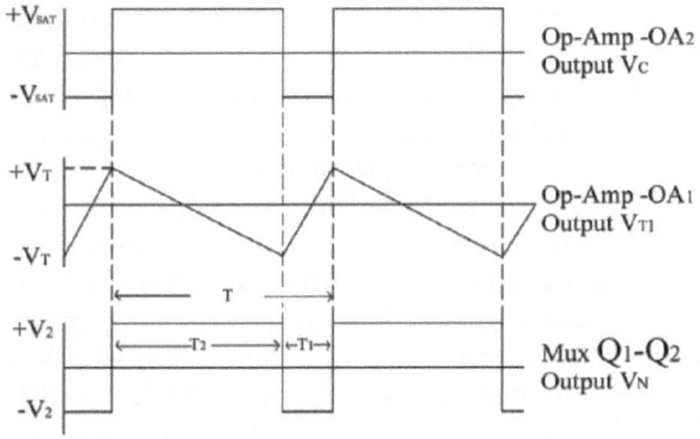

FIGURE 11.4 Associated waveforms of Figure 11.3.

$\pm V_2$ as maximum values. The R_4C_2 lowpass filter gives the average value of this pulse train V_N and is given as

$$V_O = \frac{1}{T}\left[\int_O^{T_2} V_2\,dt + \int_{T_2}^{T_1+T_2} (-V_2)dt\right] = \frac{V_2}{T}\left[T_2 - T_1\right] \tag{11.13}$$

Equation (11.12) in (11.13) gives

$$V_O = \frac{V_1V_2}{V_{SAT}} \tag{11.14}$$

The output voltage depends on V_{SAT}; a highly stable precision power supply must be used (as $V_{SAT} \sim 0.76V_{CC}$). The output voltage is also proportional to attenuation constant α of the lowpass filter. The attenuation constant α will be in the range of 0.80 to 0.90. The polarity of both input voltages V_1 and V_2 are of any polarity. Hence, this multiplier is of the four-quadrant type.

DESIGN EXERCISES

1. The transistor multiplexers Q_1–Q_2 in Figures 11.3(a) and 11.3(b) are to be replaced with JFET transistor multiplexers and MOSFET multiplexers. In each, (i) draw circuit diagrams, (ii) explain their working operation, (iii) draw waveforms at appropriate places, and (iv) deduce expression for the output voltages.

PRACTICAL EXERCISES

1. Verify the multiplier circuits shown in Figure 11.3 by experimenting with the AFC trainer kit.

2. As discussed later in Chapter 18 and Section 18.2, convert the analog multipliers of Figure 11.1 into analog dividers by both type I and II methods. In each, (i) draw the circuit diagrams, (ii) explain their working operation, (iii) draw waveforms at appropriate places, and (iv) deduce expression for the output voltage. Verify this divider by experimenting with the AFC trainer kit.
3. As discussed later in Chapter 18 and Section 18.3, convert the analog multipliers of Figure 11.3 into analog square rooters by both type I and II methods. In each, (i) draw the circuit diagrams, (ii) explain their working operation (iii) draw waveforms at appropriate places, and (iv) deduce expression for the output voltage. Verify this square rooter by experimenting with the AFC trainer kit.
4. Replace the transistor multiplexers in Figure 11.3 with CD4053 analog multiplexers IC. (i) Draw the circuit diagrams, (ii) explain their working operation, (iii) draw waveforms at appropriate places, and (iv) deduce expression for the output voltage. Verify this multiplier by experimenting with the AFC trainer kit.
5. Convert the multiplier obtained in Problem 4 into a divider (as discussed in Chapter 18 and Section 18.2). (i) Draw the circuit diagrams, (ii) explain their working operation, (iii) draw waveforms at appropriate places, and (iv) deduce expression for the output voltage. Verify this analog divider by experimenting with the AFC trainer kit.
6. Convert the multiplier obtained in Problem 4 into a square rooter (as discussed in Chapter 18 and Section 18.3). (i) Draw the circuit diagrams, (ii) explain their working operation, (iii) draw waveforms at appropriate places, and (iv) deduce expression for the output voltage. Verify this square rooter by experimenting with the AFC trainer kit.

11.3 DOUBLE DUAL SLOPE PEAK RESPONDING MULTIPLIERS

The circuit diagrams of double dual slope peak responding multipliers are shown in Figure 11.5, and their associated waveforms are shown in Figure 11.6. Figure 11.5(a)

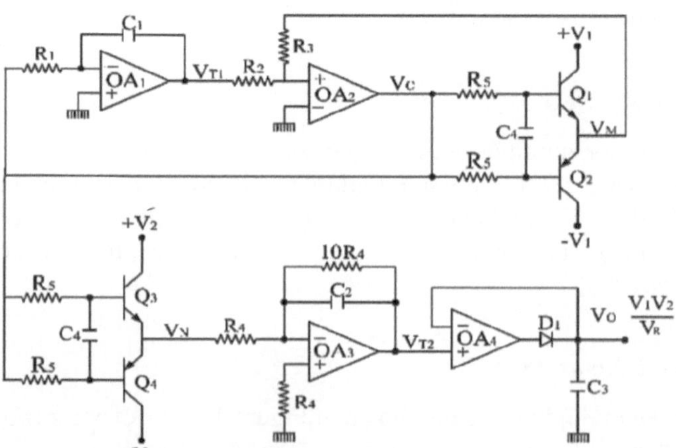

FIGURE 11.5(A) Double dual slope peak detecting multiplier.

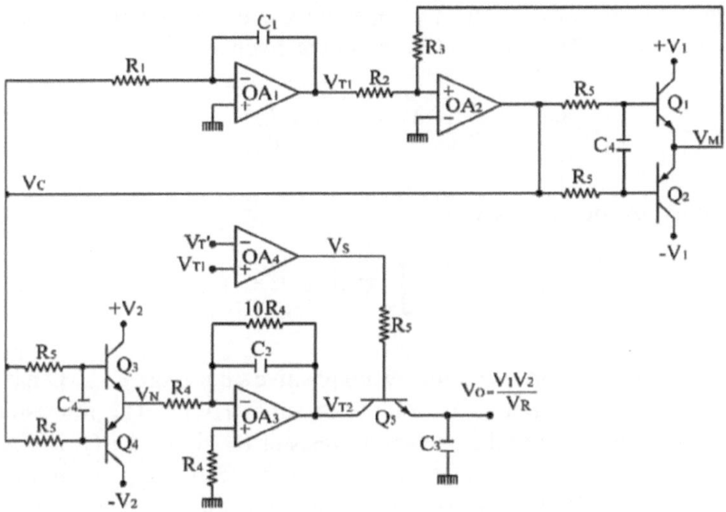

FIGURE 11.5(B) Double dual slope peak sampling multiplier.

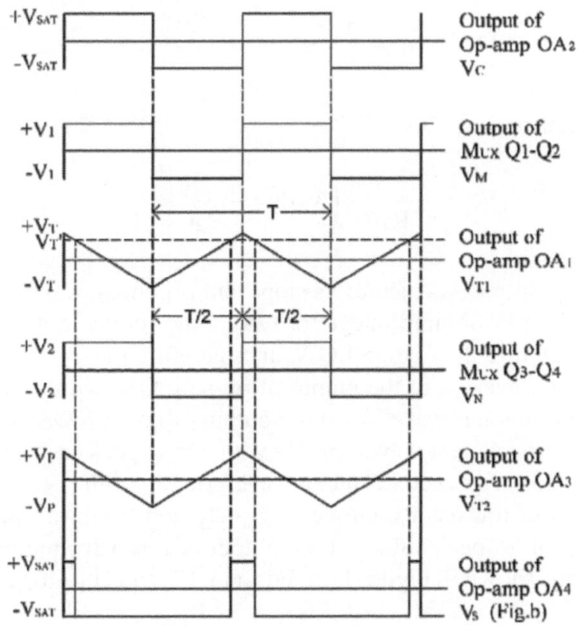

FIGURE 11.6 Associated waveforms of Figure 11.5.

shows a double dual slope peak detecting multiplier, and Figure 11.5(b) shows a double dual slope peak sampling multiplier. Let initially the comparator OA_2 output be LOW. The transistor multiplexer Q_1–Q_2 selects $(-V_1)$ to the one end of resistor R_3 (transistor Q_1 is OFF and Q_2 is ON) $(-V_{SAT})$ is given to the integrator formed by resistor R_1, capacitor C_1, and op-amp OA_1. The transistor multiplexer Q_3–Q_4 selects $(-V_2)$

to the integrator formed by resistor R_4, capacitor C_2, and op-amp OA_3 (transistor Q_3 is OFF and Q_4 is ON). The op-amp OA_1 output is given as

$$V_{T1} = -\frac{1}{R_1C_1}\int -V_{SAT}\,dt = \frac{V_{SAT}}{R_1C_1}t \tag{11.15}$$

The op-amp OA_3 output is given as

$$V_{T2} = -\frac{1}{R_4C_2}\int -V_2\,dt = \frac{V_2}{R_4C_2}t \tag{11.16}$$

The output of op-amp OA_1 is rising toward positive saturation, and when it reaches a value of $+V_T$, the comparator OA_2 output becomes HIGH. The transistor multiplexer Q_1–Q_2 selects ($+V_1$) to the one end of resistor R_3 (transistor Q_1 is ON and Q_2 is OFF), ($+V_{SAT}$) is given to the integrator formed by R_1, C_1 and op-amp OA_1. The transistor multiplexer Q_3–Q_4 selects ($+V_2$) to the integrator formed by R_4, C_2, and op-amp OA_3 (transistor Q_3 is ON and Q_4 is OFF). The output of op-amp OA_1 will now be

$$V_{T1} = -\frac{1}{R_1C_1}\int V_{SAT}\,dt = -\frac{V_{SAT}}{R_1C_1}t \tag{11.17}$$

And the output of op-amp OA_3 will be

$$V_{T2} = -\frac{1}{R_4C_2}\int V_2\,dt = -\frac{V_2}{R_4C_2}t \tag{11.18}$$

The output of op-amp OA_1 changes its slope and is going toward negative saturation. When the output of op-amp integrator OA_1 comes down to a value ($-V_T$), the comparator OA_2 output will become LOW, and therefore the cycle repeats itself to give (i) a triangular wave V_{T1} at the output of op-amp OA_1 with $\pm V_T$ peak-to-peak value, (ii) another triangular wave V_{T2} at the output of op-amp OA_3 with $\pm V_P$ peak-to-peak value, (iii) first square waveform V_C with $\pm V_{SAT}$ peak-to-peak values at the output of op-amp OA_2, (iv) second square waveform V_M with $\pm V_1$ as peak-to-peak values at the output of transistor multiplexer Q_1–Q_2, and (v) third square waveform V_N with $\pm V_2$ as peak-to-peak values at the output of transistor multiplexer Q_3–Q_4. From the waveforms shown in Figure 11.6, Eqs. (11.17) and (11.18), and the fact that at $t = T/2$, $V_{T1} = 2V_T$, $V_{T2} = 2V_P$

$$2V_T = \frac{V_{SAT}}{R_1C_1}\frac{T}{2} \tag{11.19}$$

$$2V_P = \frac{V_2}{R_4C_2}\frac{T}{2} \tag{11.20}$$

From Eqs. (11.19) and (11.20)

$$2V_P = \frac{V_2}{R_4C_2}\frac{2V_TR_1C_1}{V_{SAT}}$$

Let us assume $R_1 = R_4$, $C_1 = C_2$.

$$V_P = \frac{V_2}{V_{SAT}}V_T \tag{11.21}$$

When the comparator OA_2 output is LOW ($-V_{SAT}$), ($-V_1$) will be at the output of transistor multiplexer Q_1–Q_2, the effective voltage at the non-inverting terminal of comparator OA_2 will be by the superposition principle

$$\frac{(-V_1)}{(R_2+R_3)}R_2 + \frac{(+V_T)}{(R_2+R_3)}R_3$$

When this effective voltage at the non-inverting terminal of comparator OA_2 becomes zero

$$\frac{(-V_1)R_2+(+V_T)R_3}{(R_2+R_3)} = 0$$

$$(+V_T) = (+V_1)\frac{R_2}{R_3}$$

When the comparator OA_2 output is HIGH ($+V_{CC}$), the effective voltage at the non-inverting terminal of comparator OA_2 will be by the superposition principle

$$\frac{(+V_1)}{(R_2+R_3)}R_2 + \frac{(-V_T)}{(R_2+R_3)}R_3$$

When this effective voltage at the non-inverting terminal of comparator OA_2 becomes zero

$$\frac{(+V_1)R_2+(-V_T)R_3}{(R_2+R_3)} = 0$$

$$(-V_T) = (-V_1)\frac{R_2}{R_3}$$

$$\pm V_T = \pm V_1\frac{R_2}{R_3} \tag{11.22}$$

Equation (11.22) in (11.21) gives

$$V_P = \frac{V_1 V_2}{V_{SAT}} \frac{R_2}{R_3}, \quad \text{Let } V_R = \frac{V_{SAT}}{R_2} R_3$$

$$V_P = \frac{V_1 V_2}{V_R} \tag{11.23}$$

In the circuit shown in Figure 11.5(a), the peak detector realized by op-amp OA_4, diode D_1, and capacitor C_3 gives this peak value V_P at the output: $V_O = V_P$.

In the circuit shown in Figure 11.5(b), the peak value V_P is obtained by the sample and hold circuit realized by transistor Q_5 and capacitor C_3. The sampling pulse VS is generated by op-amp OA_4 by comparing a slightly less than voltage VT called VT' with the triangular wave VT_1. The sampled output is given as $V_O = V_P$.

From Eq. (11.23), the output voltage is given as $V_O = V_P$.

$$V_O = \frac{V_1 V_2}{V_R} \tag{11.24}$$

The output voltage depends on sharpness and linearity of triangular waveform. The output voltage depends on V_{SAT}; a highly stable precision power supply must be used (as $V_{SAT} \sim 0.76V_{CC}$). For peak detecting multiplier, the polarity of both input voltages V_1 and V_2 must be single polarity only. Hence, this multiplier is of single-quadrant type. For peak sampling multiplier, the input voltages V_1 must have positive polarity only and that of V_2 can be any polarity. Hence, this multiplier is of two-quadrant type.

DESIGN EXERCISES

1. The transistor multiplexers Q_1–Q_2 and Q_3–Q_4 in and Figure 11.5(b) are to be replaced with JFET multiplexers and MOSFET multiplexers. In each, (i) draw the circuit diagrams, (ii) explain their working operation, (iii) draw waveforms at appropriate places, and (iv) deduce expression for the output voltage.

PRACTICAL EXERCISES

1. Verify the multiplier circuits shown in Figure 11.5 by experimenting with the AFC trainer kit.
2. As discussed later in Chapter 18 and Section 18.2, convert the analog multipliers of Figure 11.5 into analog dividers by both type I and II methods. In each, (i) draw the circuit diagrams, (ii) explain their working operation, (iii) draw waveforms at appropriate places, and (iv) deduce expression for the output voltage. Verify this divider by experimenting with the AFC trainer kit.
3. As discussed later in Chapter 18 and Section 18.3, convert the analog multiplier of Figure 11.5 into analog square rooters by both type I and II methods.

In each, (i) draw the circuit diagrams, (ii) explain their working operation, (iii) draw waveforms at appropriate places, and (iv) deduce expression for the output voltage. Verify this square rooter by experimenting with the AFC trainer kit.

4. Replace the transistor multiplexers in Figure 11.5 with CD4053 analog multiplexers IC. (i) Draw the circuit diagrams, (ii) explain their working operation, (iii) draw waveforms at appropriate places, and (iv) deduce expression for the output voltage. Verify this multiplier by experimenting with the AFC trainer kit.
5. Convert the multipliers obtained in Problem 4 into dividers (as discussed in Chapter 18 and Section 18.2). (i) Draw the circuit diagrams, (ii) explain their working operation, (iii) draw waveforms at appropriate places, and (iv) deduce expression for the output voltage. Verify this analog divider by experimenting with the AFC trainer kit.
6. Convert the multiplier obtained in Problem 4 into a square rooter (as discussed later Chapter 18 and Section 18.3). (i) Draw the circuit diagrams, (ii) explain their working operation, (iii) draw waveforms at appropriate places, and (iv) deduce expression for the output voltage. Verify this square rooter by experimenting with the AFC trainer kit.

11.4 PEAK RESPONDING MULTIPLIERS WITH VOLTAGE-TO-PERIOD CONVERTER

The multiplexing type peak responding multipliers using a voltage-to-period (V/T) converter are shown in Figure 11.7, and their associated waveforms are shown in

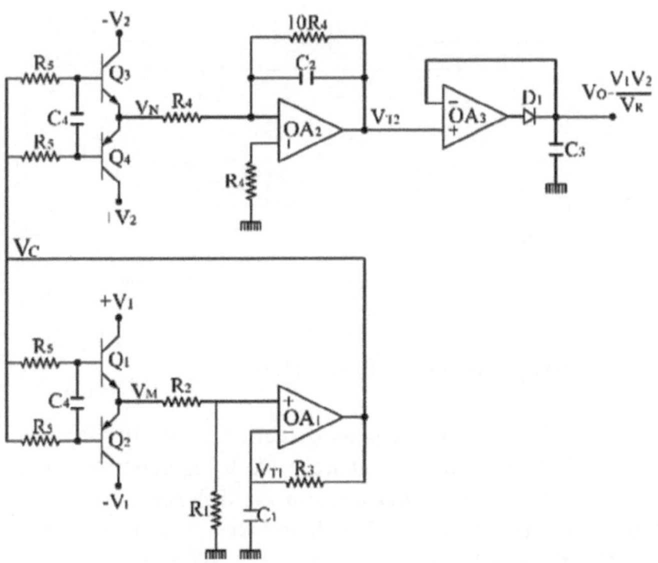

FIGURE 11.7(A) Peak detecting multiplier using voltage-to-period converter.

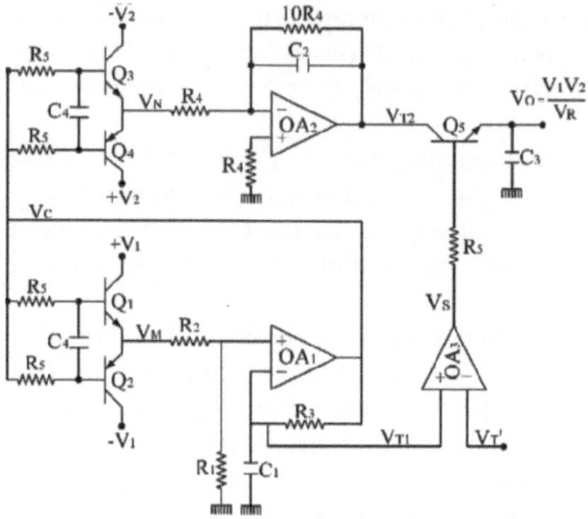

FIGURE 11.7(B) Peak sampling multiplier using voltage-to-period converter.

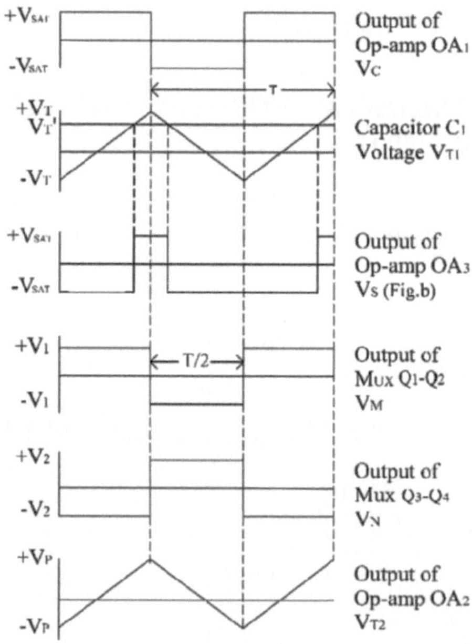

FIGURE 11.8 Associated waveforms of Figure 11.7.

Figure 11.8. Figure 11.7(a) shows a peak detecting multiplier, and Figure 11.7(b) shows a peak sampling multiplier. A square wave V_C is generated by op-amp OA_1, resistors R_1, R_2, R_3, capacitor C_1, and transistor multiplexer Q_1–Q_2. The square wave V_C controls transistor multiplexer Q_3–Q_4. The transistor multiplexer Q_3–Q_4 connects $(-V_2)$ during HIGH time of V_C and $(+V_2)$ during LOW time of V_C. Another square wave V_N with $\pm V_2$ peak-to- peak values is generated at the transistor multiplexer Q_3–Q_4 output. The integrator formed by op-amp OA_2, resistors R_4, and capacitor C_2

converts the square wave V_N into a triangular wave V_{T2} of $\pm V_P$ peak-to-peak values. V_P is proportional to V_2.

$$V_P = K_1V_2 \tag{11.25}$$

The time period 'T' of square wave V_C is proportional to V_1

$$T = K_2V_1 \tag{11.26}$$

The integrator output is given as

$$V_{T2} = -\frac{1}{R_4C_2}\int -V_2\,dt = \frac{V_2}{R_4C_2}t \tag{11.27}$$

For half time period for t = T/2. From the waveforms shown in Figure 11.8, Eq. (11.27), and the fact that at t = T/2, $V_{T2} = 2\,V_P$.

$$2V_P = \frac{V_2}{R_4C_2}\frac{T}{2};\quad V_P = \frac{V_2}{4R_4C_2}V_1K_2 \tag{11.28}$$

K_1 and K_2 are constant values. Let $V_R = \dfrac{4R_4C_2}{K_2}$

$$V_P = \frac{V_1V_2}{V_R} \tag{11.29}$$

In Figure 11.7(a), the peak detector realized by op-amp OA_3, diode D_1, and capacitor C_3 gives this peak value V_P at the output: $V_O = V_P$.

In Figure 11.7(b), the peak value V_P is sampled by the sample and hold circuit realized by transistor Q_5 and capacitor C_3 with a sampling pulse V_S. The sampling pulse V_S is generated by comparing capacitor C_1 voltage VT_1 with slightly less than its peak value VT, i.e., VT'. The sampled output is $V_O = V_P$.

From Eq. (11.29), the output voltage is given as $V_O = V_P$.

$$V_O = \frac{V_1V_2}{V_R} \tag{11.30}$$

The output voltage depends on V/T converter constant value K and V_{SAT}. For peak detecting multiplier, the polarity of both input voltages V_1 and V_2 must be single polarity only. Hence, this multiplier is of the single-quadrant type. For peak sampling multiplier, the input voltages V_1 must have positive polarity only, and that of V_2 can be any polarity. Hence, this multiplier is of the two-quadrant type.

DESIGN EXERCISES

1. The transistor multiplexers Q_1–Q_2 and Q_3–Q_4 in Figures 11.7(a) and (b) are to be replaced with JFET multiplexers and MOSFET multiplexers. In each,

(i) draw the circuit diagrams, (ii) explain their working operation, (iii) draw waveforms at appropriate places, and (iv) deduce expression for the output voltage.

PRACTICAL EXERCISES

1. Verify the multiplier circuit shown in Figure 11.7 by experimenting with the AFC trainer kit.
2. As discussed later in Chapter 18 and Section 18.2, convert the analog multiplier of Figure 11.7 into an analog divider by both type I and II methods. In each, (i) draw the circuit diagrams, (ii) explain their working operation, (iii) draw waveforms at appropriate places, and (iv) deduce expression for the output voltage. Verify this divider by experimenting with the AFC trainer kit.
3. As discussed later in Chapter 18 and Section 18.3, convert the analog multiplier of Figure 11.7 into an analog square rooter by both type I and II methods. In each, (i) draw the circuit diagrams, (ii) explain their working operation, (iii) draw waveforms at appropriate places, and (iv) deduce expression for the output voltage. Verify this square rooter by experimenting with the AFC trainer kit.
4. Replace the transistor multiplexers in Figure 11.7 with CD4053 analog multiplexers IC. (i) Draw the circuit diagrams, (ii) explain their working operation, (iii) draw waveforms at appropriate places, and (iv) deduce expression for the output voltage. Verify this multiplier by experimenting with the AFC trainer kit.
5. Convert the multipliers obtained in Problem 4 into dividers (as discussed later in Chapter 18 and Section 18.2). (i) Draw the circuit diagrams, (ii) explain their working operation, (iii) draw waveforms at appropriate places, and (iv) deduce expression for the output voltage. Verify this analog divider by experimenting with the AFC trainer kit.
6. Convert the multiplier obtained in Problem 4 into square rooter (as discussed later in Chapter 18 and Section 18.3). (i) Draw the circuit diagrams, (ii) explain their working operation, (iii) draw waveforms at appropriate places, and (iv) deduce expression for the output voltage. Verify this square rooter by experimenting with the AFC trainer kit.

12 Design of Analog Multipliers – Switching

As discussed later in Chapter 11, if the width of a pulse train is made proportional to one voltage and the amplitude of the same pulse train to a second voltage, then the average value of this pulse train is proportional to the product of two voltages, and it is called "time division multiplier," "pulse averaging multiplier," or "sigma delta multiplier." The time division multiplier (TDM) can be implemented using (1) triangular wave, (2) sawtooth wave, and (3) without using any reference clock.

As discussed in Chapter 11, peak responding multipliers are classified into (1) peak detecting multipliers and (2) peak sampling multipliers. A short pulse/sawtooth waveform is generated whose time period 'T' is proportional to one voltage. Another input voltage is integrated during the time period 'T'. The peak value of the integrated voltage is proportional to the product of the input voltages. This is called a "double single slope peak responding multiplier." A square/triangular waveform is generated whose time period 'T' is proportional to one voltage. Another input voltage is integrated during the time period 'T'. The peak value of the integrated voltage is proportional to the product of the input voltages. This is called a "double dual slope peak responding multiplier." A rectangular pulse waveform is generated whose OFF time is proportional to one voltage. Another voltage is integrated during this OFF time. The peak value of integrated output is proportional to the product of the two input voltages. This is called a "pulse width integrated peak responding multiplier." At the output of a peak responding multiplier, if peak detector is used, it is called "peak detecting multiplier" and if sample and hold is used, it is called "sampling multiplier."

A multiplier uses either analog switches or analog multiplexers for its operation. If analog switches are used, it is called a "switching multiplier," and if analog multiplexers are used, it is called a "multiplexing multiplier." Multiplexing multipliers are discussed in Chapter 11 and switching multipliers are discussed in this chapter.

12.1 SAWTOOTH WAVE–BASED TIME DIVISION MULTIPLIERS – TYPE I

The circuit diagrams of type I sawtooth wave–based TDMs are shown in Figure 12.1, and their associated waveforms are shown in Figure 12.2. Figure 12.1(a) shows a series switching TDM, and Figure 12.1(b) shows a shunt or parallel switching TDM. As discussed in Figure 4.7(c), the op-amp OA_1 along with transistor Q_1, resistors R_1, R_2, R_3, R_4, R_5, diodes D_1, D_2, and capacitor C_1 constitute a sawtooth wave generator, and a sawtooth wave V_{S1} with peak value V_R and time period 'T' is generated at its output.

DOI: 10.1201/9781003221449-15

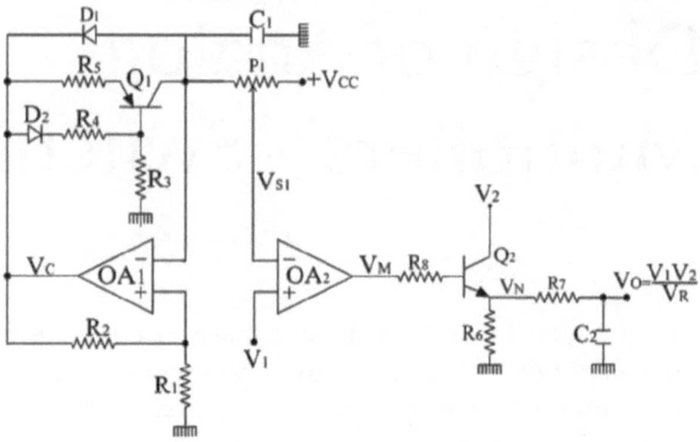

FIGURE 12.1(A) Series switching sawtooth wave–based time division multiplier – type I.

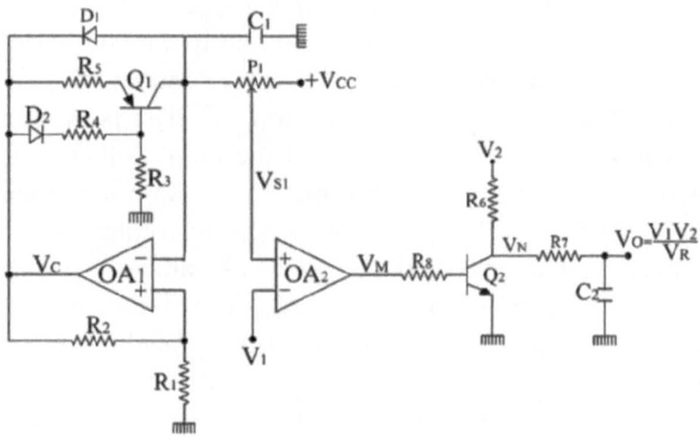

FIGURE 12.1(B) Shunt switching sawtooth wave–based time division multiplier – type I.

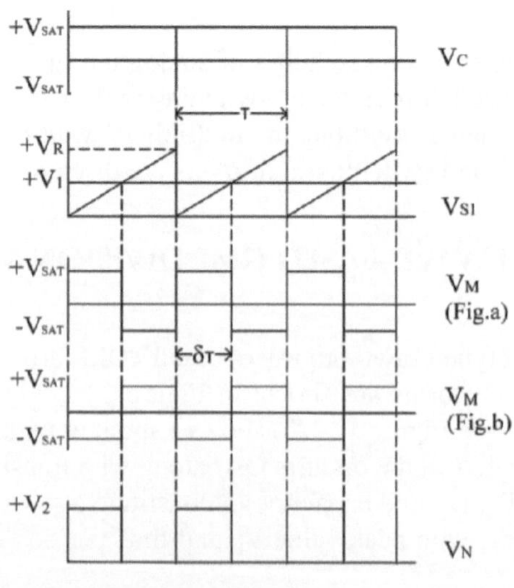

FIGURE 12.2 Associated waveforms of Figure 12.1.

The time period 'T' of this sawtooth wave V_{S1} is given as

$$T = 2R_5C_1 \ln\left(1 + 2\frac{R_1}{R_2}\right)$$

The peak value V_R is given as

$$V_R = \beta(V_{SAT}) + \frac{\beta(V_{SAT})}{1.5} \tag{12.1}$$

Where β is given as $\beta = \dfrac{R_1}{R_1 + R_2}$

The comparator OA_2 compares this sawtooth wave V_{S1} with an input voltage V_1 and produces a rectangular waveform V_M.

The ON time, Figure 12.1(a), or the OFF time, Figure 12.1(b), of V_M is given as

$$\delta_T = \frac{V_1}{V_R} T \tag{12.2}$$

The rectangular pulse V_M controls the transistor Q_2.

In Figure 12.1(a), when V_M is HIGH, the transistor Q_2 is ON, and another input voltage V_2 is connected to R_7C_2 lowpass filter. When V_M is LOW, the transistor Q_2 is OFF, and zero volts exists on R_7C_2 lowpass filter.

In Figure 12.1(b), when V_M is HIGH, the transistor Q_2 is ON, and zero volts exists on R_7C_2 lowpass filter. When V_M is LOW, the transistor Q_2 is OFF, and another input voltage V_2 is connected to R_7C_2 lowpass filter.

Another rectangular pulse V_N with maximum value V_2 is generated at the transistor Q_2 output. The R_7C_2 lowpass filter gives an average value of this pulse train V_N and is given as

$$V_O = \frac{1}{T}\int_0^{\delta_T} V_2 dt = \frac{V_2}{T}\delta_T \tag{12.3}$$

Equation (12.2) in (12.3) gives

$$V_O = \frac{V_1V_2}{V_R} \tag{12.4}$$

In order to avoid a loading effect, the value of R_7 must be greater than that of R_6. The output voltage depends on sharpness and linearity of the sawtooth waveform. The offset on the sawtooth wave will cause error at the output; hence, it is to be nulled by suitable circuitry (potentiometer P_1). Since the output voltage depends on V_R, a highly stable precision power supply must be used. The output voltage is also proportional to attenuation constant α of the lowpass filter. The attenuation constant α will be in the range of 0.80 to 0.90. The polarity of V_1 must be positive only. However, V_2 can have any polarity. Hence, this multiplier is of the two-quadrant type.

DESIGN EXERCISES

1. The transistor switch Q_2 in Figures 12.1(a) and (b) is to be replaced with an FET switch and MOSFET switch. In each, (i) draw the circuit diagrams, (ii) explain their working operations, (iii) draw waveforms at appropriate places, and (iv) deduce expression for their output voltages.
2. The sawtooth generator in part of Figures 12.1(a) and (b) is to be replaced with sawtooth generator of Figures 4.7(b) and (d). (i) Draw the circuit diagrams, (ii) explain their working operations, (iii) draw waveforms at appropriate places, and (iv) deduce expression for their output voltages.

PRACTICAL EXERCISES

1. Verify the multiplier circuits shown in Figure 12.1 by experimenting with the AFC trainer kit.
2. As discussed later in Chapter 18 and Section 18.2, convert the analog multipliers of Figure 12.1 into analog dividers by both type I and II methods. In each, (i) draw the circuit diagrams, (ii) explain their working operation, (iii) draw waveforms at appropriate places, and (iv) deduce expression for the output voltage. Verify this divider by experimenting with the AFC trainer kit.
3. As discussed later in Chapter 18 and Section 18.3, convert the analog multipliers of Figure 12.1 into analog square rooters by both type I and II methods. In each, (i) draw the circuit diagrams, (ii) explain their working operation, (iii) draw waveforms at appropriate places, and (iv) deduce expression for the output voltage. Verify this square rooter by experimenting with the AFC trainer kit.
4. Replace the transistor switch Q_2 in Figure 12.1 with CD4066 non-inverted controlled analog switch IC and DG201 inverted controlled analog switch ICs. In each, (i) draw the circuit diagrams, (ii) explain their working operation, (iii) draw waveforms at appropriate places, and (iv) deduce expression for the output voltage. Verify this multiplier by experimenting with the AFC trainer kit.
5. Convert the multipliers obtained in Problem 4 into dividers as shown in Section 18.2. (i) Draw the circuit diagrams, (ii) explain their working operation, (iii) draw waveforms at appropriate places, and (iv) deduce expression for the output voltage. Verify this divider by experimenting with the AFC trainer kit.
6. Convert the multipliers obtained in Problem 4 into square rooters as shown in Section 18.3. (i) Draw the circuit diagrams, (ii) explain their working operation, (iii) draw waveforms at appropriate places, and (iv) deduce expression for the output voltage. Verify this square rooter by experimenting with the AFC trainer kit.

12.2 SAWTOOTH WAVE–BASED TIME DIVISION MULTIPLIERS – TYPE II

The circuit diagrams of type II sawtooth wave–based TDMs are shown in Figure 12.3, and their associated waveforms are shown in Figure 12.4. Figure 12.3(a) shows a series switching TDM, and Figure 12.3(b) shows a shunt switching TDM. Initially,

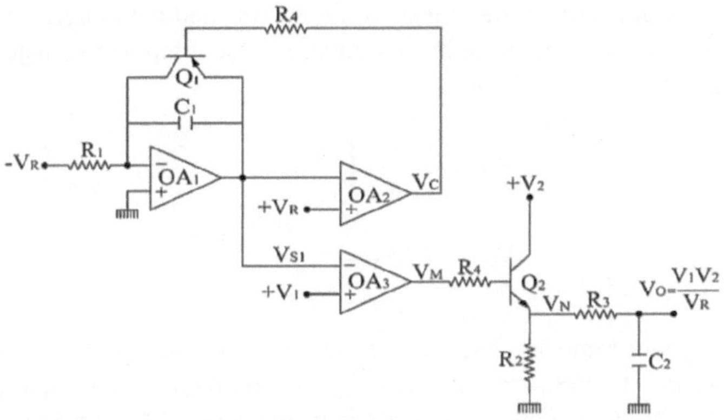

FIGURE 12.3(A) Series switching sawtooth wave–based time division multiplier – type II.

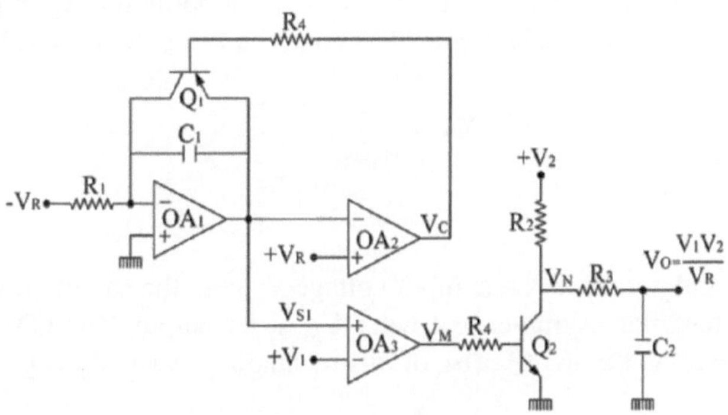

FIGURE 12.3(B) Shunt switching sawtooth wave–based time division multiplier – type II.

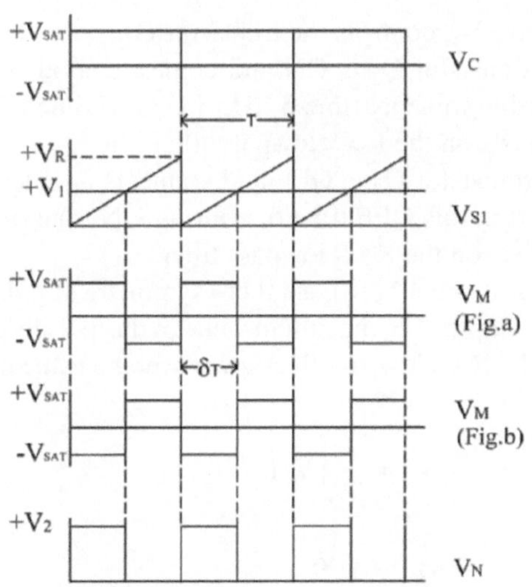

FIGURE 12.4 Associated waveforms of Figure 12.3.

op-amp OA_2 output is HIGH, the transistor Q_1 is OFF, and the integrator formed by resister R_1, capacitor C_1, and op-amp OA_1 integrates the reference voltage $(-V_R)$ and its output is given as

$$V_{S1} = -\frac{1}{R_1C_1}\int -V_R\,dt$$

$$V_{S1} = \frac{V_R}{R_1C_1}t \tag{12.5}$$

A positive-going ramp is generated at the output of op-amp OA_1, and when it reaches the value of reference voltage $+V_R$, the comparator OA_2 output becomes LOW. The transistor Q_1 is ON, the two terminals of capacitor C_1 are shorted together, and hence integrator output becomes zero. Then comparator output is HIGH, and the sequence therefore repeats to give a perfect sawtooth wave V_{S1} of peak value V_R at op-amp OA_1 output and a short pulse waveform V_C at op-amp OA_2 output as shown in Figure 12.4. From Eq. (12.5), Figure 12.4, and the fact that at $t = T$, $V_{S1} = +V_R$

$$V_R = \frac{V_R}{R_1C_1}T$$

$$T = R_1C_1 \tag{12.6}$$

Comparator OA_3 compares one input voltage V_1 with the sawtooth wave V_{S1} and produces a rectangular asymmetrical wave V_M at its output. The ON time, Figure 12.3(a), or OFF time, Figure 12.3(b), of this rectangular wave V_M is given as

$$\delta_T = \frac{V_1}{V_R}T \tag{12.7}$$

This rectangular wave V_M controls the transistor Q_2.

In Figure 12.3(a), transistor Q_2 is ON and connects another input voltage V_2 to R_3C_2 lowpass filter during this ON time δ_T. During OFF time of V_M, transistor Q_2 is OFF, and zero volts exists on the R_3C_2 lowpass filter.

In Figure 12.3(b), transistor Q_2 is OFF and connects another input voltage V_2 to R_3C_2 lowpass filter during this OFF time δ_T. During ON time of V_M, transistor Q_2 is ON, and zero volts exists on the R_3C_2 lowpass filter.

In both Figures 12.3(a) and 12.3(b), another asymmetrical pulse wave V_N is generated at the transistor Q_2 output. The maximum value of this pulse V_N is V_2 and duty cycle is proportional to V_1. The R_3C_2 lowpass filter gives average value of V_N and is given as

$$V_O = \frac{1}{T}\int_0^{\delta_T} V_2 dt = \frac{V_2}{T}\delta_T$$

$$V_O = \frac{V_1V_2}{V_R} \tag{12.8}$$

In order to avoid a loading effect, the value of R_3 must be greater than that of R_2. The output voltage depends on sharpness and linearity of the sawtooth waveform. The offset that exists on the sawtooth wave will cause error at the output; hence, it is to be nulled by suitable circuitry (potentiometer P_1, as shown in Figure 12.1). Since the output voltage depends on V_R, a highly stable precision reference voltage V_R must be used. The output voltage is also proportional to attenuation constant α of the lowpass filter. The attenuation constant α will be in the range of 0.80 to 0.90. The polarity of V_1 must be positive only. However, V_2 can have any polarity. Hence, this multiplier is of the two-quadrant type.

DESIGN EXERCISES

1. The transistor switches in Figures 12.3(a) and 12.3(b) are to be replaced with FET switches and MOSFET switches. In each, (i) draw the circuit diagrams, (ii) explain their working operations, (iii) draw waveforms at appropriate places, and (iv) deduce expression for their output voltages.

PRACTICAL EXERCISES

1. Verify the multiplier circuits shown in Figure 12.3 by experimenting with the AFC trainer kit.
2. As discussed in Chapter 18 and Section 18.2, convert the analog multipliers of Figure 12.3 into analog dividers by both type I and II methods. In each, (i) draw the circuit diagrams, (ii) explain their working operation, (iii) draw waveforms at appropriate places, and (iv) deduce expression for the output voltage. Verify this divider by experimenting with the AFC trainer kit.
3. As discussed in Chapter 18 and Section 18.3, convert the analog multipliers of Figure 12.3 into analog square rooters by both type I and II methods. In each, (i) draw the circuit diagrams, (ii) explain their working operation, (iii) draw waveforms at appropriate places, and (iv) deduce expression for the output voltage. Verify this square rooter by experimenting with the AFC trainer kit.
4. Replace the transistor switch Q_2 in Figure 12.3 with CD4066 non-inverted controlled analog switch IC and DG201 inverted controlled analog switch IC. In each, (i) draw the circuit diagrams, (ii) explain their working operation, (iii) draw waveforms at appropriate places, and (iv) deduce expression for the output voltage. Verify this multiplier by experimenting with the AFC trainer kit.
5. Convert the multipliers obtained in Problem 4 into divider as shown in Section 18.2. (i) Draw the circuit diagrams, (ii) explain their working operation, (iii) draw waveforms at appropriate places, and (iv) deduce expression for the output voltage. Verify this divider by experimenting with the AFC trainer kit.
6. Convert the multipliers obtained in Problem 4 into square rooter as shown in Section 18.3. (i) Draw the circuit diagrams, (ii) explain their working operation, (iii) draw waveforms at appropriate places, and (iv) deduce expression for the output voltage. Verify this square rooter by experimenting with the AFC trainer kit.

12.3 TRIANGULAR WAVE–BASED TIME DIVISION MULTIPLIERS – TYPE I

The circuit diagrams of type I triangular wave–based TDMs are shown in Figure 12.5, and their associated waveforms are shown in Figure 12.6. Figure 12.5(a) shows a series switching TDM, and Figure 12.5(b) shows a shunt or parallel switching TDM. As discussed in Section 4.5, Figure 4.9(a), a triangular wave V_{T1} with $\pm V_T$ peak-to-peak value and time period 'T' is generated by the op-amp OA_1 and transistors Q_1 and Q_2. The value of V_T is given as

$$V_T = \beta(V_{SAT}) \simeq \beta(0.76)(V_{CC}) \tag{12.9}$$

where β is given as $\beta = \dfrac{R_1}{R_1 + R_2}$

And time period 'T' is given as

$$T = 4R_5C_1\frac{R_1}{R_2}$$

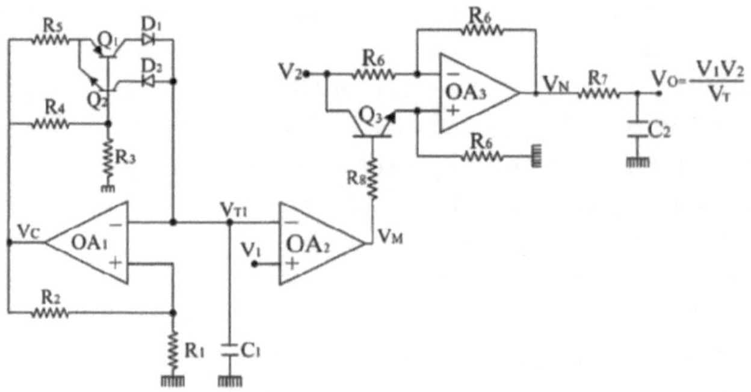

FIGURE 12.5(A) Series switching triangular wave–based time division multiplier – type I.

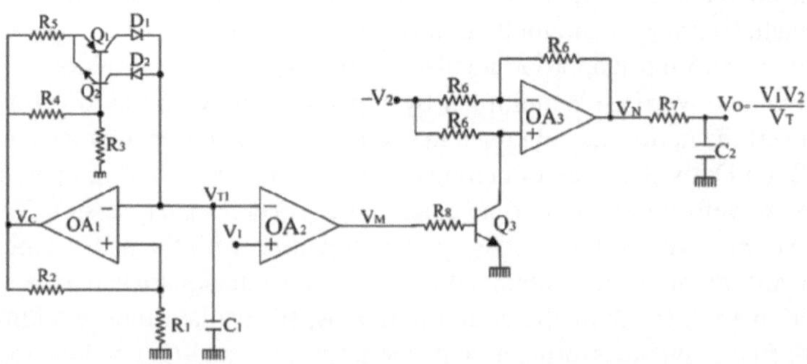

FIGURE 12.5(B) Shunt switching triangular wave–based time division multiplier – type I.

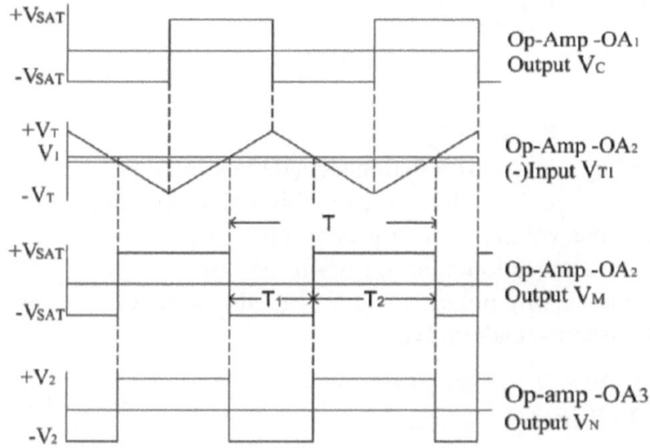

FIGURE 12.6 Associated waveforms of Figure 12.5.

One input voltage V_1 is compared with the generated triangular wave V_{T1} by the comparator OA_2. An asymmetrical rectangular waveform V_M is generated at the comparator OA_2 output. From the waveforms shown in Figure 12.6, it is observed that

$$T_1 = \frac{V_T - V_1}{2V_T}T, \qquad T_2 = \frac{V_T + V_1}{2V_T}T, \qquad T = T_1 + T_2 \tag{12.10}$$

The rectangular wave V_M controls the transistor Q_3.

In Figure 12.5(a), during HIGH(T_2) of the rectangular waveform V_M the transistor Q_3 is ON, the op-amp OA_3 along with resistors R_6 will work as a non-inverting amplifier and $+V_2$ will appear at its output ($V_N = +V_2$). During LOW(T_1) of the rectangular waveform V_M the transistor Q_3 is OFF, the op-amp OA_3 along with resistors R_6 will work as an inverting amplifier and $-V_2$ will appear at its output ($V_N = -V_2$).

In Figure 12.5(b), during HIGH(T_2) of the rectangular waveform V_M the transistor Q_3 is ON, the op-amp OA_3 along with resistors R_6 will work as an inverting amplifier and $+V_2$ will appear at its output ($V_N = +V_2$). During LOW(T_1) of the rectangular waveform V_M the transistor Q_3 is OFF, the op-amp OA_3 along with resistors R_6 will work as a non-inverting amplifier and $-V_2$ will appear at its output ($V_N = -V_2$).

In both circuits shown in Figure 12.5(a) and 12.5(b), another rectangular wave V_N with peak-to-peak values of $\pm V_2$ is generated at the output of op-amp OA_3. The R_7C_2 lowpass filter gives average value of this pulse train V_N and is given as

$$V_O = \frac{1}{T}\left[\int_{O}^{T_2} V_2\,dt + \int_{T_2}^{T_1+T_2} (-V_2)\,dt\right] = \frac{V_2}{T}(T_2 - T_1) \tag{12.11}$$

Equations (12.10) in (12.11) give

$$V_O = \frac{V_1 V_2}{V_T} \tag{12.12}$$

The output voltage depends on sharpness and linearity of the triangular waveform. Since the output voltage depends on V_T, a highly stable precision voltage V_T must be obtained. The output voltage is also proportional to attenuation constant α of the lowpass filter. The attenuation constant α will be in the range of 0.80 to 0.90. The polarity of V_1 must be any polarity, and V_2 can also have any polarity. Hence, this multiplier is of the four-quadrant type.

DESIGN EXERCISES

1. The transistor switch Q_3 in Figures 12.5(a) and 12.5(b) is to be replaced with an FET switch and MOSFET switch. In each, (i) draw the circuit diagrams, (ii) explain their working operations, (iii) draw waveforms at appropriate places, and (iv) deduce expression for their output voltages.

PRACTICAL EXERCISES

1. Verify the multiplier circuits shown in Figure 12.5 by experimenting with the AFC trainer kit.
2. As discussed in Chapter 18 and Section 18.2, convert the analog multipliers of Figure 12.5 into analog dividers by both type I and II methods. In each, (i) draw the circuit diagrams, (ii) explain their working operation, (iii) draw waveforms at appropriate places, and (iv) deduce expression for the output voltage. Verify this divider by experimenting with the AFC trainer kit.
3. As discussed in Chapter 18 and Section 18.3, convert the analog multipliers of Figure 12.5 into analog square rooters by both type I and II methods. In each, (i) draw the circuit diagrams, (ii) explain their working operation, (iii) draw waveforms at appropriate places, and (iv) deduce expression for the output voltage. Verify this square rooter by experimenting with the AFC trainer kit.
4. Replace the transistor switch Q_3 in Figure 12.5 with CD4066 non-inverted controlled analog switch IC and DG201 inverted controlled analog switch ICs. In each, (i) draw the circuit diagrams, (ii) explain their working operation, (iii) draw waveforms at appropriate places, and (iv) deduce expression for the output voltage. Verify this multiplier by experimenting with the AFC trainer kit.
5. Convert the multipliers obtained in Problem 4 into dividers as shown in Section 18.2. (i) Draw the circuit diagrams, (ii) explain their working operation, (iii) draw waveforms at appropriate places, and (iv) deduce expression for the output voltage. Verify this divider by experimenting with the AFC trainer kit.
6. Convert the multipliers obtained in Problem 4 into square rooters as shown in Section 18.3. (i) Draw the circuit diagrams, (ii) explain their working operation, (iii) draw waveforms at appropriate places, and (iv) deduce expression for the output voltage. Verify this square rooter by experimenting with the AFC trainer kit.

12.4 TRIANGULAR WAVE–BASED TIME DIVISION MULTIPLIERS – TYPE II

The circuit diagrams of type II triangular wave–based TDMs are shown in Figure 12.7, and their associated waveforms are shown in Figure 12.8. Figure 12.7(a) shows a series switching TDM, and Figure 12.7(b) shows a shunt or parallel switching TDM. As discussed in Section 4.5, Figure 4.9(b), op-amps OA_1 and OA_2 constitute a triangular/square wave generator. The output of op-amp OA_1 is triangular wave V_{T1} with $\pm V_T$ peak values and time period 'T'. The output of OA_2 is a square waveform with $\pm V_{SAT}$ peak-to-peak values.

$$\pm V_T = \pm V_{SAT} \frac{R_2}{R_3} \tag{12.13}$$

$$T = 4R_1C_1 \frac{R_2}{R_3}$$

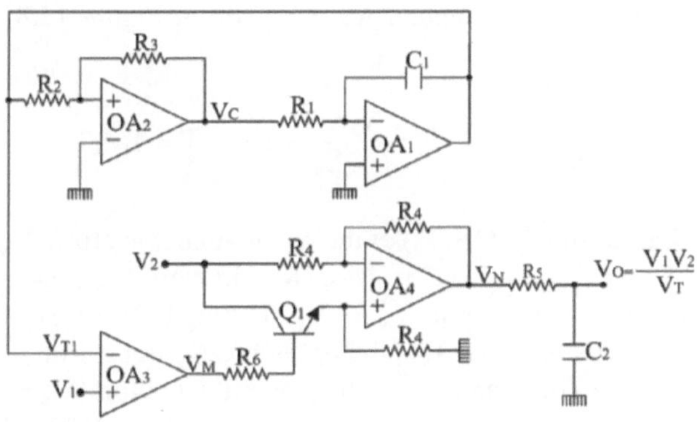

FIGURE 12.7(A) Series switching triangular wave–based time division multiplier – type II.

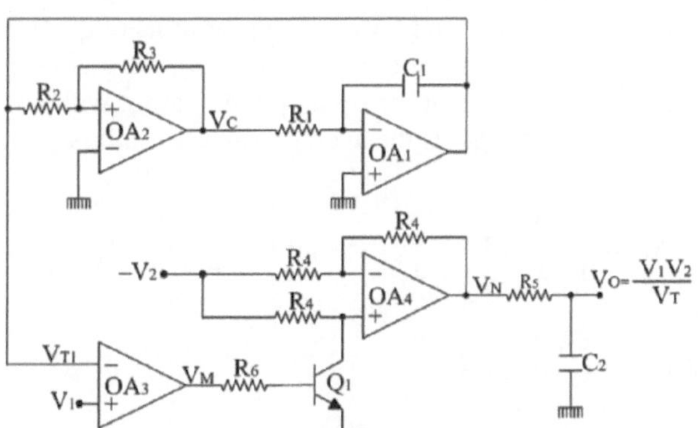

FIGURE 12.7(B) Shunt switching triangular wave–based time division multiplier – type II.

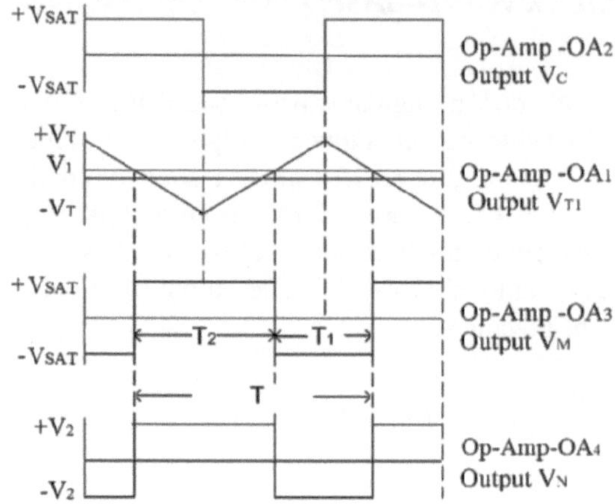

FIGURE 12.8 Associated waveforms of Figure 12.7.

The comparator OA_3 compares this triangular wave V_{T1} with one input voltage V_1 and produces asymmetrical rectangular wave V_M. From Figure 12.8, it is observed that

$$T_1 = \frac{V_T - V_1}{2V_T} T \qquad T_2 = \frac{V_T + V_1}{2V_T} T \qquad T = T_1 + T_2 \tag{12.14}$$

In Figure 12.7(a), during HIGH(T_2) of the rectangular waveform V_M the transistor Q_1 is ON, the op-amp OA_4 along with resistors R_4 will work as a non-inverting amplifier and $+V_2$ will appear at its output ($V_N = +V_2$). During LOW(T_1) of the rectangular waveform V_M the transistor Q_1 is OFF, the op-amp OA_4 along with resistors R_4 will work as an inverting amplifier and $-V_2$ will appear at its output($V_N = -V_2$).

In Figure 12.7(b), during HIGH (T_2) of the rectangular waveform V_M the transistor Q_1 is ON, the op-amp OA_4 along with resistors R_4 will work as an inverting amplifier and $+V_2$ will appear at its output ($V_N = +V_2$). During LOW (T_1) of the rectangular waveform V_M the transistor Q1 is OFF, the op-amp OA_4 along with resistors R_4 will work as a non-inverting amplifier and $-V_2$ will appear at its output ($V_N = -V_2$).

In both circuits shown in Figures 12.7(a) and 12.7(b), another asymmetrical rectangular waveform V_N is generated at the op-amp OA_4 output with $\pm V_2$ peak-to-peak values. The R_5C_2 lowpass filter gives the average value V_O and is given as

$$V_O = \frac{1}{T}\left[\int_{O}^{T_2} V_2\,dt + \int_{T_2}^{T_1+T_2} (-V_2)\,dt\right] = \frac{V_2}{T}\left[T_2 - T_1\right] \tag{12.15}$$

Equation (12.14) in (12.15) gives

$$V_O = \frac{V_1 V_2}{V_T} \tag{12.16}$$

The output voltage depends on sharpness and linearity of the triangular waveform. Since the output voltage depends on V_T, a highly stable precision voltage V_T must be obtained. The output voltage is also proportional to attenuation constant α of the lowpass filter. The attenuation constant α will be in the range of 0.80 to 0.90. The polarity of V_1 must be any polarity, and V_2 can also have any polarity. Hence this multiplier is of the four-quadrant type.

DESIGN EXERCISES

1. The transistor switch Q_1 in Figures 12.7(a) and 12.7(b) is to be replaced with an FET switch and MOSFET switch. In each, (i) draw the circuit diagrams, (ii) explain their working operations, (iii) draw waveforms at appropriate places, and (iv) deduce expression for their output voltages.

PRACTICAL EXERCISES

1. Verify the multiplier circuits shown in Figure 12.7 by experimenting with the AFC trainer kit.
2. As discussed in Chapter 18 and Section 18.2, convert the analog multipliers of Figure 12.7 into analog dividers by both type I and II methods. In each, (i) draw the circuit diagrams, (ii) explain their working operation, (iii) draw waveforms at appropriate places, and (iv) deduce expression for the output voltage. Verify this divider by experimenting with the AFC trainer kit.
3. As discussed in Chapter 18 and Section 18.3, convert the analog multipliers of Figure 12.7 into analog square rooters by both type I and II methods. In each, (i) draw the circuit diagrams, (ii) explain their working operation, (iii) draw waveforms at appropriate places, and (iv) deduce expression for the output voltage. Verify this square rooter by experimenting with the AFC trainer kit.
4. Replace the transistor switch in Figure 12.7 with CD4066 non-inverted controlled analog switch IC and DG201 inverted controlled analog switch ICs. In each, (i) draw the circuit diagrams, (ii) explain their working operation, (iii) draw waveforms at appropriate places, and (iv) deduce expression for the output voltage. Verify this multiplier by experimenting with the AFC trainer kit.
5. Convert the multipliers obtained in Problem 4 into dividers as shown in Section 18.2. (i) Draw the circuit diagrams, (ii) explain their working operation, (iii) draw waveforms at appropriate places, and (iv) deduce expression for the output voltage. Verify this divider by experimenting with the AFC trainer kit.
6. Convert the multipliers obtained in Problem 4 into square rooters as shown in Section 18.3. (i) Draw the circuit diagrams, (ii) explain their working operation, (iii) draw waveforms at appropriate places, and (iv) deduce expression for the output voltage. Verify this square rooter by experimenting with the AFC trainer kit.

12.5 TIME DIVISION MULTIPLIERS WITH NO REFERENCE – TYPE I

The circuit diagrams of TDMs without using either a sawtooth wave or triangular wave as reference are shown in Figure 12.9, and their associated waveforms are shown

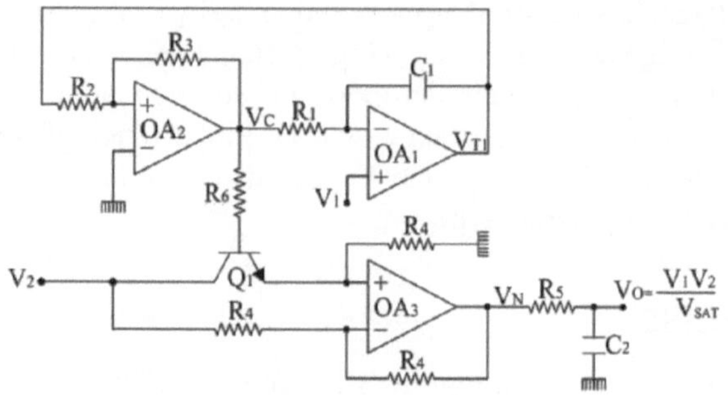

FIGURE 12.9(A) Series switching time division multiplier with no reference – type I.

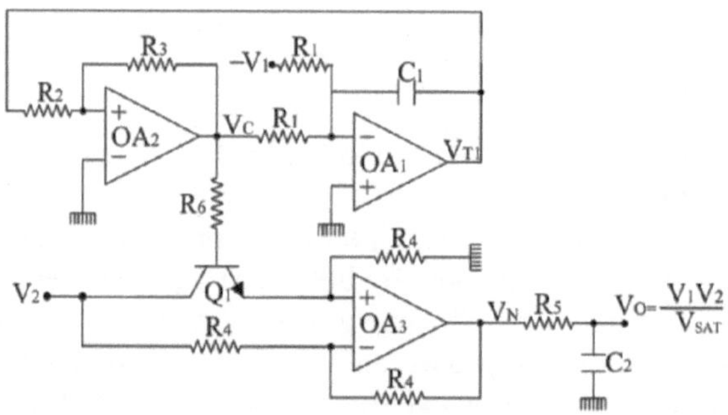

FIGURE 12.9(B) Equivalent circuit of Figure 12.9(a).

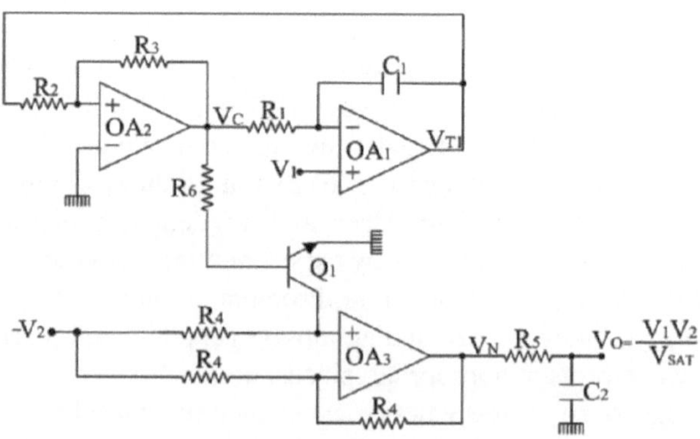

FIGURE 12.9(C) Shunt switching time division multiplier with no reference – type I.

in Figure 12.10. Figures 12.9(a) and 12.9(b) show a series switching TDM, and Figures 12.9(c) and 12.9(d) show a shunt or parallel switching TDM. The op-amps OA_1 and OA_2 along with R_1, C_1, R_2 and R_3 constitute an asymmetrical rectangular wave generator, and a rectangular wave V_C is generated at the output of op-amp OA_2.

$$\pm V_T = \frac{R_2}{R_3} V_{SAT} \tag{12.17}$$

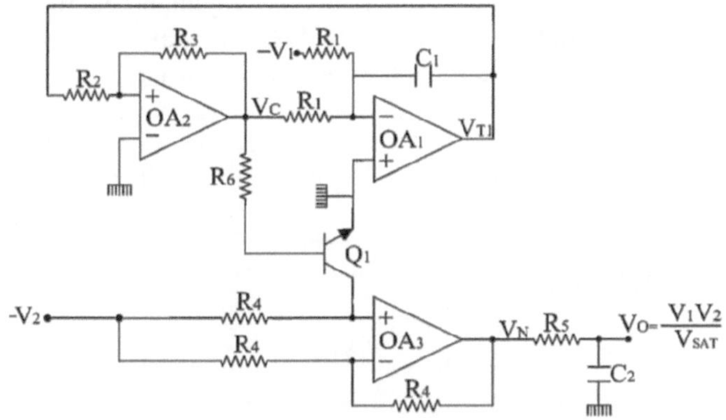

FIGURE 12.9(D) Equivalent circuit of Figure 12.9(c).

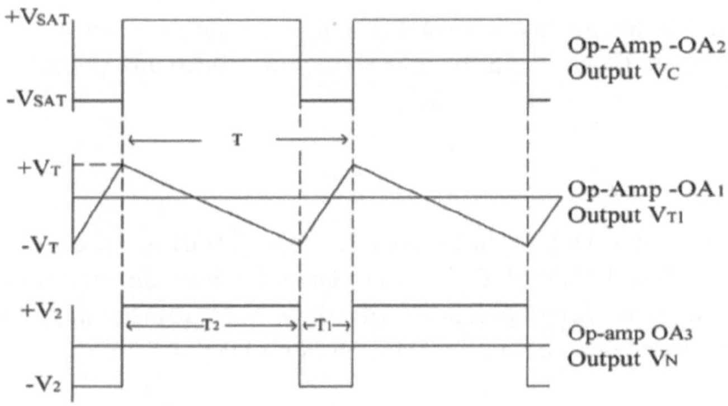

FIGURE 12.10 Associated waveforms of Figure 12.9.

From the waveforms shown in Figure 12.10, it is observed that

$$T_1 = \frac{V_{SAT} - V_1}{2V_{SAT}} T \qquad T_2 = \frac{V_{SAT} + V_1}{2V_{SAT}} T \qquad T = T_1 + T_2 \tag{12.18}$$

Let us consider Figures 12.9(a) and 12.9(b). During HIGH (T_2) of the rectangular waveform V_C the transistor Q_1 is ON, the op-amp OA_3 along with resistors R_4 will work as a non-inverting amplifier and $+V_2$ will appear at its output ($V_N = +V_2$). During LOW (T_1) of the rectangular waveform V_C the transistor Q_1 is OFF, the op-amp OA_3 along with resistors R_4 will work as an inverting amplifier and $-V_2$ will appear at its output ($V_N = -V_2$).

Let us consider Figures 12.9(c) and 12.9(d). During HIGH (T_2) of the rectangular waveform V_C the transistor Q_1 is ON, the op-amp OA_3 along with resistors R_4 will work as an inverting amplifier and $+V_2$ will appear at its output ($V_N = +V_2$). During LOW (T_1) of the rectangular waveform V_C the transistor Q_1 is OFF, the op-amp OA_3 along with resistors R_4 will work as a non-inverting amplifier and $-V_2$ will appear at its output ($V_N = -V_2$).

In all circuits of Figures 12.9(a)–12.9(d), another asymmetrical rectangular wave V_N is generated at the output of op-amp OA_3 with $\pm V_2$ as maximum values. The R_5C_2 lowpass filter gives an average value of this pulse train V_N and is given as

$$V_O = \frac{1}{T}\left[\int_{O}^{T_2} V_2\,dt + \int_{T_2}^{T_1+T_2} (-V_2)\,dt\right] = \frac{V_2}{T}\left[T_2 - T_1\right] \tag{12.19}$$

Equation (12.18) in (12.19) gives

$$V_O = \frac{V_1V_2}{V_{SAT}} \tag{12.20}$$

The output voltage depends on sharpness and linearity of the integrator waveform. Since the output voltage depends on V_{SAT}, a highly stable precision power supply is to be used. The output voltage is also proportional to attenuation constant α of the lowpass filter. The attenuation constant α will be in the range of 0.80 to 0.90. The polarity of V_1 must be any polarity, and V_2 can also have any polarity. Hence, this multiplier is of the four-quadrant type.

DESIGN EXERCISES

1. The transistor switch Q_1 in Figures 12.9(a)–12.9(d) is to be replaced with an FET switch and MOSFET switch. In each, (i) draw the circuit diagrams, (ii) explain their working operations, (iii) draw waveforms at appropriate places, and (iv) deduce expression for their output voltages.

PRACTICAL EXERCISES

1. Verify the multiplier circuits shown in Figure 12.9 by experimenting with the AFC trainer kit.
2. As discussed in Chapter 18 and Section 18.2, convert the analog multipliers of Figure 12.9 into analog dividers by both type I and II methods. In each, (i) draw the circuit diagrams, (ii) explain their working operation, (iii) draw waveforms at appropriate places, and (iv) deduce expression for the output voltage. Verify this divider by experimenting with the AFC trainer kit.
3. As discussed in Chapter 18 and Section 18.3, convert the analog multipliers of Figure 12.9 into analog square rooters by both type I and II methods. In each, (i) draw the circuit diagrams, (ii) explain their working operation, (iii) draw waveforms at appropriate places, and (iv) deduce expression for the output voltage. Verify this square rooter by experimenting with the AFC trainer kit.
4. Replace the transistor switch in Figure 12.9 with CD4066 non-inverted controlled analog switch IC and DG201 inverted controlled analog switch IC. In each, (i) draw the circuit diagrams, (ii) explain their working operation, (iii) draw waveforms at appropriate places, and (iv) deduce expression for the output voltage. Verify this multiplier by experimenting with the AFC trainer kit.

5. Convert the multipliers obtained in Problem 4 into dividers as shown in Section 18.2. (i) Draw the circuit diagrams, (ii) explain their working operation, (iii) draw waveforms at appropriate places, and (iv) deduce expression for the output voltage. Verify this divider by experimenting with the AFC trainer kit.
6. Convert the multipliers obtained in Problem 4 into square rooters as shown in Section 18.3. (i) Draw the circuit diagrams, (ii) explain their working operation, (iii) draw waveforms at appropriate places, and (iv) deduce expression for the output voltage. Verify this square rooter by experimenting with the AFC trainer kit.

12.6 DOUBLE DUAL SLOPE PEAK RESPONDING MULTIPLIERS

The circuit diagrams of double dual slope peak responding multipliers are shown in Figure 12.11 and their associated waveforms are shown in Figure 12.12. Figure 12.11(a) shows a series switching peak detecting multiplier, Figure 12.11(b) shows a shunt or parallel switching peak detecting multiplier, Figure 12.11(c) shows a series switching peak sampling multiplier, and Figure 12.11(d) shows a shunt or parallel switching peak sampling multiplier. Let initially the comparator OA_2 output be LOW ($-V_{SAT}$).

In Figures 12.11(a) and 12.11(c), the transistor Q_1 is OFF, the op-amp OA_5 along with resistors R_5 will work as an inverting amplifier and $-V_1$ will appear at its output ($V_M = -V_1$). The transistor Q_2 is OFF, the op-amp OA_6 along with resistors R_6 will work as an inverting amplifier and $-V_2$ will appear at its output ($V_N = -V_2$).

In Figures 12.11(b) and 12.11(d), the transistor Q_1 is OFF, the op-amp OA_5 along with resistors R_5 will work as a non-inverting amplifier and $-V_1$ will appear at its output ($V_M = -V_1$). The transistor Q_2 is OFF, the op-amp OA_6 along with resistors R_6 will work as a non-inverting amplifier and $-V_2$ will appear at its output ($V_N = -V_2$).

In all the circuits of Figures 12.11(a)–12.11(d), ($-V_{SAT}$) is given to the integrator formed by resistor R_1, capacitor C_1, and op-amp OA_1.

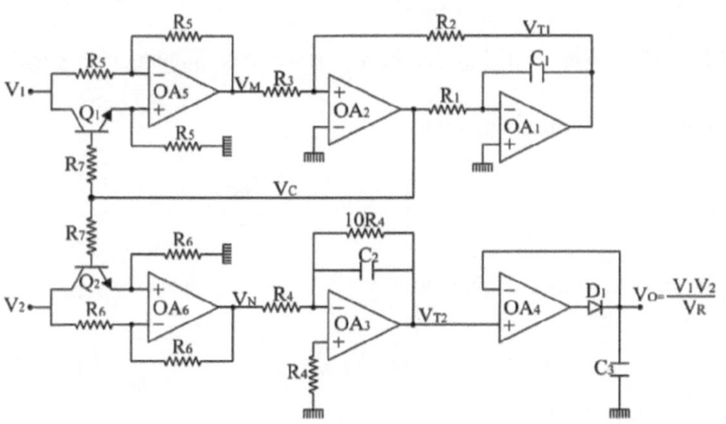

FIGURE 12.11(A) Series switching double dual slope peak detecting multiplier.

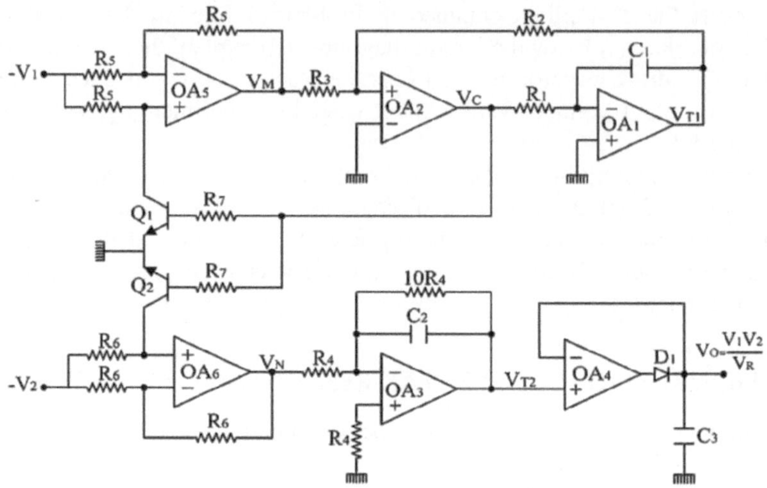

FIGURE 12.11(B) Parallel switching double dual slope peak detecting multiplier.

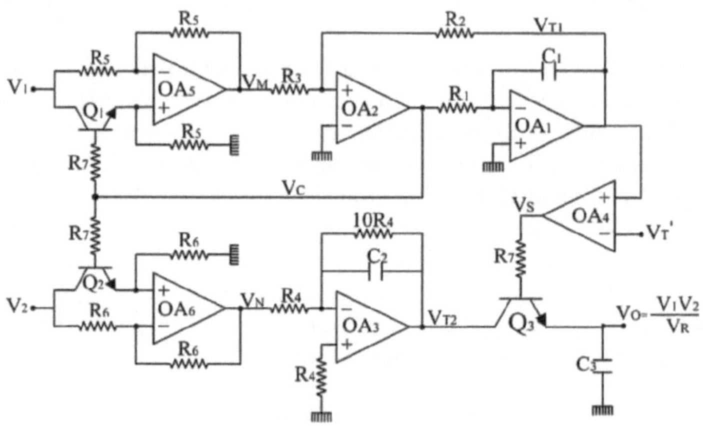

FIGURE 12.11(C) Series switching double dual slope peak sampling multiplier.

The op-amp OA_1 output is given as

$$V_{T1} = -\frac{1}{R_1C_1}\int -V_{SAT}\,dt = \frac{V_{SAT}}{R_1C_1}t \tag{12.21}$$

The op-amp OA_3 output is given as

$$V_{T2} = -\frac{1}{R_4C_2}\int -V_2\,dt = \frac{V_2}{R_4C_2}t \tag{12.22}$$

The output of op-amp OA_1 is rising toward positive saturation, and when it reaches a value of $+V_T$, the comparator OA_2 output becomes HIGH ($+V_{SAT}$).

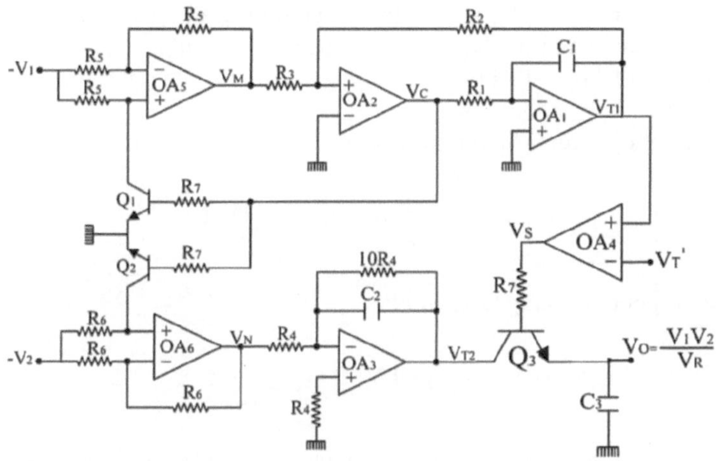

FIGURE 12.11(D) Parallel switching double dual slope sampling multiplier.

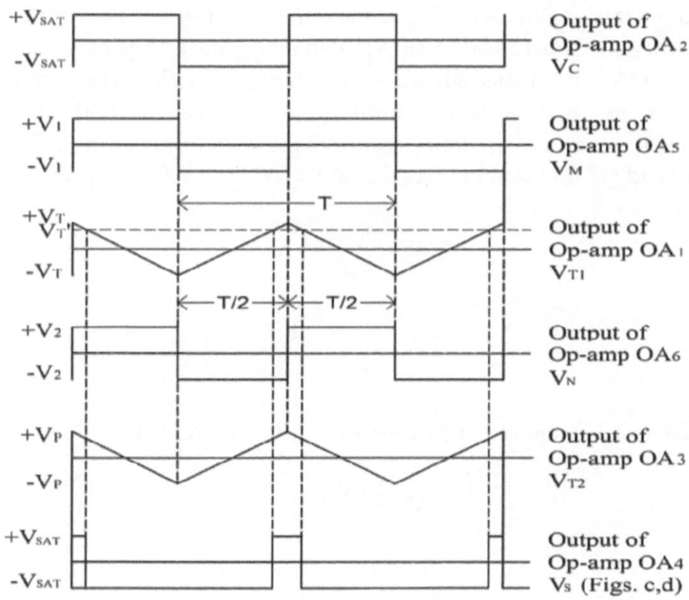

FIGURE 12.12 Associated waveforms of Figures 12.11(a)–12.11(d).

In Figures 12.11(a) and 12.11(c), the transistor Q_1 is ON, the op-amp OA_5 along with resistors R_5 will work as a non-inverting amplifier and $+V_1$ will appear at its output ($V_M = +V_1$). The transistor Q_2 is ON, the op-amp OA_6 along with resistors R_6 will work as a non-inverting amplifier and $+V_2$ will appear at its output ($V_N = +V_2$).

In Figures 12.11(b) and 12.11(d), the transistor Q_1 is ON, the op-amp OA_5 along with resistors R_5 will work as an inverting amplifier and $+V_1$ will appear at its output

($V_M = +V_1$). The transistor Q_2 is ON, the op-amp OA_6 along with resistors R_6 will work as an inverting amplifier and $+V_2$ will appear at its output ($V_N = +V_2$).

In all Figures 12.11(a)–12.11(d), $+V_{SAT}$ is given to the integrator formed by R_1, C_1 and op-amp OA_1.

The output of op-amp OA_1 will now be

$$V_{T1} = -\frac{1}{R_1C_1}\int V_{SAT}\,dt = -\frac{V_{SAT}}{R_1C_1}t \tag{12.23}$$

And the output of op-amp OA_3 will be

$$V_{T2} = -\frac{1}{R_4C_2}\int V_2\,dt = -\frac{V_2}{R_4C_2}t \tag{12.24}$$

The output of op-amp OA_1 changes its slope and is going toward negative saturation. When the output of op-amp integrator OA_1 comes down to a value ($-V_T$), the comparator OA_2 output will become LOW, and therefore the cycle repeats itself to give (i) a triangular wave V_{T1} at the output of op-amp OA_1 with $\pm V_T$ peak-to-peak value, (ii) another triangular wave V_{T2} at the output of op-amp OA_3 with $\pm V_P$ peak-to-peak value, (iii) first square waveform V_C with $\pm V_{SAT}$ peak-to-peak values at the output of op-amp OA_2, (iv) second square waveform V_M with $\pm V_1$ as peak-to-peak values at the output of op-amp OA_5, and (v) third square waveform V_N with $\pm V_2$ as peak-to-peak values at the output of op-amp OA_6. From the waveforms shown in Figure 12.12, Eqs. (12.21) and (12.22), and the fact that at $t = T/2$, $V_{T1} = 2V_T$, $V_{T2} = 2V_P$

$$2V_T = \frac{V_{SAT}}{R_1C_1}\frac{T}{2} \tag{12.25}$$

$$2V_P = \frac{V_2}{R_4C_2}\frac{T}{2} \tag{12.26}$$

From Eqs. (12.25) and (12.26) by assuming $R_1 = R_4$, $C_1 = C_2$

$$V_P = \frac{V_2}{V_{SAT}}V_T$$

$$\pm V_T = \pm\frac{R_2}{R_3}V_1$$

$$V_P = \frac{V_1V_2}{V_{SAT}}\frac{R_2}{R_3}$$

Let

$$V_R = \frac{V_{SAT}}{R_2}R_3$$

$$V_P = \frac{V_1 V_2}{V_R} \tag{12.27}$$

In the circuits shown in Figures 12.11(a) and 12.11(b), the peak detector realized by op-amp OA_4, diode D_1 and capacitor C_3 gives this peak value V_P at its output. $V_O = V_P$.

In the circuits shown in Figures 12.11(c) and 12.11(d), the peak value V_P is obtained by the sample and hold circuit realized by transistor Q_3 and capacitor C_3. The sampling pulse V_S is generated by op-amp OA_4 by comparing a slightly less than voltage of V_T called V_T' with the triangular wave V_{T1}. The sample and hold operation is illustrated in Figure 12.12. The sampled output is given as $V_O = V_P$.

From Eq. (12.27) the output voltage will be $V_O = V_P$

$$V_O = \frac{V_1 V_2}{V_R} \tag{12.28}$$

The output voltage depends on sharpness and linearity of the triangular waveform. Since the output voltage depends on V_R, a highly stable precision voltage V_R must be obtained. The output voltage is also proportional to detecting response of peak detector in peak detecting multiplier and ON time of short sampling pulse V_S in peak sampling multiplier. For peak detecting multipliers, the polarity of V_1 must be single polarity, and V_2 can also have single polarity; hence, these multipliers are of the single-quadrant type. For peak sampling multipliers, the polarity of V_1 must be single polarity, and V_2 can also have any polarity; hence, these multipliers are of the two-quadrant type.

DESIGN EXERCISES

1. The transistor switches Q_1 and Q_2 in Figure 12.11 are to be replaced with FET switches and MOSFET switches. In each, (i) draw the circuit diagrams, (ii) explain their working operations, (iii) draw waveforms at appropriate places, and (iv) deduce expression for their output voltages.

PRACTICAL EXERCISES

1. Verify the multiplier circuits shown in Figure 12.11 by experimenting with the AFC trainer kit.
2. As discussed in Chapter 18 and Section 18.2, convert the analog multipliers of Figure 12.11 into analog dividers by both type I and II methods. In each, (i) draw the circuit diagrams, (ii) explain their working operation, (iii) draw waveforms at appropriate places, and (iv) deduce expression for the output voltage. Verify this divider by experimenting with the AFC trainer kit.
3. As discussed in Chapter 18 and Section 18.3, convert the analog multipliers of Figure 12.11 into analog square rooters by both type I and II methods. In each, (i) draw the circuit diagrams, (ii) explain their working operation, (iii) draw

waveforms at appropriate places, and (iv) deduce expression for the output voltage. Verify this square rooter by experimenting with the AFC trainer kit.

4. Replace the transistor switches in Figure 12.11 with CD4066 non-inverted controlled analog switch IC and DG201 inverted controlled analog switch ICs. In each, (i) draw the circuit diagrams, (ii) explain their working operation, (iii) draw waveforms at appropriate places, and (iv) deduce expression for the output voltage. Verify this multiplier by experimenting with the AFC trainer kit.
5. Convert the multipliers obtained in Problem 4 into dividers as shown in Section 18.2. (i) Draw the circuit diagrams, (ii) explain their working operation, (iii) draw waveforms at appropriate places, and (iv) deduce expression for the output voltage. Verify this divider by experimenting with the AFC trainer kit.
6. Convert the multipliers obtained in Problem 4 into square rooters as shown in Section 18.3. (i) Draw the circuit diagrams, (ii) explain their working operation, (iii) draw waveforms at appropriate places, and (iv) deduce expression for the output voltage. Verify this square rooter by experimenting with the AFC trainer kit.

12.7 PEAK RESPONDING MULTIPLIERS WITH VOLTAGE-TO-PERIOD CONVERTER – SWITCHING TYPE

The switching type peak responding multipliers using a V/T converter are shown in Figure 12.13, and their associated waveforms are shown in Figure 12.14. Figure 12.13(a) shows a series switching peak detecting multiplier, Figure 12.13(b) shows a shunt or parallel switching peak detecting multiplier, Figure 12.13(c) shows a series switching peak sampling multiplier, and Figure 12.13(d) shows a shunt or parallel switching peak sampling multiplier. A square wave V_C is generated by op-amp OA_1, resistors R_1, R_2, R_3, and capacitor C_1.

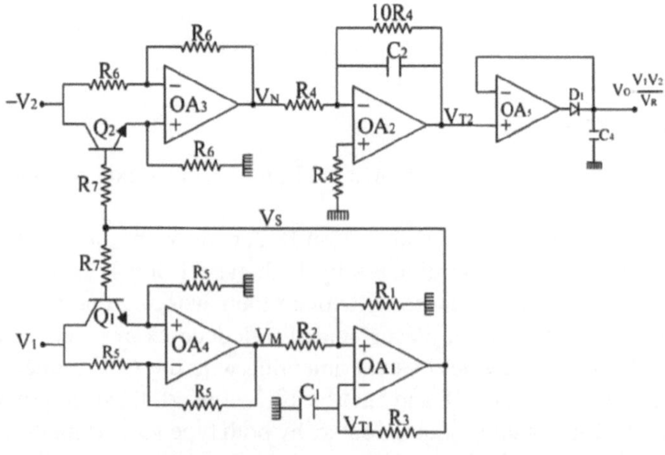

FIGURE 12.13(A) Peak detecting multiplier using voltage-to-period converter.

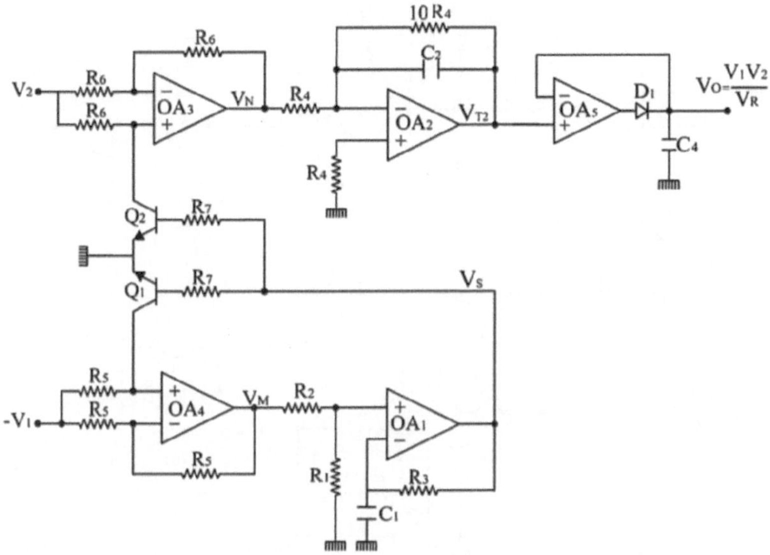

FIGURE 12.13(B) Shunt switching peak detecting multiplier using voltage-to-period converter.

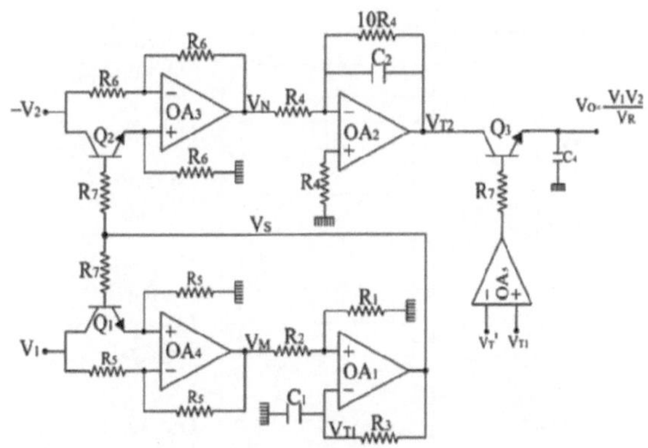

FIGURE 12.13(C) Peak sampling multiplier using voltage-to-period converter.

In Figures 12.13(a) and 12.13(c), during HIGH of the square waveform V_C, (i) the transistor Q_1 is ON, the op-amp OA_4 along with resistors R_5 will work as a non-inverting amplifier and $+V_1$ will appear at its output ($V_M = +V_1$), and (ii) transistor Q_2 is ON, the op-amp OA_3 along with resistors R_6 will work as a non-inverting amplifier and $-V_2$ will appear at its output ($V_N = -V_2$). During LOW of the square waveform V_C, (i) transistor Q_1 is OFF, the op-amp OA_4 along with resistors R_5 will work as an inverting amplifier and $-V_1$ will appear at its output ($V_M = -V_1$), and (ii) transistor Q_2 is OFF, the op-amp OA_3 along with resistors R_6 will work as an inverting amplifier and $+V_2$ will appear at its output ($V_N = +V_2$).

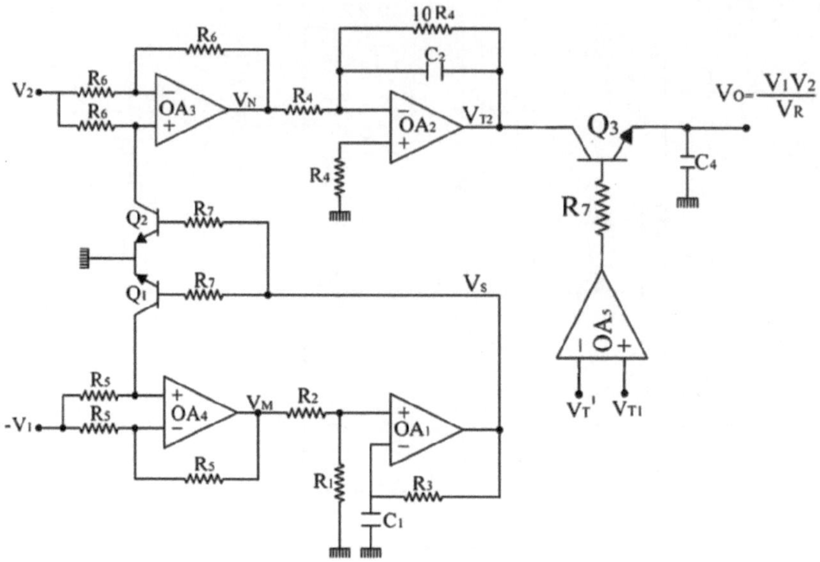

FIGURE 12.13(D) Shunt switching peak sampling multiplier using voltage-to-period converter.

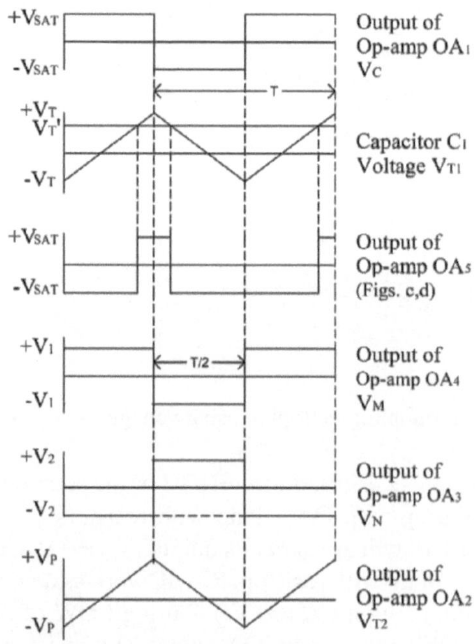

FIGURE 12.14 Associated waveforms of Figures 12.13(a)–12.13(d).

In Figures 12.13(b) and 12.13(d), during HIGH of the square waveform V_C, (i) transistor Q_1 is ON, the op-amp OA_4 along with resistors R_5 will work as an inverting amplifier and $+V_1$ will appear at its output ($V_M = +V_1$), and (ii) transistor Q_2 is ON, the op-amp OA_3 along with resistors R_6 will work as an inverting amplifier and $-V_2$ will appear at its output ($V_N = -V_2$). During LOW of the square waveform V_C, (i) transistor Q_1 is OFF, the op-amp OA_4 along with resistors R_5 will work as a non-inverting amplifier and $-V_1$ will appear at its output ($V_M = -V_1$), and (ii) transistor Q_2 is OFF, the op-amp OA_3 along with resistors R_6 will work as a non-inverting amplifier and $+V_2$ will appear at its output ($V_N = +V_2$).

In all the circuits of Figure 12.13(a)–12.13(d), two square waves, (i) V_M with $\pm V_1$ peak-to-peak at the output of op-amp OA_4, and (ii) V_N with $\pm V_2$ peak-to-peak at the output of op-amp OA_3, are generated from the astable clock V_C.

The integrator formed by op-amp OA_2, resistors R_4, and capacitor C_2 converts the square wave V_N into a triangular wave V_{T2} of $\pm V_P$ peak-to-peak values. V_P is proportional to V_2

$$V_P = K_1 V_2 \tag{12.29}$$

The time period 'T' of square wave V_S is proportional to V_1

$$T = K_2 V_1 \tag{12.30}$$

The integrator output is given as

$$V_{T2} = \frac{1}{R_4 C_2} \int V_2 \, dt = \frac{V_2}{R_4 C_2} t \tag{12.31}$$

For half time period, from the waveforms shown in Figure 12.14 and from Eq. (12.31) at $t = T/2$, $V_{T2} = 2\,V_P$.

$$2V_P = \frac{V_2}{R_4 C_2} \frac{T}{2}; \; V_P = \frac{V_2}{4 R_4 C_2} V_1 K_2 \tag{12.32}$$

K_1 and K_2 are constant values

Let

$$V_R = \frac{4 R_4 C_2}{K_2}$$

$$V_P = \frac{V_1 V_2}{V_R} \tag{12.33}$$

In Figures 12.13(a)and 12.13(b), the peak detector realized by op-amp OA_5, diode D_1, and capacitor C_4 gives this peak value V_P at the output: $V_O = V_P$.

In Figures 12.13(c) and 12.13(d), the peak value V_P is sampled by the sample and hold circuit realized by transistor Q_3 and capacitor C_3 with a sampling pulse V_S. The sampling pulse V_S is generated by comparing capacitor C_1 voltage V_{T1} with slightly less than its peak value V_T, i.e., V_T' by the comparator OA_5. The sampled output V_O is the peak value V_P.

From Eq. (12.33), the output voltage will be $V_O = V_P$

$$V_O = \frac{V_1 V_2}{V_R} S \tag{12.34}$$

The output voltage depends on V_R; a highly stable precision voltage V_R must be obtained. The output voltage also depends on the proportional constant of V/T converter. The output voltage is also proportional to detecting response of peak detector in peak detecting multiplier and ON time of short sampling pulse V_S in peak sampling multiplier. For peak detecting multipliers, the polarity of V_1 must be of single polarity, and V_2 can also have single polarity; hence, these multipliers are of the single-quadrant type. For peak sampling multipliers, the polarity of V_1 must be of single polarity and V_2 can also have any polarity; hence, these multipliers are of the two-quadrant type.

DESIGN EXERCISES

1. The transistor switches Q_1 and Q_2 in Figure 12.13 are to be replaced with FET switches and MOSFET switches. In each, (i) draw the circuit diagrams, (ii) explain their working operations, (iii) draw waveforms at appropriate places, and (iv) deduce expression for their output voltages.

PRACTICAL EXERCISES

1. Verify the multiplier circuits shown in Figure 12.13 by experimenting with the AFC trainer kit.
2. As discussed in Chapter 18 and Section 18.2, convert the analog multipliers of Figure 12.13 into analog dividers by both type I and II methods. In each, (i) draw the circuit diagrams, (ii) explain their working operation, (iii) draw waveforms at appropriate places, and (iv) deduce expression for the output voltage. Verify this divider by experimenting with the AFC trainer kit.
3. As discussed in Chapter 18 and Section 18.3, convert the analog multipliers of Figure 12.13 into analog square rooters by both type I and II methods. In each, (i) draw the circuit diagrams, (ii) explain their working operation, (iii) draw waveforms at appropriate places, and (iv) deduce expression for the output voltage. Verify this square rooter by experimenting with the AFC trainer kit.
4. Replace the transistor switches in Figure 12.13 with CD4066 non-inverted controlled analog switch IC and DG201 inverted controlled analog switch ICs. In each, (i) draw the circuit diagrams, (ii) explain their working operation, (iii) draw waveforms at appropriate places, and (iv) deduce expression for the output voltage. Verify this multiplier by experimenting with the AFC trainer kit.

5. Convert the multiplier obtained in Problem 4 into dividers as shown in Section 18.2. (i) Draw the circuit diagrams, (ii) explain their working operation, (iii) draw waveforms at appropriate places, and (iv) deduce expression for the output voltage. Verify this divider by experimenting with the AFC trainer kit.
6. Convert the multipliers obtained in Problem 4 into square rooters as shown in Section 18.3. (i) Draw the circuit diagrams, (ii) explain their working operation, (iii) draw waveforms at appropriate places, and (iv) deduce expression for the output voltage. Verify this square rooter by experimenting with the AFC trainer kit.

12.8 PULSE POSITION PEAK RESPONDING MULTIPLIER

The circuit diagrams of pulse position peak responding multipliers are shown in Figure 12.15, and their associated waveforms are shown in Figure 12.16. Figure 12.15(a) shows a pulse position peak detecting multiplier, and Figure 12.15(b) shows a pulse position peak sampling multiplier.

The op-amp OA_1 along with transistor Q_1, resistors R_1, R_6, R_3, R_4, R_5, and capacitor C_1 constitutes a sawtooth wave generator. A sawtooth wave V_{S1} with peak value

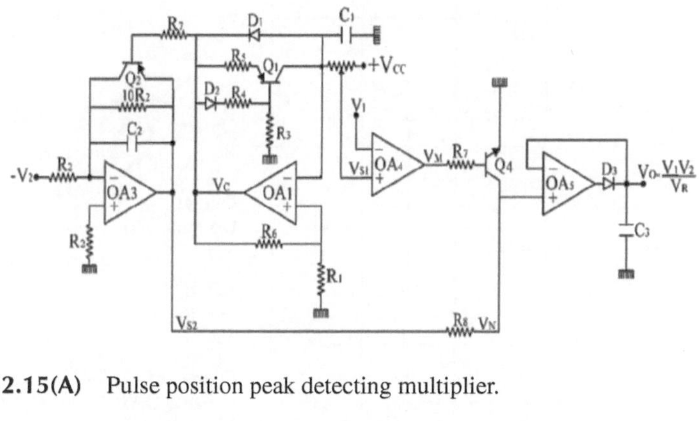

FIGURE 12.15(A) Pulse position peak detecting multiplier.

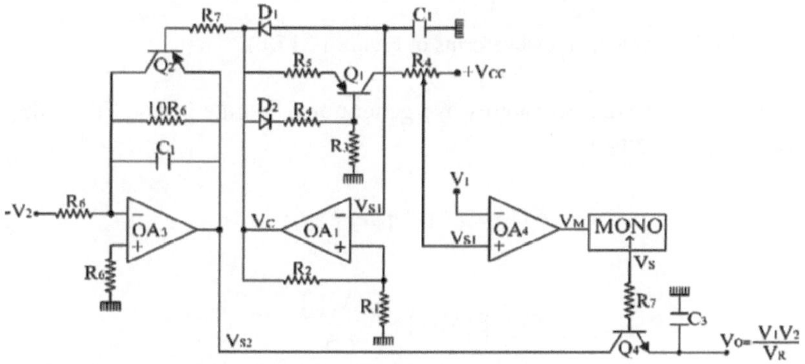

FIGURE 12.15(B) Pulse position peak sampling multiplier.

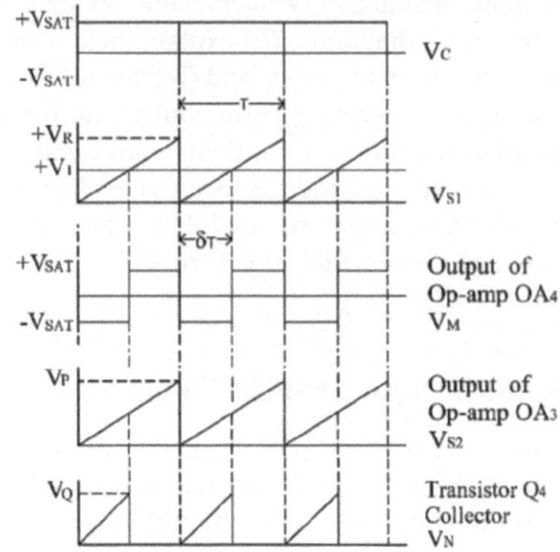

FIGURE 12.16(A) Associated waveforms of Figure 12.15(a).

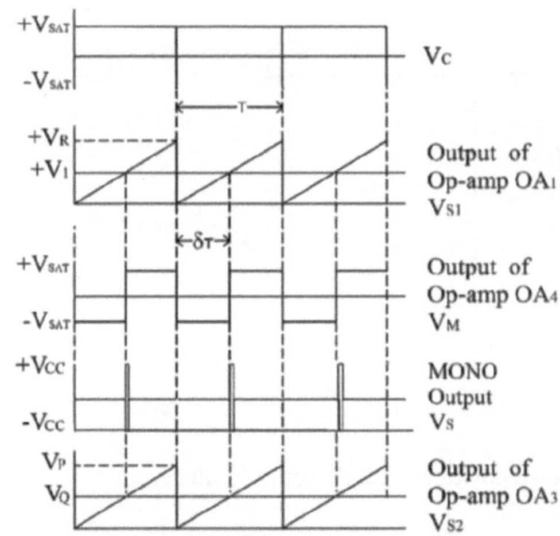

FIGURE 12.16(B) Associated waveforms of Figure 12.15(b).

of V_R is generated by this sawtooth wave generator. The time period 'T' of this sawtooth wave V_{S1} is given as

$$T = 2R_5C_1 \ln\left(1 + 2\frac{R_1}{R_6}\right)$$

$$V_R = \beta(V_{CC}) + \frac{\beta(V_{CC})}{1.5}$$

where β is given as $\beta = \dfrac{R_1}{R_1 + R_6}$

The short pulse waveform V_C from the sawtooth wave generator controls transistor Q_2. During HIGH of V_C, transistor Q_2 is OFF, and an integrator is formed by op-amp OA_3, resistor R_2, and capacitor C_2. The integrator output is given as

$$V_{S2} = -\frac{1}{R_2C_2}\int -V_2\,dt = \frac{V_2}{R_2C_2}t \tag{12.35}$$

During the short LOW of V_C, the transistor Q_2 is ON, and the capacitor C_2 is short-circuited so that the integrator output becomes zero volts. Another sawtooth wave V_{S2} with peak value of V_P is generated at the output of op-amp OA_3.

From Eq. (12.35), waveforms in Figure 12.16, and the fact that at $t = T$, $V_{S2} = V_P$

$$V_P = \frac{V_2}{R_2C_2}T \tag{12.36}$$

Two sawtooth waveforms V_{S1} and V_{S2} are of same time period 'T' and are generated in all circuits of Figure 12.15. The sawtooth wave V_{S1} is compared with first input voltage V_1 by the comparator OA_4, and a rectangular wave V_M is generated at the output of comparator OA_4. The OFF time of this rectangular waveform V_M is given as

$$\delta_T = \frac{V_1}{V_R}T \tag{12.37}$$

The rectangular pulse V_M controls transistor Q_4. During LOW of V_M, the transistor Q_4 is OFF, and the sawtooth wave V_{S2} is existing at the collector of transistor Q_4. During HIGH of V_M, the transistor Q_4 is ON, and zero volt exists on collector of transistor Q_4. A semi-sawtooth wave V_N with peak value V_Q is generated at the collector of transistor Q_4. The peak value V_Q is given as

$$V_Q = \frac{V_P}{T}\delta_T \tag{12.38}$$

In Figure 12.15(a), the peak detector realized by op-amp OA_5, diode D_1, and capacitor C_3 gives this peak value V_Q at its output. Hence, $V_O = V_Q$.

In Figure 12.15(b), the rectangular pulse V_M is given to monostable multivibrator which gives narrow spikes V_S during every rising edge of V_M. This spike V_S is acting as a sampling pulse to the sample and hold circuit. The sample and hold circuit realized by transistor Q_4, capacitor C_3 gives the peak value of V_{S2} at the time of δ_T. From Figure 12.16(b), the sampled output is V_Q, i.e., $V_O = V_Q$.

Equations (12.36) and (12.37) in (12.38) give

$$V_O = \frac{V_1V_2}{V_R}\frac{T}{R_2C_2} \tag{12.39}$$

Let $T = R_2C_2$

$$V_O = \frac{V_1V_2}{V_R} \tag{12.40}$$

The output voltage depends on sharpness and linearity of the sawtooth waveform. Since the output voltage depends on V_R, a highly stable precision voltage V_R must be obtained. The output voltage also depends on the proportional constant of V/T period converter. The output voltage is also proportional to detecting response of peak detector in peak detecting multiplier and ON time of short-sampling pulse V_S in peak sampling multiplier. For peak detecting multipliers, the polarity of V_1 must be positive and V_2 must have negative polarity; hence, this multiplier is of the single-quadrant type. For peak sampling multipliers, the polarity of V_1 must be positive and V_2 can have any polarity; hence, this multiplier is of the two-quadrant type.

PRACTICAL EXERCISES

1. Verify the multiplier circuits shown in Figure 12.15 by experimenting with the AFC trainer kit.
2. As discussed in Chapter 18 and Section 18.2, convert the analog multipliers of Figure 12.15 into analog dividers by both type I and II methods. In each, (i) draw the circuit diagrams, (ii) explain their working operation, (iii) draw waveforms at appropriate places, and (iv) deduce expression for the output voltage. Verify this divider by experimenting with the AFC trainer kit.
3. As discussed in Chapter 18 and Section 18.3, convert the analog multipliers of Figure 12.15 into analog square rooters by both type I and II methods. In each, (i) draw the circuit diagrams, (ii) explain their working operation, (iii) draw waveforms at appropriate places, and (iv) deduce expression for the output voltage. Verify this square rooter by experimenting with the AFC trainer kit.
4. Replace the transistor switches in Figure 12.15 with CD4066 non-inverted controlled analog switch IC and DG201 inverted controlled analog switch ICs. In each, (i) draw the circuit diagrams, (ii) explain their working operation, (iii) draw waveforms at appropriate places and (iv) deduce expression for the output voltage. Verify this multiplier by experimenting with the AFC trainer kit.
5. Convert the multipliers obtained in Problem 4 into dividers as shown in Section 18.2. (i) Draw the circuit diagrams, (ii) explain their working operation, (iii) draw waveforms at appropriate places, and (iv) deduce expression for the output voltage. Verify this divider by experimenting with the AFC trainer kit.
6 Convert the multipliers obtained in Problem 4 into square rooters as shown in Section 18.3. (i) Draw the circuit diagrams, (ii) explain their working operation, (iii) draw waveforms at appropriate places, and (iv) deduce expression for the output voltage. Verify this square rooter by experimenting with the AFC trainer kit.

TUTORIAL EXERCISES

1. Design series and shunt switching time division multipliers with a sawtooth wave generator.
2. You are given a block of triangular wave generator. Design series and shunt type time division multipliers from this triangular wave generator. (i) Draw waveforms at appropriate places and (ii) deduce expression for the output voltage.
3. In the multiplier circuit shown in Figure 12.7(b), the comparator OA_3 terminals are interchanged and $-V_2$ is changed to $+V_2$. (i) Draw waveforms at appropriate places and (ii) deduce expression for the output voltage.
4. In the multiplier circuits shown in Figures 12.1(a), 12.1(b), 12.3(a), 12.3(b), 12.5(a), 12.5(b), 12.7(a), 12.7(b), 12.9(a)–12.9(d), and 12.11(a), 12.11(b), if the polarity of input voltage V_2 is reversed, (i) draw waveforms at appropriate places and (ii) deduce expression for the output voltage.

13 Design of Analog Dividers

Analog dividers are classified into (i) time division dividers (TDDs) and (ii) peak responding dividers (PRDs). Peak responding dividers are further classified into (i) peak detecting dividers and (ii) peak sampling dividers. A short pulse/sawtooth waveform whose time period 'T' is inversely proportional to one voltage is generated. Another input voltage is integrated during the time period 'T'. The peak value of the integrated voltage is proportional to the division of the input voltages. This is called "double single slope peak responding divider." A square/triangular waveform whose time period 'T' is inversely proportional to one voltage is generated. Another input voltage is integrated during the time period 'T'. The peak value of the integrated voltage is proportional to the division of the input voltages. This is called "double dual slope peak responding divider." A rectangular pulse waveform is generated whose OFF time is proportional to one voltage and time period 'T' is inversely proportional to another voltage. Reference voltage is integrated during this OFF time. The peak value of integrated output is proportional to the division of the two input voltages. This is called "pulse width integrated peak responding divider." At the output stage, if peak detector is used, it is called "peak detecting divider," and if sample and hold is used, it is called "sampling divider." A PRD uses either analog switches or analog multiplexers for its operation. If analog switches are used, it is called a "switching peak responding divider," and if analog multiplexers are used, it is called "multiplexing peak responding divider." The designs of all types of analog dividers are described in this chapter.

13.1 TIME DIVISION DIVIDER WITH NO REFERENCE – MULTIPLEXING

The circuit diagrams of TDDs without using either sawtooth wave or triangular wave as reference are shown in Figures 13.1(a) and 13.1(b), and their associated waveforms are shown in Figure 13.2. The op-amps OA_1 and OA_2 along with R_1, C_1, R_2 and R_2 constitute an asymmetrical rectangular wave generator, and a rectangular wave V_C is generated at the output of op-amp OA_2.

When the output of comparator OA_2 is LOW, transistor Q_1 is OFF and Q_2 is ON, $(-V_1)$ is given to the differential integrator formed by resistor R_1, capacitor C_1, and op-amp OA_1. When the output of comparator OA_2 is HIGH, transistor Q_1 is ON and Q_2 is OFF, $(+V_1)$ is given to the differential integrator formed by resistor R_1, capacitor C_1, and op-amp OA_1.

$$\pm V_T = \frac{R_2}{R_3} V_{SAT} \tag{13.1}$$

DOI: 10.1201/9781003221449-16

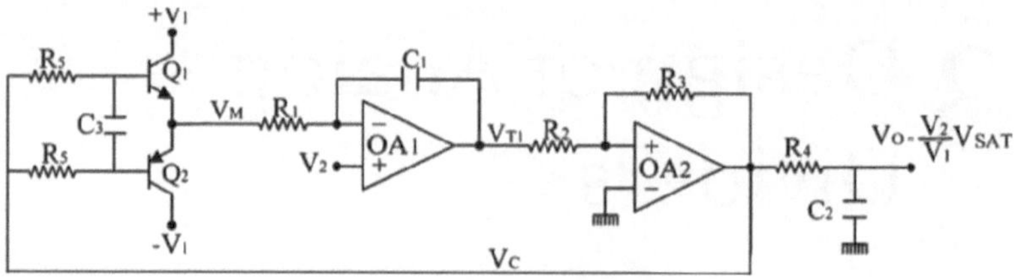

FIGURE 13.1(A) Circuit diagram of time division dividers without reference.

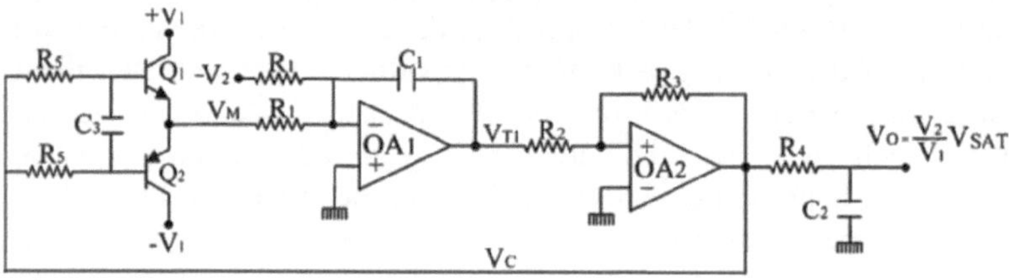

FIGURE 13.1(B) Equivalent circuit diagram of Figure 13.1(a).

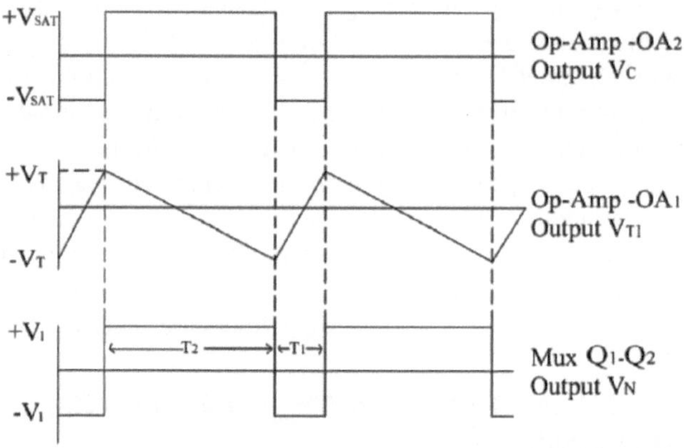

FIGURE 13.2 Associated waveforms of Figure 13.1(a).

From the waveforms shown in Figure 13.2, it is observed that

$$T_1 = \frac{V_1 - V_2}{2V_1} T \qquad T_2 = \frac{V_1 + V_2}{2V_1} T \qquad T = T_1 + T_2 \tag{13.2}$$

The R_4C_2 lowpass filter gives an average value of the rectangular wave V_C and is given as

$$V_O = \frac{1}{T}\left[\int_O^{T_2} V_{SAT}\,dt + \int_{T_2}^{T_1+T_2} \left(-V_{SAT}\right)dt\right] \tag{13.3}$$

$$= \frac{V_{SAT}}{T}\left[T_2 - T_1\right]$$

$$V_O = \frac{V_2}{V_1} V_{SAT} \tag{13.4}$$

The above equation is valid only when V_1 is greater than V_2, because that is the condition for oscillation. Since the output voltage depends on the supply voltage ($V_{SAT} \sim 0.76V_{CC}$), highly stable precision power supply is required. Since the circuit filters the waveform V_C, it must be a perfect rectangular wave with fast rising and falling edges; if there are any deviations in these rising and falling edges, then that will cause error in the output. Hence, op-amp bandwidth should be high, and the recommended op-amp IC is LF356.

DESIGN EXERCISES

1. The transistor multiplexer in Figures 13.1(a) and 13.1(b) is to be replaced with an FET multiplexer and MOSFET multiplexer. In each, (i) draw the circuit diagrams, (ii) explain their working operation, (iii) draw waveforms at appropriate places, and (iv) deduce expression for the output voltage.

PRACTICAL EXERCISES

1. Verify the divider circuits shown in Figure 13.1(a) by experimenting with the AFC trainer kit.
2. As discussed in Chapter 18 and Section 18.4, convert the analog dividers of Figure 13.1(a) into analog multipliers by both type I and II methods. In each, (i) draw the circuit diagrams, (ii) explain their working operation, (iii) draw waveforms at appropriate places, and (iv) deduce expression for the output voltage. Verify this multiplier by experimenting with the AFC trainer kit.
3. As discussed in Chapter 18 and Section 18.5, convert the analog dividers of Figure 13.1(a) into analog square rooters. In each, (i) draw the circuit diagrams, (ii) explain their working operation, (iii) draw waveforms at appropriate places, and (iv) deduce expression for the output voltage. Verify this square rooter by experimenting with the AFC trainer kit.
4. Replace the transistor multiplexers in Figure 13.1(a) with CD4053 analog multiplexers IC. (i) Draw the circuit diagrams, (ii) explain their working operation, (iii) draw waveforms at appropriate places, and (iv) deduce expression for the output voltage. Verify this divider by experimenting with the AFC trainer kit.
5. Convert the divider obtained in Problem 4 into multipliers as discussed in Section 18.4. (i) Draw the circuit diagrams, (ii) explain their working operation, (iii) draw waveforms at appropriate places, and (iv) deduce expression for the output voltage. Verify this multiplier by experimenting with the AFC trainer kit.
6. Convert the divider obtained in Problem 4 into square rooters as discussed in Section 18.5. (i) Draw the circuit diagrams, (ii) explain their working operation, (iii) draw waveforms at appropriate places, and (iv) deduce expression for the output voltage. Verify this square rooter by experimenting with the AFC trainer kit.

13.2 TIME DIVISION DIVIDER WITHOUT REFERENCE – SWITCHING

The circuit diagrams of TDDs without using triangular or sawtooth wave as reference are shown in Figure 13.3 and their associated waveforms in Figure 13.4. Figures 13.3(a) and 13.3(b) show a series switching TDD, and Figures 13.3(c) and 13.3(d) show shunt or parallel switching TDD. The op-amps OA_1 and OA_2 along with R_1, C_1, R_2 and R_3 constitute an asymmetrical rectangular wave generator, and a rectangular wave V_C is generated at the output of op-amp OA_2.

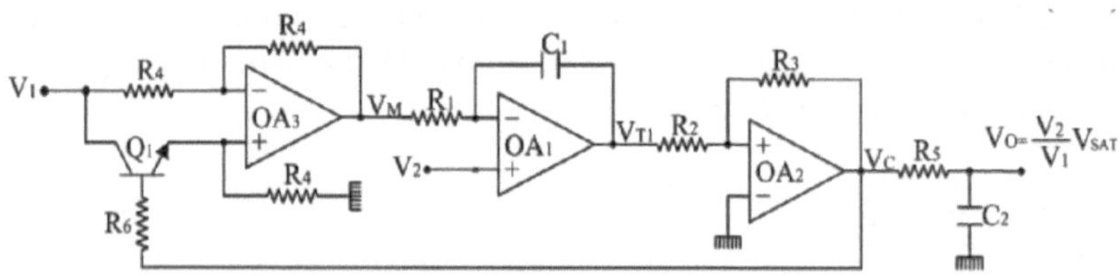

FIGURE 13.3(A) Series switching time division divider without reference.

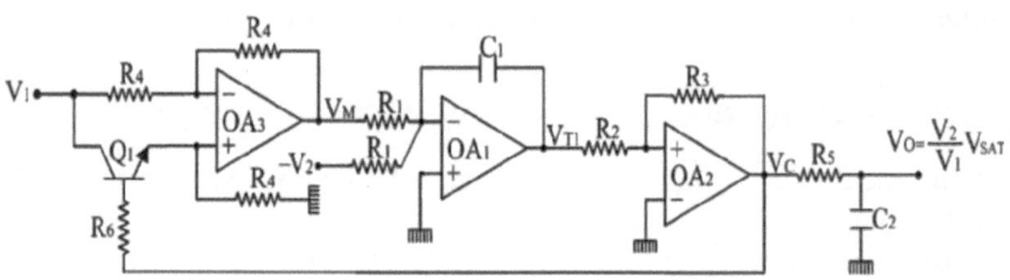

FIGURE 13.3(B) Equivalent circuit of Figure 13.3(a).

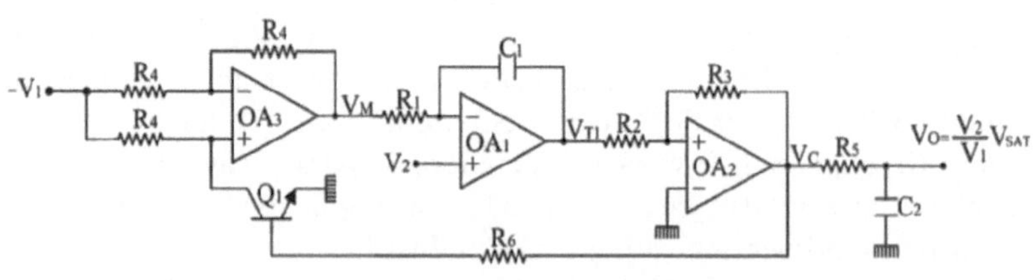

FIGURE 13.3(C) Shunt switching time division divider without reference.

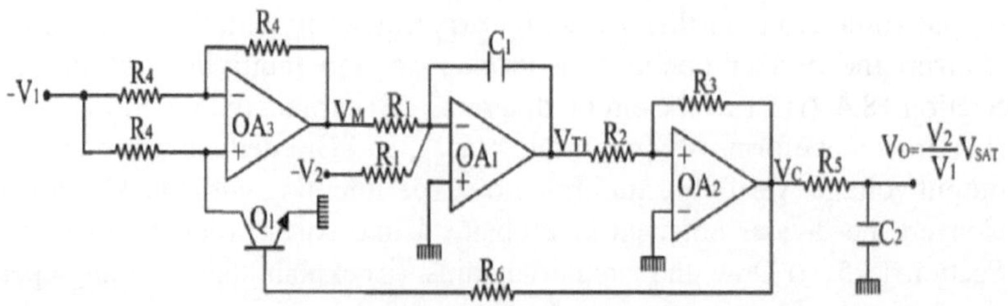

FIGURE 13.3(D) Equivalent circuit of Figure 13.3(c).

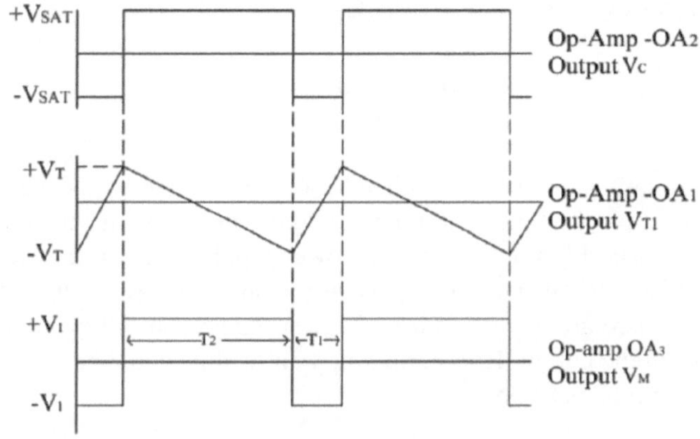

FIGURE 13.4 Associated waveforms of Figure 13.3.

In Figures 13.3(a) and 13.3(b), during HIGH of V_C, the transistor Q_1 is ON and the op-amp OA_3 along with resistors R_4 will work as a non-inverting amplifier, and $+V_1$ will exist at its output ($V_M = +V_1$). During LOW of V_C, the transistor Q_1 is OFF and the op-amp OA_3 along with resistors R_4 will work as an inverting amplifier and $-V_1$ will exist at its output ($V_M = -V_1$).

In Figures 13.3(c) and 13.3(d), during HIGH of V_C, the transistor Q_1 is ON and the op-amp OA_3 along with resistors R_4 will work as an inverting amplifier, and $+V_1$ will exist at its output ($V_M = +V_1$). During LOW of V_C, the transistor Q_1 is OFF and the op-amp OA_3 along with resistors R_4 will work as a non-inverting amplifier, and $-V_1$ will exist at its output ($V_M = -V_1$).

A rectangular waveform V_M with $\pm V_1$ peak-to-peak value is generated at the output of op-amp OA_3.

$$\pm V_T = \frac{R_2}{R_3} V_{SAT} \tag{13.5}$$

From the waveforms shown in Figure 13.4, it is observed that

$$T_1 = \frac{V_1 - V_2}{2V_1} T \qquad T_2 = \frac{V_1 + V_2}{2V_1} T \qquad T = T_1 + T_2 \tag{13.6}$$

The R_5C_2 lowpass filter gives average value of the rectangular wave V_C and is given as

$$V_O = \frac{1}{T}\left[\int_{O}^{T_2} V_{SAT}\, dt + \int_{T_2}^{T_1+T_2} (-V_{SAT})\, dt\right]$$

$$= \frac{V_{SAT}}{T}\left[T_2 - T_1\right] \tag{13.7}$$

Equations (13.5) and (13.6) in (13.7) give

$$V_O = \frac{V_2}{V_1} V_{SAT} \tag{13.8}$$

The above equation is valid only when V_1 is greater than V_2, because that is the condition for oscillation. Since the output voltage depends on the supply voltage ($V_{SAT} \sim 0.76V_{CC}$), a highly stable precision power supply is required. Since the circuit filters the waveform V_C, it must be a perfect rectangular wave with fast rising and falling edges; if there are any deviations in these rising and falling edges, then that will cause error in the output. Hence, op-amp bandwidth should be high and the recommended op-amp IC is LF356.

DESIGN EXERCISES

1. The transistor switches in Figures 13.3(a) and 13.3(b) are to be replaced with FET switches and MOSFET switches. In each, (i) draw the circuit diagrams, (ii) explain their working operations, (iii) draw waveforms at appropriate places, and (iv) deduce expression for their output voltages.

PRACTICAL EXERCISES

1. Verify the divider circuits shown in Figure 13.3 by experimenting with the AFC trainer kit.
2. As discussed in Chapter 18 and Section 18.4, convert the analog divider of Figure 13.3 into analog multipliers by both type I and II methods. In each, (i) draw the circuit diagrams, (ii) explain their working operation, (iii) draw waveforms at appropriate places, and (iv) deduce expression for the output voltage. Verify this multiplier by experimenting with the AFC trainer kit.
3. As discussed in Chapter 18 and Section 18.5, convert the analog dividers of Figure 13.3 into analog square rooters. In each, (i) draw the circuit diagrams, (ii) explain their working operation, (iii) draw waveforms at appropriate places, and (iv) deduce expression for the output voltage. Verify this square rooter by experimenting with the AFC trainer kit.
4. Replace the transistor switches in Figure 13.3 with CD4066 and DG201 analog switch ICs. (i) Draw the circuit diagrams, (ii) explain their working operation, (iii) draw waveforms at appropriate places, and (iv) deduce expression for the output voltage. Verify this square rooter by experimenting with the AFC trainer kit.
5. Convert the divider obtained in Problem 4 into a multiplier as discussed in Section 18.4. (i) Draw the circuit diagrams, (ii) explain their working operation, (iii) draw waveforms at appropriate places, and (iv) deduce expression for the output voltage. Verify this multiplier by experimenting with the AFC trainer kit.
6. Convert the divider obtained in Problem 4 into square rooter as discussed in Section 18.5. (i) Draw the circuit diagrams, (ii) explain their working operation, (iii) draw waveforms at appropriate places, and (iv) deduce expression

for the output voltage. Verify this square rooter by experimenting with the AFC trainer kit.

13.3 DOUBLE DUAL SLOPE PEAK RESPONDING DIVIDERS – MULTIPLEXING

The circuit diagrams of double dual slope PRDs are shown in Figure 13.5, and their associated waveforms are shown in Figure 13.6. Figure 13.5(a) shows a peak detecting divider, and Figure 13.5(b) shows a peak sampling divider. Let initially op-amp

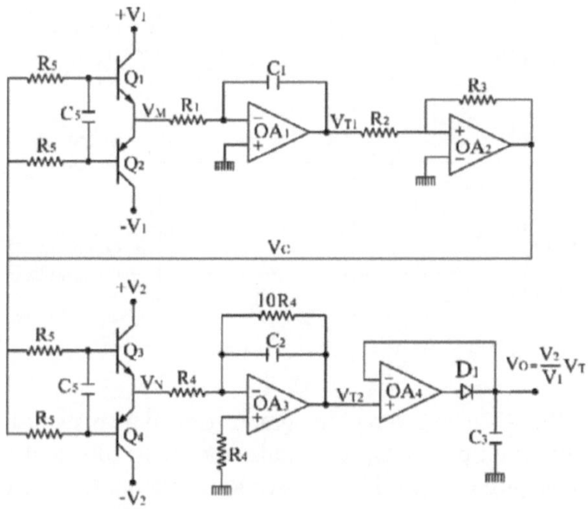

FIGURE 13.5(A) Double dual slope peak detecting divider.

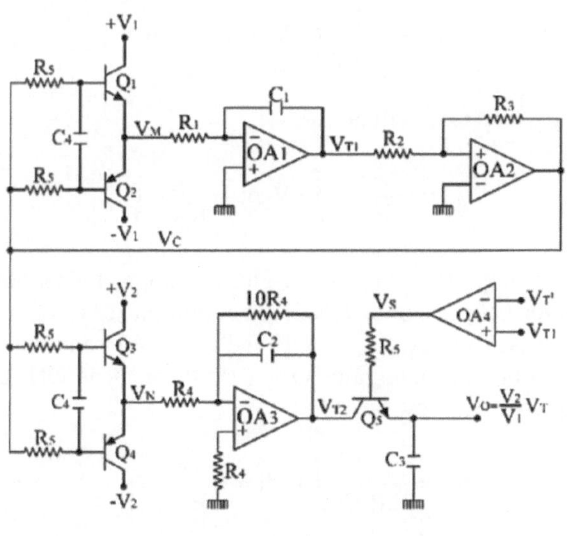

FIGURE 13.5(B) Double dual slope peak sampling divider.

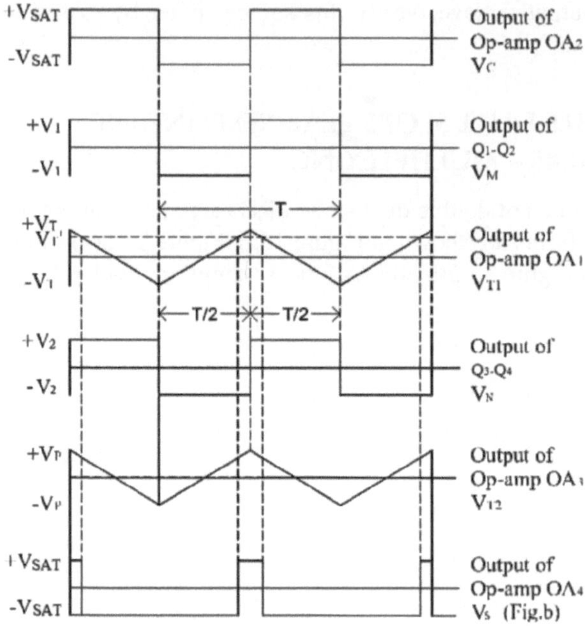

FIGURE 13.6 Associated waveforms of Figure 13.5.

OA_2 output be LOW, ($-V_1$) be given to integrator formed by resistor R_1, capacitor C_1, op-amp OA_1 by the multiplexer Q_1–Q_2 (transistor Q_1 is OFF and Q_2 is ON), and ($-V_2$) be given to integrator formed by resistor R_4, capacitor C_2, and op-amp OA_3 by the multiplexer Q_3–Q_4 (transistor Q_3 is OFF and Q_4 is ON). The output of op-amp OA_1 will be

$$V_{T1} = -\frac{1}{R_1C_1}\int -V_1\,dt = \frac{V_1}{R_1C_1}t \tag{13.9}$$

The output of op-amp OA_3 will be

$$V_{T2} = -\frac{1}{R_4C_2}\int -V_2\,dt = \frac{V_2}{R_4C_2}t \tag{13.10}$$

The output of op-amp OA_1 is a positive-going ramp, and when it reaches a value $+V_T$, the comparator OA_2 becomes HIGH. The multiplexer Q_1–Q_2 selects ($+V_1$) to op-amp integrator OA_1 (transistor Q_1 is ON and Q_2 is OFF) and the multiplexer Q_3–Q_4 selects ($+V_2$) to op-amp integrator OA_3 (transistor Q_3 is ON and Q_4 is OFF). Output of op-amp OA_1 will now be

$$V_{T1} = -\frac{1}{R_1C_1}\int V_1\,dt = -\frac{V_1}{R_1C_1}t \tag{13.11}$$

The output of op-amp OA_3 will now be

$$V_{T2} = -\frac{1}{R_4C_2}\int V_2\,dt = -\frac{V_2}{R_4C_2}t \tag{13.12}$$

The output of op-amp OA_1 changes slope from $(+V_T)$ toward $(-V_T)$, and when it reaches a value of $(-V_T)$, the comparator OA_2 becomes LOW and the cycle therefore repeats to give (i) a triangular wave V_{T1} at the output of op-amp OA_1 with $\pm V_T$ peak-to-peak value, (ii) another triangular wave V_{T2} at the output of op-amp OA_3 with $\pm V_P$ peak-to-peak value, (iii) first square waveform V_C with $\pm V_{SAT}$ peak-to-peak values at the output of op-amp OA_2, (iv) second square waveform V_M with $\pm V_1$ as peak-to-peak values at the output of multiplexer Q_1–Q_2, and (v) third square waveform V_N with $\pm V_2$ as peak-to-peak values at the output of multiplexer Q_3–Q_4.

From Eqs. (13.9) and (13.10) and the fact at $t = T/2$, $V_{T1} = 2V_T$, $V_{T2} = 2V_P$

$$2V_T = \frac{V_1}{R_1C_1}\frac{T}{2} \tag{13.13}$$

$$2V_P = \frac{V_2}{R_4C_2}\frac{T}{2} \tag{13.14}$$

From Eqs. (13.13) and (13.14)

$$2V_P = \frac{V_2}{R_4C_2}\frac{2V_TR_1C_1}{V_1}$$

Let us assume $R_1 = R_4$, $C_1 = C_2$

$$V_P = \frac{V_2}{V_1}V_T \tag{13.15}$$

$$V_T = \frac{R_2}{R_3}V_{SAT}$$

In the circuit shown in Figure 13.5(a), the peak detector realized by op-amp OA_4, diode D_1 and capacitor C_3 gives the peak value V_P of the triangular wave V_{T2}, i.e., $V_O = V_P$.

In the circuit shown in Figure 13.5(b), the peak value V_P is obtained by the sample and hold circuit realized by transistor Q_5 and capacitor C_3. The sampling pulse V_S is generated by op-amp OA_4 by comparing a slightly less than voltage of V_T called V_T' with the triangular wave V_{T1}. The sampled output is given as $V_O = V_P$.

From Eq. (13.15)

$$V_O = \frac{V_2}{V_1}V_T \tag{13.16}$$

DESIGN EXERCISES

1. The multiplexers in Figures 13.5(a) and 13.5(b) are to be replaced with FET multiplexers and MOSFET multiplexers. In each, (i) draw the circuit diagrams, (ii) explain their working operation, (iii) draw waveforms at appropriate places, and (iv) deduce expression for the output voltage.

PRACTICAL EXERCISES

1. Verify the divider circuits shown in Figure 13.5 by experimenting with the AFC trainer kit.
2. As discussed in Chapter 18 and Section 18.4, convert the analog dividers of Figure 13.5 into analog multipliers by both type I and II methods. In each, (i) draw the circuit diagrams, (ii) explain their working operation, (iii) draw waveforms at appropriate places, and (iv) deduce expression for the output voltage. Verify this multiplier by experimenting with the AFC trainer kit.
3. As discussed in Chapter 18 and Section 18.5, convert the analog dividers of Figure 13.5 into analog square rooters. In each, (i) draw the circuit diagrams, (ii) explain their working operation, (iii) draw waveforms at appropriate places, and (iv) deduce expression for the output voltage. Verify this square rooter by experimenting with the AFC trainer kit.
4. Replace the transistor multiplexers in Figure 13.5 with CD4053 analog multiplexers IC. (i) Draw the circuit diagrams, (ii) explain their working operation, (iii) draw waveforms at appropriate places, and (iv) deduce expression for the output voltage. Verify this divider by experimenting with the AFC trainer kit.
5. Convert the divider obtained in Problem 4 into multipliers as discussed in Section 18.4. (i) Draw the circuit diagrams, (ii) explain their working operation, (iii) draw waveforms at appropriate places, and (iv) deduce expression for the output voltage. Verify this multiplier by experimenting with the AFC trainer kit.
6. Convert the divider obtained in Problem 4 into square rooters as discussed in Section 18.5. (i) Draw the circuit diagrams, (ii) explain their working operation, (iii) draw waveforms at appropriate places, and (iv) deduce expression for the output voltage. Verify this square rooter by experimenting with the AFC trainer kit.

13.4 DOUBLE DUAL SLOPE PEAK RESPONDING DIVIDERS – SWITCHING

The circuit diagrams of double dual slope PRDs are shown in Figure 13.7, and their associated waveforms are shown in Figure 13.8. Figure 13.7(a) shows a series switching double dual slope peak detecting divider, Figure 13.7(b) shows a shunt switching double dual slope peak detecting divider, Figure 13.7(c) shows a series switching double dual slope peak sampling divider, and Figure 13.7(d) shows a shunt switching double dual slope peak sampling divider. Let initially comparator OA_2 output be LOW.

In Figures 13.7(a) and 13.7(c), the transistor Q_1 is OFF, the op-amp OA_5 along with resistors R_5 will work as an inverting amplifier, and $-V_1$ will appear at its output

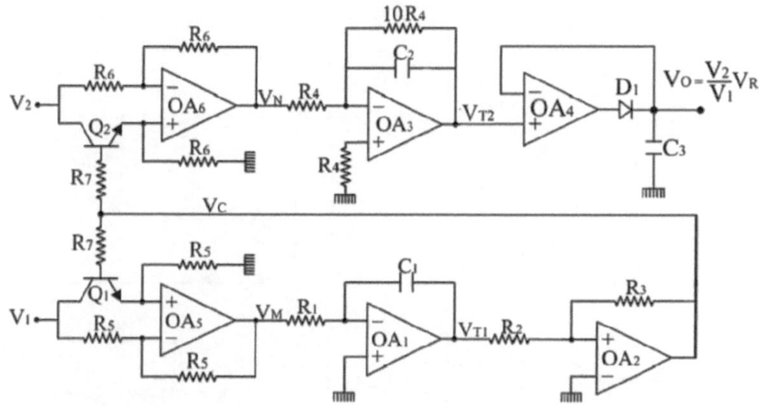

FIGURE 13.7(A) Series switching double dual slope peak detecting divider.

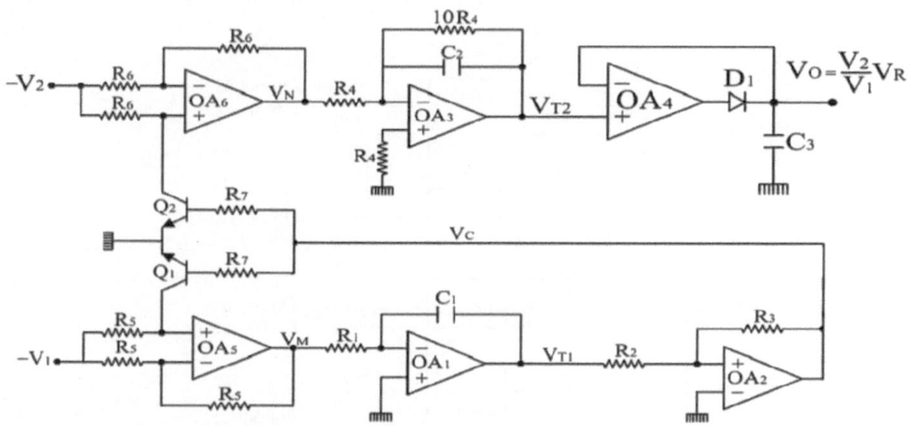

FIGURE 13.7(B) Shunt switching double dual slope peak detecting divider.

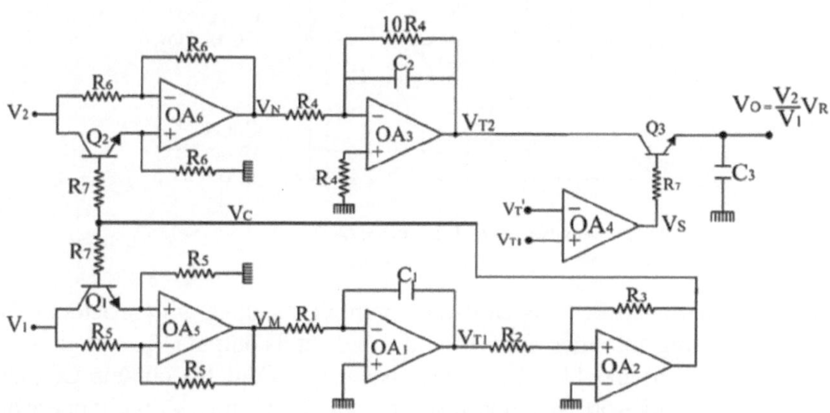

FIGURE 13.7(C) Series switching double dual slope peak sampling divider.

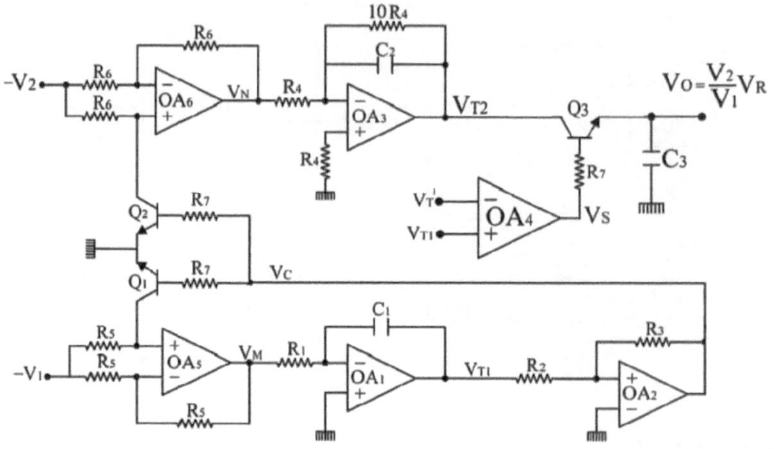

FIGURE 13.7(D) Shunt switching double dual slope peak sampling divider.

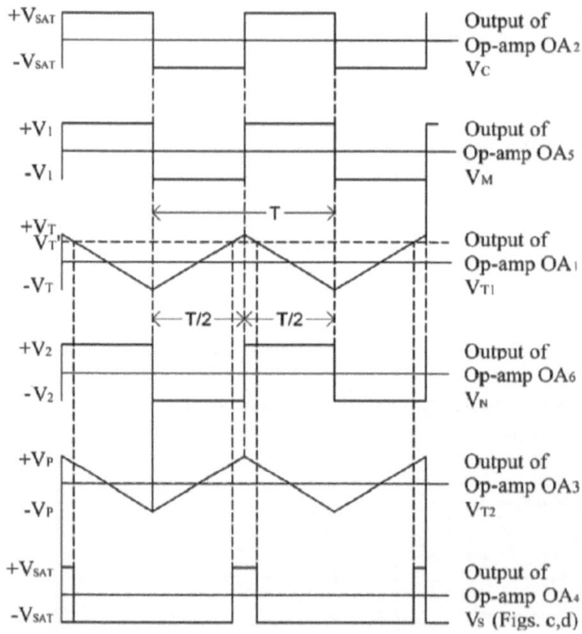

FIGURE 13.8 Associated waveforms of Figure 13.7.

($V_M = -V_1$). The transistor Q_2 is OFF, the op-amp OA_6 along with resistors R_6 will work as an inverting amplifier, and $-V_2$ will appear at its output ($V_N = -V_2$).

In Figures 13.7(b) and 13.7(d), the transistor Q_1 is OFF, the op-amp OA_5 along with resistors R_5 will work as a non-inverting amplifier, and $-V_1$ will appear at its output ($V_M = -V_1$). The transistor Q_2 is OFF, the op-amp OA_6 along with resistors R_6 will work as a non-inverting amplifier, and $-V_2$ will appear at its output ($V_N = -V_2$).

The output of op-amp OA_1 will be

$$V_{T1} = -\frac{1}{R_1C_1}\int -V_1\,dt = \frac{V_1}{R_1C_1}t \tag{13.17}$$

The output of op-amp OA_3 will be

$$V_{T2} = -\frac{1}{R_4C_2}\int -V_2\,dt = \frac{V_2}{R_4C_2}t \tag{13.18}$$

The output of op-amp OA_1 is a positive-going ramp, and when it reaches a value $+V_T$, the comparator OA_2 output becomes HIGH.

In Figures 13.7(a) and 13.7(c), the transistor Q_1 is ON, the op-amp OA_5 along with resistors R_5 will work as a non-inverting amplifier, and $+V_1$ will appear at its output ($V_M = +V_1$). The transistor Q_2 is ON, the op-amp OA_6 along with resistors R_6 will work as a non-inverting amplifier, and $+V_2$ will appear at its output ($V_N = +V_2$).

In Figures 13.7(b) and 13.7(d), the transistor Q_1 is ON, the op-amp OA_5 along with resistors R_5 will work as an inverting amplifier, and $+V_1$ will appear at its output ($V_M = +V_1$). The transistor Q_2 is ON, the op-amp OA_6 along with resistors R_6 will work as an inverting amplifier, and $+V_2$ will appear at its output ($V_N = +V_2$).

The output of op-amp OA_1 will now be

$$V_{T1} = -\frac{1}{R_1C_1}\int V_1\,dt = -\frac{V_1}{R_1C_1}t \tag{13.19}$$

The output of op-amp OA_3 will now be

$$V_{T2} = -\frac{1}{R_4C_2}\int V_2\,dt = -\frac{V_2}{R_4C_2}t \tag{13.20}$$

The output of op-amp OA_1 changes slope from ($+V_T$) toward ($-V_T$), and when it reaches a value of ($-V_T$), the comparator OA_2 output becomes LOW, and the cycle therefore repeats to give (i) a triangular waveform V_{T1} with $\pm V_T$ peak-to-peak at the output of op-amp OA_1 and (ii) another triangular waveform V_{T2} with $\pm V_P$ peak-to-peak at the output of op-amp OA_3, and (iii) first square waveform V_C with $\pm V_{SAT}$ peak-to-peak value at the output of op-amp OA_2, (iv) second square waveform V_M with $\pm V_1$ peak-to-peak value at the output of op-amp OA_5, and (v) third square waveform V_N with $\pm V_2$ peak-to-peak value at the output of op-amp OA_6.

From the waveforms shown in Figure 13.8, Eq. (13.18), and the fact that at $t = T/2$, $V_{T1} = 2V_T$, $V_{T2} = 2V_P$

$$2V_T = \frac{V_1}{R_1C_1}\frac{T}{2} \tag{13.21}$$

$$2V_P = \frac{V_2}{R_2C_2}\frac{T}{2} \tag{13.22}$$

From Eqs. (13.21) and (13.22)

$$V_P = \frac{V_2 V_T}{V_1} \frac{R_1 C_1}{R_2 C_2}$$

Let us assume $R_1 = R_2$, $C_1 = C_2$

$$V_P = \frac{V_2}{V_1} V_T \tag{13.23}$$

$$V_T = \frac{R_2}{R_3} V_{SAT}$$

In Figures 13.7(a) and 13.7(b), the peak detector realized by op-amp OA_4, diode D_1 and capacitor C_3 gives the peak value V_P of the triangular wave V_{T2}. Hence $V_O = V_P$.

In Figures 13.7(c) and 13.7(d), the peak value V_P is obtained by the sample and hold circuit realized by transistor Q_3 and capacitor C_3. The sampling pulse is generated by op-amp OA_4 by comparing a slightly less than voltage of VT, called VT' with the triangular wave VT_1. The sample and hold output is $V_O = V_P$.

From Eq. 13.23,

$$V_O = \frac{V_2}{V_1} V_T = \frac{V_2}{V_1} \frac{R_2}{R_3} V_{SAT}$$

Let $V_R = R_2/R_3 V_{SAT}$.

$$V_O = \frac{V_2}{V_1} V_R \tag{13.24}$$

DESIGN EXERCISES

1. The transistor switches in Figures 13.7(a) and 13.7(b) are to be replaced with FET switches and MOSFET switches. In each, (i) draw the circuit diagrams, (ii) explain their working operations, (iii) draw waveforms at appropriate places, and (iv) deduce expression for their output voltages.

PRACTICAL EXERCISES

1. Verify the divider circuits shown in Figure 13.7 by experimenting with the AFC trainer kit.
2. As discussed in Chapter 18 and Section 18.4, convert the analog dividers of Figure 13.7 into analog multipliers by both type I and II methods. In each, (i) draw the circuit diagrams, (ii) explain their working operation, (iii) draw waveforms at appropriate places, and (iv) deduce expression for the output voltage. Verify this multiplier by experimenting with the AFC trainer kit.

3. As discussed in Chapter 18 and Section 18.5, convert the analog dividers of Figure 13.7 into analog square rooters. In each, (i) draw the circuit diagrams, (ii) explain their working operation, (iii) draw waveforms at appropriate places, and (iv) deduce expression for the output voltage. Verify this square rooter by experimenting with the AFC trainer kit.
4. Replace the transistor switches in Figure 13.7 with CD4066 and DG201 analog switch ICs. (i) Draw the circuit diagrams, (ii) explain their working operation, (iii) draw waveforms at appropriate places, and (iv) deduce expression for the output voltage. Verify this square rooter by experimenting with the AFC trainer kit.
.5 Convert the dividers obtained in Problem 4 into multipliers as discussed in Section 18.4. (i) Draw the circuit diagrams, (ii) explain their working operation, (iii) draw waveforms at appropriate places, and (iv) deduce expression for the output voltage. Verify this multiplier by experimenting with the AFC trainer kit.
6. Convert the dividers obtained in Problem 4 into square rooters as discussed in Section 18.5. (i) Draw the circuit diagrams, (ii) explain their working operation, (iii) draw waveforms at appropriate places, and (iv) deduce expression for the output voltage. Verify this square rooter by experimenting with the AFC trainer kit.

13.5 PEAK RESPONDING DIVIDERS USING VOLTAGE-TO-FREQUENCY CONVERTER –MULTIPLEXING

The PRDs using voltage-to-frequency (V/F) converters are shown in Figure 13.9, and their associated waveforms are shown in Figure 13.10. Figure 13.9(a) shows a peak

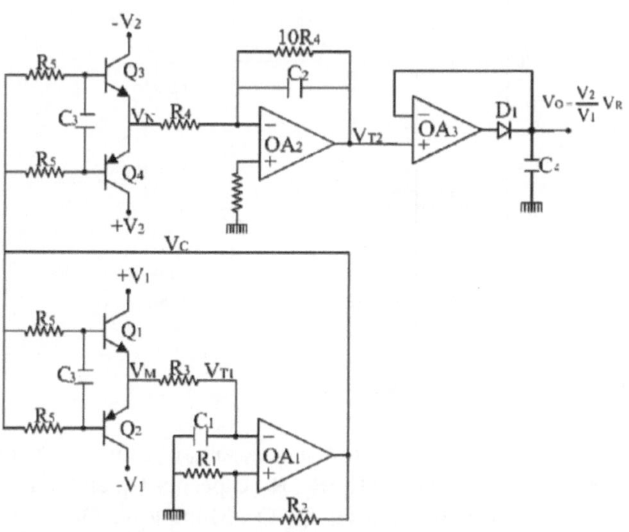

FIGURE 13.9(A) Peak detecting divider using voltage-to-frequency converter.

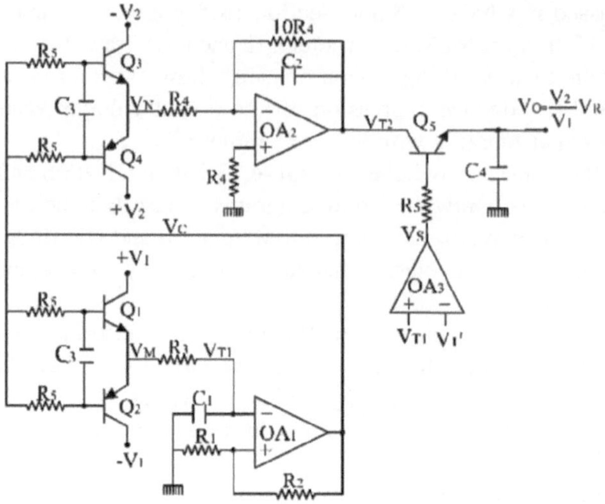

FIGURE 13.9(B) Peak sampling divider using voltage-to-frequency converter.

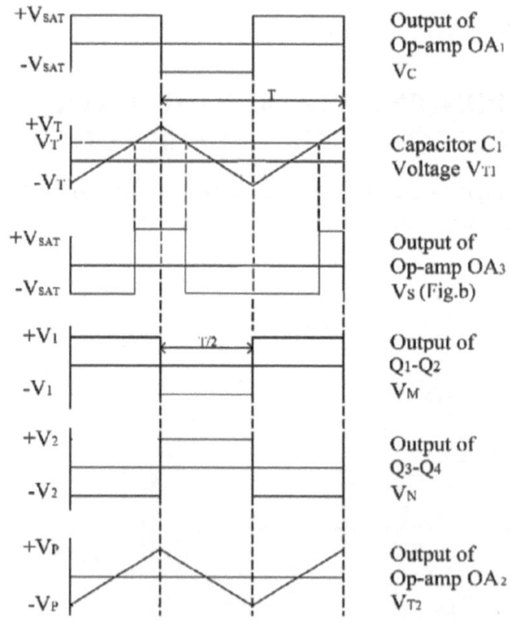

FIGURE 13.10 Associated waveforms of Figure 13.9.

detecting divider, and Figure 13.9(b) shows a peak sampling divider. A square wave V_C is generated by op-amp OA_1, resistors R_1, R_2, R_3, capacitor C_1, and multiplexer Q_1–Q_2. The square wave V_C controls multiplexer Q_3–Q_4. Multiplexer Q_3–Q_4 connects $(-V_2)$ during HIGH time of V_C and $(+V_2)$ during LOW time of V_C. Another square wave V_N with $\pm V_2$ peak-to-peak values is generated at the multiplexer Q_3–Q_4 output. The integrator formed by op-amp OA_2, resistors R_4, and capacitor C_2 converts the square wave V_N into a triangular wave V_{T2} of $\pm V_P$ peak-to-peak values. V_P is proportional to V_2.

$$V_P = K_1 V_2 \tag{13.25}$$

The time period 'T' of square wave V_C is inversely proportional to V_1

$$T = \frac{K_2}{V_1} \tag{13.26}$$

The integrator output is given as

$$V_{T2} = \frac{1}{R_4 C_2} \int V_2\, dt = \frac{V_2}{R_4 C_2} t \tag{13.27}$$

For half time period for t = T/2. From the waveforms shown in Figure 13.10, Eq. (13.27), and the fact that at $t = T/2$, $V_{T2} = 2V_P$

$$2V_P = \frac{V_2}{R_4 C_2} \frac{T}{2}; \; V_P = \frac{V_2}{4R_4 C_2} \frac{K_2}{V_1} \tag{13.28}$$

Let

$$V_R = \frac{K_2}{4R_4 C_2}$$

$$V_P = \frac{V_2 V_R}{V_1} \tag{13.29}$$

In the circuit shown in Figure 13.9(a), the peak detector realized by op-amp OA_3, diode D_1, and capacitor C_4 gives this peak value V_P at the output: $V_O = V_P$.

In the circuit shown in Figure 13.9(b), the sample and hold circuit composed by transistor Q_5 with capacitor C_4 gives the peak value V_P. The sampling pulse V_S is generated by comparator op-amp OA_3. The sampling pulse V_S is generated by comparing the capacitor C_1 voltage VT_1 of ±VT peak-to- peak value with a voltage of VT' which is slightly less than the value of VT by comparator OA_3. The sampled output V_O is the peak value V_P: $V_O = V_P$.

From Eq. (13.29)

$$V_O = \frac{V_2}{V_1} V_R \tag{13.30}$$

DESIGN EXERCISES

1. The transistor multiplexers in Figures 13.9(a) and 13.9(b) are to be replaced with FET multiplexers and MOSFET multiplexers. In each, (i) draw the

circuit diagram, (ii) explain its working operation, (iii) draw waveforms at appropriate places, and (iv) deduce expression for the output voltage.

PRACTICAL EXERCISES

1. Verify the divider circuits shown in Figure 13.9 by experimenting with the AFC trainer kit.
2. As discussed in Chapter 18 and Section 18.4, convert the analog dividers of Figure 13.9 into analog multipliers by both type I and II methods. In each, (i) draw the circuit diagrams, (ii) explain their working operation, (iii) draw waveforms at appropriate places, and (iv) deduce expression for the output voltage. Verify this multiplier by experimenting with the AFC trainer kit.
3. As discussed in Chapter 18 and Section 18.5, convert the analog dividers of Figure 13.9 into analog square rooters. In each, (i) draw the circuit diagrams, (ii) explain their working operation, (iii) draw waveforms at appropriate places, and (iv) deduce expression for the output voltage. Verify this square rooter by experimenting with the AFC trainer kit.
4. Replace the transistor multiplexers in Figure 13.9 with CD4053 analog multiplexers IC. (i) Draw the circuit diagrams, (ii) explain their working operation, (iii) draw waveforms at appropriate places, and (iv) deduce expression for the output voltage. Verify this divider by experimenting with the AFC trainer kit.
5 Convert the dividers obtained in Problem 3 into multipliers as discussed in Section 18.4. (i) Draw the circuit diagrams, (ii) explain their working operation, (iii) draw waveforms at appropriate places, and (iv) deduce expression for the output voltage. Verify this multiplier by experimenting with the AFC trainer kit.
6. Convert the divider obtained in Problem 4 into square rooters as discussed in Section 18.5. (i) Draw the circuit diagrams, (ii) explain their working operation, (iii) draw waveforms at appropriate places, and (iv) deduce expression for the output voltage. Verify this square rooter by experimenting with the AFC trainer kit.

13.6 PEAK RESPONDING DIVIDERS USING VOLTAGE-TO-FREQUENCY CONVERTER – SWITCHING TYPE

The PRDs using voltage-to-frequency converters are shown in Figure 13.11, and their associated waveforms are shown in Figure 13.12. Figure 13.11(a) shows a series switching peak detecting divider, Figure 13.11(b) shows a shunt switching peak detecting divider, Figure 13.11(c) shows a series switching peak sampling divider, and Figure 13.11(d) shows a shunt switching peak sampling divider. A square wave V_C is generated by op-amp OA_1, resistors R_1, R_2, R_3, capacitor C_1, and control amplifier realized by op-amp OA_3 and resistors R_6. The square wave V_C controls transistor switches Q_1 and Q_2.

Let us first consider the circuit shown in Figures 13.11(a) and 13.11(c). During HIGH of the square waveform V_C, (i) the transistor Q_1 is ON, the op-amp OA_3 along with resistors R_6 will work as a non-inverting amplifier, and $+V_1$ will appear at its output ($V_M = +V_1$); (ii) the transistor Q_2 is ON, the op-amp OA_4 along with resistors

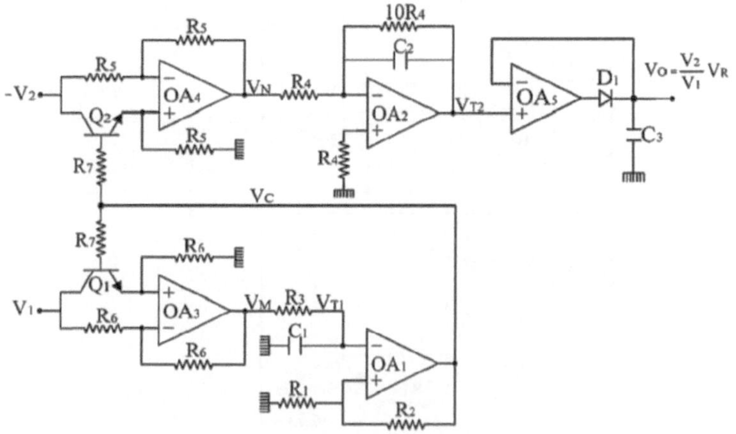

FIGURE 13.11(A) Series switching peak detecting divider using voltage-to-frequency converter.

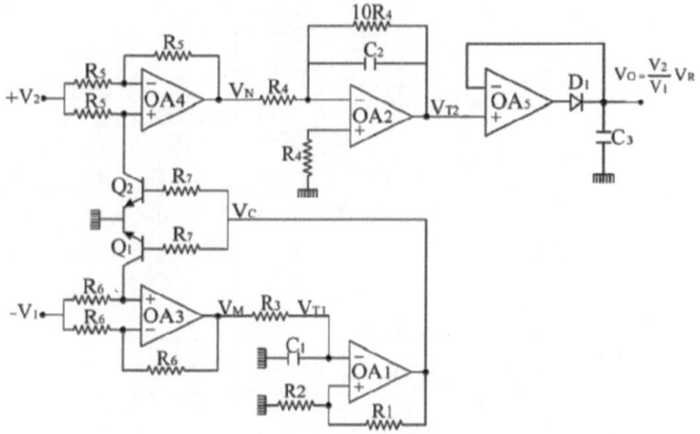

FIGURE 13.11(B) Shunt switching peak detecting divider using voltage-to-frequency converter.

R_5 will work as a non-inverting amplifier, and $-V_2$ will appear at its output ($V_N = -V_2$). During LOW of the square waveform V_C, (i) the transistor Q_1 is OFF, the op-amp OA_3 along with resistors R_6 will work as an inverting amplifier, and $-V_1$ will appear at its output ($V_M = -V_1$); (ii) the transistor Q_2 is OFF, the op-amp OA_4 along with resistors R_5 will work as an inverting amplifier and $+V_2$ will appear at its output ($V_N = +V_2$).

Next, consider the circuits shown in Figures 13.11(b) and 13.11(d). During HIGH of the square waveform V_C, (i) the transistor Q_1 is ON, the op-amp OA_3 along with resistors R_6 will work as an inverting amplifier, and $+V_1$ will appear at its output ($V_M = +V_1$), (ii) the transistor Q_1 is ON, the op-amp OA_4 along with resistors R_5 will work as inverting amplifier and $-V_2$ will appear at its output ($V_N = -V_2$). During

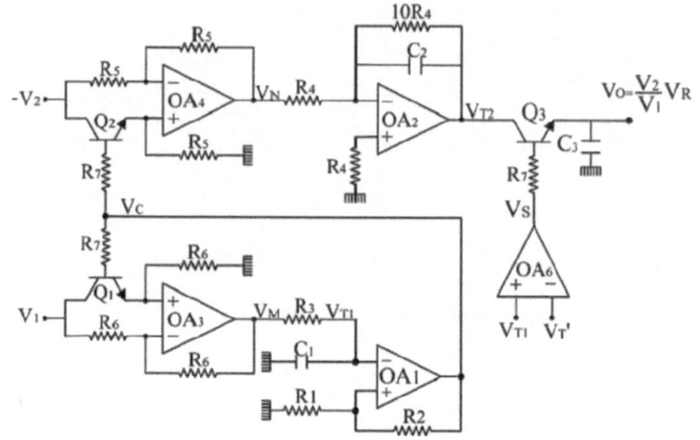

FIGURE 13.11(C) Series switching peak sampling divider using voltage-to-frequency converter.

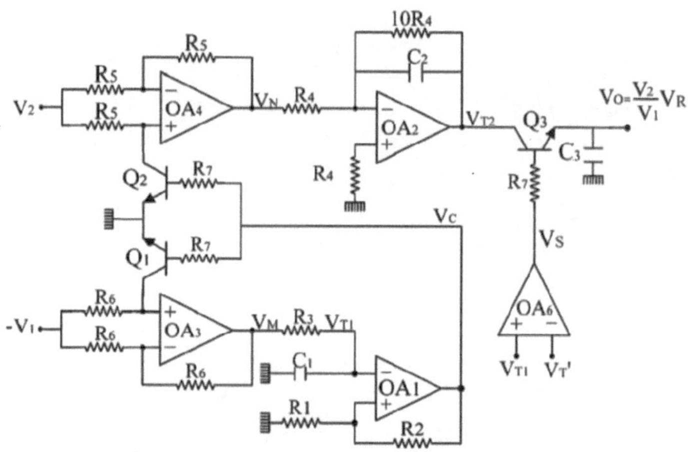

FIGURE 13.11(D) Shunt switching peak sampling divider using voltage-to-frequency converter.

LOW of the square waveform V_C; (i) the transistor Q_1 is OFF, the op-amp OA_3 along with resistors R_6 will work as a non-inverting amplifier, and $-V_1$ will appear at its output ($V_M = -V_1$), (ii) the transistor Q_2 is OFF, the op-amp OA_4 along with resistors R_5 will work as a non-inverting amplifier, and $+V_2$ will appear at its output ($V_N = +V_2$).

In all the circuits of Figures 13.11(a), 13.11(b), 13.11(c), and 13.11(d), two square waves are generated: (i) V_M with $\pm V_1$ peak-to-peak at the output of op-amp OA_3, and (ii) V_N with $\pm V_2$ peak-to-peak at the output of op-amp OA_4.

The integrator formed by op-amp OA_2, resistors R_4, and capacitor C_2 converts the square wave V_N into a triangular wave V_{T2} of $\pm V_P$ peak-to-peak values. V_P is proportional to V_2

$$V_P = K_1 V_2 \tag{13.31}$$

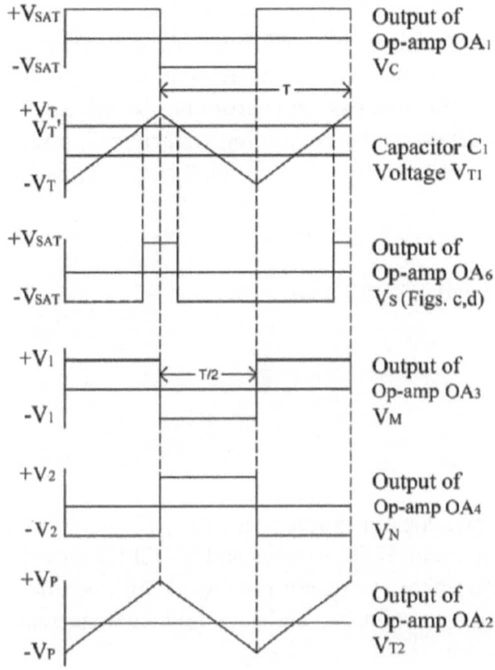

FIGURE 13.12 Associated waveforms of Figures 13.11(a), 13.11(b), 13.11(c), and 13.11(d).

The time period 'T' of square wave V_S is inversely proportional to V_1

$$T = \frac{K_2}{V_1} \tag{13.32}$$

The integrator output is given as

$$V_{T2} = \frac{1}{R_4C_2}\int V_2\,dt = \frac{V_2}{R_4C_2}t \tag{13.33}$$

For half time period for t = T/2. From the waveforms shown in Figure 13.12, Eq. (13.33), and the fact that at $t = T/2$, $V_{T2} = 2\,V_P$

$$2V_P = \frac{V_2}{R_4C_2}\frac{T}{2};\ V_P = \frac{V_2}{4R_4C_2}\frac{K_2}{V_1}$$

Let

$$V_R = \frac{K_2}{4R_4C_2}$$

$$V_P = \frac{V_2}{V_1}V_R \tag{13.34}$$

In Figures 13.11(a) and 13.11(b), the peak detector realized by op-amp OA_5, diode D_1, and capacitor C_3 gives this peak value V_P at the output: $V_O = V_P$.

In Figures 13.11(c) and 13.11(d), this peak value V_P is sampled by the sample and hold circuit realized by transistor Q_3 and capacitor C_3 with a sampling pulse V_S. The sampling pulse V_S is generated by comparing capacitor C_1 voltage V_{T1} with slightly less than its peak value V_T, i.e., V_T'. The sampled output V_O is the peak value V_P.

$$V_O = V_P \tag{13.35}$$

From Eqs. (13.34) and (13.35)

$$V_O = \frac{V_2}{V_1} V_R \tag{13.36}$$

DESIGN EXERCISES

1. The transistor switches in Figures 13.11(a), 13.11(b), 13.11(c), and 13.11(d) are to be replaced with FET switches and MOSFET switches. In each, (i) draw the circuit diagrams, (ii) explain their working operations, (iii) draw waveforms at appropriate places, and (iv) deduce expression for their output voltages.

PRACTICAL EXERCISES

1. Verify the divider circuits shown in Figure 13.11 by experimenting with the AFC trainer kit.
2. As discussed in Chapter 18 and Section 18.4, convert the analog dividers of Figure 13.11 into analog multipliers by both type I and II methods. In each, (i) draw the circuit diagrams, (ii) explain their working operation, (iii) draw waveforms at appropriate places, and (iv) deduce expression for the output voltage. Verify this multiplier by experimenting with the AFC trainer kit.
3. As discussed in Chapter 18 and Section 18.5, convert the analog dividers of Figure 13.11 into analog square rooters. In each, (i) draw the circuit diagrams, (ii) explain their working operation, (iii) draw waveforms at appropriate places, and (iv) deduce expression for the output voltage. Verify this square rooter by experimenting with the AFC trainer kit.
4. Replace the transistor switches in Figure 13.11 with CD4066 and DG 201 analog switch ICs. In each, (i) draw the circuit diagrams, (ii) explain their working operation, (iii) draw waveforms at appropriate places, and (iv) deduce expression for the output voltage. Verify this divider by experimenting with the AFC trainer kit.
5. Convert the dividers obtained in Problem 3 into multipliers as discussed in Section 18.4. (i) Draw the circuit diagrams, (ii) explain their working operation, (iii) draw waveforms at appropriate places, and (iv) deduce expression for the output voltage. Verify this multiplier by experimenting with the AFC trainer kit.

6. Convert the divider obtained in Problem 4 into square rooters as discussed in Section 18.5. (i) Draw the circuit diagrams, (ii) explain their working operation, (iii) draw waveforms at appropriate places, and (iv) deduce expression for the output voltage. Verify this square rooter by experimenting with the AFC trainer kit.

14 Design of Time Division Multiplier-Cum-Divider – Multiplexing

Like multipliers, multipliers-cum-dividers (MCDs) are also classified into (i) time division MCDs, (ii) peak responding MCDs, and (iii) pulse position responding MCDs. The peak responding MCD is further classified into (i) peak detecting MCD and (ii) peak sampling MCD.

A pulse train whose maximum value is proportional to one voltage (V_3) is generated. If the width of this pulse train is made proportional to one voltage (V_2) and inversely proportional to another voltage (V_1), then the average value of the pulse train is proportional to $\frac{V_2 V_3}{V_1}$. This is called "time division multiplier-cum-divider (TDMCD)." TDMCDs using multiplexers are described in this chapter and using switches are described next in Chapter 15.

14.1 SAWTOOTH WAVE–BASED MULTIPLIER-CUM-DIVIDER – TYPE I

The circuit diagram of double multiplexing–averaging TDMCD is shown in Figure 14.1, and its associated waveforms are shown in Figure 14.2. A sawtooth wave V_{S1} of peak value V_R and time period 'T' is generated by the op-amp OA_1. The time period 'T' is given as

$$T = 2R_5C_1 \ln\left(1 + 2\frac{R_1}{R_2}\right) \tag{14.1}$$

Comparator OA_2 compares the sawtooth wave V_{S1} with the voltage V_Y and produces a rectangular waveform V_K. The ON time δ_T of V_K is given as

$$\delta_T = \frac{V_Y}{V_R} T \tag{14.2}$$

The rectangular pulse V_K controls the transistor multiplexer Q_1–Q_2. When V_K is HIGH, input voltage V_1 is connected to R_6C_2 lowpass filter (transistor Q_1 is ON and Q_2 is OFF). When V_K is LOW, zero volts is connected to R_6C_2 lowpass filter (transistor Q_1 is OFF and Q_2 is ON). Another rectangular pulse V_M with maximum value of

DOI: 10.1201/9781003221449-17

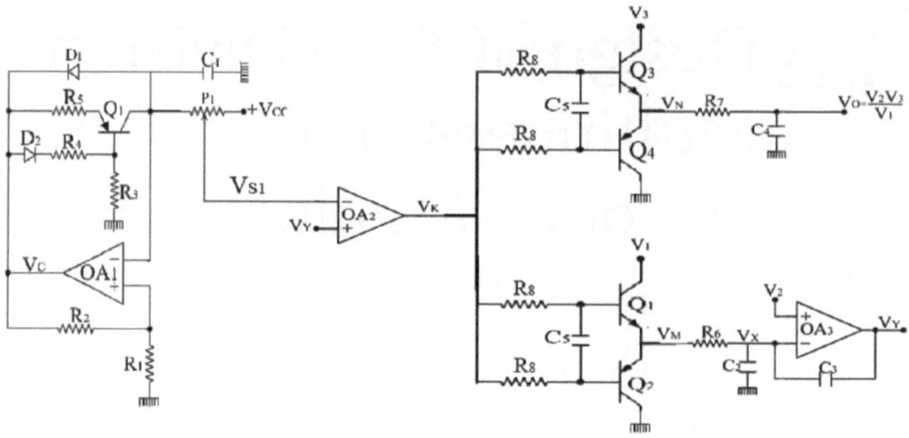

FIGURE 14.1 Double multiplexing–averaging time division multiplier-cum-divider.

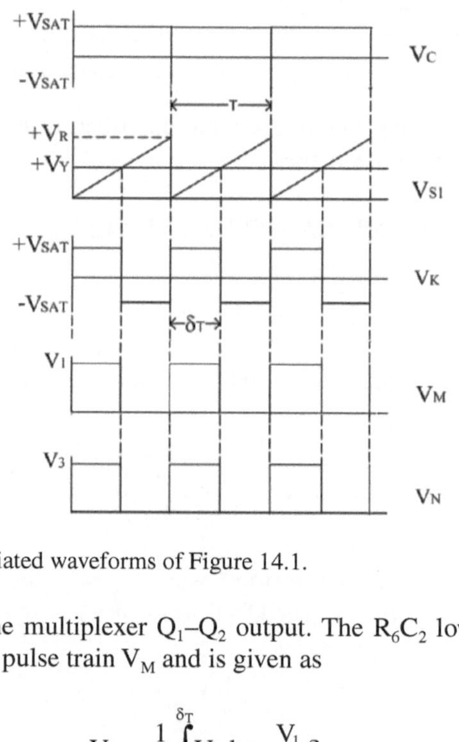

FIGURE 14.2 Associated waveforms of Figure 14.1.

V_1 is generated at the multiplexer Q_1–Q_2 output. The R_6C_2 lowpass filter gives an average value of this pulse train V_M and is given as

$$V_X = \frac{1}{T}\int_0^{\delta_T} V_1\, dt = \frac{V_1}{T}\delta_T \tag{14.3}$$

$$V_X = \frac{V_1 V_Y}{V_R} \tag{14.4}$$

The op-amp OA_3 is configured in a negative closed loop feedback and a positive DC voltage is ensured in the feedback loop. Hence, its inverting terminal voltage must equal to its non-inverting terminal voltage.

$$V_2 = V_X \tag{14.5}$$

From Eqs. (14.4) and (14.5)

$$V_Y = \frac{V_2 V_R}{V_1} \tag{14.6}$$

The rectangular pulse V_K also controls the second multiplexer Q_3–Q_4. When V_K is HIGH, another input voltage V_3 is connected to R_7C_4 lowpass filter (transistor Q_3 is ON and Q_4 is OFF). When V_K is LOW, zero volts is connected to R_7C_4 lowpass filter (transistor Q_3 is OFF and Q_4 is ON). Another rectangular pulse V_N with maximum value of V_3 is generated at the multiplexer Q_3–Q_4 output. The R_7C_4 lowpass filter gives an average value of this pulse train V_N and is given as

$$V_O = \frac{1}{T}\int_0^{\delta_T} V_3 dt = \frac{V_3}{T}\delta_T \tag{14.7}$$

Equations (14.2) and (14.6) in (14.7) give

$$V_O = \frac{V_2 V_3}{V_1} \tag{14.8}$$

The output voltage depends on sharpness and linearity of the sawtooth waveform. The offset that exists on the sawtooth wave will cause an error at the output; hence, it is to be nulled by suitable circuitry (potentiometer P_1). The output voltage is proportional to attenuation constant α of the lowpass filter. The attenuation constant α will be in the range of 0.80 to 0.90. The polarities of V_1 and V_2 are only positive. However, V_3 can have any polarity.

DESIGN EXERCISES

1. In the MCD circuit shown in Figure 14.1, the sawtooth wave generator shown in Figure 4.7(c) is used. Replace this sawtooth wave generator with other sawtooth generators. In each, (i) draw circuit diagrams, (ii) draw waveforms at appropriate places, (iii) explain their working operation, and (iv) deduce expression for their output voltages.
2. Replace transistor multiplexers Q_1–Q_2 and Q_3–Q_4 in Figure 14.1 with JFET multiplexers and MOSFET multiplexers. In each, (i) draw circuit diagrams, (ii) draw waveforms at appropriate places, (iii) explain their working operation, and (iv) deduce expression for their output voltages.

PRACTICAL EXERCISES

1. Verify the MCD circuit shown in Figure 14.1 by experimenting with the AFC trainer kit.

2. Replace the transistor multiplexers in Figure 14.1 with CD4053 analog multiplexer. (i) Draw the circuit diagrams, (ii) explain their working operation, (iii) draw waveforms at appropriate places, and (iv) deduce expression for the output voltage. Verify this MCD by experimenting with the AFC trainer kit.
3. As discussed in Chapter 18 and Section 18.9, convert the MCDs of Figure 14.1 into square root of multiplication (SRM). (i) Draw the circuit diagrams, (ii) explain their working operation, (iii) draw waveforms at appropriate places, and (iv) deduce expression for the output voltage. Verify this SRM by experimenting with the AFC trainer kit.
4. As discussed in Chapter 19 and Section 19.6, convert the MCDs of Figure 14.1 into vector magnitude circuits (VMCs). (i) Draw the circuit diagrams, (ii) explain their working operation, (iii) draw waveforms at appropriate places, and (iv) deduce expression for the output voltage. Verify this VMC by experimenting with the AFC trainer kit.
5. Convert the MCD obtained in Problem 2 into SRM as discussed in Section 18.9. (i) Draw the circuit diagrams, (ii) explain their working operation, (iii) draw waveforms at appropriate places, and (iv) deduce expression for the output voltage. Verify this SRM by experimenting with the AFC trainer kit.
6. Convert the MCD obtained in Problem 2 into VMC as discussed in Section 19.6. (i) Draw the circuit diagrams, (ii) explain their working operation, (iii) draw waveforms at appropriate places, and (iv) deduce expression for the output voltage. Verify this VMC by experimenting with the AFC trainer kit.

14.2 SAWTOOTH WAVE–BASED MULTIPLIER-CUM-DIVIDER – TYPE II

The circuit diagram of sawtooth wave–based TDMCD is shown in Figure 14.3, and its associated waveforms are shown in Figure 14.4. A sawtooth wave V_{S1} with peak value V_R and time period 'T' is generated by around op-amp OA_1. The comparator OA_2 compares the sawtooth wave V_{S1} with the third input voltage V_3 and produces a rectangular waveform V_M. The ON time δ_{T1} of V_M is given as

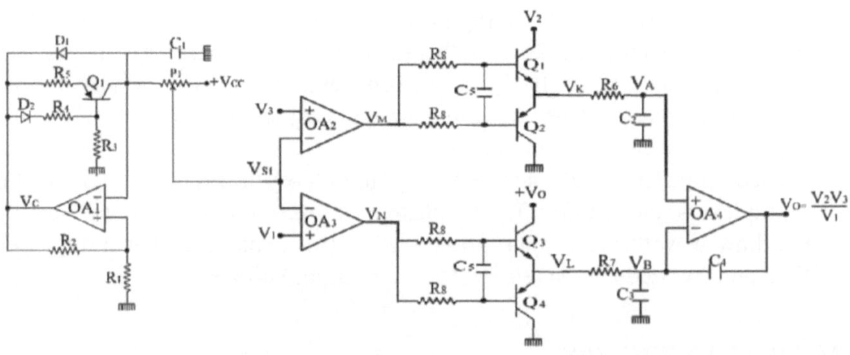

FIGURE 14.3 Sawtooth wave–based time division multiplier-cum-divider – type II.

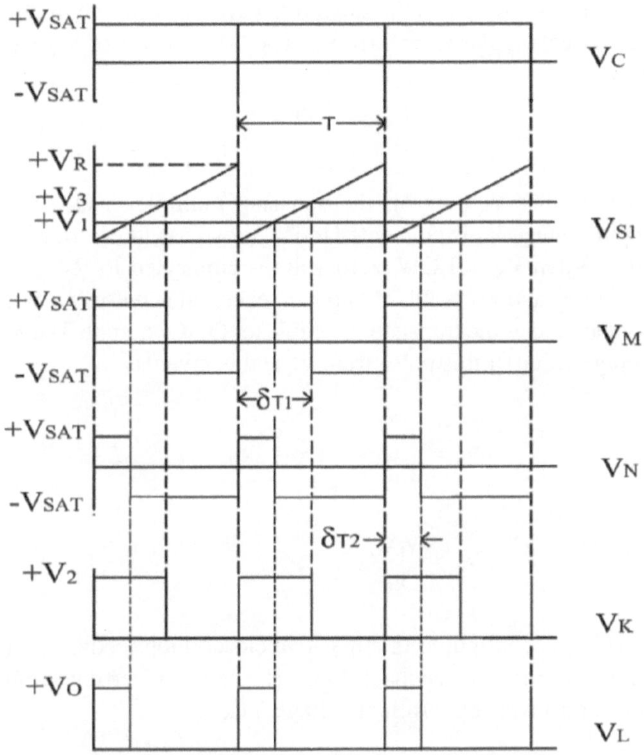

FIGURE 14.4 Associated waveforms of Figure 14.3.

$$\delta_{T1} = \frac{V_3}{V_R} T \tag{14.9}$$

The rectangular pulse V_M controls the transistor multiplexer Q_1–Q_2. When V_M is HIGH, the second input voltage V_2 is connected to R_6C_2 lowpass filter (transistor Q_1 is ON and Q_2 is OFF). When V_M is LOW, zero volts is connected to R_6C_2 lowpass filter (transistor Q_1 is OFF and Q_2 is ON). Another rectangular pulse V_K with maximum value of V_2 is generated at the transistor multiplexer Q_1–Q_2 output. The R_6C_2 lowpass filter gives average value of this pulse train V_K and is given as

$$V_A = \frac{1}{T}\int_0^{\delta_{T1}} V_2\,dt \tag{14.10}$$

$$= \frac{V_2}{T}\delta_{T1} \tag{14.11}$$

$$V_A = \frac{V_2 V_3}{V_R} \tag{14.12}$$

The comparator OA_3 compares the sawtooth wave V_{S1} with the first input voltage V_1 and produces a rectangular waveform V_N. The ON time δ_{T2} of V_N is given as

$$\delta_{T2} = \frac{V_1}{V_R} T \tag{14.13}$$

The rectangular pulse V_N controls the transistor multiplexer Q_3–Q_4. When V_N is HIGH, the output voltage V_O is connected to R_7C_3 lowpass filter (transistor Q_3 is ON and Q_4 is OFF). When V_N is LOW, zero volts is connected to R_7C_3 lowpass filter (transistor Q_3 is OFF and Q_4 is ON). Another rectangular pulse V_L with maximum value of V_O is generated at the transistor multiplexer Q_3–Q_4 output. The R_7C_3 lowpass filter gives average value of this pulse train V_L and is given as

$$V_B = \frac{1}{T} \int_0^{\delta_{T2}} V_O \, dt = \frac{V_O}{T} \delta_{T2}$$

$$V_B = \frac{V_1 V_O}{V_R} \tag{14.14}$$

The op-amp OA_4 is configured in a negative closed loop feedback and a positive DC voltage is ensured in the feedback loop. Hence, its inverting terminal voltage must equal to its non-inverting terminal voltage, i.e.,

$$V_A = V_B \tag{14.15}$$

Equations (14.12) and (14.14) in (14.15) give

$$V_O = \frac{V_2 V_3}{V_1} \tag{14.16}$$

The output voltage depends on sharpness and linearity of the sawtooth waveform. The offset that exists on the sawtooth wave will cause an error at the output; hence, it is to be nulled by suitable circuitry (potentiometer P_1). The output voltage is proportional to attenuation constant α of the lowpass filter. The attenuation constant α will be in the range of 0.80 to 0.90. The polarity of V_1, V_2, and V_3 will be only positive.

DESIGN EXERCISES

1. In the MCD circuit shown in Figure 14.3, replace this sawtooth wave generator with other sawtooth generators shown in Section 4.4. In each, (i) draw circuit diagrams, (ii) draw waveforms at appropriate places, (iii) explain their working operation, and (iv) deduce expression for their output voltages.
2. Replace transistor multiplexers Q_2–Q_3 and Q_4–Q_5 in Figure 14.3 with FET multiplexers and MOSFET multiplexers. In each, (i) draw circuit diagrams, (ii) draw waveforms at appropriate places, (iii) explain their working operation, and (iv) deduce expression for their output voltages.

PRACTICAL EXERCISES

1. Verify the MCD circuit shown in Figure 14.3 by experimenting with the AFC trainer kit.
2. Replace the transistor multiplexers in Figure 14.3 with a CD4053 analog multiplexer. (i) Draw the circuit diagrams, (ii) explain their working operation, (iii) draw waveforms at appropriate places, and (iv) deduce expression for the output voltage. Verify this MCD by experimenting with the AFC trainer kit.
3. As discussed in Chapter 18 and Section 18.9, convert the MCDs of Figure 14.3 into SRM. (i) Draw the circuit diagrams, (ii) explain their working operation, (iii) draw waveforms at appropriate places, and (iv) deduce expression for the output voltage. Verify this SRM by experimenting with the AFC trainer kit.
4. As discussed in Chapter 19 and Section 19.6, convert the MCDs of Figure 14.3 into VMC. (i) Draw the circuit diagrams, (ii) explain their working operation, (iii) draw waveforms at appropriate places, and (iv) deduce expression for the output voltage. Verify this VMC by experimenting with the AFC trainer kit.
5. Convert the MCD obtained in Problem 2 into SRM as discussed in Section 18.9. (i) Draw the circuit diagrams, (ii) explain their working operation, (iii) draw waveforms at appropriate places, and (iv) deduce expression for the output voltage. Verify this SRM by experimenting with the AFC trainer kit.
6. Convert the MCD obtained in Problem 2 into VMC as discussed in Section 19.6. (i) Draw the circuit diagrams, (ii) explain their working operation, (iii) draw waveforms at appropriate places, and (iv) deduce expression for the output voltage. Verify this VMC by experimenting with the AFC trainer kit.

14.3 SAWTOOTH WAVE–BASED MULTIPLIER-CUM-DIVIDER – TYPE III

The circuit diagram of sawtooth wave–based TDMCD is shown in Figure 14.5, and its associated waveforms are shown in Figure 14.6. A sawtooth wave V_{S1} with peak value V_R and time period 'T' is generated by around op-amp OA_1. Comparator OA_2

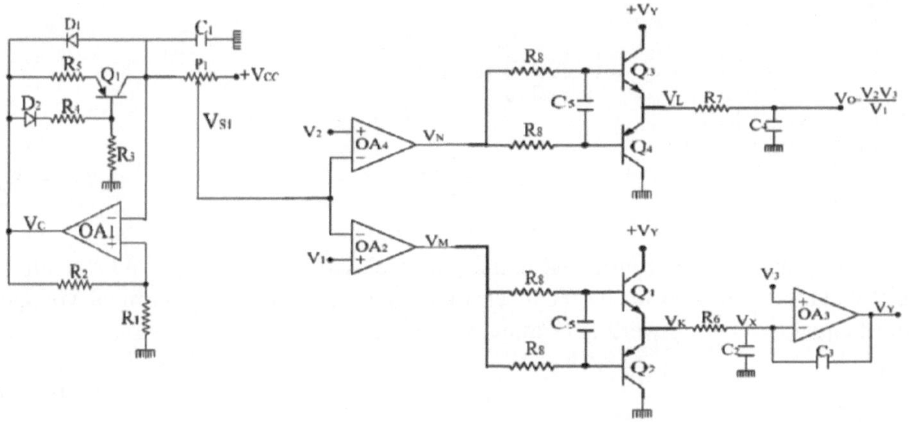

FIGURE 14.5 Sawtooth wave–based time division multiplier-cum-divider – type III.

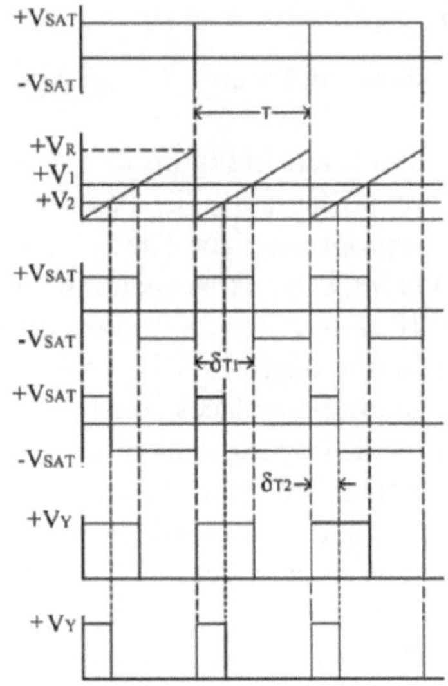

FIGURE 14.6 Associated waveforms of Figure 14.5.

compares the sawtooth wave with first input voltage V_1 and produces a rectangular waveform V_M. The ON time δ_{T1} of V_M is given as

$$\delta_{T1} = \frac{V_1}{V_R} T \tag{14.17}$$

The rectangular pulse V_M controls the transistor multiplexer Q_1–Q_2. When V_M is HIGH, the voltage V_Y is connected to R_6C_2 lowpass filter (transistor Q_1 is ON and Q_2 is OFF). When V_M is LOW, zero volts is connected to R_6C_2 lowpass filter (transistor Q_1 is OFF and Q_2 is ON). Another rectangular pulse V_K with maximum value of V_Y is generated at the multiplexer Q_1–Q_2 output. The R_6C_2 lowpass filter gives an average value of this pulse train V_K and is given as

$$V_X = \frac{1}{T} \int_0^{\delta_{T1}} V_Y \, dt = \frac{V_Y}{T} \delta_{T1}$$

$$V_X = \frac{V_1 V_Y}{V_R} \tag{14.18}$$

The op-amp OA_3 is configured in a negative closed loop feedback and a positive DC voltage is ensured in the feedback loop. Hence, its inverting terminal voltage must equal to its non-inverting terminal voltage, i.e.,

$$V_X = V_3 \tag{14.19}$$

From Eqs. (14.18) and (14.19)

$$V_Y = \frac{V_3 V_R}{V_1} \tag{14.20}$$

Comparator OA_4 compares the sawtooth wave V_{S1} with the second input voltage V_2 and produces a rectangular waveform V_N. The ON time δ_{T2} of V_N is given as

$$\delta_{T2} = \frac{V_2}{V_R} T \tag{14.21}$$

The rectangular pulse V_N controls the transistor multiplexer Q_3–Q_4. When V_N is HIGH, the voltage V_Y is connected to R_7C_4 lowpass filter (transistor Q_3 is ON and Q_4 is OFF). When V_N is LOW, zero volts is connected to R_7C_4 lowpass filter (transistor Q_3 is OFF and Q_4 is ON). Another rectangular pulse V_L with maximum value of V_Y is generated at the multiplexer Q_3–Q_4 output. The R_7C_4 lowpass filter gives an average value of this pulse train V_L and is given as

$$V_O = \frac{1}{T} \int_0^{\delta_{T2}} V_Y \, dt = \frac{V_Y}{T} \delta_{T2} \tag{14.22}$$

$$V_O = \frac{V_2 V_Y}{V_R} \tag{14.23}$$

Equation (14.20) in (14.23) gives

$$V_O = \frac{V_2 V_3}{V_1} \tag{14.24}$$

The output voltage depends on sharpness and linearity of the sawtooth waveform. The offset that exists on the sawtooth wave will cause an error at the output; hence, it is to be nulled by suitable circuitry (potentiometer P_1). The output voltage is proportional to attenuation constant α of the lowpass filter. The attenuation constant α will be in the range of 0.80 to 0.90. The polarities of V_1, V_2, and V_3 are only positive.

DESIGN EXERCISES

1. In the MCD circuit shown in Figure 14.5, replace the sawtooth wave generator with other saw tooth generators shown in Section 4.4. In each, (i) draw circuit diagrams, (ii) draw waveforms at appropriate places, (iii) explain their working operation, and (iv) deduce expression for their output voltages.
2. Replace transistor multiplexers in Figure 14.5 with FET multiplexers and MOSFET multiplexers. In each, (i) draw circuit diagrams, (ii) draw waveforms at appropriate places, (iii) explain their working operation, and (iv) deduce expression for their output voltages.

PRACTICAL EXERCISES

1. Verify the MCD circuit shown in Figure 14.5 by experimenting with the AFC trainer kit.
2. Replace the transistor multiplexers in Figure 14.5 with CD4053 analog multiplexer. (i) Draw the circuit diagrams, (ii) explain their working operation, (iii) draw waveforms at appropriate places, and (iv) deduce expression for the output voltage. Verify this MCD by experimenting with the AFC trainer kit.
3. As discussed in Chapter 18 and Section 18.9, convert the MCDs of Figure 14.5 into SRM. (i) Draw the circuit diagrams, (ii) explain their working operation, (iii) draw waveforms at appropriate places, and (iv) deduce expression for the output voltage. Verify this SRM by experimenting with the AFC trainer kit.
4. As discussed in Chapter 19 and Section 19.6, convert the MCDs of Figure 14.5 into VMC. (i) Draw the circuit diagrams, (ii) explain their working operation, (iii) draw waveforms at appropriate places, and (iv) deduce expression for the output voltage. Verify this VMC by experimenting with the AFC trainer kit.
5. Convert the MCD obtained in Problem 2 into SRM as discussed in Section 18.9. (i) Draw the circuit diagrams, (ii) explain their working operation, (iii) draw waveforms at appropriate places, and (iv) deduce expression for the output voltage. Verify this SRM by experimenting with the AFC trainer kit.
6. Convert the MCD obtained in Problem 2 into VMC as discussed in Section 19.6. (i) Draw the circuit diagrams, (ii) explain their working operation, (iii) draw waveforms at appropriate places, and (iv) deduce expression for the output voltage. Verify this VMC by experimenting with the AFC trainer kit.

14.4 TRIANGULAR WAVE–BASED TIME DIVISION MULTIPLIER-CUM-DIVIDER – TYPE I

The circuit diagram of triangular wave–based TDMCD is shown in Figure 14.7, and its associated waveforms are shown in Figure 14.8. A triangular wave V_{T1} of $\pm V_T$

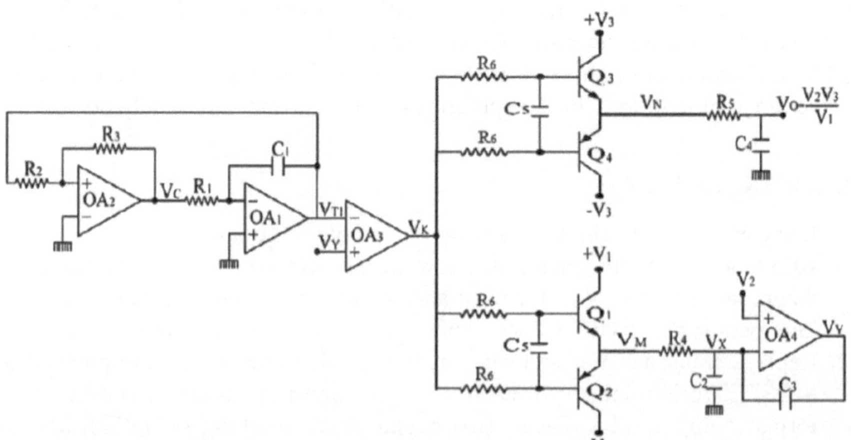

FIGURE 14.7 Triangular wave–based time division multiplier-cum-divider – type I.

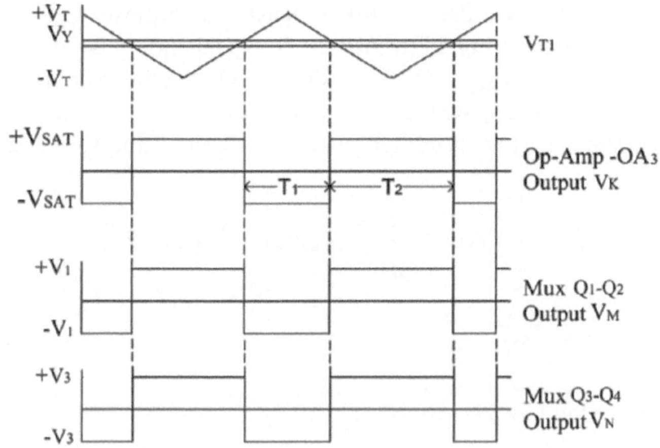

FIGURE 14.8 Associated waveforms of Figure 14.7.

peak-to-peak values and time period 'T' is generated by the op-amps OA_1 and OA_2. The comparator OA_3 compares the triangular wave V_{T1} with the voltage V_Y and produces asymmetrical rectangular wave V_K. From Figure 14.8, it is observed that

$$T_1 = \frac{V_T - V_Y}{2V_T} T \qquad T_2 = \frac{V_T + V_Y}{2V_T} T \qquad T = T_1 + T_2 \tag{14.25}$$

The rectangular wave V_K controls transistor multiplexer Q_1–Q_2 which connects $(+V_1)$ to its output during T_2 (transistor Q_1 is ON and Q_2 is OFF) and $(-V_1)$ to its output during T_1 (transistor Q_1 is OFF and Q_2 is ON). Another asymmetrical rectangular waveform V_M is generated at the transistor multiplexer Q_1–Q_2 output with $\pm V_1$ peak-to-peak values. The R_4C_2 lowpass filter gives the average value of V_M and is given as

$$V_X = \frac{1}{T}\left[\int_{O}^{T_2} V_1\,dt + \int_{T_2}^{T_1+T_2} (-V_1)\,dt\right] = \frac{V_1}{T}\left[T_2 - T_1\right]$$
$$V_X = \frac{V_1 V_Y}{V_T} \tag{14.26}$$

The op-amp OA_4 is configured in a negative closed loop feedback and a positive DC voltage is ensured in the feedback loop. Hence, its inverting terminal voltage must equal to its non-inverting terminal voltage, i.e.,

$$V_X = V_2 \tag{14.27}$$

From Eqs. (14.26) and (14.27)

$$V_Y = \frac{V_2 V_T}{V_1} \tag{14.28}$$

The rectangular wave V_K also controls transistor multiplexer Q_3–Q_4 which connects $(+V_3)$ to its output during T_2 (transistor Q_3 is ON and Q_4 is OFF) and $(-V_3)$ to output during T_1 (transistor Q_3 is OFF and Q_4 is ON). Another asymmetrical rectangular wave V_N is generated at the transistor multiplexer Q_3–Q_4 output with $\pm V_3$ peak-to-peak values. The R_5C_4 lowpass filter gives the average value V_O and is given as

$$V_O = \frac{1}{T}\left[\int_{O}^{T_2} V_3\,dt + \int_{T_2}^{T_1+T_2} (-V_3)\,dt\right] = \frac{V_3}{T}\left[T_2 - T_1\right] \tag{14.29}$$

$$V_O = \frac{V_3 V_Y}{V_T}$$

Equation (14.28) in (14.29) gives

$$V_O = \frac{V_2 V_3}{V_1} \tag{14.30}$$

DESIGN EXERCISES

1. In the MCD circuit shown in Figure 14.7, the triangular wave generator shown in Figure 4.9(b) is used. Replace this triangular wave generator with other triangular wave generator shown in Figure 4.9(a). (i) Draw circuit diagrams, (ii) draw waveform at appropriate places, (iii) explain their working operation, and (iv) deduce expression for their output voltages.
2. Replace transistor multiplexers Q_1–Q_2 and Q_3–Q_4 in Figure 14.7 with FET multiplexers and MOSFET multiplexers. In each, (i) draw circuit diagrams, (ii) draw waveforms at appropriate places, (iii) explain their working operation, and (iv) deduce expression for their output voltages.

PRACTICAL EXERCISES

1. Verify the MCD circuit shown in Figure 14.7 by experimenting with the AFC trainer kit.
2. Replace the transistor multiplexers in Figure 14.7 with CD4053 analog multiplexer IC. (i) Draw the circuit diagrams, (ii) explain their working operation, (iii) draw waveforms at appropriate places, and (iv) deduce expression for the output voltage. Verify this MCD by experimenting with the AFC trainer kit.
3. As discussed in Chapter 18 and Section 18.9, convert the MCDs of Figure 14.7 into SRM. (i) Draw the circuit diagrams, (ii) explain their working operation, (iii) draw waveforms at appropriate places, and (iv) deduce expression for the output voltage. Verify this SRM by experimenting with the AFC trainer kit.
4. As discussed in Chapter 19 and Section 19.6, convert the MCDs of Figure 14.7 into VMC. (i) Draw the circuit diagrams, (ii) explain their working operation, (iii) draw waveforms at appropriate places, and (iv) deduce expression for the output voltage. Verify this VMC by experimenting with the AFC trainer kit.

5. Convert the MCD obtained in Problem 2 into SRM as discussed in Section 18.9. (i) Draw the circuit diagrams, (ii) explain their working operation, (iii) draw waveforms at appropriate places, and (iv) deduce expression for the output voltage. Verify this SRM by experimenting with the AFC trainer kit.
6. Convert the MCD obtained in Problem 2 into VMC as discussed in Section 19.6. (i) Draw the circuit diagrams, (ii) explain their working operation, (iii) draw waveforms at appropriate places, and (iv) deduce expression for the output voltage. Verify this VMC by experimenting with the AFC trainer kit.

14.5 TRIANGULAR WAVE–BASED TIME DIVISION MULTIPLIER-CUM-DIVIDER – TYPE II

The circuit diagram of triangular wave–based TDMCD is shown in Figure 14.9, and its associated waveforms are shown in Figure 14.10. A triangular wave V_{T1} of $\pm V_T$ peak-to-peak values and time period 'T' is generated by around the op-amp OA_1.

The first input voltage V_1 is compared with the triangular wave V_{T1} by the comparator on OA_2. An asymmetrical rectangular waveform V_N is generated at the comparator OA_2 output. From the waveforms shown in Figure 14.10, it is observed that

$$T_1 = \frac{V_T - V_1}{2V_T} T \qquad T_2 = \frac{V_T + V_1}{2V_T} T \qquad T = T_1 + T_2 \tag{14.31}$$

This rectangular wave V_N is given as control input to the transistor multiplexer Q_1–Q_2. The transistor multiplexer Q_1–Q_2 connects the voltage ($+V_Y$) during T_2 (transistor Q_1 is ON and Q_2 is OFF), and ($-V_Y$) during T_1 (transistor Q_1 is OFF and Q_2 is ON). Another rectangular asymmetrical wave V_K with peak-to-peak values of $\pm V_Y$ is

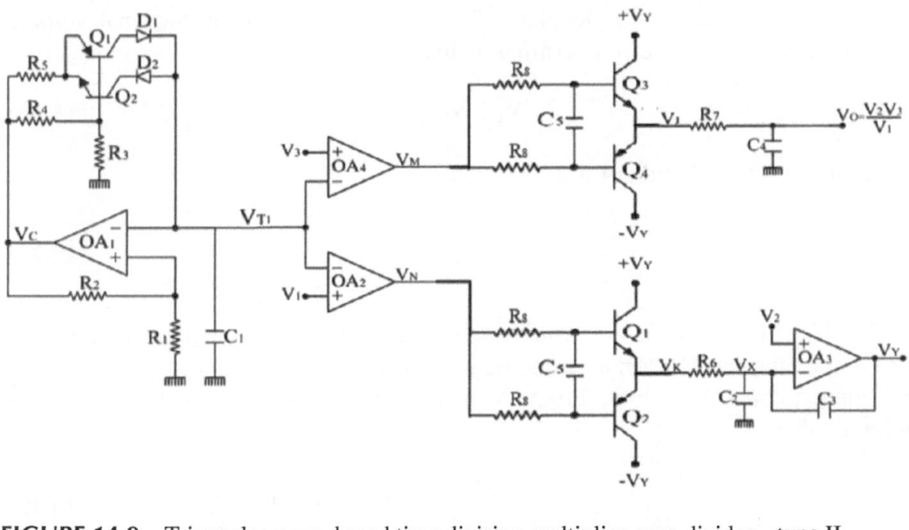

FIGURE 14.9 Triangular wave–based time division multiplier-cum-divider – type II.

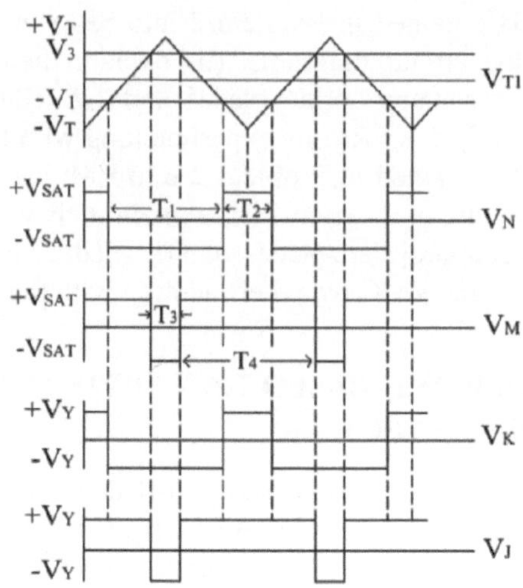

FIGURE 14.10 Associated waveforms of Figure 14.9.

generated at the transistor multiplexer Q_1–Q_2 output. The R_6C_2 lowpass filter gives average value of the pulse train V_N which is given as

$$V_X = \frac{1}{T}\left[\int_{O}^{T_2} V_Y\,dt + \int_{T_2}^{T_1+T_2} (-V_Y)\,dt\right] = \frac{V_Y}{T}(T_2 - T_1) \tag{14.32}$$

$$V_X = \frac{V_1 V_Y}{V_T}$$

The op-amp OA_3 is configured in a negative closed loop feedback and a positive DCvoltage is ensured in the feedback loop. Hence, its inverting terminal voltage must equal to its non-inverting terminal voltage, i.e.,

$$V_2 = V_X \tag{14.33}$$

From Eqs. (14.32) and (14.33)

$$V_Y = \frac{V_2 V_T}{V_1} \tag{14.34}$$

The third input voltage V_3 is compared with the generated triangular wave V_{T1} by the comparator on OA_4. An asymmetrical rectangular waveform V_M is generated at the comparator OA_4 output. From the waveforms shown in Figure 14.10, it is observed that

$$T_3 = \frac{V_T - V_3}{2V_T}T \qquad T_4 = \frac{V_T + V_3}{2V_T}T \qquad T = T_3 + T_4 \tag{14.35}$$

This rectangular wave V_M is given as control input to the transistor multiplexer Q_3–Q_4. The transistor multiplexer Q_3–Q_4 connects the voltage ($+V_Y$) during T_4 (transistor Q_3 is ON and Q_4 is OFF), and ($-V_Y$) during T_3 (transistor Q_3 is OFF and Q_4 is ON). Another rectangular asymmetrical wave V_J with peak-to-peak values of $\pm V_Y$ is generated at the transistor multiplexer Q_3–Q_4 output. The R_7C_4 lowpass filter gives an average value of the pulse train V_J and is given as

$$V_O = \frac{1}{T}\left[\int_{O}^{T_4} V_Y \, dt + \int_{T_4}^{T_3+T_4} (-V_Y) dt\right] = \frac{V_Y}{T}(T_4 - T_3) \tag{14.36}$$

$$V_O = \frac{V_Y V_3}{V_T} \tag{14.37}$$

Equation (14.34) in (14.37) gives

$$V_O = \frac{V_2 V_3}{V_1} \tag{14.38}$$

DESIGN EXERCISES

1. In the MCD circuit shown in Figure 14.9, the triangular wave generator shown in Figure 4.9(a) is used. Replace it with Figure 4.9(b). (i) Draw circuit diagrams, (ii) draw waveforms at appropriate places, (iii) explain their working operation, and (iv) deduce expression for their output voltages.
2. Replace transistor multiplexer Q_1–Q_2 and Q_3–Q_4 in Figure 14.9 with FET multiplexers and MOSFET multiplexers. In each, (i) draw circuit diagrams, (ii) draw waveforms at appropriate places, (iii) explain their working operation, and (iv) deduce expression for their output voltages.

PRACTICAL EXERCISES

1. Verify the MCD circuit shown in Figure 14.9 by experimenting with the AFC trainer kit.
2. Replace the transistor multiplexers in Figure 14.9 with CD4053 analog multiplexer IC. (i) Draw the circuit diagrams, (ii) explain their working operation, (iii) draw waveforms at appropriate places, and (iv) deduce expression for the output voltage. Verify this MCD by experimenting with the AFC trainer kit.
3. As discussed in Chapter 18 and Section 18.9, convert the MCDs of Figure 14.9 into SRM. (i) Draw the circuit diagrams, (ii) explain their working operation, (iii) draw waveforms at appropriate places, and (iv) deduce expression for the output voltage. Verify this SRM by experimenting with the AFC trainer kit.
4. As discussed in Chapter 19 and Section 19.6, convert the MCDs of Figure 14.9 into VMC. (i) Draw the circuit diagrams, (ii) explain their working operation, (iii) draw waveforms at appropriate places, and (iv) deduce expression for the output voltage. Verify this VMC by experimenting with the AFC trainer kit.

5. Convert the MCD obtained in Problem 2 into SRM as discussed in Section 18.9. (i) Draw the circuit diagrams, (ii) explain their working operation, (iii) draw waveforms at appropriate places, and (iv) deduce expression for the output voltage. Verify this SRM by experimenting with the AFC trainer kit.
6. Convert the MCD obtained in Problem 2 into VMC as discussed in Section 19.6. (i) Draw the circuit diagrams, (ii) explain their working operation, (iii) draw waveforms at appropriate places, and (iv) deduce expression for the output voltage. Verify this VMC by experimenting with the AFC trainer kit.

14.6 TRIANGULAR WAVE–BASED MULTIPLIER-CUM-DIVIDER – TYPE III

The circuit diagrams of triangular wave–based TDMCD is shown in Figure 14.11, and their associated waveforms are shown in Figure 14.12. A triangular wave of $\pm V_T$ peak-to-peak value and time period 'T' is generated around op-amp OA_1.

The second input voltage V_2 is compared with the triangular wave V_{T1} by the comparator on OA_2. An asymmetrical rectangular waveform V_M is generated at the comparator OA_2 output. From the waveforms shown in Figure 14.12, it is observed that

$$T_3 = \frac{V_T - V_2}{2V_T} T \qquad T_4 = \frac{V_T + V_2}{2V_T} T \qquad T = T_3 + T_4 \tag{14.39}$$

This rectangular wave V_M is given as control input to the transistor multiplexer Q_1–Q_2. The transistor multiplexer Q_1–Q_2 connects the third input voltage ($+V_3$) during T_4 (transistor Q_1 is ON and Q_2 is OFF), and ($-V_3$) during T_3 (transistor Q_1 is OFF and Q_2 is ON). Another rectangular asymmetrical square wave V_J with peak-to-peak values of $\pm V_3$ is generated at the transistor multiplexer Q_1–Q_2 output. The R_6C_2 low-pass filter gives average value of the pulse train V_J and is given as

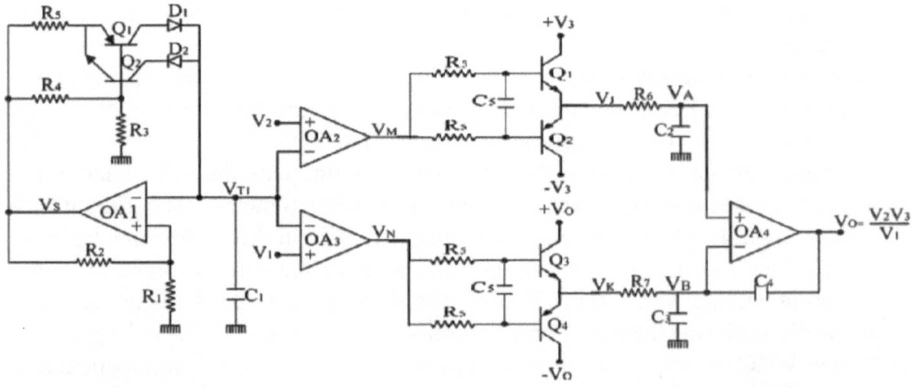

FIGURE 14.11 Triangular wave–based time division multiplier-cum-divider – type III.

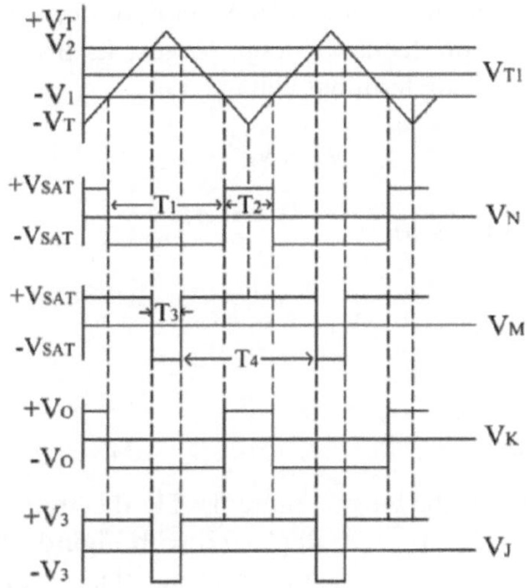

FIGURE 14.12 Associated waveforms of Figure 14.11.

$$V_A = \frac{1}{T}\left[\int_{O}^{T_4} V_3\,dt + \int_{T_4}^{T_3+T_4} (-V_3)\,dt\right] = \frac{V_3}{T}(T_4 - T_3)$$
$$V_A = \frac{V_2 V_3}{V_T} \tag{14.40}$$

The first input voltage V_1 is compared with the generated triangular wave V_{T1} by the comparator on OA_3. An asymmetrical rectangular waveform V_N is generated at the comparator OA_3 output. From the waveforms shown in Figure 14.12, it is observed that

$$T_1 = \frac{V_T - V_1}{2V_T}T \qquad T_2 = \frac{V_T + V_1}{2V_T}T \qquad T = T_1 + T_2 \tag{14.41}$$

This rectangular wave V_N is given as control input to the transistor multiplexer Q_3–Q_4. The transistor multiplexer Q_3–Q_4 connects the output voltage ($+V_O$) during T_2 (transistor Q_3 is ON and Q_4 is OFF), and ($-V_O$) during T_1 (transistor Q_3 is OFF and Q_4 is ON). Another rectangular asymmetrical wave V_K with peak-to-peak values of $\pm V_O$ is generated at the transistor multiplexer Q_3–Q_4 output. The R_7C_3 lowpass filter gives average value of the pulse train V_N and is given as

$$V_B = \frac{1}{T}\left[\int_{O}^{T_2} V_O\,dt + \int_{T_2}^{T_1+T_2} (-V_O)\,dt\right] = \frac{V_O}{T}(T_2 - T_1)$$
$$V_B = \frac{V_1 V_O}{V_T} \tag{14.42}$$

The op-amp OA_4 is configured in a negative closed loop feedback and a positive DC voltage is ensured in the feedback loop. Hence, its inverting terminal voltage must equal to its non-inverting terminal voltage, i.e.,

$$V_A = V_B \tag{14.43}$$

From Eqs. (14.42) and (14.40)

$$V_O = \frac{V_2 V_3}{V_1} \tag{14.44}$$

DESIGN EXERCISES

1. In the MCD circuit shown in Figure 14.11, the triangular wave generator shown in Figure 4.9(a) is used. Replace it with Figure 4.9(b). (i) Draw circuit diagrams, (ii) draw waveforms at appropriate places, (iii) explain their working operation, and (iv) deduce expression for their output voltages.
2. Replace transistor multiplexers Q_1–Q_2 and Q_3–Q_4 in Figure 14.11 with FET multiplexers and MOSFET multiplexers. In each, (i) draw circuit diagrams, (ii) draw waveforms at appropriate places, (iii) explain their working operation, and (iv) deduce expression for their output voltages.

PRACTICAL EXERCISES

1. Verify the MCD circuit shown in Figure 14.11 by experimenting with the AFC trainer kit.
2. Replace the transistor multiplexers in Figure 14.11 with CD 4053 analog multiplexer IC, (i) draw the circuit diagrams (ii) explain their working operation (iii) draw waveforms at appropriate places and (iv) deduce expression for the output voltage. Verify this MCD by experimenting with the AFC trainer kit.
3. As discussed in Chapter 18 and Section 18.9, convert the MCDs of Figure 14.11 into SRM. (i) Draw the circuit diagrams, (ii) explain their working operation, (iii) draw waveforms at appropriate places, and (iv) deduce expression for the output voltage. Verify this SRM by experimenting with the AFC trainer kit.
4. As discussed in Chapter 19 and Section 19.6, convert the MCDs of Figure 14.11 into VMC. (i) Draw the circuit diagrams, (ii) explain their working operation, (iii) draw waveforms at appropriate places, and (iv) deduce expression for the output voltage. Verify this VMC by experimenting with the AFC trainer kit.
5. Convert the MCD obtained in Problem 2 into SRM as discussed in Section 18.9. (i) Draw the circuit diagrams, (ii) explain their working operation, (iii) draw waveforms at appropriate places, and (iv) deduce expression for the output voltage. Verify this SRM by experimenting with the AFC trainer kit.
6. Convert the MCD obtained in Problem 2 into VMC as discussed in Section 19.6. (i) Draw the circuit diagrams, (ii) explain their working operation, (iii) draw waveforms at appropriate places, and (iv) deduce expression for the output voltage. Verify this VMC by experimenting with the AFC trainer kit.

14.7 TIME DIVISION MULTIPLIER-CUM-DIVIDER WITH NO REFERENCE – TYPE I

The MCD using the time division principle without any reference clock is shown in Figure 14.13, and its associated waveforms are shown in Figure 14.14. Let initially the comparator OA_2 output be LOW. The transistor multiplexer Q_1–Q_2 connects ($-V_1$) to the differential integrator composed by resistor R_1, capacitor C_1, and op-amp OA_1 (transistor Q_1 is OFF and Q_2 is ON). The output of differential integrator will be

$$V_{T1} = \frac{1}{R_1C_1}\int (V_2 + V_1)dt - V_T$$

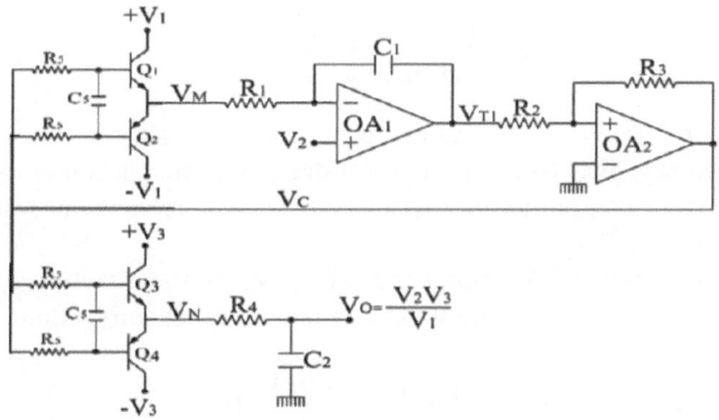

FIGURE 14.13 Time division multiplier-cum-divider with no reference – type I.

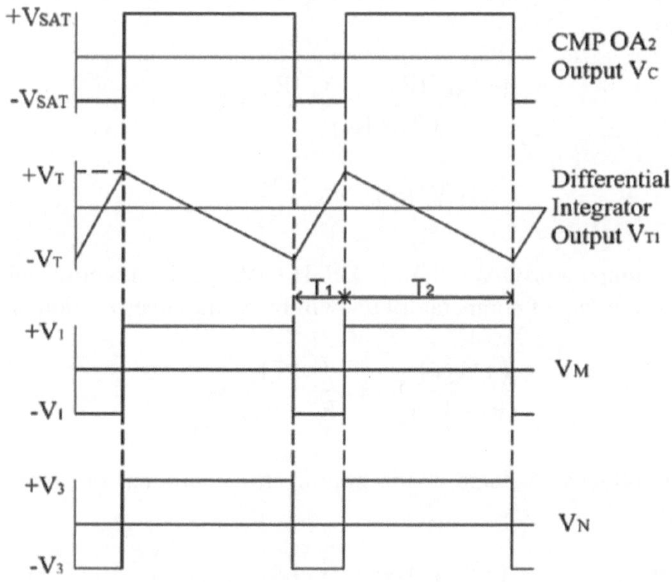

FIGURE 14.14 Associated waveforms of Figure 14.13.

$$V_{T1} = \frac{(V_1 + V_2)}{R_1C_1} t - V_T \tag{14.45}$$

The output of differential integrator is rising toward positive saturation, and when it reaches the voltage level of '$+V_T$', the comparator OA_2 output becomes HIGH. The transistor multiplexer Q_1–Q_2 connects ($+V_1$) to the differential integrator composed by resistor R_1, capacitor C_1, and op-amp OA_1 (transistor Q_1 is ON and Q_2 is OFF). Now, the output of differential integrator will be

$$V_{T1} = \frac{1}{R_1C_1} \int (V_2 - V_1) dt + V_T$$

$$V_{T1} = -\frac{(V_1 - V_2)}{R_1C_1} t + V_T \tag{14.46}$$

The output of differential integrator reverses toward negative saturation, and when it reaches the voltage level ($-V_T$), the comparator OA_2 output becomes LOW, and the cycle therefore repeats in order to give an asymmetrical rectangular wave V_C at the output of comparator OA_2.

When the comparator OA_2 output is LOW ($-V_{SAT}$), the effective voltage at the non-inverting terminal of comparator OA_2 will be by the superposition principle

$$\frac{(-V_{SAT})}{(R_2 + R_3)} R_2 + \frac{(+V_T)}{(R_2 + R_3)} R_3$$

When this effective voltage at the non-inverting terminal of comparator OA_2 becomes zero

$$\frac{(-V_{SAT})R_2 + (+V_T)R_3}{(R_2 + R_3)} = 0$$

$$(+V_T) = (+V_{SAT}) \frac{R_2}{R_3}$$

When the comparator OA_2 output is HIGH ($+V_{SAT}$), the effective voltage at the non-inverting terminal of comparator OA_2 will be by the superposition principle

$$\frac{(+V_{SAT})}{(R_2 + R_3)} R_2 + \frac{(-V_T)}{(R_2 + R_3)} R_3$$

When this effective voltage at the non-inverting terminal of comparator OA_2 becomes zero

$$\frac{(+V_{SAT})R_2 + (-V_T)R_3}{(R_2 + R_3)} = 0$$

$$(-V_T) = (-V_{SAT})\frac{R_2}{R_3}$$

$$\pm V_T = \pm V_{SAT}\frac{R_2}{R_3};\ 0.76(\pm V_{CC})\frac{R_2}{R_3} \tag{14.47}$$

From the waveforms shown in Figure 14.14, it is observed that

$$T_1 = \frac{V_1 - V_2}{2V_1}T,\quad T_2 = \frac{V_1 + V_2}{2V_1}T,\quad T = T_1 + T_2 \tag{14.48}$$

The asymmetrical rectangular wave V_C controls another transistor multiplexer Q_3–Q_4. The transistor multiplexer Q_3–Q_4 selects ($+V_3$) during ON time T_2 (transistor Q_3 is ON and Q_4 is OFF), and ($-V_3$) during OFF time T_1 of the rectangular wave V_C (transistor Q_3 is OFF and Q_4 is ON). Another rectangular wave V_N with $\pm V_3$ peak-to-peak values is generated at the transistor multiplexer Q_3–Q_4 output. The R_4C_2 low-pass filter gives an average value of this pulse train V_N and is given as

$$V_O = \frac{1}{T}\left[\int_O^{T_2} V_3\,dt + \int_{T_2}^{T_1+T_2} (-V_3)\,dt\right],\quad V_O = \frac{V_3(T_2 - T_1)}{T} \tag{14.49}$$

Equations (14.48) in (14.49) give

$$V_O = \frac{V_2 V_3}{V_1} \tag{14.50}$$

The output voltage is also proportional to the attenuation constant α of the lowpass filter. The attenuation constant α will be in the range of 0.80 to 0.90. The polarity of all input voltages may be any polarity.

DESIGN EXERCISES

1. Replace transistor multiplexers Q_1–Q_2 and Q_3–Q_4 in Figure 14.13 with FET multiplexers and MOSFET multiplexers. In each, (i) draw circuit diagrams, (ii) draw waveforms at appropriate places, (iii) explain their working operation, and (iv) deduce expression for their output voltages.
2. The inputs of the transistor multiplexers Q_3–Q_4 in Figure 14.13 are interchanged. (i) Draw waveforms at appropriate places, (ii) explain their working operation, and (iii) deduce expression for their output voltages.

PRACTICAL EXERCISES

1. Verify the MCD circuit shown in Figure 14.13 by experimenting with the AFC trainer kit.

2. Replace the transistor multiplexers in Figure 14.13 with CD4053 analog multiplexer IC. (i) Draw the circuit diagrams, (ii) explain their working operation, (iii) draw waveforms at appropriate places, and (iv) deduce expression for the output voltage. Verify this MCD by experimenting with the AFC trainer kit.
3. As discussed in Chapter 18 and Section 18.9, convert the MCDs of Figure 14.13 into SRM. (i) Draw the circuit diagrams, (ii) explain their working operation, (iii) draw waveforms at appropriate places, and (iv) deduce expression for the output voltage. Verify this SRM by experimenting with the AFC trainer kit.
4. As discussed in Chapter 19 and Section 19.6, convert the MCDs of Figure 14.13 into VMC. (i) Draw the circuit diagrams, (ii) explain their working operation, (iii) draw waveforms at appropriate places, and (iv) deduce expression for the output voltage. Verify this VMC by experimenting with the AFC trainer kit.
5. Convert the MCD obtained in Problem 2 into SRM as discussed in Section 18.9. (i) Draw the circuit diagrams, (ii) explain their working operation, (iii) draw waveforms at appropriate places, and (iv) deduce expression for the output voltage. Verify this SRM by experimenting with the AFC trainer kit.
6. Convert the MCD obtained in Problem 2 into VMC as discussed in Section 19.6. (i) Draw the circuit diagrams, (ii) explain their working operation, (iii) draw waveforms at appropriate places, and (iv) deduce expression for the output voltage. Verify this VMC by experimenting with the AFC trainer kit.

14.8 TIME DIVISION MULTIPLIER-CUM-DIVIDER WITH NO REFERENCE – TYPE II

The MCD using the time division principle without any reference clock is shown in Figure 14.15, and its associated waveforms are shown in Figure 14.16. Let initially the comparator OA_2 output be LOW. The transistor multiplexer Q_1–Q_2 connects

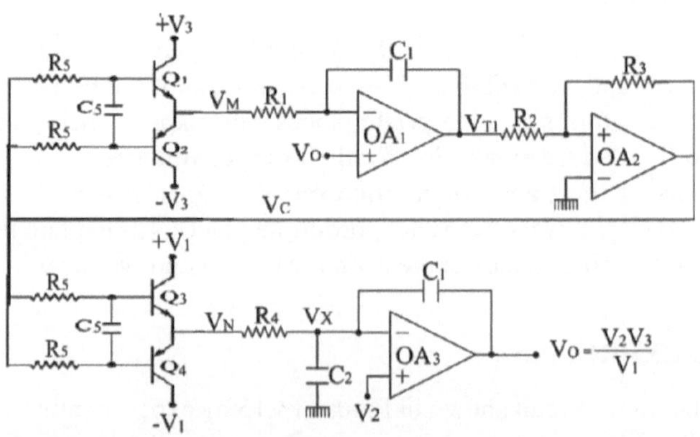

FIGURE 14.15 Time division multiplier-cum-divider with no reference – type II.

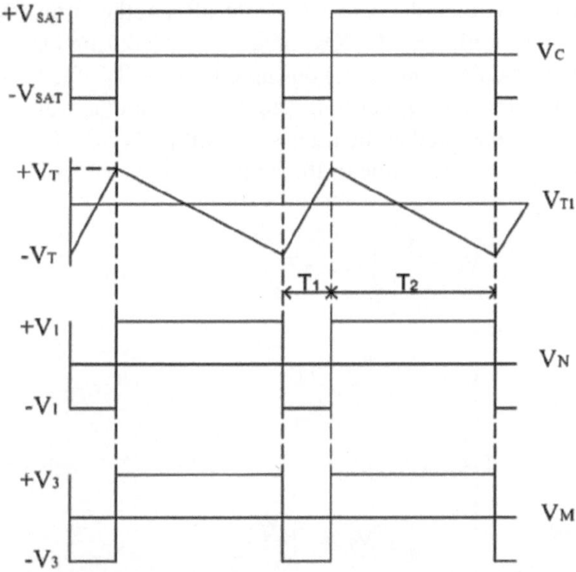

FIGURE 14.16 Associated waveforms of Figure 14.15.

$(-V_3)$ to one end of resistor R_1 (transistor Q_1 is OFF and Q_2 is ON). The output of differential integrator composed by resistor R_1, capacitor C_1, and op-amp OA_1 will be

$$V_{T1} = \frac{1}{R_1C_1}\int (V_O + V_3)dt - V_T, \quad V_{T1} = \frac{(V_O + V_3)}{R_1C_1}t - V_T \tag{14.51}$$

The output of differential integrator OA_1 rises toward positive saturation, and when it reaches the voltage level $(+V_T)$, the comparator OA_2 output becomes HIGH. The transistor multiplexer Q_1–Q_2 connects $(+V_3)$ to one end of resistor R_1 (transistor Q_1 is ON and Q_2 is OFF). Now, the output of differential integrator OA_1 will be

$$V_{T1} = \frac{1}{R_1C_1}\int (V_O - V_3)dt + V_T, \quad V_{T1} = -\frac{(V_3 - V_O)}{R_1C_1}t + V_T \tag{14.52}$$

The output of differential integrator OA_1 reverses toward negative saturation, and when it reaches the voltage level $(-V_T)$, the comparator OA_2 output becomes LOW, and the cycle therefore repeats in order to give an asymmetrical rectangular wave V_C at the output of comparator OA_2.

From the waveforms shown in Figure 14.16, it is observed that

$$T_1 = \frac{V_3 - V_O}{2V_3}T, \quad T_2 = \frac{V_3 + V_O}{2V_3}T, \quad T = T_1 + T_2 \tag{14.53}$$

The asymmetrical rectangular wave V_C controls another transistor multiplexer Q_3–Q_4. The transistor multiplexer Q_3–Q_4 connects ($+V_1$) during ON time T_2 (transistor Q_3 is ON and Q_4 is OFF), and ($-V_1$) during OFF time T_1 of the rectangular wave V_C (transistor Q_3 is OFF and Q_4 is ON). Another rectangular wave V_N with $\pm V_1$ as peak-to-peak value is generated at the transistor multiplexer Q_3–Q_4 output. The R_4C_2 lowpass filter gives an average value of this pulse train V_N and is given as

$$V_X = \frac{1}{T}\left[\int_{O}^{T_2} V_1\,dt + \int_{T_2}^{T_1+T_2} (-V_1)\,dt\right]$$

$$V_X = \frac{V_1(T_2 - T_1)}{T} \tag{14.54}$$

Equations (14.53) in (14.54) give

$$V_X = \frac{V_O V_1}{V_3} \tag{14.55}$$

The op-amp OA_3 is at negative closed loop configuration and a positive DC voltage is ensured in the feedback loop. Hence, its non-inverting terminal voltage is equal to its inverting terminal voltage, i.e., $V_2 = V_X$

$$V_O = \frac{V_2 V_3}{V_1} \tag{14.56}$$

The output voltage is also proportional to attenuation constant α of the lowpass filter. The attenuation constant α will be in the range of 0.80 to 0.90. The polarity of all input voltages may be any polarity.

DESIGN EXERCISES

1. Replace transistor multiplexers Q_1–Q_2 and Q_3–Q_4 in Figure 14.15 with FET multiplexers and MOSFET multiplexers. In each, (i) draw circuit diagrams, (ii) draw waveforms at appropriate places, (iii) explain their working operation, and (iv) deduce expression for their output voltages.

PRACTICAL EXERCISES

1. Verify the MCD circuit shown in Figure 14.15 by experimenting with the AFC trainer kit.
2. Replace the transistor multiplexers in Figure 14.15 with CD4053 analog multiplexer IC. (i) Draw the circuit diagrams, (ii) explain their working operation, (iii) draw waveforms at appropriate places, and (iv) deduce expression for the output voltage. Verify this MCD by experimenting with the AFC trainer kit.

3. As discussed in Chapter 18 and Section 18.9, convert the MCDs of Figure 14.15 into SRM. (i) Draw the circuit diagrams, (ii) explain their working operation, (iii) draw waveforms at appropriate places, and (iv) deduce expression for the output voltage. Verify this SRM by experimenting with the AFC trainer kit.
4. As discussed in Chapter 19 and Section 19.6, convert the MCDs of Figure 14.15 into VMC. (i) Draw the circuit diagrams, (ii) explain their working operation, (iii) draw waveforms at appropriate places, and (iv) deduce expression for the output voltage. Verify this square rooter by experimenting with the AFC trainer kit.
5. Convert the MCD obtained in Problem 2 into SRM as discussed in Section 18.9. (i) Draw the circuit diagrams, (ii) explain their working operation, (iii) draw waveforms at appropriate places, and (iv) deduce expression for the output voltage. Verify this SRM by experimenting with the AFC trainer kit.
6. Convert the MCD obtained in Problem 2 into VMC as discussed in Section 19.6. (i) Dwraw the circuit diagrams, (ii) explain their working operation, (iii) draw waveforms at appropriate places, and (iv) deduce expression for the output voltage. Verify this VMC by experimenting with the AFC trainer kit.

15 Design of Time Division Multiplier-Cum-Divider – Switching

Like multipliers, multipliers-cum-dividers (MCDs) also are classified into (i) time division MCDs, (ii) peak responding MCDs, and (iii) pulse position responding MCDs. The peak responding MCD is further classified into a (i) peak detecting MCD and (ii) peak sampling MCD.

A pulse train whose maximum value is proportional to one voltage (V_3) is generated. If the width of this pulse train is made proportional to one voltage (V_2) and inversely proportional to another voltage (V_1), then the average value of pulse train is proportional to $\frac{V_2V_3}{V_1}$. This is called "time division multiplier-cum-divider (TDMCD). TDMCDs using multiplexers are described in the previous Chapter 14 and using switches are described in this chapter.

15.1 SAWTOOTH WAVE–BASED DOUBLE SWITCHING–AVERAGING TIME DIVISION MULTIPLIER-CUM-DIVIDER

The circuit diagrams of double switching–averaging TDMCDs are shown in Figure 15.1, and their associated waveforms are shown in Figure 15.2. Figure 15.1(a) shows a series switching MCD, and Figure 15.1(b) shows a shunt switching MCD. As discussed in Section 4.4, a sawtooth wave V_{S1} of peak value V_R time period 'T' is generated by op-amps OA_1, OA_2, and transistor Q_1.

The comparator OA_3 compares the sawtooth wave with the voltage V_Y and produces a first rectangular waveform V_K. The ON time, Figure 15.1(a), or OFF time, Figure 15.1(b), δ_T of V_K is given as

$$\delta_T = \frac{V_Y}{V_R}T \tag{15.1}$$

The rectangular pulse V_K controls the transistor switches Q_2 and Q_3.

In Figure 15.1(a), when V_K is HIGH, the transistor Q_2 is ON, and third input voltage V_3 is connected to R_3C_2 lowpass filter, the transistor Q_3 is ON, and the first input voltage V_1 is connected to R_4C_3 lowpass filter. When V_K is LOW, the transistor Q_2 is OFF, and zero volts exists on R_3C_2 lowpass filter, the transistor Q_3 is OFF, and zero volts exists on R_4C_3 lowpass filter.

In Figure 15.1(b), when V_K is HIGH, the transistor Q_2 is ON, and zero volts exists on R_3C_2 lowpass filter, the transistor Q_3 is ON, and zero volts exists on R_4C_3 lowpass

DOI: 10.1201/9781003221449-18

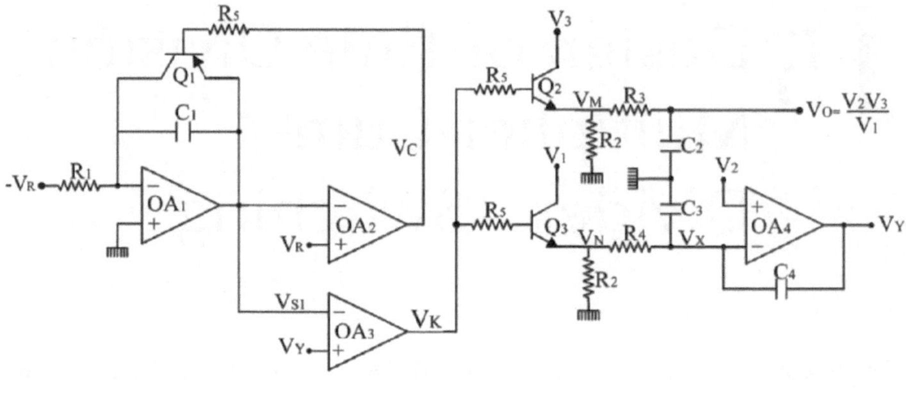

FIGURE 15.1(A) Double series switching–averaging time division multiplier-cum-divider.

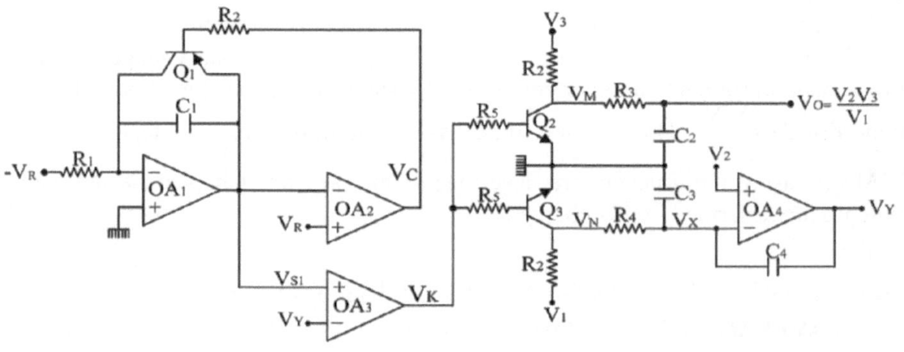

FIGURE 15.1(B) Double shunt switching–averaging time division multiplier-cum-divider.

filter. When V_K is LOW, the transistor Q_2 is OFF, and third input voltage V_3 is connected to R_3C_2 lowpass filter, the transistor Q_3 is OFF, and the first input voltage V_1 is connected to the R_4C_3 lowpass filter.

Second rectangular pulse V_N with maximum value of V_1 is generated at the transistor Q_3 output. The R_4C_3 lowpass filter gives average value of this pulse train V_N and is given as

$$V_X = \frac{1}{T}\int_0^{\delta_T} V_1\,dt = \frac{V_1}{T}\delta_T$$

$$V_X = \frac{V_1 V_Y}{V_R} \tag{15.2}$$

The op-amp OA_4 is configured in a negative closed loop feedback, and a positive DC voltage is ensured in the feedback loop. Hence, its inverting terminal voltage must equal to its non-inverting terminal voltage.

$$V_2 = V_X \tag{15.3}$$

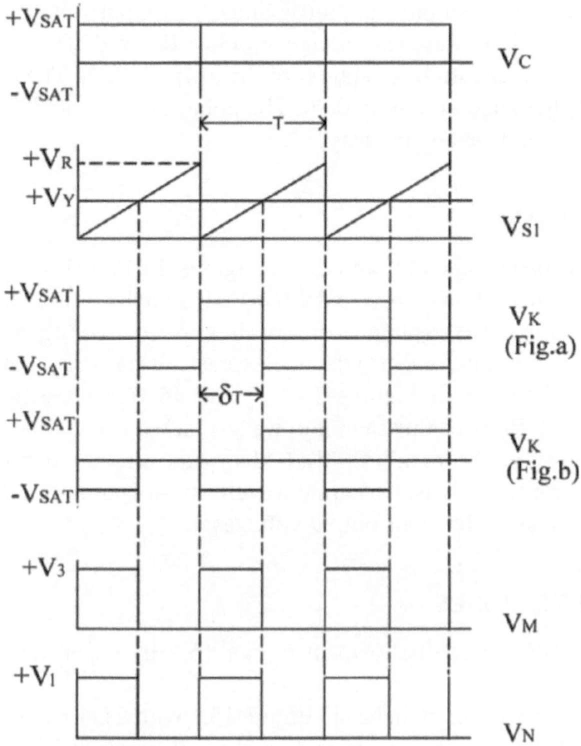

FIGURE 15.2 Associated waveforms of Figure 15.1.

From Eqs. (15.2) and (15.3),

$$V_Y = \frac{V_2 V_R}{V_1} \tag{15.4}$$

Third rectangular pulse V_M with maximum value V_3 is generated at the transistor Q_2 output. The R_3C_2 lowpass filter gives average value of this pulse train V_M and is given as

$$V_O = \frac{1}{T}\int_0^{\delta_T} V_3 \, dt = \frac{V_3}{T}\delta_T \tag{15.5}$$

Equations (15.1) and (15.4) in (15.5) give

$$V_O = \frac{V_2 V_3}{V_1} \tag{15.6}$$

The output voltage depends on the sharpness and linearity of the sawtooth waveform. The offset that exists on the sawtooth wave will cause an error at the output;

hence, it is to be nulled by suitable circuitry. Since the output voltage depends on V_R, a highly stable precision reference voltage V_R must be used. The output voltage is also proportional to attenuation constant α of the lowpass filter. The attenuation constant α will be in the range of 0.80 to 0.90. The polarities of V_1 and V_2 are only positive. However, V_3 can have any polarity.

DESIGN EXERCISES

1. The transistor switches Q_2 and Q_3 in Figures 15.1(a) and 15.1(b) are to be replaced with FET switches and MOSFET switches. In each, (i) draw the circuit diagrams, (ii) explain their working operations, (iii) draw waveforms at appropriate places, and (iv) deduce expression for their output voltages.
2. The circuits shown in Figures 15.1(a) and 15.1(b) use sawtooth wave of Figure 4.7(e). Replace this sawtooth wave generator with other sawtooth generators shown in Figure 4.7(a)–(d). (i) Draw the circuit diagrams, (ii) explain their working operations, (iii) draw waveforms at appropriate places, and (iv) deduce expression for their output voltages.

PRACTICAL EXERCISES

1. Verify the MCD circuit shown in Figure 15.1 by experimenting with AFC trainer kit.
2. Replace the transistor switches in Figure 15.1 with CD4066 non-inverter controlled switches and DG201 inverter controlled switches. In each, (i) draw the circuit diagrams, (ii) explain their working operation, (iii) draw waveforms at appropriate places, and (iv) deduce expression for the output voltage. Verify these MCDs by experimenting with the AFC trainer kit.
3. As discussed in Chapter 18 and Section 18.9, convert the MCDs of Figure 15.1 into SRM. (i) Draw the circuit diagrams, (ii) explain their working operation, (iii) draw waveforms at appropriate places, and (iv) deduce expression for the output voltage. Verify this SRM by experimenting with the AFC trainer kit.
4. As discussed in Chapter 19 and Section 19.6, convert the MCDs of Figure 15.1 into VMC. (i) Draw the circuit diagrams, (ii) explain their working operation, (iii) draw waveforms at appropriate places, and (iv) deduce expression for the output voltage. Verify this square rooter by experimenting with the AFC trainer kit.
5. Convert the MCD circuits obtained from Problem 2 into SRM. (i) Draw the circuit diagrams, (ii) explain their working operation, (iii) draw waveforms at appropriate places, and (iv) deduce expression for the output voltage. Verify this SRM by experimenting with the AFC trainer kit.
6. Convert the MCD circuits obtained from Problem 2 into VMC. (i) Draw the circuit diagrams, (ii) explain their working operation, (iii) draw waveforms at appropriate places, and (iv) deduce expression for the output voltage. Verify this VMC by experimenting with the AFC trainer kit.

15.2 SAWTOOTH WAVE–BASED TIME DIVISION MULTIPLY–DIVIDE MULTIPLIER-CUM-DIVIDER

The circuit diagrams of the sawtooth wave–based time division multiply–divide MCDs are shown in Figure 15.3 and their associated waveforms are shown in Figure 15.4. Figure 15.3(a) shows a series switching MCD, and Figure 15.3(b) shows a shunt switching MCD. As discussed in Section 4.4, Figure 4.7(d), a sawtooth wave V_{S1} of period 'T' with peak value V_R is generated around operational amplifier OA_1. The comparator OA_2 compares the sawtooth wave with an input voltage V_3 and

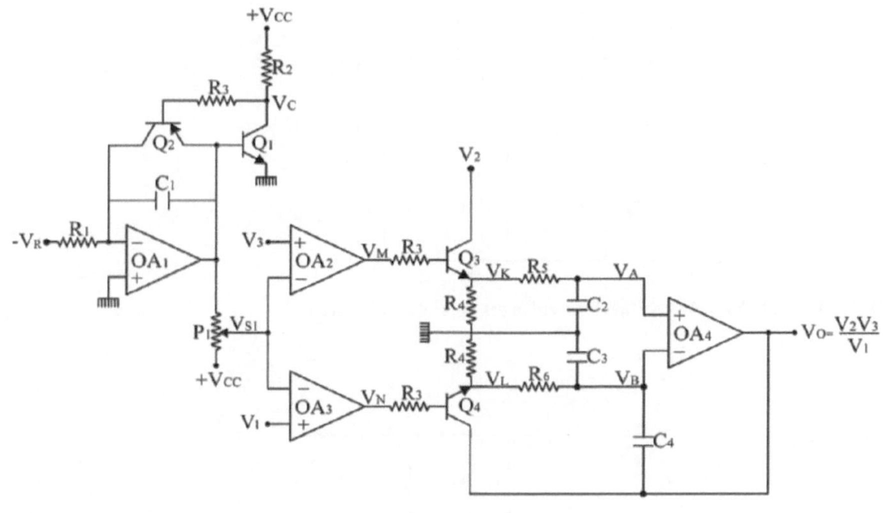

FIGURE 15.3(A) Series switching time division multiply–divide multiplier-cum-divider.

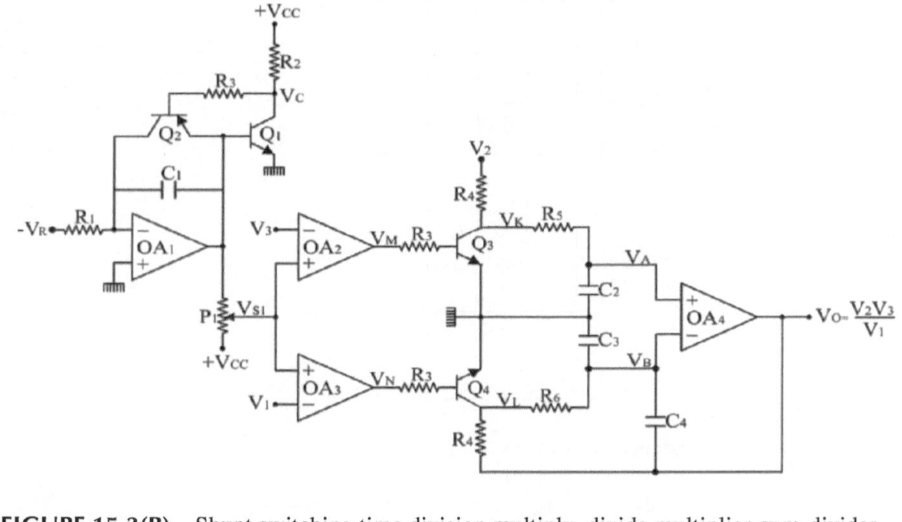

FIGURE 15.3(B) Shunt switching time division multiply–divide multiplier-cum-divider.

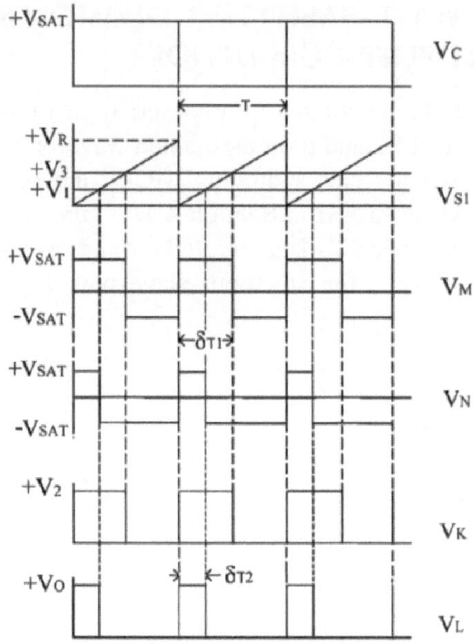

FIGURE 15.4(A) Associated waveforms of Figure 15.3(a).

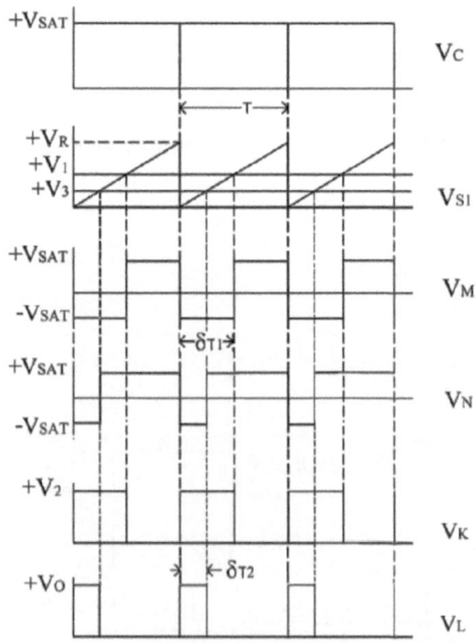

FIGURE 15.4(B) Associated waveforms of Figure 15.3(b).

produces a rectangular waveform V_M. The ON time δ_{T1} of V_M, Figure 15.3(a), or OFF time δ_{T1} of V_M, Figure 15.3(b), is given as

$$\delta_{T1} = \frac{V_3}{V_R} T \tag{15.7}$$

The rectangular pulse V_M controls the transistor Q_3.

In Figure 15.3(a), when V_M is HIGH, transistor Q_3 is ON, another input voltage V_2 is connected to R_5C_2 lowpass filter. When V_M is LOW, transistor Q_3 is OFF, zero volts is connected to R_5C_2 lowpass filter.

In Figure 15.3(b), when V_M is HIGH, transistor Q_3 is ON, zero volts is connected to R_5C_2 lowpass filter. When V_M is LOW, transistor Q_3 is OFF, another input voltage V_2 is connected to R_5C_2 lowpass filter.

Another rectangular pulse V_K with maximum value of V_2 is generated at the transistor Q_3 output. The R_5C_2 lowpass filter gives an average value of this pulse train V_K and is given as

$$V_A = \frac{1}{T}\int_0^{\delta_{T1}} V_2\, dt = \frac{V_2}{T}\delta_{T1}$$

$$V_A = \frac{V_2 V_3}{V_R} \tag{15.8}$$

Comparator OA_3 compares the sawtooth wave V_{S1} with the first input voltage V_1 and produces a rectangular waveform V_N. The ON time $\delta_{T2}$of V_N, Figure 15.9(a), or OFF time $\delta_{T2}$of V_N, Figure 15.9(b), is given as

$$\delta_{T2} = \frac{V_1}{V_R} T \tag{15.9}$$

The rectangular pulse V_N controls the second transistor Q_4.

In Figure 15.3(a), when V_N is HIGH, the output voltage V_O is connected to R_6C_3 lowpass filter (transistor Q_4 is ON). When V_N is LOW, zero volts is connected to R_5C_3 lowpass filter (transistor Q_4 is OFF).

In Figure 15.3(b), when V_N is HIGH, zero volts is connected to R_6C_3 lowpass filter (transistor Q_4 is ON). When V_N is LOW, output voltage V_O is connected to R_6C_3 lowpass filter (transistor Q_4 is OFF).

Another rectangular pulse V_L with maximum value of V_O is generated at the transistor Q_4 output. The R_6C_3 lowpass filter gives an average value of this pulse train V_L and is given as

$$V_B = \frac{1}{T}\int_0^{\delta_{T2}} V_O\, dt = \frac{V_O}{T}\delta_{T2}$$

$$V_B = \frac{V_1 V_O}{V_R} \tag{15.10}$$

The op-amp OA_4 is configured in a negative closed loop feedback and a positive DC voltage is ensured in the feedback loop. Hence, its inverting terminal voltage must equal to its non-inverting terminal voltage, i.e.,

$$V_A = V_B \tag{15.11}$$

From Eqs. (15.8) and (15.10)

$$V_O = \frac{V_2 V_3}{V_1} \tag{15.12}$$

The output voltage depends on the sharpness and linearity of the sawtooth waveform. The offset that exists on the sawtooth wave will cause error at the output; hence, it is to be nulled by suitable circuitry (potentiometer P_1). The output voltage is also proportional to attenuation constant α of the lowpass filter. The attenuation constant α will be in the range of 0.80 to 0.90. The polarities of V_1, V_2 and V_3 will be only positive.

DESIGN EXERCISES

1. The transistor switches Q_3 and Q_4 in Figures 15.3(a) and 15.3(b) are to be replaced with FET switches and MOSFET switches. In each, (i) draw the circuit diagrams, (ii) explain their working operations, (iii) draw waveforms at appropriate places, and (iv) deduce expression for their output voltages.
2. The circuits shown in Figures 15.3(a) and 15.3(b) use sawtooth wave of Figure 4.7(d). Replace this sawtooth wave generator with other sawtooth generator shown in Figures 4.7(a), (b), and (c). (i) Draw the circuit diagrams, (ii) explain their working operations, (iii) draw waveforms at appropriate places and (iv) deduce expression for their output voltages.

PRACTICAL EXERCISES

1. Verify the MCD circuit shown in Figure 15.3 by experimenting with the AFC trainer kit.
2. Replace the transistor switches in Figure 15.3 with CD4066 non-inverter controlled switches and DG201 inverter controlled switches. In each, (i) draw the circuit diagrams, (ii) explain their working operation, (iii) draw waveforms at appropriate places, and (iv) deduce expression for the output voltage. Verify these MCDs by experimenting with the AFC trainer kit.
3. As discussed in Chapter 18 and Section 18.9, convert the MCDs of Figure 15.3 into SRM. (i) Draw the circuit diagrams, (ii) explain their working operation, (iii) draw waveforms at appropriate places, and (iv) deduce expression for the output voltage. Verify this SRM by experimenting with the AFC trainer kit.

4. As discussed in Chapter 19 and Section 19.6, convert the MCDs of Figure 15.3 into VMC. (i) Draw the circuit diagrams, (ii) explain their working operation, (iii) draw waveforms at appropriate places, and (iv) deduce expression for the output voltage. Verify this square rooter by experimenting with the AFC trainer kit.
5. Convert the MCD circuits obtained from Problem 2 into SRM. (i) Draw the circuit diagrams, (ii) explain their working operation, (iii) draw waveforms at appropriate places, and (iv) deduce expression for the output voltage. Verify this SRM by experimenting with the AFC trainer kit.
6. Convert the MCD circuits obtained from Problem 2 into VMC. (i) Draw the circuit diagrams, (ii) explain their working operation, (iii) draw waveforms at appropriate places, and (iv) deduce expression for the output voltage. Verify this VMC by experimenting with the AFC trainer kit.

15.3 SAWTOOTH WAVE–BASED TIME DIVISION DIVIDE–MULTIPLY MULTIPLIER-CUM-DIVIDER

The circuit diagrams of sawtooth wave–based time division divide–multiply MCD are shown in Figure 15.5, and their associated waveforms are shown in Figure 15.6. Figure 15.5(a) shows a series switching MCD, and Figure 15.5(b) shows a shunt switching MCD. A sawtooth wave V_{S1} of V_R peak value and time period 'T' is generated by around the op-amp OA_1. The comparator OA_2 compares the sawtooth wave with an input voltage V_1 and produces a rectangular waveform V_M. The ON time $\delta_{T1}$of V_M, Figure 15.5(a), or OFF time δ_{T1} of V_M, Figure 15.5(b), is given as

$$\delta_{T1} = \frac{V_1}{V_R} T \tag{15.13}$$

The rectangular pulse V_M controls the transistor Q_3.

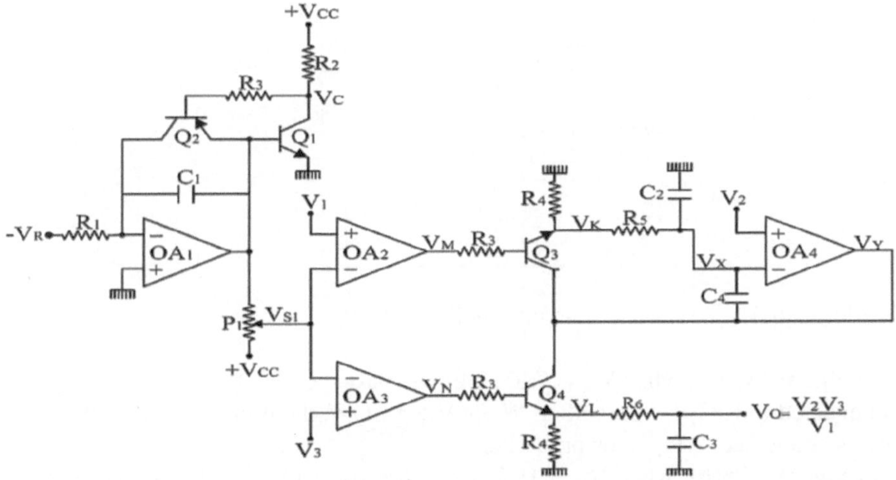

FIGURE 15.5(A) Series switching time division divide–multiply multiplier-cum-divider.

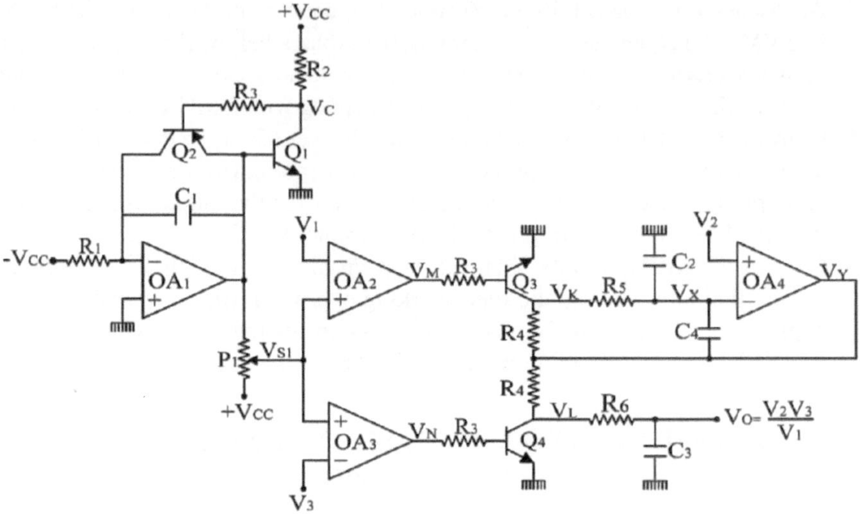

FIGURE 15.5(B) Shunt switching time division divide–multiply multiplier-cum-divider.

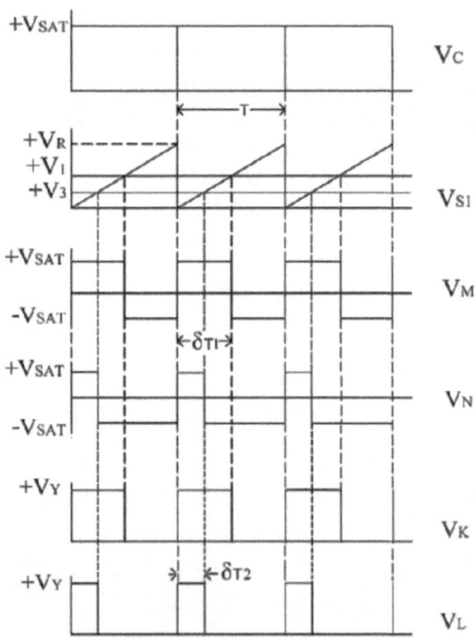

FIGURE 15.6(A) Associated waveforms of Figure 15.5(a).

In Figure 15.5(a), when V_M is HIGH, the transistor Q_3 is ON, and the voltage V_Y is connected to R_5C_2 lowpass filter. When V_M is LOW, the transistor Q_3 is OFF, zero volts is connected to R_5C_2 lowpass filter.

In Figure 15.5(b), when V_M is HIGH, the transistor Q_3 is ON, and zero volts is connected to R_5C_2 lowpass filter. When V_M is LOW, the transistor Q_3 is OFF, input voltage V_Y is connected to R_5C_2 lowpass filter.

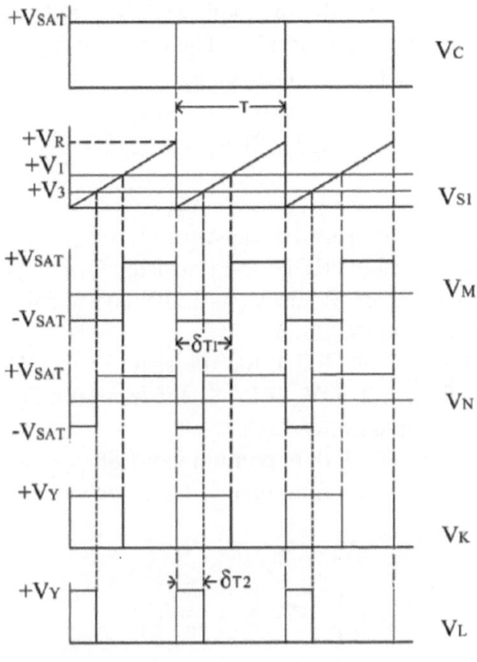

FIGURE 15.6(B) Associated waveforms of Figure 15.5(b).

Another rectangular pulse V_K with maximum value of V_Y is generated at the transistor Q_3 output. The R_5C_2 lowpass filter gives average value of this pulse train V_K and is given as

$$V_X = \frac{1}{T}\int_0^{\delta_{T1}} V_Y \, dt = \frac{V_Y}{T}\delta_{T1}$$

$$V_X = \frac{V_1 V_Y}{V_R} \quad (15.14)$$

The op-amp OA_4 is configured in a negative closed loop feedback and a positive DC voltage is ensured in the feedback loop. Hence, its inverting terminal voltage must equal to its non-inverting terminal voltage, i.e.,

$$V_X = V_2 \quad (15.15)$$

From Eqs. (15.14) and (15.15)

$$V_Y = \frac{V_2 V_R}{V_1} \quad (15.16)$$

The comparator OA_3 compares the sawtooth wave V_{S1} with another input voltage V_3 and produces a rectangular waveform V_N. The ON time $\delta_{T2}$of V_N, Figure 15.5(a), or OFF time $\delta_{T2}$of V_N, Figure 15.5(b), is given as

$$\delta_{T2} = \frac{V_3}{V_R} T \tag{15.17}$$

The rectangular pulse V_N controls the transistor Q_4.

In Figure 15.5(a), when V_N is HIGH, the transistor Q_4 is ON, the voltage V_Y is connected to R_6C_3 lowpass filter. When V_N is LOW, the transistor Q_4 is OFF, zero volts is connected to R_6C_3 lowpass filter.

In Figure 15.5(b), when V_N is HIGH, the transistor Q_4 is ON, zero volts is connected to R_6C_3 lowpass filter. When V_N is LOW, the transistor Q_4 is OFF, the voltage V_Y is connected to R_6C_3 lowpass filter.

Another rectangular pulse V_L with maximum value of V_Y is generated at the transistor Q_4 output. The R_6C_3 lowpass filter gives an average value of this pulse train V_L and is given as

$$V_O = \frac{1}{T}\int_0^{\delta_{T2}} V_Y \, dt = \frac{V_Y}{T}\delta_{T2} \tag{15.18}$$

$$V_O = \frac{V_3 V_Y}{V_R} \tag{15.19}$$

Equation (15.16) in (15.19) gives

$$V_O = \frac{V_2 V_3}{V_1} \tag{15.20}$$

The output voltage depends on the sharpness and linearity of the sawtooth waveform. The offset that exists on the sawtooth wave will cause an error at the output; hence, it is to be nulled by suitable circuitry (potentiometer P_1). The output voltage is also proportional to attenuation constant α of the lowpass filter. The attenuation constant α will be in the range of 0.80 to 0.90. The polarities of V_1, V_2 and V_3 are only positive.

DESIGN EXERCISES

1. The transistors Q_3 and Q_4 in Figures 15.5(a) and 15.5(b) are to be replaced with FET switches and MOSFET switches. In each, (i) draw the circuit diagrams, (ii) explain their working operations, (iii) draw waveforms at appropriate places, and (iv) deduce expression for their output voltages.
2 The circuits shown in Figures 15.5(a) and 15.5(b) use sawtooth wave of Figure 4.7(d). Replace this sawtooth wave generator with other sawtooth generators shown in Figures 4.7(a)–(c). In each, (i) draw the circuit diagrams, (ii) explain

their working operations, (iii) draw waveforms at appropriate places, and (iv) deduce expression for their output voltages.

PRACTICAL EXERCISES

1. Verify the MCD circuit shown in Figure 15.5 by experimenting with AFC trainer kit.
2. Replace the transistor switches in Figure 15.5 with CD4066 non-inverter controlled switches and DG201 inverter controlled switches. In each, (i) draw the circuit diagrams, (ii) explain their working operation, (iii) draw waveforms at appropriate places, and (iv) deduce expression for the output voltage. Verify these MCDs by experimenting with the AFC trainer kit.
3. As discussed in Chapter 18 and Section 18.9, convert the MCDs of Figure 15.5 into SRM. (i) Draw the circuit diagrams, (ii) explain their working operation, (iii) draw waveforms at appropriate places, and (iv) deduce expression for the output voltage. Verify this SRM by experimenting with the AFC trainer kit.
4. As discussed in Chapter 19 and Section 19.6, convert the MCDs of Figure 15.5 into VMC. (i) Draw the circuit diagrams, (ii) explain their working operation, (iii) draw waveforms at appropriate places, and (iv) deduce expression for the output voltage. Verify this VMC by experimenting with the AFC trainer kit.
5. Convert the MCD circuits obtained from Problem 2 into SRM. (i) Draw the circuit diagrams, (ii) explain their working operation, (iii) draw waveforms at appropriate places, and (iv) deduce expression for the output voltage. Verify this SRM by experimenting with the AFC trainer kit.
6. Convert the MCD circuits obtained from Problem 2 into VMC. (i) Draw the circuit diagrams, (ii) explain their working operation, (iii) draw waveforms at appropriate places, and (iv) deduce expression for the output voltage. Verify this VMC by experimenting with the AFC trainer kit.

15.4 TRIANGULAR WAVE–BASED TIME DIVISION MULTIPLIER-CUM-DIVIDER

The circuit diagram of triangular wave-based time division MCD is shown in Figure 15.7, and its associated waveforms are shown in Figure 15.8. Figure 15.7(a) shows a series switching MCD, and Figure 15.7(b) shows a shunt switching MCD. As discussed in Section 4.5, Figure 4.9(a), the output of op-amp OA_1 is triangular wave V_{T1} with $\pm V_T$ peak values and time period 'T'.

Comparator OA_2 compares this triangular wave V_{T1} with the voltage V_Y and produces an asymmetrical rectangular wave V_K. From Figure 15.8, it is observed that

$$T_1 = \frac{V_T - V_Y}{2V_T}T, \quad T_2 = \frac{V_T + V_Y}{2V_T}T \qquad T = T_1 + T_2 \tag{15.21}$$

The rectangular wave V_K controls transistors Q_3 and Q_4.

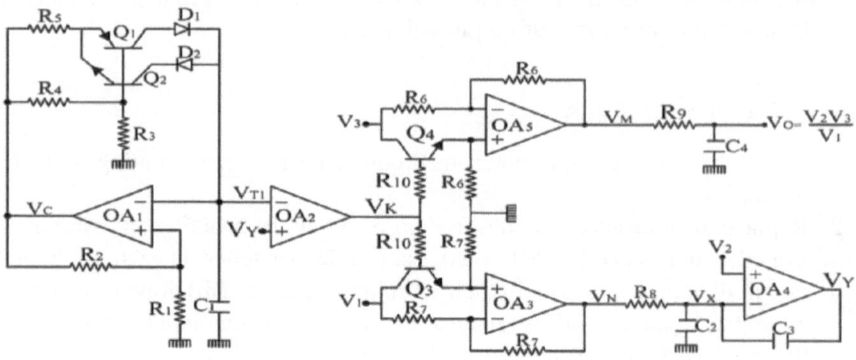

FIGURE 15.7(A) Series switching time division multiplier-cum-divider.

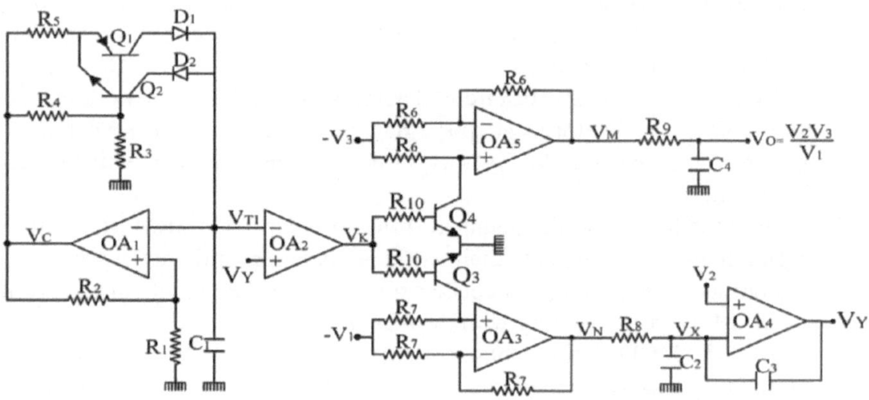

FIGURE 15.7(B) Shunt switching time division multiplier-cum-divider.

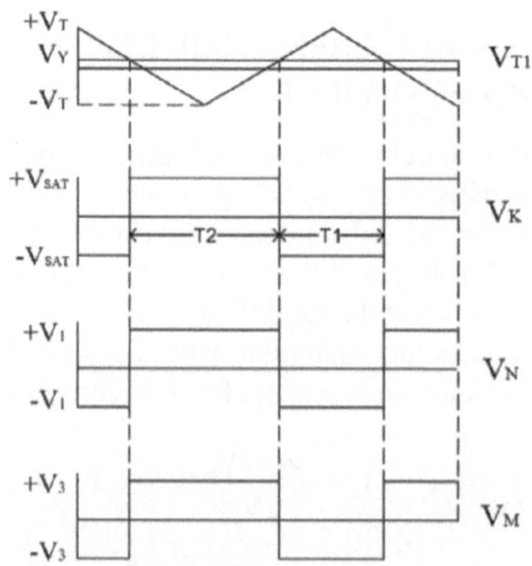

FIGURE 15.8 Associated waveforms of Figure 15.7.

During ON time T_2 of this rectangular wave V_K, in Figure 15.7(a), the transistor Q_4 is ON, the op-amp OA_5 along with resistors R_6 will work as non-inverting amplifiers and $+V_3$ will exist on its output ($V_M = +V_3$). The transistor Q_3 is ON, the op-amp OA_3 along with resistors R_7 will work as a non-inverting amplifier, and $+V_1$ will exist on its output ($V_N = +V_1$). In Figure 15.7(b), the transistor Q_4 is ON, the op-amp OA_5 along with resistors R_6 will work as inverting amplifiers and $+V_3$ will exist on its output ($V_M = +V_3$). The transistor Q_3 is ON, the op-amp OA_3 along with resistors R_7 will work as an inverting amplifier and $+V_1$ will exist on its output ($V_N = +V_1$).

During OFF time T_1 of this rectangular wave V_K, in Figure 15.7(a), the transistor Q_4 is OFF, the op-amp OA_5 along with resistors R_6 will work as inverting amplifiers and $-V_3$ will exist on its output ($V_M = -V_3$). The transistor Q_3 is OFF, the op-amp OA_3 along with resistors R_7 will work as an inverting amplifier, and $-V_1$ will exist on its output ($V_N = -V_1$). In Figure 15.7(b), the transistor Q_4 is OFF, the op-amp OA_5 along with resistors R_6 will work as non-inverting amplifiers and $-V_3$ will exist on its output ($V_M = -V_3$). The transistor Q_3 is OFF, the op-amp OA_3 along with resistors R_7 will work as a non-inverting amplifier and $-V_1$ will exist on its output ($V_N = -V_1$).

Two asymmetrical rectangular waveforms are generated: (i) V_M at the output of op-amp OA_5 and (ii) V_N at the output of op-amp OA_3. The R_8C_2 lowpass filter gives the average value of V_N and is given as

$$V_X = \frac{1}{T}\left[\int_{O}^{T_2} V_1\,dt + \int_{T_2}^{T_1+T_2} (-V_1)\,dt\right] = \frac{V_1}{T}\left[T_2 - T_1\right] \tag{15.22}$$

$$V_X = \frac{V_1 V_Y}{V_T}$$

The op-amp OA_4 is configured in a negative closed loop feedback and a positive DC voltage is ensured in the feedback loop. Hence, its inverting terminal voltage must equal to its non-inverting terminal voltage, i.e.,

$$V_X = V_2 \tag{15.23}$$

From Eqs. (15.22) and (15.20)

$$V_Y = \frac{V_2 V_T}{V_1} \tag{15.24}$$

The R_9C_4 lowpass filter gives the average value of the rectangular waveform V_M and is given as

$$V_O = \frac{1}{T}\left[\int_{O}^{T_2} V_3\,dt + \int_{T_2}^{T_1+T_2} (-V_3)\,dt\right] = \frac{V_3}{T}\left[T_2 - T_1\right] \tag{15.25}$$

$$V_O = \frac{V_3 V_Y}{V_T}$$

Equation (15.24) in (15.25) gives

$$V_O = \frac{V_2 V_3}{V_1} \tag{15.26}$$

The output voltage depends on the sharpness and linearity of the triangular waveform. The output voltage is also proportional to attenuation constant α of the lowpass filter. The attenuation constant α will be in the range of 0.80 to 0.90. The polarities of V_1 and V_2 are only positive. However, V_3 can have any polarity.

DESIGN EXERCISES

1. In the MCD circuits shown in Figures 15.7(a) and 15.7(b), the triangular wave generator shown in Figure 4.9(a) is used. Replace this triangular wave generator with another triangular wave generator shown in Figure 4.9(b). (i) Draw circuit diagrams, (ii) draw waveforms at appropriate places, (iii) explain their working operation, and (iv) deduce expression for their output voltages.
2. Replace the transistor switches Q_3 and Q_4 in Figures 15.7(a) and 15.7(b) with FET switches and MOSFET switches. In each, (i) draw circuit diagrams, (ii) draw waveforms at appropriate places, (iii) explain their working operation, and (iv) deduce expression for their output voltages.

PRACTICAL EXERCISES

1. Verify the MCD circuit shown in Figure 15.7 by experimenting with the AFC trainer kit.
2. Replace the transistor switches in Figure 15.7 with CD4066 non-inverter controlled switches and DG201 inverter controlled switches. In each, (i) draw the circuit diagrams, (ii) explain their working operation, (iii) draw waveforms at appropriate places, and (iv) deduce expression for the output voltage. Verify these MCDs by experimenting with the AFC trainer kit.
3. As discussed in Chapter 18 and Section 18.9, convert the MCDs of Figure 15.7 into SRM. (i) Draw the circuit diagrams, (ii) explain their working operation, (iii) draw waveforms at appropriate places, and (iv) deduce expression for the output voltage. Verify this SRM by experimenting with the AFC trainer kit.
4. As discussed in Chapter 19 and Section 19.6, convert the MCDs of Figure 15.7 into VMC. (i) Draw the circuit diagrams, (ii) explain their working operation, (iii) draw waveforms at appropriate places, and (iv) deduce expression for the output voltage. Verify this VMC by experimenting with the AFC trainer kit.
5. Convert the MCD circuits obtained from Problem 2 into SRM. (i) Draw the circuit diagrams, (ii) explain their working operation, (iii) draw waveforms at appropriate places, and (iv) deduce expression for the output voltage. Verify this SRM by experimenting with the AFC trainer kit.
6. Convert the MCD circuits obtained from Problem 2 into VMC. (i) Draw the circuit diagrams, (ii) explain their working operation, (iii) draw waveforms at appropriate places, and (iv) deduce expression for the output voltage. Verify this VMC by experimenting with the AFC trainer kit.

15.5 TRIANGULAR WAVE–BASED DIVIDE–MULTIPLY TIME DIVISION MULTIPLIER-CUM-DIVIDER

The circuit diagrams of triangular wave-based divide–multiply TDMCD are shown in Figures 15.9, and their associated waveforms are shown in Figure 15.10. Figure 15.9(a) shows a series switching MCD, and Figure 15.9(b) shows a shunt switching MCD. As discussed in Section 4.5, Figure 4.9(a), a triangular wave V_{T1} with $\pm V_T$ peak-to-peak value is generated by around the op-amp OA_1 transistors Q_1 and Q_2.

The first input voltage V_1 is compared with the generated triangular wave V_{T1} by the comparator on OA_2. An asymmetrical rectangular waveform V_N is generated at the comparator OA_2 output. From the waveforms shown in Figure 15.10, it is observed that

$$T_1 = \frac{V_T - V_1}{2V_T} T \qquad T_2 = \frac{V_T + V_1}{2V_T} T \qquad T = T_1 + T_2 \tag{15.27}$$

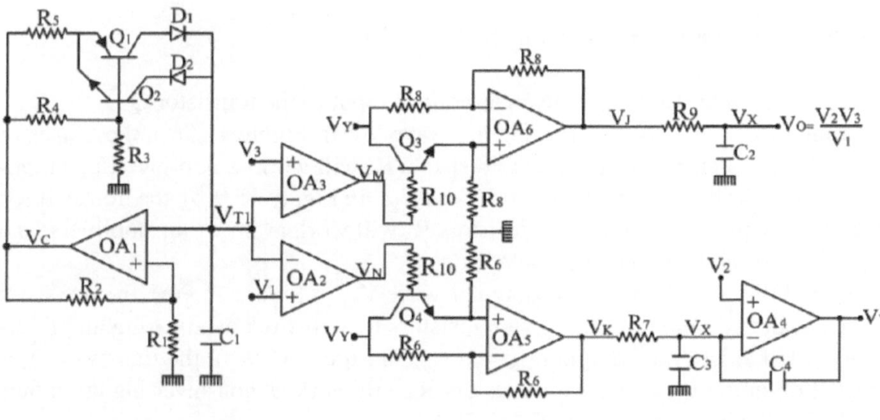

FIGURE 15.9(A) Series switching divide–multiply time division multiplier-cum-divider.

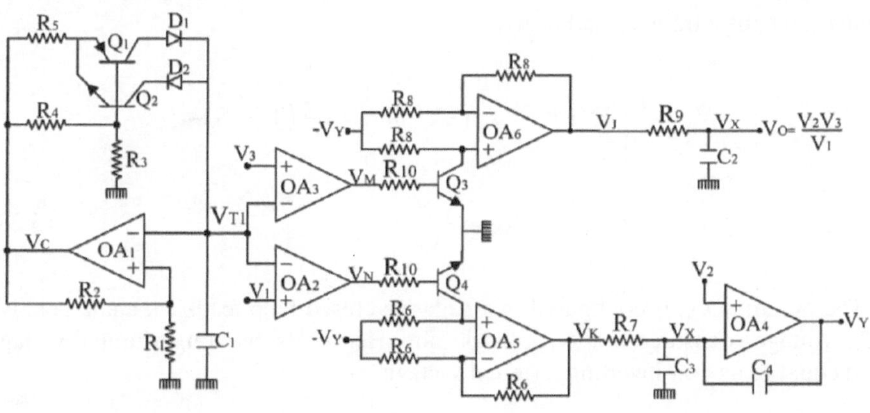

FIGURE 15.9(B) Shunt switching divide–multiply time division multiplier-cum-divider.

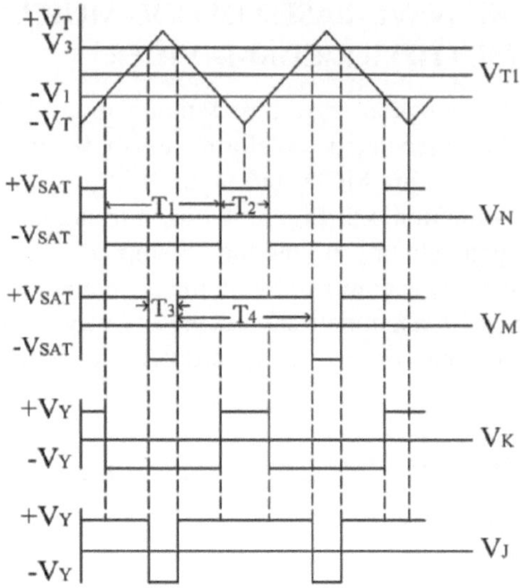

FIGURE 15.10 Associated waveforms of Figure 15.9.

This rectangular wave V_N is given as control input to the transistor Q_4.

During ON time T_2 of this rectangular wave V_N, in Figure 15.9(a), the transistor Q_4 is ON, the op-amp OA_5 along with resistors R_6 will work as non-inverting amplifiers, and $+V_Y$ will exist on its output ($V_K = +V_Y$). In Figure 15.9(b), the transistor Q_4 is ON, the op-amp OA_5 along with resistors R_6 will work as inverting amplifiers, and $+V_Y$ will exist on its output ($V_K = +V_Y$).

During OFF time T_1 of this rectangular wave V_N, in Figure 15.9(a), the transistor Q_4 is OFF, the op-amp OA_5 along with resistors R_6 will work as inverting amplifiers and $-V_Y$ will exist on its output ($V_K = -V_Y$). In Figure 15.9(b), the transistor Q_4 is OFF, the op-amp OA_5 along with resistors R_6 will work as non-inverting amplifiers and $-V_Y$ will exist on its output ($V_K = -V_Y$).

Another rectangular asymmetrical wave V_K with peak-to-peak values of $\pm V_Y$ is generated at the output of op-amp OA_5. The R_7C_3 lowpass filter gives an average value of the pulse train V_K and is given as

$$V_X = \frac{1}{T}\left[\int_{O}^{T_2} V_Y\,dt + \int_{T_2}^{T_1+T_2} (-V_Y)\,dt\right] = \frac{V_Y}{T}(T_2 - T_1)$$

$$V_X = \frac{V_1 V_Y}{V_T} \tag{15.28}$$

The op-amp OA_4 is configured in a negative closed loop feedback and a positive DCC voltage is ensured in the feedback loop. Hence, its inverting terminal voltage must equal to its non-inverting terminal voltage, i.e.,

$$V_2 = V_X \tag{15.29}$$

From Eqs. (15.28) and (15.29)

$$V_Y = \frac{V_2 V_T}{V_1} \tag{15.30}$$

The third input voltage V_3 is compared with the generated triangular wave V_{T1} by the comparator on OA_3. An asymmetrical rectangular waveform V_M is generated at the comparator OA_3 output. From the waveforms shown in Figure 15.10, it is observed that

$$T_3 = \frac{V_T - V_3}{2V_T} T \qquad T_4 = \frac{V_T + V_3}{2V_T} T \qquad T = T_3 + T_4 \tag{15.31}$$

This rectangular wave V_M is given as control input to the transistor Q_3.

During ON time T_4 of this rectangular wave V_M, in Figure 15.9(a), the transistor Q_3 is ON, the op-amp OA_6 along with resistors R_8 will work as non-inverting amplifiers, and $+V_Y$ will exist on its output ($V_J = +V_Y$). In Figure 15.9(b), the transistor Q_3 is ON, the op-amp OA_6 along with resistors R_8 will work as an inverting amplifier, and $+V_Y$ will exist on its output ($V_J = +V_Y$).

During OFF time T_3 of this rectangular wave V_M, in Figure 15.9(a), the transistor Q_3 is OFF, the op-amp OA_6 along with resistors R_8 will work as inverting amplifiers, and $-V_Y$ will exist on its output ($V_J = -V_Y$). In Figure 15.9(b), the transistor Q_3 is OFF, the op-amp OA_6 along with resistors R_8 will work as non-inverting amplifiers, and $-V_Y$ will exist on its output ($V_J = -V_Y$).

Another rectangular asymmetrical wave V_J with peak-to-peak values of $\pm V_Y$ is generated at the output of op-amp OA_6. The R_9C_2 lowpass filter gives average value of the pulse train V_J and is given as

$$V_O = \frac{1}{T}\left[\int_O^{T_4} V_Y\,dt + \int_{T_4}^{T_3+T_4} (-V_Y)\,dt\right] = \frac{V_Y}{T}(T_4 - T_3) \tag{15.32}$$

$$V_O = \frac{V_Y V_3}{V_T} \tag{15.33}$$

Equation (15.30) in (15.33) gives

$$V_O = \frac{V_2 V_3}{V_1} \tag{15.34}$$

The output voltage depends on the sharpness and linearity of the triangular waveform. The output voltage is proportional to attenuation constant α of the lowpass filter. The attenuation constant α will be in the range of 0.80 to 0.90. The polarities of V_1 and V_2 are only positive. However, V_3 can have any polarity.

DESIGN EXERCISES

1. In the MCD circuits shown in Figures 15.9(a) and 15.9(b), the triangular wave generator shown in Figure 4.9(a) is used. Replace this triangular wave generator with other triangular wave generator shown in Figure 4.9(b). (i) Draw circuit diagrams, (ii) draw waveforms at appropriate places, (iii) explain their working operation and (iv) deduce expression for their output voltages
2. Replace the transistor switches Q_3 & Q_4 in Figures 15.9(a), 15.9(b) with FET switches and MOSFET switches. In each, (i) Draw circuit diagrams (ii) draw waveforms at appropriate places (iii) explain their working operation, and (iv) deduce expression for their output voltages.

PRACTICAL EXERCISES

1. Verify the MCD circuit shown in Figure 15.9 by experimenting with the AFC trainer kit.
2. Replace the transistor switches in Figure 15.9 with CD4066 non-inverter controlled switches and DG201 inverter controlled switches. In each, (i) draw the circuit diagrams, (ii) explain their working operation, (iii) draw waveforms at appropriate places, and (iv) deduce expression for the output voltage. Verify these MCDs by experimenting with the AFC trainer kit.
3. As discussed in Chapter 18 and Section 18.9, convert the MCDs of Figure 15.9 into SRM. (i) Draw the circuit diagrams, (ii) explain their working operation, (iii) draw waveforms at appropriate places, and (iv) deduce expression for the output voltage. Verify this SRM by experimenting with the AFC trainer kit.
4. As discussed in Chapter 19 and Section 19.6, convert the MCDs of Figure 15.9 into VMC. (i) Draw the circuit diagrams, (ii) explain their working operation, (iii) draw waveforms at appropriate places, and (iv) deduce expression for the output voltage. Verify this VMC by experimenting with the AFC trainer kit.
5. Convert the MCD circuits obtained from Problem 2 into SRM. (i) Draw the circuit diagrams, (ii) explain their working operation, (iii) draw waveforms at appropriate places, and (iv) deduce expression for the output voltage. Verify this SRM by experimenting with the AFC trainer kit.
6. Convert the MCD circuits obtained from Problem 2 into VMC. (i) Draw the circuit diagrams, (ii) explain their working operation, (iii) draw waveforms at appropriate places, and (iv) deduce expression for the output voltage. Verify this VMC by experimenting with the AFC trainer kit.

15.6 TRIANGULAR WAVE–BASED MULTIPLY–DIVIDE TIME DIVISION MULTIPLIER-CUM-DIVIDER

The circuit diagrams of triangular wave-based multiply–divide TDMCD are shown in Figure 15.11, and its associated waveforms are shown in Figure 15.12. Figure 15.11(a) shows a series switching MCD, and Figure 15.11(b) shows a shunt switching MCD. A triangular wave V_{T1} with $\pm V_T$ peak-to-peak value is generated by the op-amp OA_1 and transistors Q_1 and Q_2.

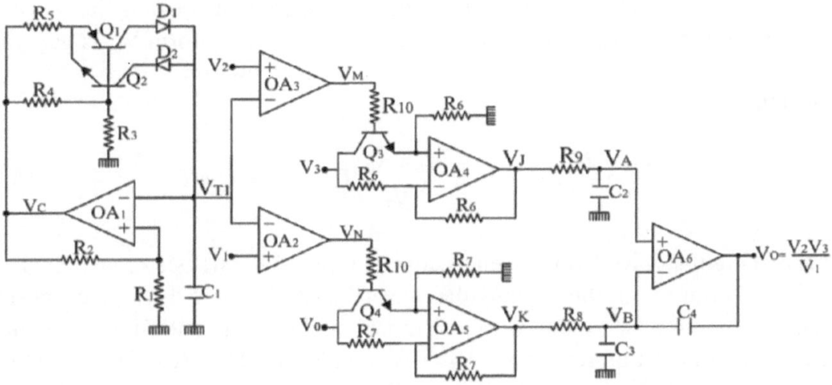

FIGURE 15.11(A) Series switching multiply–divide time division multiplier-cum-divider.

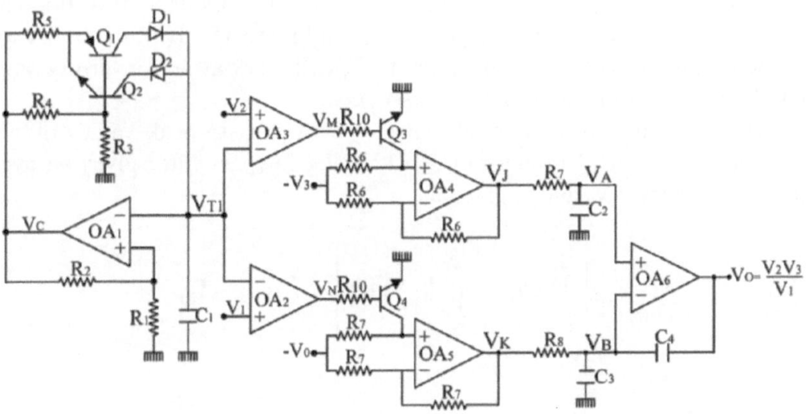

FIGURE 15.11(B) Shunt switching multiply–divide time division multiplier-cum-divider.

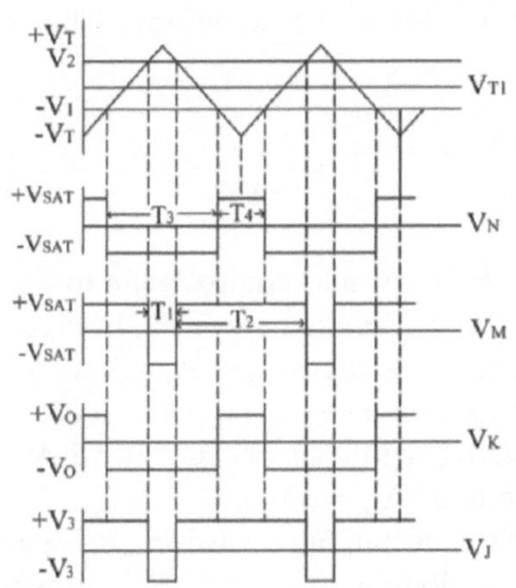

FIGURE 15.12 Associated waveforms of Figure 15.11.

The second input voltage V_2 is compared with the generated triangular wave V_{T1} by the comparator on OA_3. An asymmetrical rectangular waveform V_M is generated at the comparator OA_3 output. From the waveforms shown in Figure 15.12, it is observed that

$$T_1 = \frac{V_T - V_2}{2V_T} T \qquad T_2 = \frac{V_T + V_2}{2V_T} T \qquad T = T_1 + T_2 \tag{15.35}$$

This rectangular wave V_M is given as control input to the transistor switch Q_3.

During ON time T_2 of this rectangular wave V_M, in Figure 15.11(a), the transistor Q_3 is ON, the op-amp OA_4 along with resistors R_6 will work as a non-inverting amplifier, and $+V_3$ will exist on its output ($V_J = +V_3$). In Figure 15.11(b), the transistor Q_3 is ON, the op-amp OA_4 along with resistors R_6 will work as inverting amplifiers, and $+V_3$ will exist on its output ($V_J = +V_3$).

During OFF time T_1 of this rectangular wave V_M, in Figure 15.11(a), the transistor Q_3 is OFF, the op-amp OA_4 along with resistors R_6 will work as an inverting amplifier, and $-V_3$ will exist on its output ($V_J = -V_3$). In Figure 15.11(b), the transistor Q_3 is OFF, the op-amp OA_4 along with resistors R_6 will work as non-inverting amplifiers, and $-V_3$ will exist on its output ($V_J = -V_3$).

Another rectangular asymmetrical wave V_J with peak-to-peak values of $\pm V_3$ is generated at the output of op-amp OA_4. The R_9C_2 lowpass filter gives an average value of the pulse train V_J and is given as

$$V_A = \frac{1}{T}\left[\int_O^{T_2} V_3\,dt + \int_{T_2}^{T_1+T_2} (-V_3)\,dt\right] = \frac{V_3}{T}(T_2 - T_1)$$
$$V_A = \frac{V_2 V_3}{V_T} \tag{15.36}$$

The first input voltage V_1 is compared with the generated triangular wave V_{T1} by the comparator on OA_2. An asymmetrical rectangular waveform V_N is generated at the comparator OA_2 output. From the waveforms shown in Figure 15.12, it is observed that

$$T_3 = \frac{V_T - V_1}{2V_T} T \qquad T_4 = \frac{V_T + V_1}{2V_T} T \qquad T = T_3 + T_4 \tag{15.37}$$

This rectangular wave V_N is given as control input to the transistor switch Q_4.

During ON time T_4 of this rectangular wave V_N, in Figure 15.11(a), the transistor Q_4 is ON, the op-amp OA_5 along with resistors R_7 will work as a non-inverting amplifier, and $+V_O$ will exist on its output ($V_K = +V_O$). In Figure 15.11(b), the transistor Q_4 is ON, the op-amp OA_5 along with resistors R_7 will work as an inverting amplifier, and $+V_O$ will exist on its output ($V_K = +V_O$).

During OFF time T_3 of this rectangular wave V_N, in Figure 15.11(a), the transistor Q_4 is OFF, the op-amp OA_5 along with resistors R_7 will work as an inverting amplifier, and $-V_O$ will exist on its output ($V_K = -V_O$). In Figure 15.11(b), the transistor Q_4

is OFF, the op-amp OA_5 along with resistors R_7 will work as a non-inverting amplifier, and $-V_O$ will exist on its output ($V_K = -V_O$).

Another rectangular asymmetrical wave V_K with peak-to-peak values of $\pm V_O$ is generated at the output of op-amp OA_5. The R_8C_3 lowpass filter gives average value of the pulse train V_K and is given as

$$V_B = \frac{1}{T}\left[\int_{O}^{T_4} V_O\,dt + \int_{T_4}^{T_3+T_4} (-V_O)\,dt\right] = \frac{V_O}{T}(T_4 - T_3)$$

$$V_B = \frac{V_1 V_O}{V_T} \tag{15.38}$$

The op-amp OA_6 is configured in a negative closed loop feedback, and a positive DC voltage is ensured in the feedback loop. Hence, its inverting terminal voltage must equal to its non-inverting terminal voltage, i.e.,

$$V_A = V_B \tag{15.39}$$

From Eqs. (15.36) and (15.38)

$$V_O = \frac{V_2 V_3}{V_1} \tag{15.40}$$

The output voltage depends on the sharpness and linearity of the triangular waveform. The output voltage is also proportional to attenuation constant α of the lowpass filter. The attenuation constant α will be in the range of 0.80 to 0.90. The polarities of V_1, V_2, and V_3 are to be only positive.

DESIGN EXERCISES

1. In the MCD circuits shown in Figures 15.11(a) and 15.11(b), the triangular wave generator shown in Figure 4.9(a) is used. Replace this triangular wave generator with another triangular wave generator shown in Figure 4.9(b). (i) Draw circuit diagrams, (ii) draw waveforms at appropriate places, (iii) explain their working operation, and (iv) deduce expression for their output voltages.
2. Replace the transistor switches Q_3 and Q_4 in Figures 15.11(a) and 15.11(b) with FET switches and MOSFET switches. In each, (i) draw circuit diagrams, (ii) draw waveforms at appropriate places, (iii) explain their working operation, and (iv) deduce expression for their output voltages.

PRACTICAL EXERCISES

1. Verify the MCD circuit shown in Figure 15.11 by experimenting with the AFC trainer kit.

2. Replace the transistor switches in Figure 15.11 with CD4066 non-inverter controlled switches and DG201 inverter controlled switches. In each, (i) draw the circuit diagrams, (ii) explain their working operation, (iii) draw waveforms at appropriate places, and (iv) deduce expression for their output voltage. Verify these MCDs by experimenting with the AFC trainer kit.
3. As discussed in Chapter 18 and Section 18.9, convert the MCDs of Figure 15.11 into SRM. (i) Draw the circuit diagrams, (ii) explain their working operation, (iii) draw waveforms at appropriate places, and (iv) deduce expression for their output voltage. Verify this SRM by experimenting with the AFC trainer kit.
4. As discussed in Chapter 19 and Section 19.6, convert the MCDs of Figure 15.11 into VMC. (i) Draw the circuit diagrams, (ii) explain their working operation, (iii) draw waveforms at appropriate places, and (iv) deduce expression for their output voltage. Verify this VMC by experimenting with the AFC trainer kit.
5. Convert the MCD circuits obtained from Problem 2 into SRM. (i) Draw the circuit diagrams, (ii) explain their working operation, (iii) draw waveforms at appropriate places, and (iv) deduce expression for their output voltage. Verify this SRM by experimenting with the AFC trainer kit.
6. Convert the MCD circuits obtained from Problem 2 into VMC. (i) Draw the circuit diagrams, (ii) explain their working operation, (iii) draw waveforms at appropriate places, and (iv) deduce expression for their output voltage. Verify this VMC by experimenting with the AFC trainer kit.

15.7 TIME DIVISION MULTIPLIER-CUM-DIVIDER WITH NO REFERENCE – TYPE I – SWITCHING

The MCDs using the time division principle without any reference clock are shown in Figure 15.13, and their associated waveforms are shown in Figure 15.14. Figures 15.13(a) and 15.13(b) show series switching MCDs, and Figures 15.13(c)

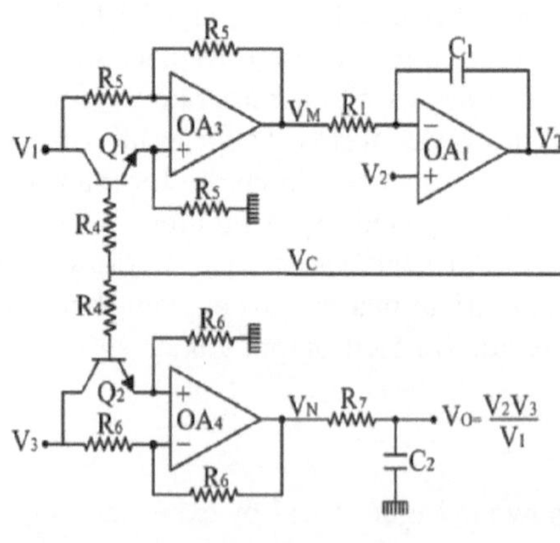

FIGURE 15.13(A) Series switching time division multiplier-cum-divider with no reference.

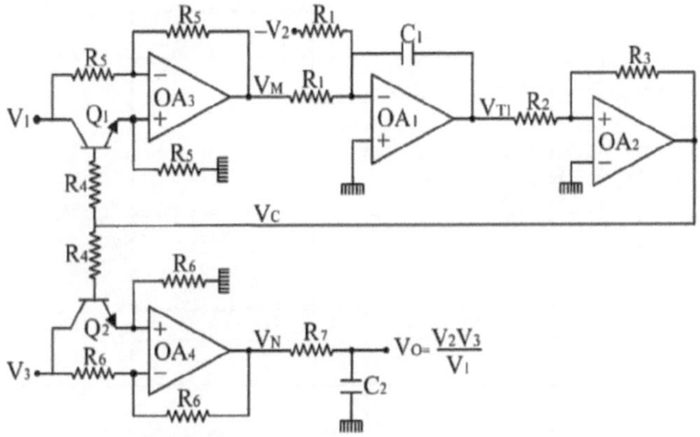

FIGURE 15.13(B) Equivalent circuit of Figure 15.13(a).

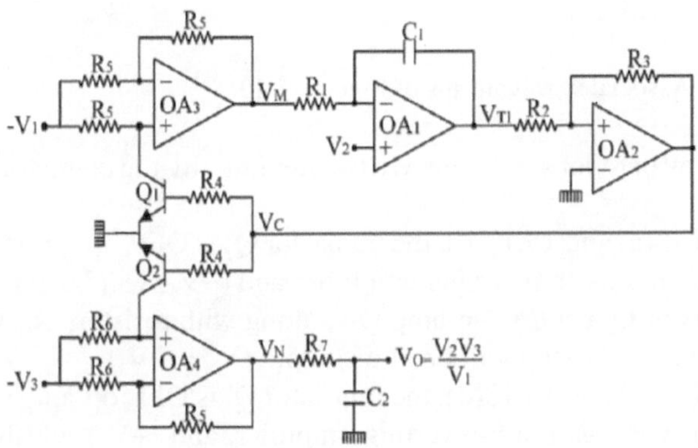

FIGURE 15.13(C) Shunt switching time division multiplier-cum-divider with no reference.

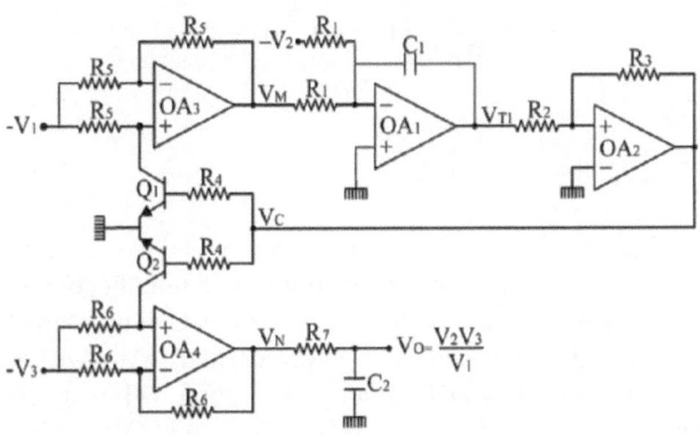

FIGURE 15.13(D) Equivalent circuit of Figure 15.13(c).

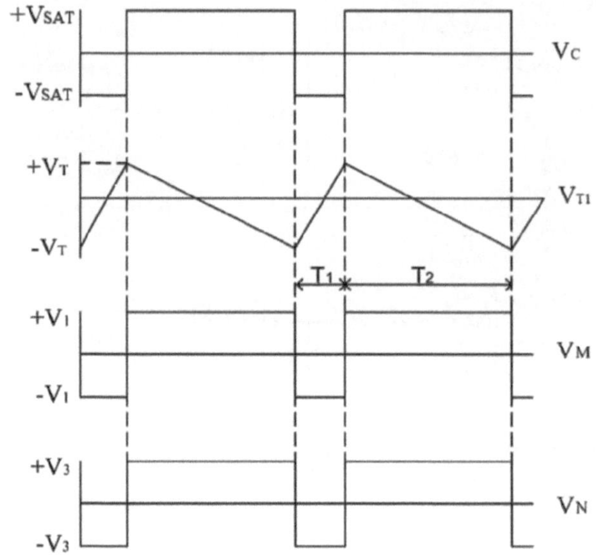

FIGURE 15.14 Associated waveforms of Figure 15.13.

and 15.13(d) show parallel switching MCDs. Let initially the comparator OA_2 output be LOW.

In Figures 15.13(a) and 15.13(b), the transistor Q_1 is OFF, op-amp OA_3 along with resistors R_5 will work as an inverting amplifier, and $(-V_1)$ will be at its output ($V_M = -V_1$). The transistor Q_2 is OFF, op-amp OA_4 along with resistors R_6 will work as an inverting amplifier, and $(-V_3)$ will be at its output ($V_N = -V_3$).

In Figures 15.13(c) and 15.13(d), the transistor Q_1 is OFF, op-amp OA_3 along with resistors R_5 will work as a non-inverting amplifier, and $(-V_1)$ will be at its output ($V_M = -V_1$). The transistor Q_2 is OFF, op-amp OA_4 along with resistors R_6 will work as a non-inverting amplifier, and $(-V_3)$ will be at its output ($V_N = -V_3$).

The output of differential integrator will be

$$V_{T1} = \frac{1}{R_1C_1}\int (V_2 + V_1)dt - V_T$$

$$V_{T1} = \frac{(V_2 + V_1)}{R_1C_1}t - V_T \tag{15.41}$$

The output of differential integrator rises toward positive saturation, and when it reaches the voltage level of $+V_T$, the comparator OA_2 output becomes HIGH.

In Figures 15.13(a) and 15.13(b), the transistor Q_1 is ON, op-amp OA_3 along with resistors R_5 will work as a non-inverting amplifier, and $(+V_1)$ will be at its output ($V_M = +V_1$). The transistor Q_2 is ON, op-amp OA_4 along with resistors R_6 will work as a non-inverting amplifier, and $(+V_3)$ will be at its output ($V_N = +V_3$).

In Figures 15.13(c) and 15.13(d), the transistor Q_1 is ON, op-amp OA_3 along with resistors R_5 will work as an inverting amplifier, and $(+V_1)$ will be at its output

($V_M = +V_1$). The transistor Q_2 is ON, op-amp OA_4 along with resistors R_6 will work as an inverting amplifier, and ($+V_3$) will be at its output ($V_N = +V_3$).

Now, the output of differential integrator will be

$$V_{T1} = \frac{1}{R_1C_1}\int (V_2 - V_1)dt + V_T$$

$$V_{T1} = -\frac{(V_1 - V_2)}{R_1C_1}t + V_T \tag{15.42}$$

The output of differential integrator reverses toward negative saturation, and when it reaches the voltage level ($-V_T$), the comparator OA_2 output becomes LOW, and the cycle therefore repeats in order to give an asymmetrical rectangular wave V_C at the output of comparator OA_2.

$$V_T = \frac{R_2}{R_3}V_{SAT} \tag{15.43}$$

From the waveforms shown in Figure 15.14, it is observed that

$$T_1 = \frac{V_1 - V_2}{2V_1}T, \quad T_2 = \frac{V_1 + V_2}{2V_1}T, \quad T = T_1 + T_2 \tag{15.44}$$

Another rectangular wave V_N is generated at the output of op-amp OA_4. The R_7C_2 lowpass filter gives an average value of this pulse train V_N and is given as

$$V_O = \frac{1}{T}\left[\int_0^{T_2} V_3 dt + \int_{T_2}^{T_1+T_2} (-V_3)dt\right], \quad V_O = \frac{V_3(T_2 - T_1)}{T} \tag{15.45}$$

Equation (15.44) in (15.45) give

$$V_O = \frac{V_2V_3}{V_1} \tag{15.46}$$

The output voltage is also proportional to attenuation constant α of the lowpass filter. The attenuation constant α will be in the range of 0.80 to 0.90. The polarities of all input voltages must be single-polarity only.

DESIGN EXERCISES

1. The transistor switches Q_1 and Q_2 in Figures 15.13(a)–15.13(d) are to be replaced with FET switches and MOSFET switches. In each, (i) draw the circuit diagrams, (ii) explain their working operations, (iii) draw waveforms at appropriate places, and (iv) deduce expression for their output voltages.

PRACTICAL EXERCISES

1. Verify the MCD circuit shown in Figure 15.13 by experimenting with the AFC trainer kit.
2. Replace the transistor switches in Figure 15.13 with CD4066 non-inverter controlled switches and DG201 inverter controlled switches. In each, (i) draw the circuit diagrams, (ii) explain their working operation, (iii) draw waveforms at appropriate places, and (iv) deduce expression for the output voltage. Verify these MCDs by experimenting with the AFC trainer kit.

.3 As discussed in Chapter 18 and Section 18.9, convert the MCDs of Figure 15.13 into SRM. (i) Draw the circuit diagrams, (ii) explain their working operation, (iii) draw waveforms at appropriate places, and (iv) deduce expression for the output voltage. Verify this SRM by experimenting with the AFC trainer kit.

4. As discussed in Chapter 19 and Section 19.6, convert the MCDs of Figure 15.13 into VMC. (i) Draw the circuit diagrams, (ii) explain their working operation, (iii) draw waveforms at appropriate places, and (iv) deduce expression for the output voltage. Verify this square rooter by experimenting with the AFC trainer kit.
5. Convert the MCD circuits obtained from Problem 2 into SRM. (i) Draw the circuit diagrams, (ii) explain their working operation, (iii) draw waveforms at appropriate places, and (iv) deduce expression for the output voltage. Verify this SRM by experimenting with the AFC trainer kit.
6. Convert the MCD circuits obtained from Problem 2 into VMC. (i) Draw the circuit diagrams, (ii) explain their working operation, (iii) draw waveforms at appropriate places, and (iv) deduce expression for the output voltage. Verify this VMC by experimenting with the AFC trainer kit.

15.8 TIME DIVISION MULTIPLIER-CUM-DIVIDER WITH NO REFERENCE – TYPE II – SWITCHING

The MCDs using the time division principle without any reference clock are shown in Figure 15.15, and their associated waveforms are shown in Figure 15.16. Figures

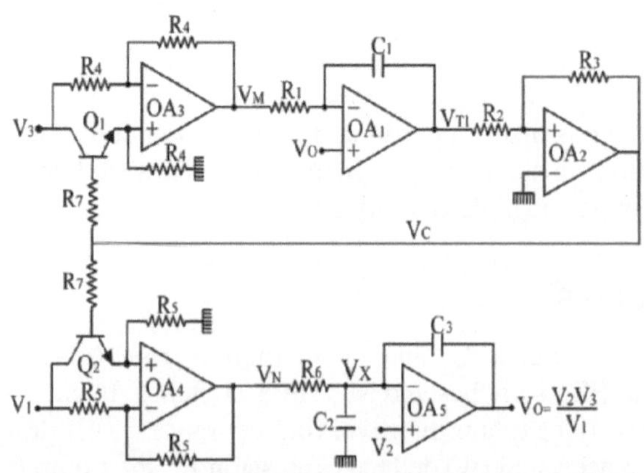

FIGURE 15.15(A) Series switching time division multiplier-cum-divider.

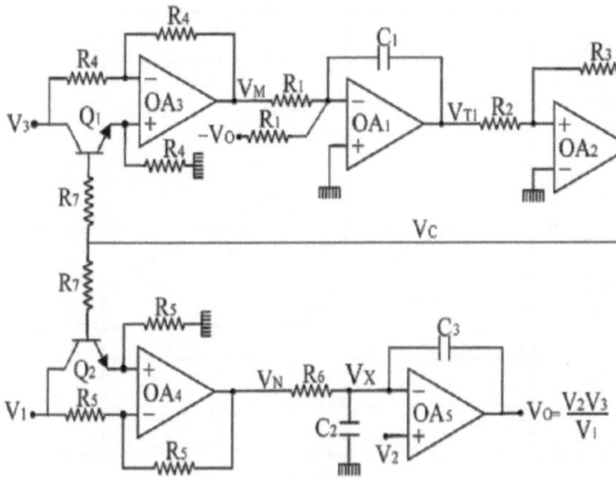

FIGURE 15.15(B) Equivalent circuit of Figure 15.15(a).

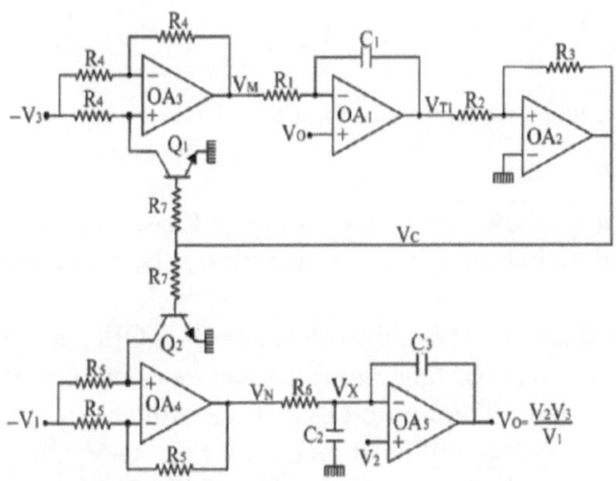

FIGURE 15.15(C) Shunt switching time division multiplier-cum-divider with no reference.

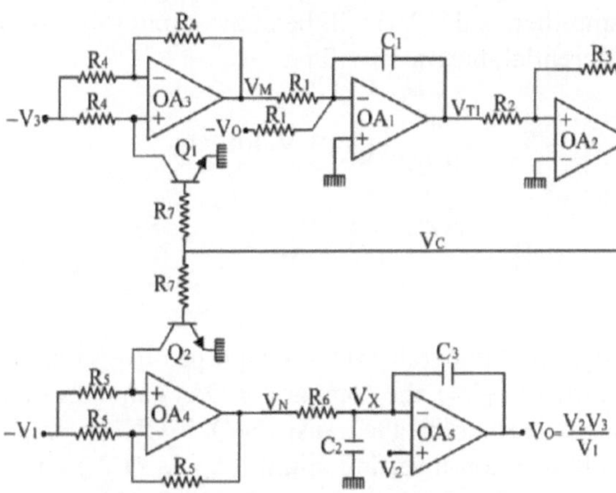

FIGURE 15.15(D) Equivalent circuit of Figure 15.15(c).

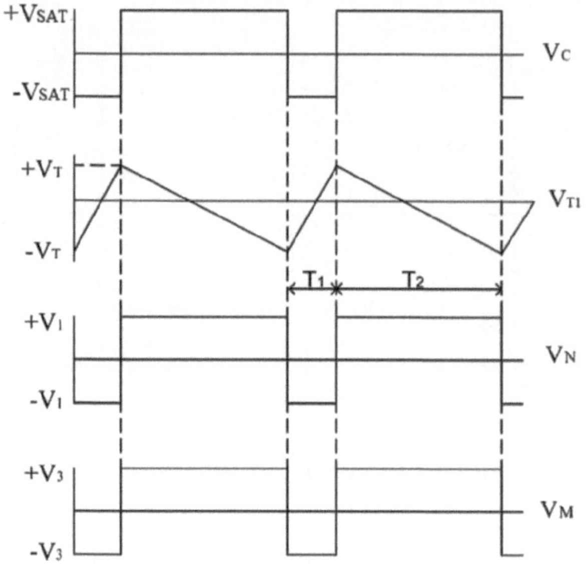

FIGURE 15.16 Associated waveforms of Figure 15.15.

15.15(a) and 15.15(b) show series switching MCDs, and Figures 15.15(c) and 15.15(d) show shunt switching MCDs. Let initially the comparator OA_2 output be LOW.

In Figures 15.15(a) and 15.15(b), the transistor Q_1 is OFF, op-amp OA_3 along with resistors R_4 will work as an inverting amplifier, and $(-V_3)$ will be at its output ($V_M = -V_3$). The transistor Q_2 is OFF, op-amp OA_4 along with resistors R_5 will work as an inverting amplifier, and $(-V_1)$ will be at its output ($V_N = -V_1$).

In Figures 15.15(c) and 15.15(d), the transistor Q_1 is OFF, op-amp OA_3 along with resistors R_4 will work as a non-inverting amplifier, and $(-V_3)$ will be at its output ($V_M = -V_3$). The transistor Q_2 is OFF, op-amp OA_4 along with resistors R_5 will work as a non-inverting amplifier, and $(-V_1)$ will be at its output ($V_N = -V_1$).

The output of differential integrator will be

$$V_{T1} = \frac{1}{R_1C_1}\int (V_O + V_3)dt - V_T$$

$$V_{T1} = \frac{(V_3 + V_O)}{R_1C_1}t - V_T \tag{15.47}$$

The output of differential integrator rises toward positive saturation, and when it reaches the voltage level of $(+V_T)$, the comparator OA_2 output becomes HIGH.

In Figures 15.15(a) and 15.15(b), the transistor Q_1 is ON, op-amp OA_3 along with resistors R_4 will work as a non-inverting amplifier, and $(+V_3)$ will be at its output ($V_M = +V_3$). The transistor Q_2 is ON, op-amp OA_4 along with resistors R_5 will work as a non-inverting amplifier, and $(+V_1)$ will be at its output ($V_N = +V_1$).

In Figures 15.15(c) and 15.15(d), the transistor Q_1 is ON, op-amp OA_3 along with resistors R_4 will work as an inverting amplifier, and $(+V_3)$ will be at its output $(V_M = +V_3)$. The transistor Q_2 is ON, op-amp OA_4 along with resistors R_5 will work as an inverting amplifier, and $(+V_1)$ will be at its output $(V_N = +V_1)$.

Now the output of the differential integrator will be

$$V_{T1} = \frac{1}{R_1C_1}\int (V_O - V_3)\,dt + V_T$$

$$V_{T1} = -\frac{(V_3 - V_O)}{R_1C_1}t + V_T \tag{15.48}$$

The output of the differential integrator reverses toward negative saturation, and when it reaches the voltage level $(-V_T)$, the comparator OA_2 output becomes LOW, and the cycle therefore repeats in order to give an asymmetrical rectangular wave V_C at the output of comparator OA_2.

$$V_T = \frac{R_2}{R_3}V_{SAT} \tag{15.49}$$

From the waveforms shown in Figure 15.16, it is observed that

$$T_1 = \frac{V_3 - V_O}{2V_3}T, \quad T_2 = \frac{V_3 + V_O}{2V_3}T, \quad T = T_1 + T_2 \tag{15.50}$$

Another rectangular wave V_N is generated at the output of op-amp OA_4. The R_6C_2 lowpass filter gives average value of this pulse train V_N and is given as

$$V_X = \frac{1}{T}\left[\int_0^{T_2} V_1\,dt + \int_{T_2}^{T_1+T_2} (-V_1)dt\right], \quad V_X = \frac{V_1(T_2 - T_1)}{T} \tag{15.51}$$

Equations (15.50) in (15.51) give

$$V_X = \frac{V_1V_O}{V_3} \tag{15.52}$$

The op-amp OA_5 is at negative closed loop feedback configuration and a positive DC voltage is ensured in the feedback. Its non-inverting terminal voltage will equal to its inverting terminal voltage. Hence,

$$V_X = V_2 \tag{15.53}$$

From Eqs. (15.52) and (15.53)

$$V_O = \frac{V_2V_3}{V_1} \tag{15.54}$$

The output voltage is also proportional to attenuation constant α of the lowpass filter. The attenuation constant α will be in the range of 0.80 to 0.90. The polarity of all input voltages must be single polarity only.

DESIGN EXERCISES

1. The transistor switches Q_1 and Q_2 in Figures 15.15(a)–15.15(d) are to be replaced with FET switches and MOSFET switches. In each, (i) draw the circuit diagrams, (ii) explain their working operations, (iii) draw waveforms at appropriate places, and (iv) deduce expression for their output voltages.

PRACTICAL EXERCISES

1. Verify the MCD circuit shown in Figure 15.15 by experimenting with the AFC trainer kit
2. Replace the transistor switches in Figure 15.15 with CD4066 non-inverter controlled switches and DG201 inverter controlled switches. In each, (i) draw the circuit diagrams, (ii) explain their working operation, (iii) draw waveforms at appropriate places, and (iv) deduce expression for the output voltage. Verify these MCDs by experimenting with the AFC trainer kit.
3. As discussed in Chapter 18 and Section 18.9, convert the MCDs of Figure 15.1 into SRM. (i) Draw the circuit diagrams, (ii) explain their working operation, (iii) draw waveforms at appropriate places, and (iv) deduce expression for the output voltage. Verify this SRM by experimenting with the AFC trainer kit.
4. As discussed in Chapter 19 and Section 19.6, convert the MCDs of Figure 15.1 into VMC. (i) Draw the circuit diagrams, (ii) explain their working operation, (iii) draw waveforms at appropriate places, and (iv) deduce expression for the output voltage. Verify this square rooter by experimenting with the AFC trainer kit.
5. Convert the MCD circuits obtained from Problem 2 into SRM. (i) Draw the circuit diagrams, (ii) explain their working operation, (iii) draw waveforms at appropriate places, and (iv) deduce expression for the output voltage. Verify this SRM by experimenting with the AFC trainer kit.
6. Convert the MCD circuits obtained from Problem 2 into VMC. (i) Draw the circuit diagrams, (ii) explain their working operation, (iii) draw waveforms at appropriate places, and (iv) deduce expression for the output voltage. Verify this VMC by experimenting with the AFC trainer kit.

16 Design of Peak Responding Multipliers-Cum-Dividers

Peak responding MCDs are classified into (i) peak detecting MCDs and (ii) peak sampling MCDs. A square wave/triangular waveform is generated whose time period 'T' is (i) proportional to one voltage (V_2) and (ii) inversely proportional to one voltage (V_1). The third input voltage V_3 is integrated during this time period 'T'. The peak value of the integrated output is proportional to $\frac{V_2V_3}{V_1}$. This is called a "double dual slope peak responding MCD." A short pulse/sawtooth waveform is generated whose time period 'T' is (i) proportional to one voltage (V_2) and (ii) inversely proportional to one voltage (V_1). The third input voltage V_3 is integrated during this time period 'T'. The peak value of the integrated output is proportional to $\frac{V_2V_3}{V_1}$. This is called a "double single slope peak responding MCD." At the output of a peak responding MCD, if peak detector is used, it is called "peak detecting MCD," and if sample and hold is used, it is called "peak sampling MCD." The peak responding MCDs of both peak detecting MCDs and peak sampling MCDs are described in this chapter.

16.1 DOUBLE SINGLE SLOPE MULTIPLIERS-CUM-DIVIDERS – SWITCHING

The circuit diagrams of double single slope peak responding MCDs are shown in Figure 16.1, and their associated waveforms are shown in Figure 16.2. Figure 16.1(a) shows a double single slope peak detecting multiplier, and Figure 16.1(b) shows a double single slope peak sampling multiplier. Let initially comparator OA_2 output be HIGH, transistor Q_1 be OFF, and an integrator formed by resistor R_1, capacitor C_1, and op-amp OA_1 integrate the input voltage ($-V_1$). The integrated output is given as

$$V_{S1} = -\frac{1}{R_1C_1}\int -V_1 dt = \frac{V_1}{R_1C_1}t \quad (16.1)$$

A positive-going ramp Vs_1 is generated at the output of op-amp OA_1. When the output of OA_1 reaches the voltage level of V_2, comparator OA_2 output becomes LOW. Transistor Q_1 is ON, and hence capacitor C_1 is shorted so that op-amp OA_1 output becomes ZERO. Then op-amp OA_2 output goes to HIGH, transistor Q_1 is OFF, and the integrator composed by R_1, C_1, and op-amp OA_1 integrates the input voltage ($-V_1$), and the cycle therefore repeats to provide (i) a sawtooth wave of peak value V_2

DOI: 10.1201/9781003221449-19

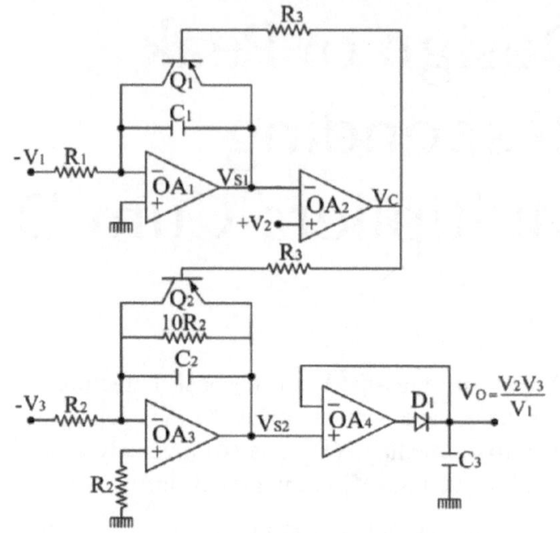

FIGURE 16.1(A) Switching double single slope peak detecting MCD.

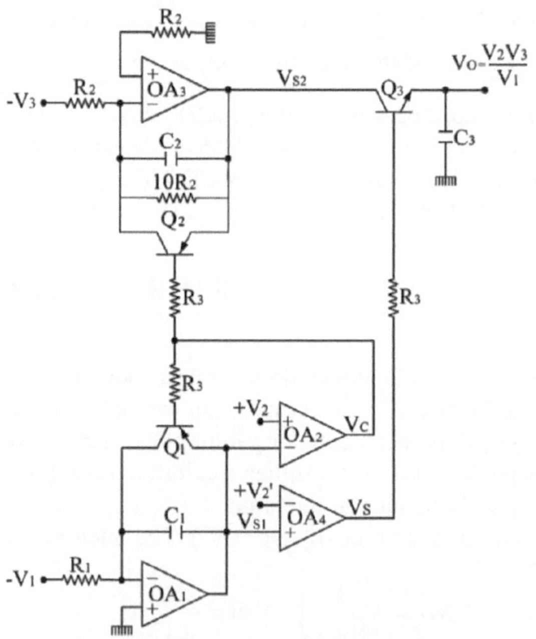

FIGURE 16.1(B) Switching double single slope peak sampling MCD.

at the output of op-amp OA_1 and (ii) a short pulse waveform V_C at the output of comparator OA_2. The short pulse V_C also controls transistor Q_2. During the short LOW time of V_C, transistor Q_2 is ON, and capacitor C_2 is short-circuited so that op-amp OA_3 output is zero volts. During HIGH time of V_C, transistor Q_2 is OFF, the

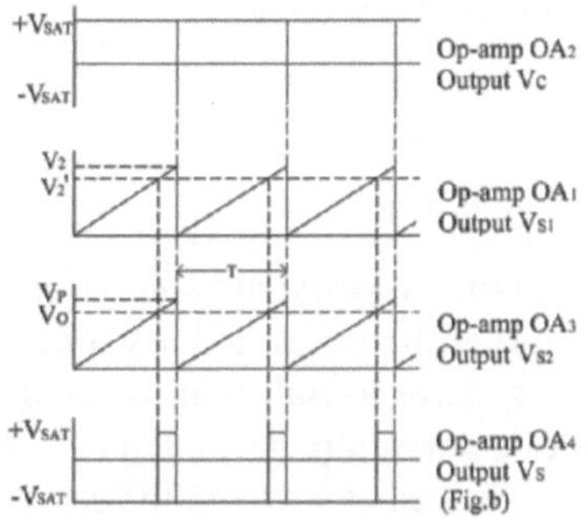

FIGURE 16.2 Associated waveforms of Figure 16.1.

integrator formed by resistor R_2, capacitor C_2, and op-amp OA_3 integrates its input voltage ($-V_3$) and its output is given as

$$V_{S2} = -\frac{1}{R_2C_2}\int -V_3 dt = \frac{V_3}{R_2C_2}t \tag{16.2}$$

Another sawtooth waveform V_{S2} with peak value V_P is generated at the output of op-amp OA_3. From the waveforms shown in Figure 16.2, Eqs. (16.1) and (16.2), and the fact that at $t = T$, $V_{S1} = V_2$, $V_{S2} = V_P$

$$V_2 = \frac{V_1}{R_1C_1}T \tag{16.3}$$

$$V_P = \frac{V_3}{R_2C_2}T \tag{16.4}$$

From Eqs. (16.3) and (16.4)

$$V_P = \frac{V_3}{R_2C_2}\frac{V_2}{V_1}R_1C_1$$

Let us assume $R_1 = R_2$ and $C_1 = C_2$, then

$$V_P = \frac{V_2V_3}{V_1} \tag{16.5}$$

In the circuit shown in Figure 16.1(a), the peak detector realized by op-amp OA_4, diode D_1, and capacitor C_3 gives this peak value V_P at its output V_O: $V_O = V_P$.

In the circuit shown in Figure 16.1(b), the peak value VP is obtained by the sample and hold circuit realized by transistor Q_3 and capacitor C_3. The sampling pulse is generated by op-amp OA_4 by comparing a slightly less than voltage of V_2 called V_2' with the sawtooth wave V_{S1}. The sample and hold operation is illustrated in Figure 16.2. The sample and hold output is $V_O = V_P$.

From Eq. (16.5), the output will be $V_O = V_P$.

$$V_O = \frac{V_2 V_3}{V_1} \tag{16.6}$$

The output voltage depends on the linearity and sharpness of the generated sawtooth waveforms. The offset voltages at the generated sawtooth waveforms will cause an error in the output; hence, they have to null with suitable external circuit (connect integrator output to one end of a potentiometer; the other end of this potentiometer is given to $+V_{CC}$). The sawtooth wave can take from the center pin of the potentiometer. A perfect sawtooth wave without any offset can be obtained by adjusting the potentiometer. (The AFC trainer kit has the offset adjustment facility.) The polarities of V_1 and V_2 must be opposite. In Figure 16.1(a) V_3 must be negative only. However, in Figure 16.1(b), V_3 may have any polarity.

DESIGN EXERCISES

1. The transistor switches in Figures 16.1(a) and 16.1(b) are to be replaced with FET switches and MOSFET switches. In each, (i) draw the circuit diagrams, (ii) explain their working operations, (iii) draw waveforms at appropriate places, and (iv) deduce expression for their output voltages.

PRACTICAL EXERCISES

1. Verify the MCD circuits shown in Figure 16.1 by experimenting with the AFC trainer kit.
2. As discussed in Chapter 18 and Section 18.6, convert the MCDs of Figure 16.1 into analog multipliers. (i) Draw the circuit diagrams, (ii) explain their working operation, (iii) draw waveforms at appropriate places, and (iv) deduce expression for the output voltage. Verify this multiplier by experimenting with the AFC trainer kit.
3. As discussed in Chapter 18 and Section 18.7, convert the MCDs of Figure 16.1 into analog dividers. (i) Draw the circuit diagrams, (ii) explain their working operation, (iii) draw waveforms at appropriate places, and (iv) deduce expression for the output voltage. Verify this divider by experimenting with the AFC trainer kit.
4. Replace the transistor multiplexers in Figure 16.1 with CD 4053 analog multiplexer IC (i) draw the circuit diagrams, (ii) explain their working operation, (iii) draw waveforms at appropriate places, and (iv) deduce expression for the output voltage. Verify these MCDs by experimenting with the AFC trainer kit.
5. As discussed in Chapter 18 and Section 18.9, convert the MCDs of Figure 16.1 into SRMs. (i) Draw the circuit diagrams, (ii) explain their working operation, (iii) draw waveforms at appropriate places, and (iv) deduce expression for the output voltage. Verify this SRM by experimenting with the AFC trainer kit.
6. As discussed in Chapter 19 and Section 19.6, convert the MCDs of Figure 16.1 into VMCs. (i) Draw the circuit diagrams, (ii) explain their working operation,

(iii) draw waveforms at appropriate places, and (iv) deduce expression for the output voltage. Verify this VMC by experimenting with the AFC trainer kit.

16.2 DOUBLE DUAL SLOPE PEAK RESPONDING MULTIPLIERS-CUM-DIVIDERS USING FEEDBACK COMPARATORS – MULTIPLEXING

The circuit diagrams of double dual slope peak responding MCDs using FBCs are shown in Figure 16.3, and their associated waveforms are shown in Figure 16.4. Figure 16.3(a) shows a peak detecting MCD, and Figure 16.3(b) shows a peak sampling MCD. Let initially comparator OA_2 output be LOW. The transistor multiplexer Q_1–Q_2 selects $(-V_1)$ to integrator I composed by resistor R_1, capacitor C_1, and op-amp OA_1 (transistor Q_1 is OFF and Q_2 is ON). The integrator I output is given as

$$V_{T1} = -\frac{1}{R_1C_1}\int(-V_1)dt = \frac{V_1}{R_1C_1}t \tag{16.7}$$

The output of integrator I is going toward positive saturation, and when it reaches a value $(+V_T)$, the comparator OA_2 output becomes HIGH. The transistor multiplexer Q_1–Q_2 selects $(+V_1)$ to integrator I composed by resistor R_1, capacitor C_1, and op-amp OA_1 (transistor Q_1 is ON and Q_2 is OFF). The integrator I output is given as

$$V_{T1} = -\frac{1}{R_1C_1}\int +V_1dt = -\frac{V_1}{R_1C_1}t \tag{16.8}$$

The output of integrator I is reversing toward negative saturation, and when it reaches a value $(-V_T)$, the comparator OA_2 output becomes LOW. The transistor

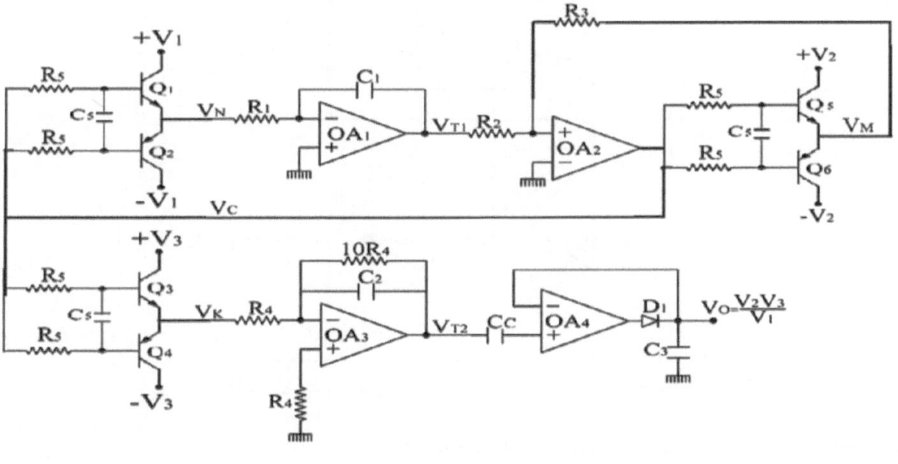

FIGURE 16.3(A) Double dual slope peak detecting multiplier-cum-divider with feedback comparator.

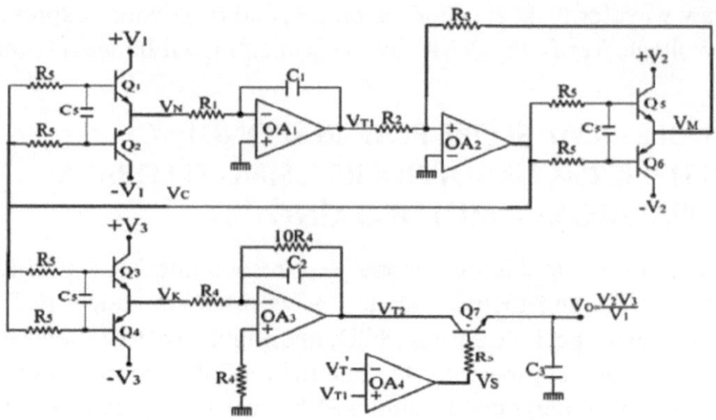

FIGURE 16.3(B) Double dual slope peak sampling multiplier-cum-divider with feedback comparator.

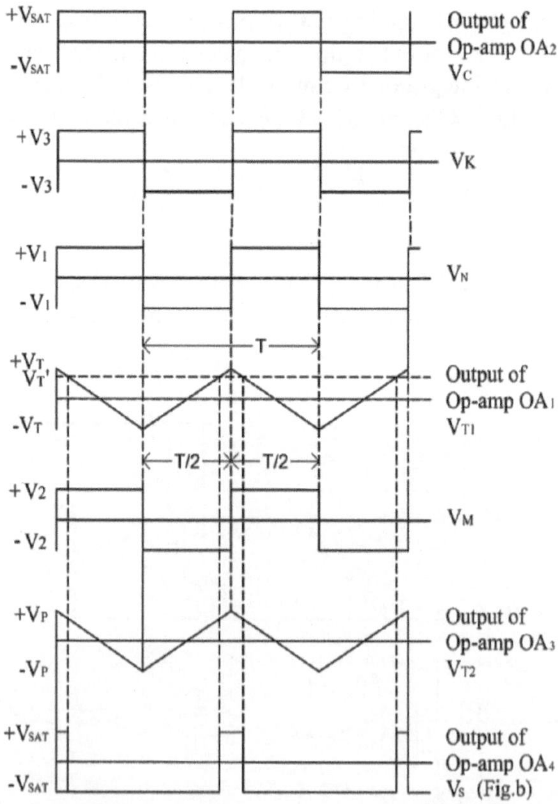

FIGURE 16.4 Associated waveforms of Figure 16.3.

multiplexer Q_1–Q_2 selects ($-V_1$), and the sequence repeats to give a triangular waveform V_{T1} of $\pm V_T$ peak-to-peak values with a time period 'T' at the output of op-amp OA_1, and a square waveform V_C at the output of comparator OA_2. From the waveforms shown in Figure 16.4, Eq. (16.7), and the fact that at $t = T/2$, $V_{T1} = 2V_T$

$$2V_T = \frac{V_1}{R_1C_1}\frac{T}{2}$$

$$T = \frac{4R_1C_1V_T}{V_1} \tag{16.9}$$

When the comparator OA_2 output is LOW ($-V_{SAT}$), ($-V_2$) will be at the output of transistor multiplexer Q_5–Q_6 (transistor Q_5 is OFF and Q_6 is ON); the effective voltage at the non-inverting terminal of comparator OA_2 will be by the superposition principle.

$$\frac{(-V_2)}{(R_2+R_3)}R_2 + \frac{(+V_T)}{(R_2+R_3)}R_3$$

When this effective voltage at the non-inverting terminal of comparator OA_2 becomes zero

$$\frac{(-V_2)R_2 + (+V_T)R_3}{(R_2+R_3)} = 0$$

$$(+V_T) = (+V_2)\frac{R_2}{R_3}$$

When the comparator OA_2 output is HIGH ($+V_{SAT}$), ($+V_2$) will be at the output of transistor multiplexer Q_5–Q_6 (transistor Q_5 is ON and Q_6 is OFF); the effective voltage at the non-inverting terminal of comparator OA_2 will be by the superposition principle

$$\frac{(+V_2)}{(R_2+R_3)}R_2 + \frac{(-V_T)}{(R_2+R_3)}R_3$$

When this effective voltage at the non-inverting terminal of comparator OA_2 becomes zero

$$\frac{(+V_2)R_2 + (-V_T)R_3}{(R_2+R_3)} = 0$$

$$(-V_T) = (-V_2)\frac{R_2}{R_3}$$

$$\pm V_T = \pm V_2 \frac{R_2}{R_3} \tag{16.10}$$

The transistor multiplexer Q_3–Q_4 connects ($+V_3$) during HIGH of the square waveform V_C (transistor Q_3 is ON and Q_4 is OFF) and ($-V_3$) during LOW of V_C (transistor Q_3 is OFF and Q_4 is ON). Another square waveform V_K with $\pm V_3$ peak-to-peak value is generated at the output of transistor multiplexer Q_3–Q_4. This square wave V_K is converted into a triangular wave V_{T2} by the integrator-II composed by resistor R_4, capacitor C_2, and op-amp OA_3 with $\pm V_P$ as peak-to-peak values of same time period 'T'. For one transition, the integrator-II output is given as

$$V_{T2} = -\frac{1}{R_4 C_2}\int(-V_3)\,dt = \frac{V_3}{R_4 C_2} t \tag{16.11}$$

From the waveforms shown in Figure 16.4, Eq. (16.11), and the fact that at $t = T/2$, $V_{T2} = 2V_p$

$$2V_P = \frac{V_3}{R_4 C_2}\frac{T}{2} \tag{16.12}$$

Equations (16.9) and (16.10) in (16.12) give

$$V_P = \frac{V_2 V_3}{V_1}\frac{R_1 C_1}{R_4 C_2}\frac{R_2}{R_3}$$

Let $R_1 = R_4$ and $C_1 = C_2$.

$$V_P = \frac{V_2 V_3}{V_1}\frac{R_2}{R_3}$$

Let $R_2/R_3 = 1$(but in practical $R_2/R_3 < 1$).

$$V_P = \frac{V_2 V_3}{V_1} \tag{16.13}$$

In Figure 16.3(a), the peak detector at output stage gives peak value 'V_P' of triangular wave V_{T2} and hence $V_O = V_P$.

In Figure 16.3(b), the peak value V_P of the triangular waveform V_{T2} is obtained by the sample and hold circuit realized by transistor Q_7 and capacitor C_3. The sampling pulse V_S is generated by op-amp OA_4 by comparing a slightly less than voltage of V_T called V_T' with the triangular wave V_{T1}. The sampled output is given as $V_O = V_P$.

From Eq. (16.13), $V_O = V_P$

$$V_O = \frac{V_2 V_3}{V_1} \tag{16.14}$$

The output voltage depends on the sharpness and linearity of generated triangular waveforms. The offset voltage occurred in the second triangular wave V_{T2} will create an error in the output voltage. In Figure 16.3(a), the capacitor CC is used to block DC offset voltage and allows only a triangular waveform to a peak detector. Its value is 0.1 μF disk. In Figure 16.3(b), the offset adjustment is to be done externally by connecting a 0.1 μF disk between output of the second integrator and a buffer. The buffer output is to give to sample and hold circuit. This offset removing arrangement is available in the AFC trainer kit. In Figure 16.3(a), all inputs should have single polarity only. In Figure 16.3(b), V_1, V_2 should have single polarity, and V_3 may have any polarity.

DESIGN EXERCISES

1. The transistor multiplexers Q_1–Q_2, Q_3–Q_4, and Q_5–Q_6 in Figures 16.3(a) and 16.3(b) are to be replaced with FET multiplexers and MOSFET multiplexers. In each, (i) draw the circuit diagrams, (ii) explain their working operation, (iii) draw waveforms at appropriate places, and (iv) deduce expression for the output voltage.

PRACTICAL EXERCISES

1. Verify the MCD circuits shown in Figure 16.3 by experimenting with the AFC trainer kit.
2. As discussed in Chapter 18 and Section 18.9, convert the MCDs of Figure 16.1 into analog multipliers. (i) Draw the circuit diagrams, (ii) explain their working operation, (iii) draw waveforms at appropriate places, and (iv) deduce expression for the output voltage. Verify this multiplier by experimenting with the AFC trainer kit.
3. As discussed in Chapter 18 and Section 18.7, convert the MCDs of Figure 16.1 into analog dividers. (i) Draw the circuit diagrams, (ii) explain their working operation, (iii) draw waveforms at appropriate places, and (iv) deduce expression for the output voltage. Verify this divider by experimenting with the AFC trainer kit.
4. Replace the transistor multiplexers in Figure 16.3 with CD 4053 analog multiplexer IC. In each, (i) draw the circuit diagrams, (ii) explain their working operation, (iii) draw waveforms at appropriate places, and (iv) deduce expression for the output voltage. Verify these MCDs by experimenting with the AFC trainer kit.
5. As discussed in Chapter 18 and Section 18.9, convert the MCDs of Figure 16.1 into SRMs. (i) Draw the circuit diagrams, (ii) explain their working operation, (iii) draw waveforms at appropriate places, and (iv) deduce expression for the output voltage. Verify this SRM by experimenting with the AFC trainer kit.
6. As discussed in Chapter 19 and Section 19.6, convert the MCDs of Figure 16.1 into VMCs. (i) Draw the circuit diagrams, (ii) explain their working operation,

(iii) draw waveforms at appropriate places, and (iv) deduce expression for the output voltage. Verify this VMC by experimenting with the AFC trainer kit.

16.3 DOUBLE DUAL SLOPE PEAK RESPONDING MULTIPLIERS-CUM-DIVIDERS WITH FLIP FLOP – MULTIPLEXING

The circuit diagram of double dual slope peak responding MCDs are shown in Figure 16.5, and their associated waveforms are shown in Figure 16.6. Figure 16.5(a) shows a peak detecting MCD, and Figure 16.5(b) shows a peak sampling MCD. Let initially flip flop output be LOW. The transistor multiplexer Q_1–Q_2 selects ($-V_1$) to integrator-I composed by resistor R_1, capacitor C_1, and op-amp OA_1 (transistor Q_1 is OFF and Q_2 is ON). The output of op-amp OA_1 is given as

$$V_{T1} = -\frac{1}{R_1C_1}\int(-V_1)dt = \frac{V_1}{R_1C_1}t \qquad (16.15)$$

The output of op-amp OA_1 is going toward positive saturation, and when it reaches the value $+V_2$, the comparator OA_2 output becomes HIGH, and it sets the flip flop output to HIGH. The transistor multiplexer Q_1–Q_2 selects ($+V_1$) to the integrator composed by resistor R_1, capacitor C_1, and op-amp OA_1 (transistor Q_1 is ON and Q_2 is OFF). The output of op-amp OA_1 is given as

$$V_{T1} = -\frac{1}{R_1C_1}\int(+V_1)dt = -\frac{V_1}{R_1C_1}t \qquad (16.16)$$

The output of op-amp OA_1 is reversing toward negative saturation, and when it reaches the value ($-V_2$), the comparator OA_3 output becomes HIGH and resets the

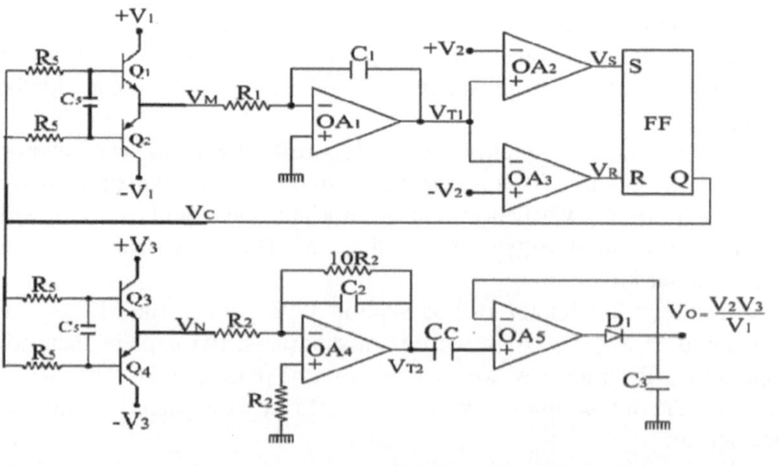

FIGURE 16.5(A) Double dual slope peak detecting MCD with flip flop.

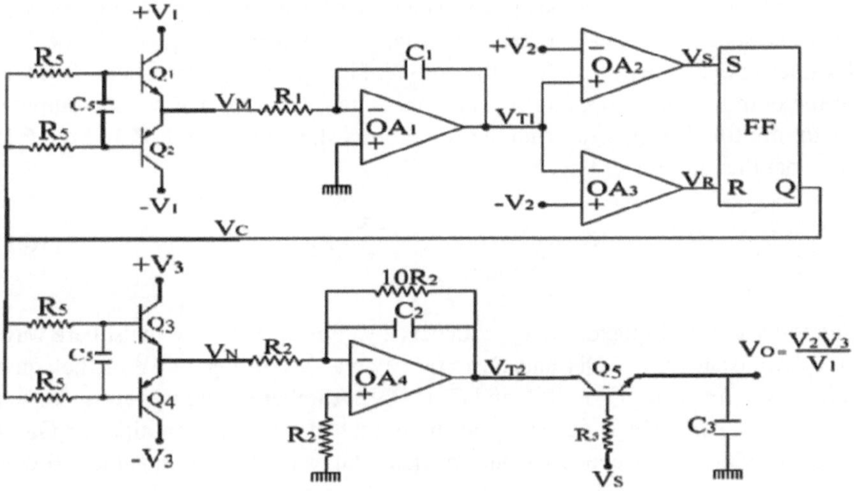

FIGURE 16.5(B) Double dual slope peak sampling MCD with flip flop.

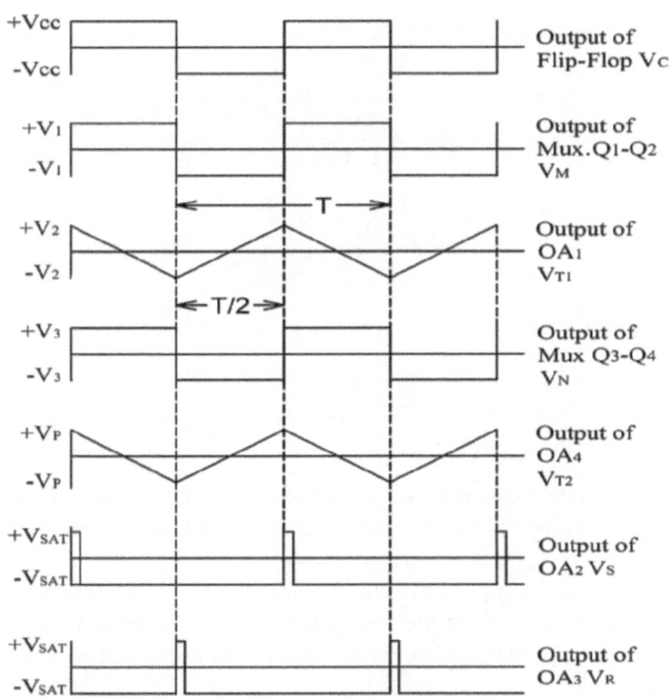

FIGURE 16.6 Associated waveforms of Figure 16.5.

flip flop so that it output becomes LOW. The transistor multiplexer Q_1–Q_2 selects (–V_1), and the sequence repeats to give (i) a triangular waveform V_{T1} of $\pm V_2$ peak-to-peak values with a time period 'T' at the output of integrator-I, (ii) a square waveform V_C at the output of flip flop, and (iii) another square waveform V_M at the output of transistor multiplexer Q_1–Q_2. From the waveforms shown in Figure 16.6, Eq. 16.15, and the fact that at $t = T/2$, $V_{T1} = 2V_2$

$$2V_2 = \frac{V_1}{R_1C_1}\frac{T}{2}, \quad T = \frac{4V_2}{V_1}R_1C_1 \tag{16.17}$$

The transistor multiplexer Q_3–Q_4 connects $+V_3$ during HIGH of the square waveform V_C (transistor Q_3 is ON and Q_4 is OFF) and $-V_3$ during LOW of the square waveform V_C (transistor Q_3 is OFF and Q_4 is ON). Another square waveform V_N with $\pm V_3$ peak-to-peak value is generated at the output of transistor multiplexer Q_3–Q_4. This square wave V_N is converted into a triangular wave V_{T2} by integrator-II composed by resistor R_2, capacitor C_2, and op-amp OA_4 with $\pm V_P$ as peak-to-peak values of the same time period 'T.' For one transition, the op-amp OA_4 output is given as

$$V_{T2} = -\frac{1}{R_2C_2}\int(-V_3)\,dt = \frac{V_3}{R_2C_2}t \tag{16.18}$$

From the waveforms shown in Figure 16.6, Eq. (16.18), and the fact that at $t = T/2$, $V_{T2} = 2V_p$

$$2V_P = \frac{V_3}{R_2C_2}\frac{T}{2}$$

$$V_P = \frac{V_2V_3}{V_1}\frac{R_1C_1}{R_2C_2}$$

Let $R_1 = R_2$ and $C_1 = C_2$.

$$V_P = \frac{V_2V_3}{V_1} \tag{16.19}$$

In Figure 16.5(a), the peak detector realized by op-amp OA_5, diode D_1, and capacitor C_3 gives peak value V_P of the triangular wave V_{T2} at its output and, hence, $V_O = V_P$.

In Figure 16.5(b), the sample and hold circuit realized by transistor Q_5 and capacitor C_3 gives the peak value V_P of the triangular wave V_{T2}. The short pulse V_S generated at the output of comparator OA_2 is acting as a sampling pulse. The sample and hold output is $V_O = V_P$.

From Eq. (16.19), $V_O = V_P$

$$V_O = \frac{V_2V_3}{V_1} \tag{16.20}$$

The output voltage depends on the sharpness and linearity of the generated triangular waveforms. The offset voltage that occurred in the second triangular wave V_{T2} will create an error in the output voltage. In Figure 16.5(a), the capacitor CC is used to block DC offset voltage and allows only a triangular waveform to a peak detector. Its value is 0.1 μF disk. In Figure 16.5(b), the offset adjustment is to be done externally by connecting a 0.1 μF disk between output of the second integrator and a buffer. The buffer output is to give to the sample and hold circuit. This offset removing arrangement is available in the AFC trainer kit. In Figure 16.5(a), all inputs should have single polarity only. In Figure 16.5(b), V_1 and V_2 should have single polarity, and V_3 may have any polarity.

DESIGN EXERCISES

1. The transistor multiplexers Q_1–Q_2 and Q_3–Q_4 in Figures 16.5(a) and 16.5(b) are to be replaced with FET multiplexers and MOSFET multiplexers. In each, (i) draw the circuit diagrams, (ii) explain their working operation, (iii) draw waveforms at appropriate places, and (iv) deduce expression for their output voltage.
2. In the MCD circuit shown in Figure 16.5(a) añd 16.5(b), if the input terminals of transistor multiplexers Q_3–Q_4 are reversed, and the sampling pulse V_S is replaced by V_R, (i) draw the circuit diagrams, (ii) explain their working operation, (iii) draw waveforms at appropriate places, and (iv) deduce expression for their output voltage.

PRACTICAL EXERCISES

1. Verify the MCD circuits shown in Figure 16.5 by experimenting with the AFC trainer kit.
2. As discussed in Chapter 18 and Section 18.6, convert the MCDs of Figure 16.5 into analog multipliers. (i) Draw the circuit diagrams, (ii) explain their working operation, (iii) draw waveforms at appropriate places, and (iv) deduce expression for the output voltage. Verify this multiplier by experimenting with the AFC trainer kit.
3. As discussed in Chapter 18 and Section 18.7, convert the MCDs of Figure 16.5 into analog dividers. (i) Draw the circuit diagrams, (ii) explain their working operation, (iii) draw waveforms at appropriate places, and (iv) deduce expression for the output voltage. Verify this dividers by experimenting with the AFC trainer kit.
4. Replace the transistor multiplexers in Figure 16.5 with CD4053 analog multiplexer IC. (i) Draw the circuit diagrams, (ii) explain their working operation, (iii) draw waveforms at appropriate places, and (iv) deduce expression for the output voltage. Verify these MCDs by experimenting with the AFC trainer kit.
5. As discussed in Chapter 18 and Section 18.9, convert the MCDs of Figure 16.5 into SRMs. (i) Draw the circuit diagrams, (ii) explain their working operation, (iii) draw waveforms at appropriate places, and (iv) deduce expression for the output voltage. Verify this SRM by experimenting with the AFC trainer kit.

6. As discussed in Chapter 19 and Section 19.6, convert the MCDs of Figure 16.5 into VMCs. (i) Draw the circuit diagrams, (ii) explain their working operation, (iii) draw waveforms at appropriate places, and (iv) deduce expression for the output voltage. Verify this VMC by experimenting with the AFC trainer kit.

16.4 DOUBLE DUAL SLOPE MULTIPLIERS-CUM-DIVIDERS WITH FLIP FLOP – SWITCHING

The MCDs using double dual slope principle with flip flop are shown in Figure 16.7, and their associated waveforms are shown in Figure 16.8. Figure 16.7(a) shows a series switching peak detecting MCD, Figure 16.7(b) shows a shunt switching peak detecting MCD, Figure 16.7(c) shows a series switching peak sampling MCD, and Figure 16.7(d) shows a shunt switching peak sampling MCD.

Let initially flip flop output V_C be LOW.

In Figures 16.7(a) and 16.7(c), transistor Q_1 is OFF, op-amp OA_6 along with resistors R_5 will work as an inverting amplifier, and $(-V_1)$ will be at its output $(V_M = -V_1)$. Transistor Q_2 is OFF, op-amp OA_7 along with resistors R_6 will work as an inverting amplifier, and $(-V_3)$ will be at its output $(V_N = -V_3)$.

In Figures 16.7(b) and 16.7(d), transistor Q_1 is OFF, op-amp OA_6 along with resistors R_5 will work as a non-inverting amplifier, and $(-V_1)$ will be at its output $(V_M = -V_1)$. Transistor Q_2 is OFF, op-amp OA_7 along with resistors R_6 will work as a non-inverting amplifier, and $(-V_3)$ will be at its output $(V_N = -V_3)$.

The output of integrator-I composed by resistor R_1, capacitor C_1, and op-amp OA_1 will be

$$V_{T1} = -\frac{1}{R_1C_1}\int(-V_1)dt = \frac{V_1}{R_1C_1}t \tag{16.21}$$

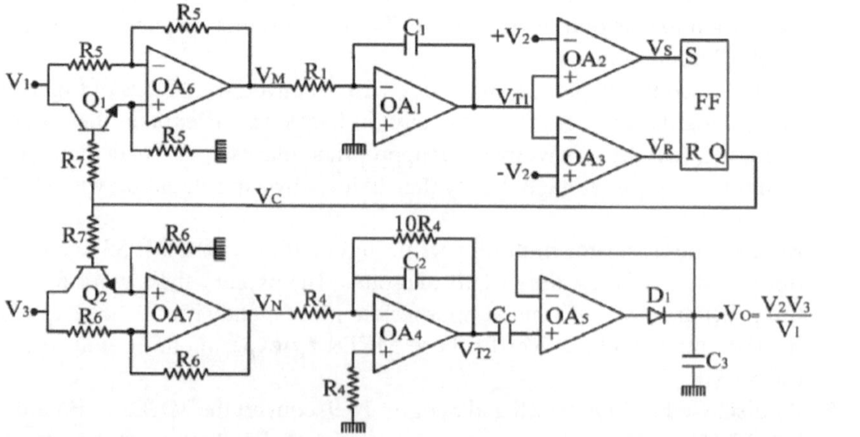

FIGURE 16.7(A) Series switching double dual slope peak detecting multiplier-cum-divider.

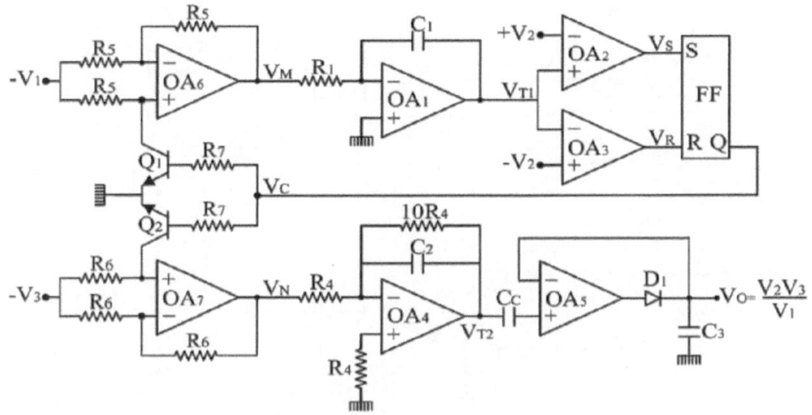

FIGURE 16.7(B) Shunt switching double dual slope peak detecting multiplier-cum-divider.

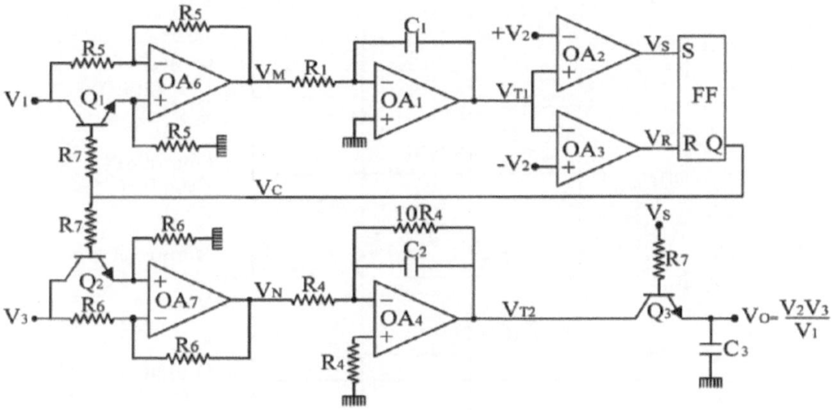

FIGURE 16.7(C) Series switching double dual slope peak sampling multiplier-cum-divider.

The output of integrator-I is going toward positive saturation, and when it reaches the value $+V_2$, the comparator OA_2 output becomes HIGH and it sets the flip flop output to HIGH.

In Figures 16.7(a) and 16.7(c), transistor Q_1 is ON, op-amp OA_6 along with resistors R_5 will work as a non-inverting amplifier, and $(+V_1)$ will be at its output $(V_M = +V_1)$. Transistor Q_2 is ON, op-amp OA_7 along with resistors R_6 will work as a non-inverting amplifier, and $(+V_3)$ will be at its output $(V_N = +V_3)$.

In Figures 16.7(b) and 16.7(d), transistor Q_1 is ON, op-amp OA_6 along with resistors R_5 will work as an inverting amplifier, and $(+V_1)$ will be at its output $(V_M = +V_1)$. Transistor Q_2 is ON, op-amp OA_7 along with resistors R_6 will work as an inverting amplifier, and $(+V_3)$ will be at its output $(V_N = +V_3)$.

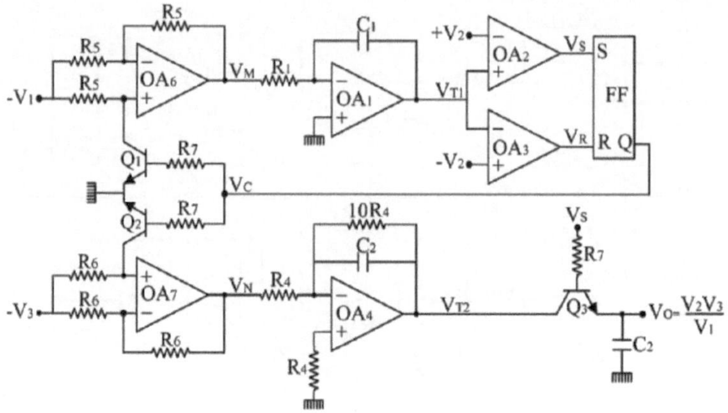

FIGURE 16.7(D) Shunt switching double dual slope peak sampling multiplier-cum-divider.

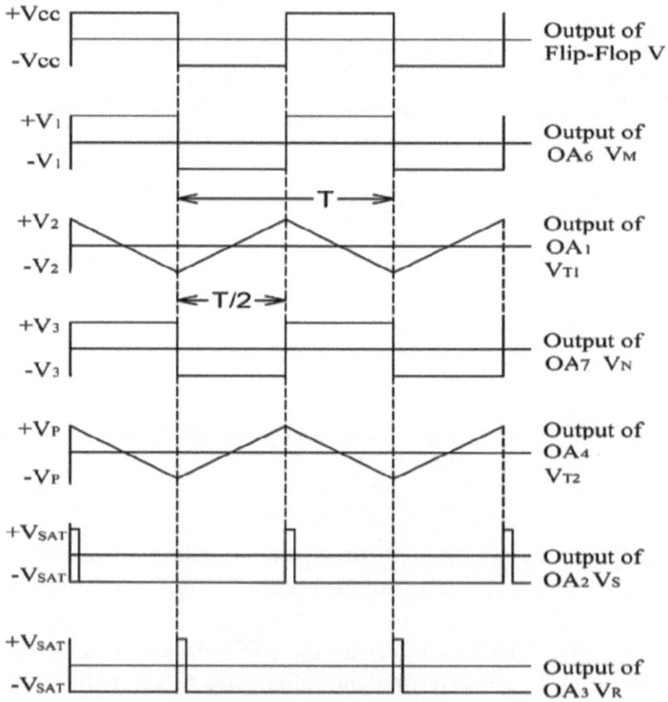

FIGURE 16.8 Associated waveforms of Figure 16.7.

The output of integrator-I composed by resistor R_1, capacitor C_1, and op-amp OA_1 will now be

$$V_{T1} = -\frac{1}{R_1C_1}\int(+V_1)dt = -\frac{V_1}{R_1C_1}t \tag{16.22}$$

The output of integrator OA_1 is reversing toward negative saturation, and when it reaches the value $(-V_2)$, the comparator OA_3 output becomes HIGH and resets the flip flop so that its output V_C becomes LOW, and the sequence repeats to give (i) a triangular waveform V_{T1} of $\pm V_2$ peak-to-peak values with a time period 'T' at the output of integrator-I, (ii) a square waveform V_C at the output of flip flop, (iii) another square waveform V_M with $\pm V_1$ peak-to-peak value at the output of op-amp OA_6, and (iv) third square waveform V_N with $\pm V_3$ peak-to-peak value is generated at the output of op-amp OA_7. From the waveforms shown in Fig 16.8, Eq. (16.21) and the fact that at $t = T/2$, $V_{T1} = 2V_2$

$$2V_2 = \frac{V_1}{R_1C_1}\frac{T}{2}, \quad T = \frac{4V_2}{V_1}R_1C_1 \tag{16.23}$$

The third square wave V_N is converted into triangular wave V_{T2} by integrator-II composed by resistor R_4, capacitor C_2, and op-amp OA_4 with $\pm V_P$ as peak-to-peak values of the same time period 'T'. For one transition, the integrator-II output is given as

$$V_{T2} = -\frac{1}{R_4C_2}\int(-V_3)dt = \frac{V_3}{R_4C_2}t \tag{16.24}$$

From the waveforms shown in Figure 16.8, Eq. (16.24), and the fact that at $t = T/2$, $V_{T2} = 2V_p$

$$2V_P = \frac{V_3}{R_4C_2}\frac{T}{2}$$

$$V_P = \frac{V_2V_3}{V_1}\frac{R_1C_1}{R_4C_2}$$

Let $R_1 = R_4$ and $C_1 = C_2$.

$$V_P = \frac{V_2V_3}{V_1} \tag{16.25}$$

In the MCD circuits shown in Figures 16.7(a) and 16.7(b), the peak detector realized by op-amp OA_5, diode D_1, and capacitor C_3 gives peak value V_P of the triangular wave V_{T2} and, hence, $V_O = V_P$.

In the MCD circuits shown in Figures 16.7(c) and 16.7(d), the peak value V_P of the triangular waveform V_{T2} is obtained by the sample and hold circuit realized by transistor Q_3 and capacitor C_3. The comparator OA_2 output V_S is acting as a sampling pulse. The sampled output is given as $V_O = V_{P.}$

$$V_O = \frac{V_2 V_3}{V_1} \tag{16.26}$$

The output voltage depends on the sharpness and linearity of the generated triangular waveforms. The offset voltage that occurred in the second triangular wave V_{T2} will create an error in the output voltage. In Figures 16.7(a) and 16.7(b) the capacitor CC is used to block DC offset voltage and allows only triangular waveform to peak detector. Its value is 0.1 μF disk. In Figures 16.7(a) and 16.7(b), the offset adjustment is to be done externally by connecting a 0.1 μF disk between output of the second integrator and a buffer. The buffer output is to give to the sample and hold circuit. This offset removing arrangement is available in the AFC trainer kit. In Figures 16.7(a) and 16.7(b), all inputs should have single polarity only. In Figure 16.7(b), V_1 and V_2 should have single polarity, and V_3 may have any polarity.

DESIGN EXERCISES

1. The transistor switches in Figures 16.7(a)–(d) are to be replaced with FET switches and MOSFET switches. In each, (i) draw the circuit diagrams, (ii) explain their working operations, (iii) draw waveforms at appropriate places, and (iv) deduce expression for their output voltages.

PRACTICAL EXERCISES

1. Verify the MCD circuits shown in Figure 16.7 by experimenting with the AFC trainer kit.
2. As discussed in Chapter 18 and Section 18.6, convert the MCDs of Figure 16.7 into analog multipliers. (i) Draw the circuit diagrams, (ii) explain their working operation, (iii) draw waveforms at appropriate places, and (iv) deduce expression for the output voltage. Verify this MCD by experimenting with the AFC trainer kit.
3. As discussed in Chapter 18 and Section 18.7, convert the MCDs of Figure 16.5 into analog divider. (i) Draw the circuit diagrams, (ii) explain their working operation, (iii) draw waveforms at appropriate places, and (iv) deduce expression for the output voltage. Verify this divider by experimenting with the AFC trainer kit.
4. Replace the transistor switches in Figure 16.7 with CD4066 non-inverted controlled analog switches and DG201 inverted controlled analog switches. In each, (i) draw the circuit diagrams, (ii) explain their working operation, (iii) draw waveforms at appropriate places, and (iv) deduce expression for the output voltage. Verify these MCDs by experimenting with the AFC trainer kit.

5. As discussed in Chapter 18 and Section 18.9, convert the MCDs of Figure 16.7 into SRMs. (i) Draw the circuit diagrams, (ii) explain their working operation, (iii) draw waveforms at appropriate places, and (iv) deduce expression for the output voltage. Verify this SRM by experimenting with the AFC trainer kit.
6. As discussed in Chapter 19 and Section 19.6, convert the MCDs of Figure 16.7 into VMCs. (i) Draw the circuit diagrams, (ii) explain their working operation, (iii) draw waveforms at appropriate places, and (iv) deduce expression for the output voltage. Verify this VMC by experimenting with the AFC trainer kit.

16.5 PULSE WIDTH INTEGRATED PEAK RESPONDING MULTIPLIERS-CUM-DIVIDERS – SWITCHING

The circuit diagrams of pulse width integrated peak responding MCDs are shown in Figure 16.9, and their associated waveforms are shown in Figure 16.10. Figure 16.9(a) shows a pulse width integrated peak detecting MCD, and Figure 16.9(b) shows a pulse width integrated peak sampling MCD. Let initially the comparator OA_2 output be HIGH, the transistor Q_1 be OFF, and the integrator formed by resistor R_1, capacitor C_1, and op-amp OA_1 integrate $(-V_1)$. The integrated output is given as

$$V_{S1} = -\frac{1}{R_1C_1}\int -V_1 dt = \frac{V_1}{R_1C_1}t \tag{16.27}$$

When the output of op-amp OA_1 is rising toward positive saturation, and it reaches the value V_R, the comparator OA_2 output will become LOW, the transistor Q_1 is ON,

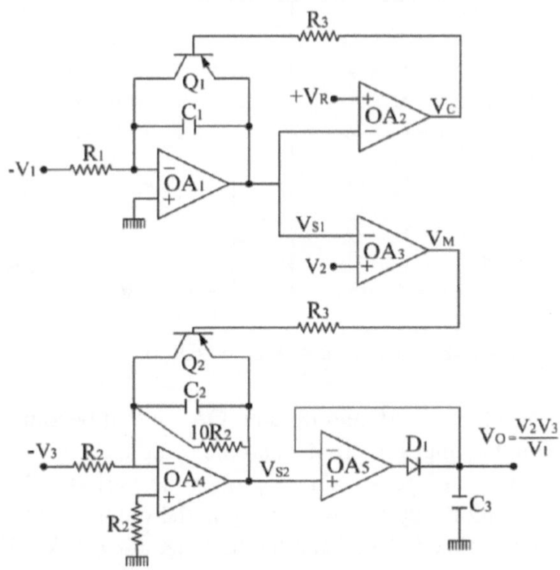

FIGURE 16.9(A) Pulse width integrated peak detecting multiplier-cum-divider.

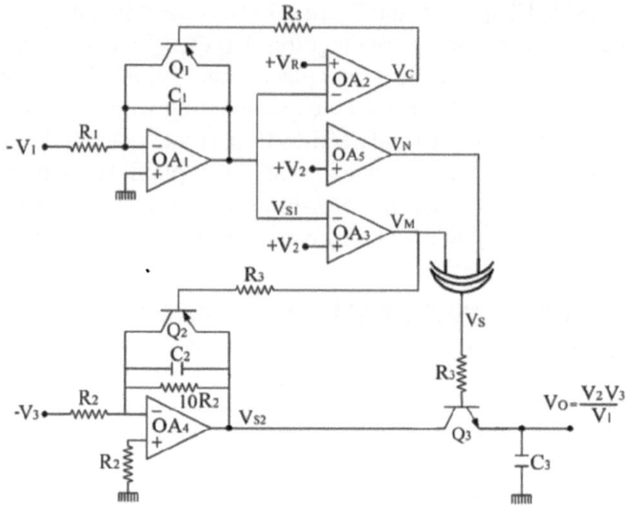

FIGURE 16.9(B) Pulse width integrated peak sampling multiplier-cum-divider.

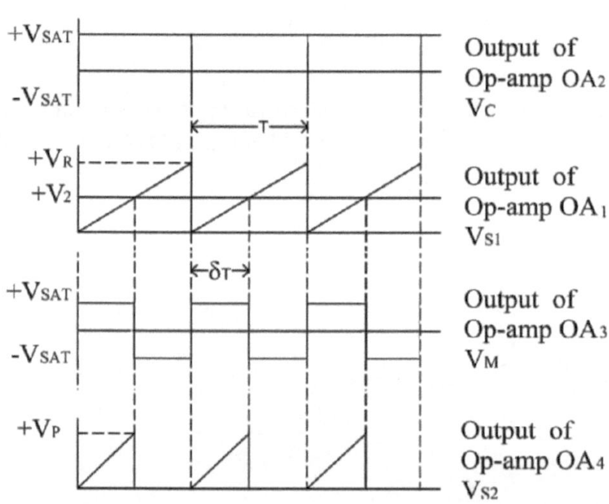

FIGURE 16.10(A) Associated waveforms of Figure 16.9(a).

the capacitor C_1 is short-circuited, and op-amp OA_1 output becomes zero. Now the comparator OA_2 output changes to HIGH, and the cycle therefore repeats to give (i) a sawtooth waveform V_{S1} of peak value V_R and time period 'T' at the output of op-amp OA_1 and (ii) a short pulse waveform V_C at the output of op-amp OA_2. From the waveforms shown in Figure 16.10 and the fact that at t = T, $V_{S1} = V_R$

$$V_R = \frac{V_1}{R_1C_1}T, T = \frac{V_R}{V_1}R_1C_1 \tag{16.28}$$

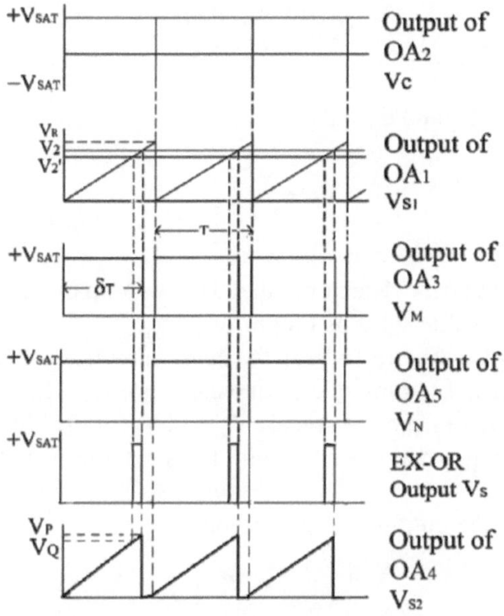

FIGURE 16.10(B) Associated waveforms of Figure 16.9(b).

The sawtooth waveform V_{S1} is compared with the second input voltage V_2 by comparator OA_3. An asymmetrical rectangular wave V_M is generated at the output of comparator OA_3. The ON time of this wave is given as

$$\delta_T = \frac{V_2}{V_R} T \tag{16.29}$$

The output of comparator OA_3 is given as control input of transistor Q_2. During OFF time of V_M, transistor Q_2 is ON and capacitor C_2 is shorted so that zero volts appears at op-amp OA_4 output. During ON time of V_M, transistor Q_2 is OFF and another integrator is formed by resistor R_2, capacitor C_2, and op-amp OA_4. This integrator integrates the input voltage $(-V_3)$ and its output is given as

$$V_{S2} = -\frac{1}{R_2C_2}\int -V_3 dt = \frac{V_3}{R_2C_2} t \tag{16.30}$$

A semi-sawtooth wave V_{S2} with peak values of V_P is generated at the output of op-amp OA_4. From the waveforms shown in Figure 16.10, Eq. (16.30), and the fact that at $t = \delta_T$, $V_{S2} = V_P$

$$V_P = \frac{V_3}{R_2C_2} \delta_T$$

$$V_P = \frac{V_2 V_3}{V_1} \frac{R_1 C_1}{R_2 C_2}$$

Let us assume $R_1 = R_2$ and $C_1 = C_2$

$$V_P = \frac{V_2 V_3}{V_1} \tag{16.31}$$

In Figure 16.9(a), the peak detector realized by op-amp $\grave{O}A_5$, diode D_1, and capacitor C_3 gives this peak value V_P at its output: $V_0 = V_P$.

In the circuit shown in Figure 16.9(b), the peak value V_P is obtained by the sample and hold circuit realized by transistor Q_3 and capacitor C_3. The sampling pulse V_S is generated by Ex-OR gate from the signals V_M and V_N. V_N is obtained by comparing slightly less than voltage of V_2, i.e., V_2' with the sawtooth waveform V_{S1}. The sampled output is given as $V_O = V_P$.

From Eq. (16.31), the output voltage will be

$$V_O = \frac{V_2 V_3}{V_1} \tag{16.32}$$

The output voltage depends on the linearity and sharpness of the generated sawtooth waveforms. The offset voltages at the generated sawtooth waveforms will cause error in the output; hence, they have to null with suitable external circuit (connect integrator output to one end of a potentiometer, the other end of this potentiometer is given to $+V_{CC}$). The sawtooth wave can take from the center pin of the potentiometer. A perfect sawtooth wave without any offset can be obtained by adjusting the potentiometer. (The AFC trainer kit has this offset adjustment facility.) The polarities of V_1 and V_2 must be opposite. In Figure 16.9(a) V_3 must be negative only. However, in Figure 16.9(b), V_3 may have any polarity.

DESIGN EXERCISES

1. The transistor switches in Figure 8.10 are to be replaced with FET switches and MOSFET switches. In each, (i) draw the circuit diagrams, (ii) explain their working operations, (iii) draw waveforms at appropriate places, and (iv) deduce expression for their output voltages.

PRACTICAL EXERCISES

1. Verify the MCD circuits shown in Figure 16.9 by experimenting with the AFC trainer kit.
2. As discussed in Chapter 18 and Section 18.6, convert the MCDs of Figure 16.9 into analog multipliers. (i) Draw the circuit diagrams, (ii) explain their working operation (iii) draw waveforms at appropriate places and (iv) deduce expression for the output voltage. Verify this MCD by experimenting with the AFC trainer kit.

3. As discussed in Chapter 18 and Section 18.7, convert the MCDs of Figure 16.9 into analog dividers. (i) Draw the circuit diagrams, (ii) explain their working operation, (iii) draw waveforms at appropriate places, and (iv) deduce expression for the output voltage. Verify this divider by experimenting with the AFC trainer kit.
4. Replace the transistor switches in Figure 16.9 with CD4066 non-inverted controlled analog switches and DG201 inverted controlled analog switches. In each, (i) draw the circuit diagrams, (ii) explain their working operation, (iii) draw waveforms at appropriate places, and (iv) deduce expression for the output voltage. Verify these MCDs by experimenting with the AFC trainer kit.
5. As discussed in Chapter 18 and Section 18.9, convert the MCDs of Figure 16.9 into SRM. (i) Draw the circuit diagrams, (ii) explain their working operation, (iii) draw waveforms at appropriate places, and (iv) deduce expression for the output voltage. Verify this SRM by experimenting with the AFC trainer kit.
6. As discussed in Chapter 19 and Section 19.6, convert the MCDs of Figure 16.9 into VMCs. (i) Draw the circuit diagrams, (ii) explain their working operation, (iii) draw waveforms at appropriate places, and (iv) deduce expression for the output voltage. Verify this VMC by experimenting with the AFC trainer kit.

16.6 MULTIPLIERS-CUM-DIVIDERS USING VOLTAGE TUNABLE ASTABLE MULTIVIBRATORS – MULTIPLEXING

The peak responding MCDs using voltage tunable astable multivibrators are shown in Figure 16.11, and their associated waveforms are shown in Figure 16.12. Figure 16.11(a) shows a peak detecting MCD and Figure 16.11(b) shows a peak sampling MCD. The op-amp OA_1 in association with transistor multiplexers Q_1–Q_2 and Q_5–Q_6 along with the components R_1, R_2, R_3, and C_1 produces a square waveform V_C. The time period 'T' of this square waveform V_C is proportional to the input voltage V_2 and inversely proportional to another input voltage V_1.

$$T = K\frac{V_2}{V_1} \tag{16.33}$$

where 'K' is constant. During LOW value of V_C, the transistor multiplexer Q_3–Q_4 selects $+V_3$ (transistor Q_3 is OFF and Q_4 is ON), and during HIGH value of V_C, $(-V_3)$ (transistor Q_3 is ON and Q_4 is OFF) is connected to the integrator OA_2. Another square wave V_K with $\pm V_3$ peak-to-peak value is generated at the transistor multiplexer Q_3–Q_4 output. This square wave V_K is converted into triangular wave V_{T2} with $\pm V_P$ peak-to-peak value by the integrator formed by resistor R_4, capacitor C_2, and op-amp OA_2. For one transition, the integrator output is given as

$$V_{T2} = -\frac{1}{R_4C_2}\int(-V_3)dt = \frac{V_3}{R_4C_2}t \tag{16.34}$$

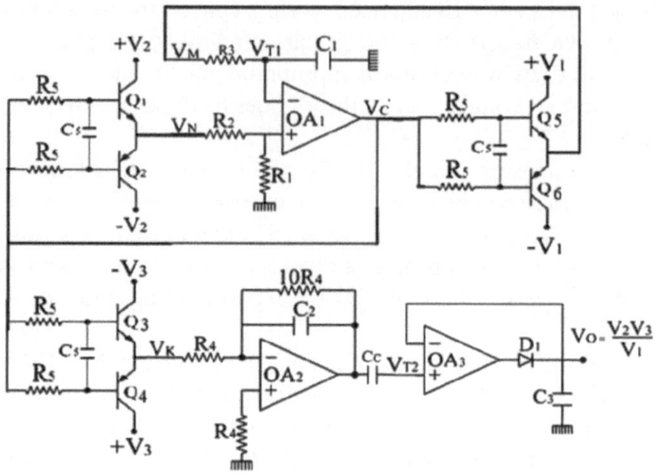

FIGURE 16.11(A) Peak detecting multiplier-cum-divider using voltage tunable astable multivibrator.

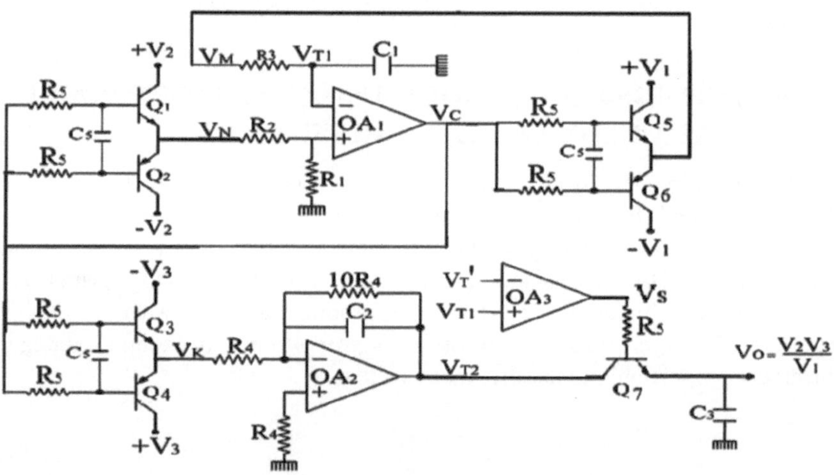

FIGURE 16.11(B) Peak sampling multiplier-cum-divider using voltage tunable astable multivibrator.

From the waveforms shown in Figure 16.12, Eq. (16.34), and the fact that at $t = T/2$, $V_{T2} = 2V_P$

$$2V_P = \frac{V_3}{R_4C_2}\frac{T}{2}$$

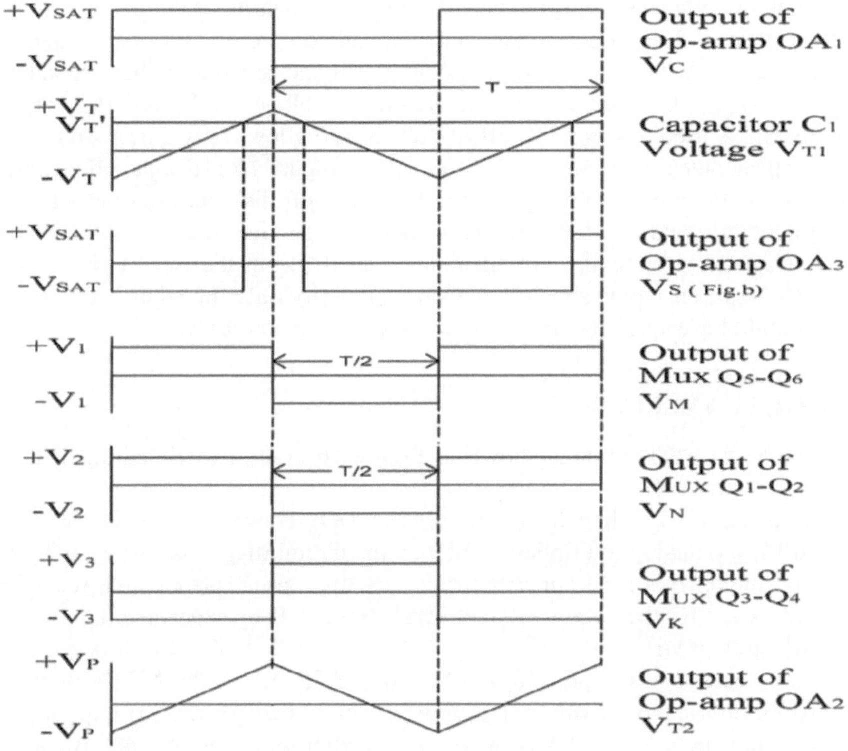

FIGURE 16.12 Associated waveforms of Figure 16.11.

$$V_P = \frac{V_2 V_3}{V_1} \frac{K}{R_4 C_2}$$

Let us assume $K = R_4 C_2$.

$$V_P = \frac{V_2 V_3}{V_1} \tag{16.35}$$

In Figure 16.11(a), the peak detector realized by op-amp OA_3, diode D_1 and capacitor C_3 gives peak value V_P of the triangular wave V_{T2} and, hence, $V_O = V_P$.

In Figure 16.11(b), the sample and hold circuit realized by transistor Q_7 and capacitor C_3 gives the peak value V_P of the triangular wave V_{T2}. The sampling pulse V_S is generated by comparing capacitor C_1 voltage V_{T1} with slightly less than its peak value (V_T'). The sample and hold output is $V_O = V_P$.

From Eq. (16.35)

$$V_O = \frac{V_2 V_3}{V_1} \tag{16.36}$$

The output voltage depends on the sharpness and linearity of the generated triangular waveforms. The output voltage also depends on the proportional constant of voltage tunable astable multivibrator. The offset voltage occurred in the second triangular wave V_{T2} will create an error in the output voltage. In Figure 16.11(a), the capacitor CC is used to block DC offset voltage and allows only a triangular waveform to a peak detector. Its value is 0.1 μF disk. In Figure 16.11(b), the offset adjustment is to be done externally by connecting a 0.1 μF disk between output of the second integrator and a buffer. The buffer output is to give to the sample and hold circuit. This offset removing arrangement is available in the AFC trainer kit. In Figure 16.11(a), all inputs should have single polarity only. In Figure 16.11(b), V_1 and V_2 should have single polarity, and V_3 may have any polarity.

PRACTICAL EXERCISES

1. Verify the MCD circuits shown in Figure 16.11 by experimenting with the AFC trainer kit.
2. As discussed in Chapter 18 and Section 18.6, convert the MCDs of Figure 16.11 into analog multipliers. (i) Draw the circuit diagrams, (ii) explain their working operation, (iii) draw waveforms at appropriate places, and (iv) deduce expression for the output voltage. Verify this MCD by experimenting with the AFC trainer kit.
3. As discussed in Chapter 18 and Section 18.7, convert the MCDs of Figure 16.11 into analog dividers. (i) Draw the circuit diagrams, (ii) explain their working operation, (iii) draw waveforms at appropriate places, and (iv) deduce expression for the output voltage. Verify this divider by experimenting with the AFC trainer kit.
4. Replace the transistor switches in Figure 16.11 with CD4053 analog multiplexer IC. (i) Draw the circuit diagrams, (ii) explain their working operation, (iii) draw waveforms at appropriate places, and (iv) deduce expression for the output voltage. Verify these MCDs by experimenting with the AFC trainer kit.
5. As discussed in Chapter 18 and Section 18.9, convert the MCDs of Figure 16.11 into SRMs. (i) Draw the circuit diagrams, (ii) explain their working operation, (iii) draw waveforms at appropriate places, and (iv) deduce expression for the output voltage. Verify this SRM by experimenting with the AFC trainer kit.
6. As discussed in Chapter 19 and Section 19.6, convert the MCDs of Figure 16.11 into VMCs. (i) Draw the circuit diagrams, (ii) explain their working operation, (iii) draw waveforms at appropriate places, and (iv) deduce expression for the output voltage. Verify this VMC by experimenting with the AFC trainer kit.

16.7 PULSE POSITION PEAK DETECTING MULTIPLIERS-CUM-DIVIDERS – SWITCHING

The circuit diagram of switching type pulse position peak detecting MCD is shown in Figure 16.13, and its associated waveforms are shown in Figure 16.14. Let initially the op-amp OA_2 output be HIGH, the transistor Q_1 be OFF, and an integrator formed by resistor R_1, capacitor C_1, and op-amp OA_1 integrate the input voltage $(-V_1)$. The integrator output is given as

$$V_{S1} = -\frac{1}{R_1C_1}\int -V_1 dt = \frac{V_1}{R_1C_1}t \tag{16.37}$$

A positive-going ramp Vs_1 is generated at the output of op-amp OA_1. When the output of OA_1 reaches the voltage level $+V_R$, the comparator OA_2 output becomes LOW. The transistor Q_1 is ON, and hence the capacitor C_1 is shorted so that op-amp

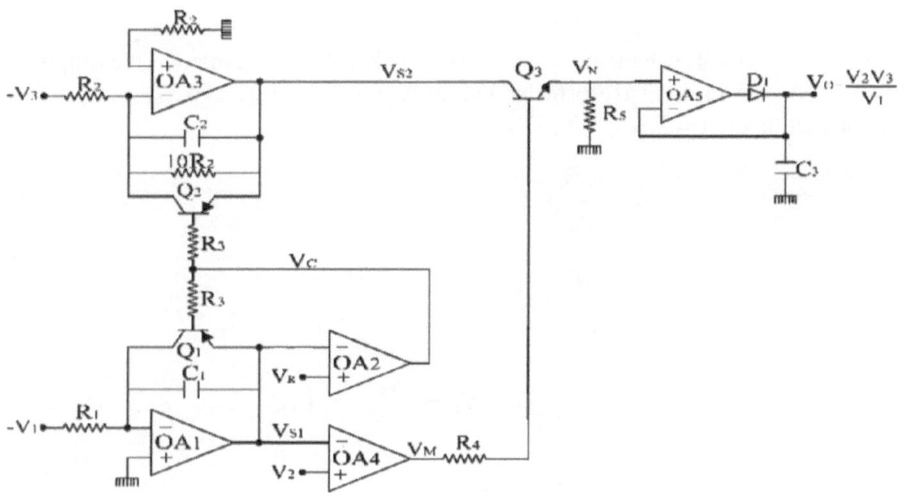

FIGURE 16.13 Pulse position peak detecting multiplier-cum-divider.

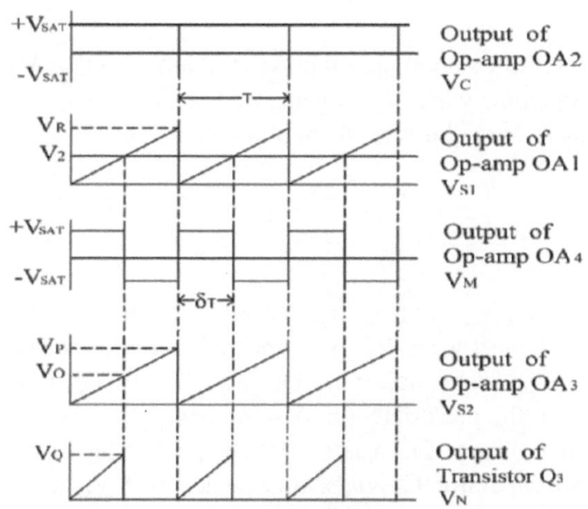

FIGURE 16.14 Associated waveforms of Figure 16.13.

OA_1 output becomes ZERO. Then op-amp OA_2 output goes to HIGH, the transistor Q_1 is OFF, and the integrator composed by R_1, C_1, and op-amp OA_1 integrates the input voltage ($-V_1$), and the cycle therefore repeats to provide (i) a sawtooth wave of peak value V_R at the output of op-amp OA_1 and (ii) a short pulse waveform V_C at the output of comparator OA_2. The short pulse V_C also controls transistor Q_2. During the short LOW time of V_C, transistor Q_2 is ON, the capacitor C_2 is short-circuited so that op-amp OA_3 output is zero volts. During HIGH time of V_C, transistor Q_2 is OFF, the integrator formed by resistor R_2, capacitor C_2, and op-amp OA_3 integrates its input voltage ($-V_3$) and its output is given as

$$V_{S2} = -\frac{1}{R_2C_2}\int -V_3 dt = \frac{V_3}{R_2C_2}t \tag{16.38}$$

Another sawtooth waveform V_{S2} with peak value V_P is generated at the output of op-amp OA_3. From the waveforms shown in Figure 16.4, Eqs. (16.37) and (16.38), and the fact that at $t = T$, $V_{S1} = V_R$, $V_{S2} = V_P$

$$V_R = \frac{V_1}{R_1C_1}T$$

$$T = \frac{V_R}{V_1}R_1C_1$$

$$V_P = \frac{V_3}{R_2C_2}T, \quad V_P = \frac{V_3}{R_2C_2}\frac{V_R}{V_1}R_1C_1$$

Let us assume $R_1 = R_2$, $C_1 = C_2$, then

$$V_P = V_3\frac{V_R}{V_1} \tag{16.39}$$

The sawtooth wave V_{S1} is compared with second input voltage V_2 by the comparator OA_4, and a rectangular wave V_M is generated at the output of comparator OA_4. The ON time of this rectangular waveform V_M is given as

$$\delta_T = \frac{V_2}{V_R}T \tag{16.40}$$

The rectangular pulse V_M controls transistor Q_3. During HIGH of V_M, transistor Q_3 is ON and the sawtooth wave V_{S2} is connected to the peak detector realized by op-amp OA_5, diode D_1, and capacitor C_3. During LOW of V_M, transistor Q_3 is OFF, and zero volt exists on the peak detector. A semi-sawtooth wave V_N with peak value V_Q is generated at the output of transistor Q_3. The peak detector realized by op-amp OA_5, diode D_1, and capacitor C_3 gives this peak value V_Q at its output. Hence, $V_O = V_Q$.

The peak value V_Q is given as

$$V_Q = \frac{V_P}{T}\delta_T \tag{16.41}$$

$$V_Q = V_O$$

Equations (16.39) and (16.40) in (16.41) give

$$V_O = \frac{V_2 V_3}{V_1} \tag{16.42}$$

The output voltage depends on the linearity and sharpness of the generated sawtooth waveforms. The offset voltages at the generated sawtooth waveforms will cause error in the output; hence, they have to null with suitable external circuit (connect integrator output to one end of a potentiometer, the other end of this potentiometer is given to $+V_{CC}$). The sawtooth wave can take from the center pin of the potentiometer. A perfect sawtooth wave without any offset can be obtained by adjusting the potentiometer. (The AFC trainer kit has this offset adjustment facility.) The polarities of V_1 and V_2 must be opposite; V_3 must be negative only.

DESIGN EXERCISES

1. The transistor switches Q_1–Q_3 in Figures 16.13 are to be replaced with FET switches and MOSFET switches. In each, (i) draw the circuit diagrams, (ii) explain their working operations, (iii) draw waveforms at appropriate places, and (iv) deduce expression for their output voltages.

PRACTICAL EXERCISES

1. Verify the MCD circuit shown in Figure 16.13 by experimenting with the AFC trainer kit.
2. As discussed in Chapter 18 and Section 18.6, convert the MCDs of Figure 16.13 into analog multiplier. (i) Draw the circuit diagrams, (ii) explain their working operation, (iii) draw waveforms at appropriate places, and (iv) deduce expression for the output voltage. Verify this MCD by experimenting with the AFC trainer kit.
3. As discussed in Chapter 18 and Section 18.7, convert the MCDs of Figure 16.13 into analog divider. (i) Draw the circuit diagrams, (ii) explain their working operation, (iii) draw waveforms at appropriate places, and (iv) deduce expression for the output voltage. Verify this divider by experimenting with the AFC trainer kit.
4. Replace the transistor switches in Figure 16.13 with CD4066 non-inverted controlled analog switches and DG201 inverted controlled analog switches. (i) Draw the circuit diagrams, (ii) explain their working operation, (iii) draw

waveforms at appropriate places, and (iv) deduce expression for the output voltage. Verify these MCDs by experimenting with the AFC trainer kit.

5. As discussed in Chapter 18 and Section 18.9, convert the MCD of Figure 16.13 into SRM. (i) Draw the circuit diagrams, (ii) explain their working operation, (iii) draw waveforms at appropriate places, and (iv) deduce expression for the output voltage. Verify this SRM by experimenting with the AFC trainer kit.
6. As discussed in Chapter 19 and Section 19.6, convert the MCD of Figure 16.13 into VMC. (i) Draw the circuit diagrams,, (ii) explain their working operation (iii) draw waveforms at appropriate places, and (iv) deduce expression for the output voltage. Verify this VMC by experimenting with the AFC trainer kit.

16.8 PULSE POSITION PEAK SAMPLING MULTIPLIERS-CUM-DIVIDERS – SWITCHING

The circuit diagram of pulse position peak sampling MCD is shown in Figure 16.15, and its associated waveforms are shown in Figure 16.16. As discussed in Section 4.4, Figure 4.7(d), op-amp OA_1 along with transistors Q_1 and Q_2 constitutes a sawtooth wave generator. The time period of the generated sawtooth wave is given as

$$T = \frac{V_R}{V_1} R_1 C_1 \tag{16.43}$$

where V_R is the V_{BE} value of transistor Q_1.

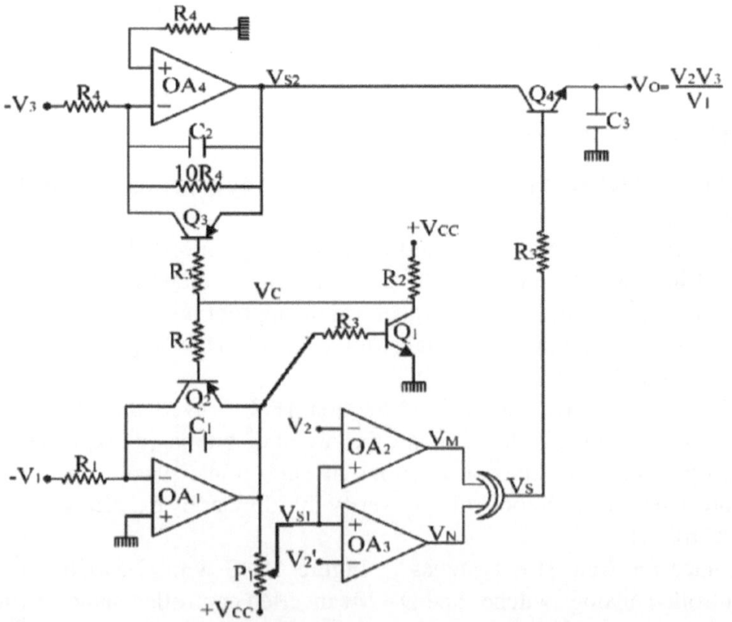

FIGURE 16.15 Pulse width integrated peak sampling multiplier-cum-divider.

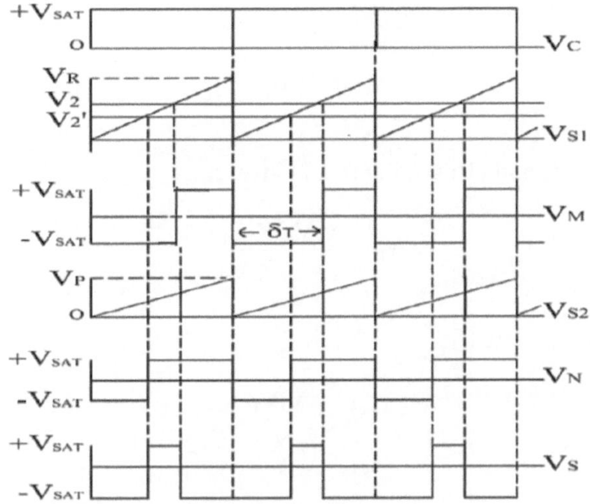

FIGURE 16.16 Associated waveforms of Figure 16.15.

This sawtooth wave V_{S1} is compared with the second input voltage V_2 by comparator OA_2 to get a rectangular pulse V_M. The OFF time of this rectangular wave V_M is given as

$$\delta_T = \frac{V_2}{V_R} T \tag{16.44}$$

The sawtooth wave V_{S1} is also compared with the slightly less than voltage of V_2, i.e., V_2' by comparator OA_3 to get another rectangular pulse V_N. The two rectangular pulses V_M and V_N are given to an Ex-OR gate. The output of Ex-OR gate is a short pulse V_S which is acting as a sampling pulse to the sample and hold circuit realized by transistor Q_4 and capacitor C_3.

The short pulse V_C in the sawtooth generator is also given to the transistor Q_3 which constitutes a controlled integrator along with the op-amp OA_4, resistor R_4, and capacitor C_2.

During ON time of V_C, transistor Q_3 is OFF, and the integrator formed around op-amp OA_4 integrates its input voltage ($-V_3$) and is given as

$$V_{S2} = -\frac{1}{R_4C_2}\int -V_3 dt = \frac{V_3}{R_4C_2} t \tag{16.45}$$

During OFF time of V_C, transistor Q_3 is ON, capacitor C_2 is short-circuited, and hence the output of the integrator becomes zero. Another sawtooth wave V_{S2} with peak value of V_P is generated at the output of op-amp OA_4. From the waveforms shown in Figure 16.16, Eq. 16.45, and the fact that at $t = T$, $V_{S2} = V_P$

$$V_P = \frac{V_3}{R_4C_2} T \tag{16.46}$$

The sampled output is given as

$$V_O = \frac{V_P}{T}\delta_T \tag{16.47}$$

Equations (16.44) and (16.46) in (16.47) give

$$V_P = \frac{V_2V_3}{V_1}\frac{R_1C_1}{R_4C_2}$$

Let $R_1 = R_4$ and $C_1 = C_2$.

$$V_P = \frac{V_2V_3}{V_1}$$

The peak value V_P of this sawtooth wave V_{S2} is obtained by the sample and hold circuit realized by transistor Q_4 and capacitor C_3. The sample and hold output $V_O = V_P$.

$$V_O = \frac{V_2V_3}{V_1} \tag{16.48}$$

The output voltage depends on the linearity and sharpness of the generated sawtooth waveforms. The offset voltages at the generated sawtooth waveforms will cause an error in the output; hence, they have to null with suitable external circuit. The polarities of V_1 and V_2 must be opposite; V_3 may have any polarity.

DESIGN EXERCISES

1. The transistor switches in Figure 16.15 are to be replaced with FET switches and MOSFET switches. In each, (i) draw the circuit diagrams, (ii) explain their working operations, (iii) draw waveforms at appropriate places, and (iv) deduce expression for their output voltages.
2. In the MCD circuit shown in Figure 16.15, the sawtooth generator is replaced with other sawtooth generators shown in Figures 4.4(a), (b) and (c). In each, (i) draw the circuit diagrams, (ii) explain their working operations, (iii) draw waveforms at appropriate places, and (iv) deduce expression for their output voltages.

PRACTICAL EXERCISES

1. Verify the MCD circuit shown in Figure 16.13 by experimenting with the AFC trainer kit.
2. As discussed in Chapter 18 and Section 18.6, convert the MCDs of Figure 16.13 into analog multiplier. (i) Draw the circuit diagrams, (ii) explain their

working operation, (iii) draw waveforms at appropriate places, and (iv) deduce expression for the output voltage. Verify this MCD by experimenting with the AFC trainer kit.

3. As discussed in Chapter 18 and Section 18.7, convert the MCDs of Figure 16.13 into analog divider. (i) Draw the circuit diagrams, (ii) explain their working operation, (iii) draw waveforms at appropriate places, and (iv) deduce expression for the output voltage. Verify this divider by experimenting with the AFC trainer kit.
4. Replace the transistor switches in Figure 16.13 with CD4066 non-inverted controlled analog switches and DG201 inverted controlled analog switches (i) Draw the circuit diagrams, (ii) explain their working operation, (iii) draw waveforms at appropriate places, and (iv) deduce expression for the output voltage. Verify these MCDs by experimenting with the AFC trainer kit.
5. As discussed in Chapter 18 and Section 18.9, convert the MCD of Figure 16.13 into SRM. (i) Draw the circuit diagrams, (ii) explain their working operation, (iii) draw waveforms at appropriate places, and (iv) deduce expression for the output voltage. Verify this SRM by experimenting with the AFC trainer kit.
6. As discussed in Chapter 19 and Section 19.6, convert the MCD of Figure 16.13 into VMC. (i) Draw the circuit diagrams, (ii) explain their working operation, (iii) draw waveforms at appropriate places, and (iv) deduce expression for the output voltage. Verify this VMC by experimenting with the AFC trainer kit.

Part D

General on Function Circuits

17 Conventional Function Circuits

Most conventional function circuits use a bipolar transistor as a predictable non-linear element. Its base emitter voltage (V_{BE}) is a logarithmic function of its collector current. By using transistors and op-amps, log amplifiers can be performed. Logarithmic amplifiers are used in many functions like multiplying, dividing, squaring, and square rooting. Log-antilog multipliers, field effect transistors (FETs) as voltage variable resistor multipliers, variable trans-conductance multipliers, Gilbert multiplier cells, triangle wave averaging multipliers, and quarter squarer multipliers are conventional multipliers. Log-antilog MCD, MCD using FETs, and MCD using MOSFETs are conventional MCDs. These conventional function circuits are described in this chapter.

17.1 LOG-ANTILOG MULTIPLIER

Figure 17.1 shows the block diagram of multiplier using log-antilog amplifiers. From Figure 17.1, the outputs of log amplifiers OA_1 and OA_2 are given as

$$V_X = V_T \ln \frac{V_1}{V_R} \tag{17.1}$$

$$V_Y = V_T \ln \frac{V_2}{V_R} \tag{17.2}$$

In the above Eqs. (17.1) and (17.2) V_T and V_R are constants of log amplifiers. The output of unity gain summer OA_3 is

$$V_Z = V_X + V_Y$$

$$V_Z = V_T \ln \left(\frac{V_1 V_2}{{V_R}^2} \right) \tag{17.3}$$

The output of unity gain summer V_Z is given to the antilog amplifier. The output voltage V_O is given as

$$V_O = V_R \ln^{-1} \left[\frac{V_Z}{V_T} \right]$$

DOI: 10.1201/9781003221449-21

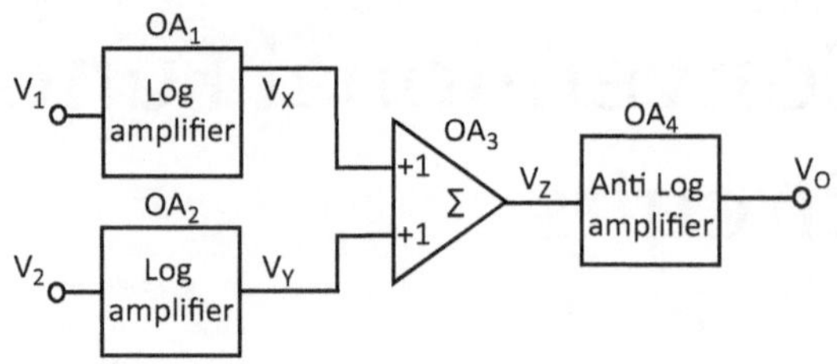

FIGURE 17.1 Block diagram of log-antilog multiplier.

$$V_O = V_R \ln^{-1\ln\left(\frac{V_1V_2}{V_R^2}\right)}$$

$$V_O = \frac{V_1V_2}{V_R} \tag{17.4}$$

Figure 17.2 shows the complete circuit diagram of log-antilog multiplier using op-amps and transistors. The logarithmic operation of bipolar transistor is

$$V_{BE} = V_T \ln\frac{I_C}{I_S}$$

where V_{BE} = base emitter voltage, V_T = KT/Q, I_C = collector current, and I_S = emitter saturation current. In Figure 17.2 the collector current of transistor Q_1 will be

$$I_{CQ1} = \frac{V_1}{R_1} \tag{17.5}$$

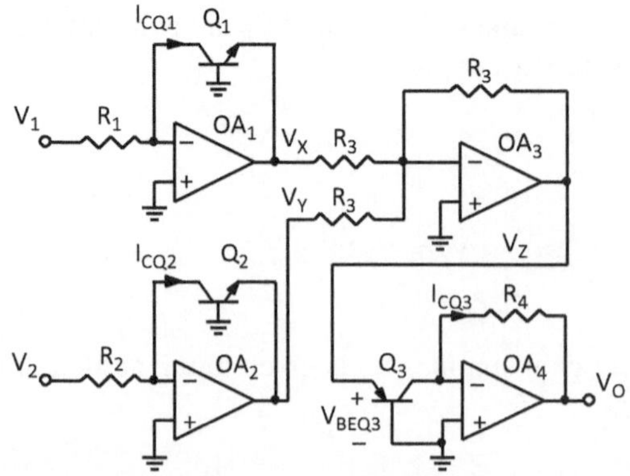

FIGURE 17.2 Circuit diagram of log-antilog multiplier.

The output of log amplifier OA_1 will be

$$V_X = -V_{BEQ1}$$

$$V_X = -V_T \ln\left(\frac{I_{CQ1}}{I_S}\right)$$

$$V_X = -V_T \ln\left(\frac{V_1}{R_1 I_S}\right) = -V_T \ln\left(\frac{V_1}{V_R}\right) \quad (17.6)$$

where $R_1 = R$ and $V_R = RI_S$.

Similarly, the output of log amplifier OA_2 will be

$$V_Y = -V_T \ln\left(\frac{V_2}{R_2 I_S}\right) = -V_T \ln\left(\frac{V_2}{V_R}\right) \quad (17.7)$$

The output of adder OA_3 will be

$$V_Z = -\left(V_X + V_Y\right) = V_T \ln\left(\frac{V_1 V_2}{V_R{}^2}\right) \quad (17.8)$$

The antilogarithmic operation of a bipolar transistor is given by

$$I_C = I_S \ln^{-1 V_{BE}/V_T}$$

In Figure 17.2 $V_{BEQ3} = V_Z$, the collector current of transistor Q_3 will be

$$I_{CQ3} = I_S \ln^{-1 V_Z/V_T}$$

The output voltage is given by

$$V_O = -I_{CQ3} R_4$$

$$V_O = -I_S R_4 \ln^{-1 V_Z/V_T}$$

Let $R_4 = R$, then

$$V_O = -V_R \ln^{-1 V_Z/V_T} \quad (17.9)$$

Equation (17.8) in (17.9) gives

$$V_O = -\frac{V_1 V_2}{V_R} \quad (17.10)$$

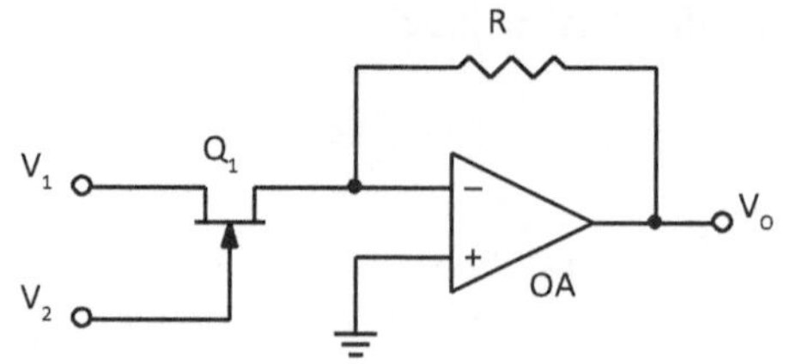

FIGURE 17.3 Basic multiplier using FET.

17.2 MULTIPLIER USING FIELD EFFECT TRANSISTORS

An ideal multiplier circuit using FET as a controlled resistor is shown in Figure 17.3. For small source-drain voltages, a FET acts as a controlled resistor whose resistance is inversely proportional to the gate source voltage. The multiplier circuit shown in Figure 17.3 is an amplifier with V_1 as input voltage whose gain is approximately proportional to V_2. Hence, the output voltage is proportional to the product of V_1 and V_2.

The circuit has several major drawbacks such as very poor linearity since the gain between V_1 and V_O is not linearly related to V_2. The properties of a FET are also very temperature sensitive causing large temperature-dependent errors. There is a restricted range of voltages for V_1 and V_2, i.e., V_1 should be in the range of −0.5 to +0.5 V, and V_2 should be 0 to 0.5 V.

17.3 VARIABLE TRANSCONDUCTANCE MULTIPLIER

The basic one-quadrant variable transconductance multiplier is shown in Figure 17.4. It uses the principle of the dependence of the transistor transconductance on the emitter current bias. Transistors Q_1 and Q_2 form a differential amplifier. For small voltages V_1, i.e., $V_1 \ll V_T$,

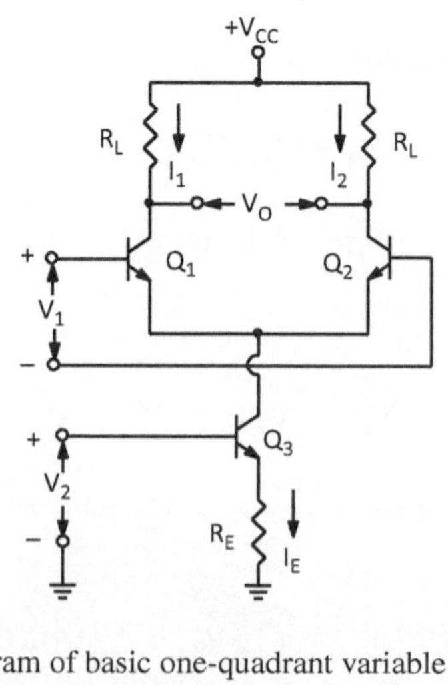

FIGURE 17.4 Circuit diagram of basic one-quadrant variable transconductance multiplier.

$$V_O = g_m R_L V_1$$

where gm = transconductance

$$g_m = \frac{I_E}{V_T}$$

In Figure 17.4, if $I_E R_E \gg V_{BE}$, then

$$V_2 = I_E R_E$$

The overall voltage transfer expression can be written as

$$V_O = \frac{I_E}{V_T} R_L V_1 = \frac{V_2}{R_E V_T} R_L V_1$$

$$V_O = \frac{R_L}{R_E V_T} V_1 V_2$$

$$V_O = \frac{V_1 V_2}{V_R} \quad (17.11)$$

where $V_R = R_E V_T / R_L$.

17.4 GILBERT'S MULTIPLIER CELL

The Gilbert cell using emitter coupled pair (Q_1–Q_2) in a series with a cross-coupled, emitter coupled pair (Q_3–Q_6) is shown in Figure 17.5. This cell is used to obtain a complete Gilbert cell basic one-quadrant multiplier circuit and is shown in Figure 17.6. V_1 and V_2 are the two input voltages. These two inputs determine the division of the total current I_E among the various branches of the circuit. As the devices are symmetrically cross-coupled and the current I_E is constant, the large common mode shift at the outputs gets eliminated.

Let us assume that all transistors are well matched and h_{fe} for all transistors is very high, i.e., $h_{fe} \gg 1$. From Figure 17.6

$$I_1 + I_2 = I_5 \quad (17.12)$$

$$I_3 + I_4 = I_6 \quad (17.13)$$

$$I_5 + I_6 = I_E \quad (17.14)$$

Let us assume $V_1 \ll V_T$. The current unbalance in the differential pairs can be expressed as

$$I_1 - I_2 = (g_m)_{12} V_1 \quad (17.15)$$

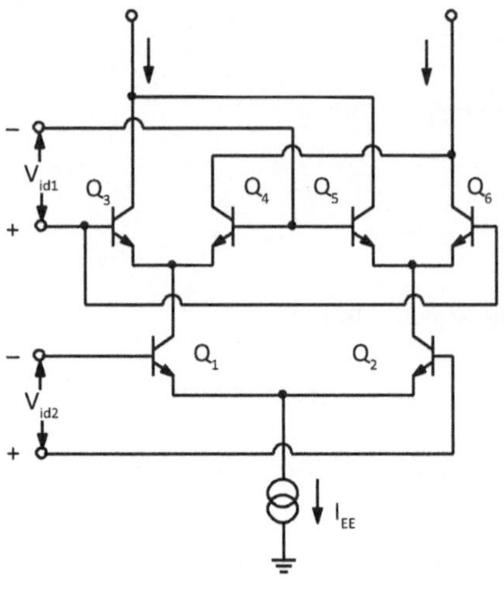

FIGURE 17.5 Circuit diagram of basic Gilbert cell.

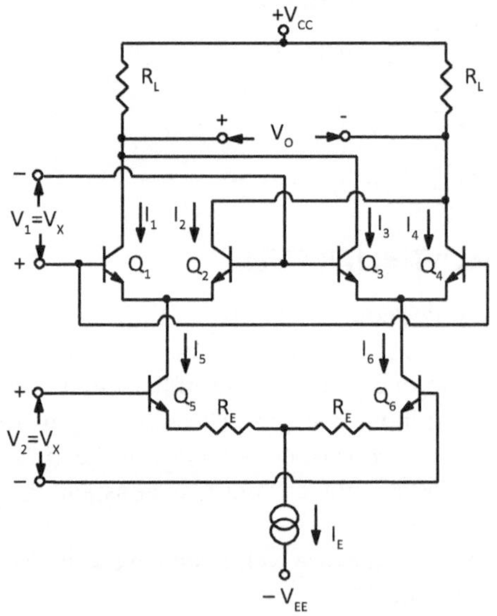

FIGURE 17.6 Basic variable transconductance multiplier using Gilbert cell.

$$I_3 - I_4 = (g_m)_{34} V_1 \tag{17.16}$$

where $(g_m)_{12}$ and $(g_m)_{34}$ are the variable transconductances of the transistor pairs Q_1–Q_2 and Q_3–Q_4, respectively. Under the absence of emitter degeneration resistance, the transconductances are directly proportional to the bias currents I_5 and I_6 and hence

$$\left(g_m\right)_{12} = \frac{I_5}{V_T} \quad (17.17)$$

$$\left(g_m\right)_{34} = \frac{I_6}{V_T} \quad (17.18)$$

The total differential output voltage V_O is given by

$$V_O = R_L\left[\left(I_1 - I_2\right) - \left(I_3 - I_4\right)\right] \quad (17.19)$$

Equations (17.15) and (17.16) in (17.19) give

$$V_O = R_L\left[\left(g_m\right)_{12} V_1 - \left(g_m\right)_{34} V_1\right] \quad (17.20)$$

Equations (17.17) and (17.18) in (17.20) give

$$V_O = \frac{R_L V_1}{V_T}\left[I_5 - I_6\right] \quad (17.21)$$

If the emitter series resistance R_E is chosen sufficiently high, such that $I_5 R_E \gg V_T$ and $I_6 R_E \gg V_T$, then

$$\left(I_5 - I_6\right) = \frac{V_2}{R_E} \quad (17.22)$$

Equation (17.22) in (17.21) gives

$$V_O = \frac{R_L}{R_E V_T} V_1 V_2 \quad (17.23)$$

$$V_O = \frac{V_1 V_2}{V_R} \quad (17.24)$$

where $V_R = R_E V_T / R_L$

17.5 TRIANGLE WAVE AVERAGING MULTIPLIER

The block diagram of a triangular wave averaging multiplier is given in Figure 17.7. It is made up of a triangle wave generator, summing amplifiers, diode rectifiers, and lowpass filters. From the block diagram shown in Figure 17.7,

$$V_X = V_T + V_1 + V_2$$

$$V_Y = V_T + V_1 - V_2$$

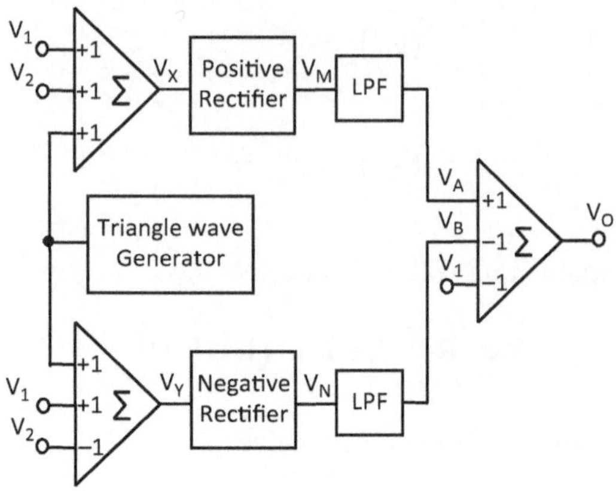

FIGURE 17.7 Block diagram of triangle wave averaging multiplier.

The voltage V_X is passed through a positive rectifier to retain a positive portion of the triangular wave. The lowpass filter output V_A will be

$$V_A = \frac{1}{T}\int_0^t V_M(t)dt$$

$$V_A = \frac{1}{T}\text{ [Area of triangle above zero level]} = \frac{1}{T}\left[\frac{1}{2}(\text{Base})(\text{peakvalue})\right]$$

Peak value = $V_1+V_2+V_T$. When peak value is V_T, Base = T/2. Therefore, if peak voltage is

$$V_1 + V_2 + V_T \text{ then base} = \frac{(V_1 + V_2 + V_T)\frac{T}{2}}{V_T}$$

$$V_A = \frac{1}{T}\left[\frac{1}{2}\left(\frac{(V_1 + V_2 + V_T)\frac{T}{2}}{V_T}\right)(V_1 + V_2 + V_T)\right]$$

$$V_A = \frac{1}{4V_T}(V_1 + V_2 + V_T)^2 \qquad (17.25)$$

The voltage V_Y is passed through a negative rectifier to retain a negative portion of the triangular wave. The lowpass filter output V_B will be

$$V_B = \frac{1}{T}\int_0^t V_N(t)dt$$

$V_B = \frac{1}{T}$ [Area of triangle below zero level] $= \frac{1}{T}\left[\frac{1}{2}(\text{Base})(\text{peakvalue})\right]$

Only the negative voltage of V_Y is passed through. Negative peak value = $V_T - (V_1 - V_2) = V_T - V_1 + V_2$. When peak value is V_T, base is T/2. Therefore, when peak value $V_T - V_1 + V_2$, then base = $\frac{(V_T - V_1 + V_2)\frac{T}{2}}{V_T}$

$$V_B = \frac{1}{T}\left[\frac{1}{2}\left(\frac{(V_T - V_1 + V_2)\frac{T}{2}}{V_T}\right)(V_T - V_1 + V_2)\right]$$

$$V_B = \frac{1}{4V_T}(-V_1 + V_2 + V_T)^2 \tag{17.26}$$

V_A, V_B and V_1 are combined through a summing amplifier in such a way that

$$V_o = V_A - V_B - V_1 \tag{17.27}$$

Equations (17.24) and (17.19) in (17.27) give

$$V_o = \frac{V_1 V_2}{V_T} \tag{17.28}$$

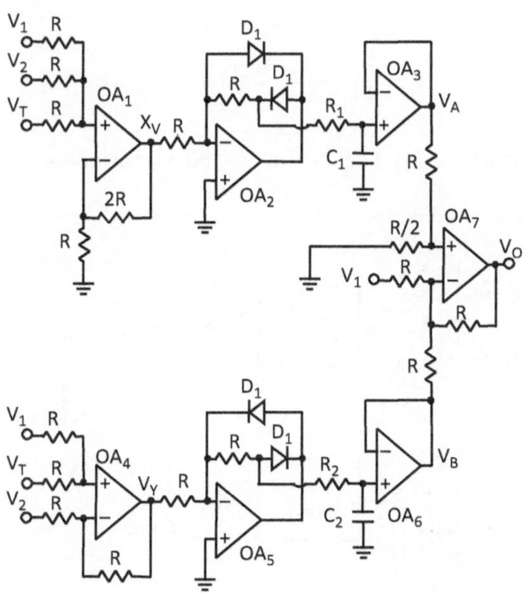

FIGURE 17.8 Circuit diagram of triangle wave averaging multiplier.

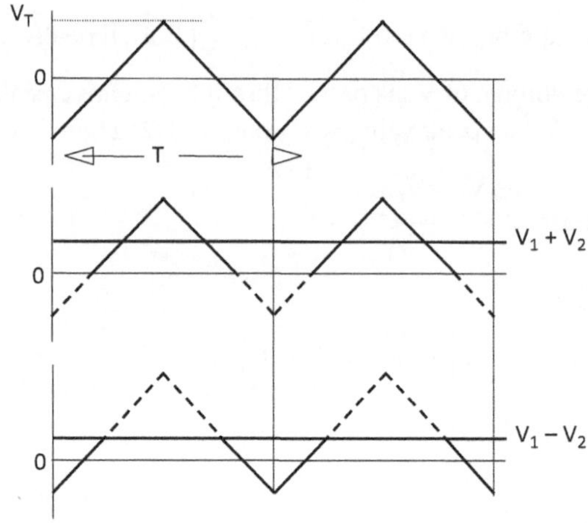

FIGURE 17.9 Associated waveforms of Figure 17.8. From top (1) to bottom (3) details: (1) reference triangular wave with peak value V_T, (2) output of positive rectifier V_M, and (3) output of negative rectifier V_N.

Figure 17.8 shows the circuit diagram of the proposed triangle averaging multiplier, and its associated waveforms are shown in Figure 17.9.

17.6 QUARTER-SQUARER MULTIPLIER

The block diagram of a quarter-squarer multiplier is shown in Figure 17.10. This multiplier is based on the algebraic equation

$$xy = \frac{1}{4}\left[(x+y)^2 - (x-y)^2\right] \tag{17.29}$$

Let us replace the above Eq. (17.29) for V_1 and V_2 variables

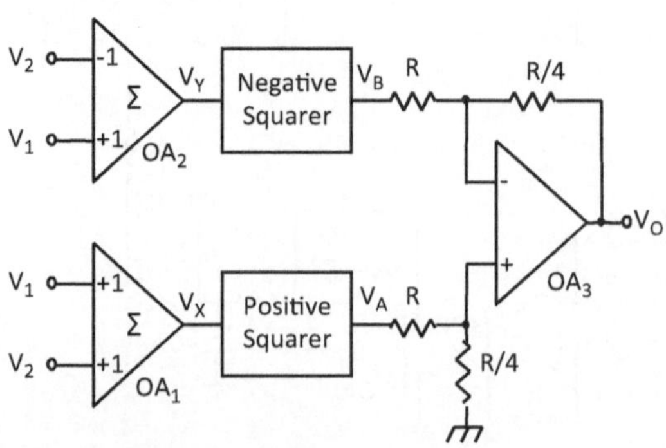

FIGURE 17.10 Block diagram of quarter-squarer multiplier.

$$V_1V_2 = \frac{1}{4}\left[(V_1+V_2)^2 - (V_1-V_2)^2\right]$$

The adder composed by op-amp OA_1 output will be

$$V_X = V_1 + V_2$$

The subtractor composed by op-amp OA_2 output will be

$$V_Y = V_1 - V_2$$

The output of the positive squarer will be

$$V_A = (V_1+V_2)^2$$

The output of the negative squarer will be

$$V_B = (V_1-V_2)^2$$

The output of the differential amplifier composed by op-amp OA_3 will be

$$V_O = \frac{1}{4}\left[(V_A)-(V_B)\right]$$

$$V_O = \frac{1}{4}\left[(V_1+V_2)^2 - (V_1-V_2)^2\right]$$

$$V_O = V_1V_2 \tag{17.30}$$

17.7 LOG-ANTILOG MULTIPLIER-CUM-DIVIDER – TYPE I

The log-antilog multiplier is shown in Figure 17.11. The logarithmic operation of bipolar transistor is

$$V_{BE} = V_T \ln\frac{I_C}{I_S} \tag{17.31}$$

where V_{BE} = Base emitter voltage, V_T = KT/Q, I_C = collector current, and I_S = emitter saturation current.

Considering log-amp OA_1, the collector current of Q_1 will be the current through the resistor R_1.

$$I_{CQ1} = \frac{V_2}{R_1}$$

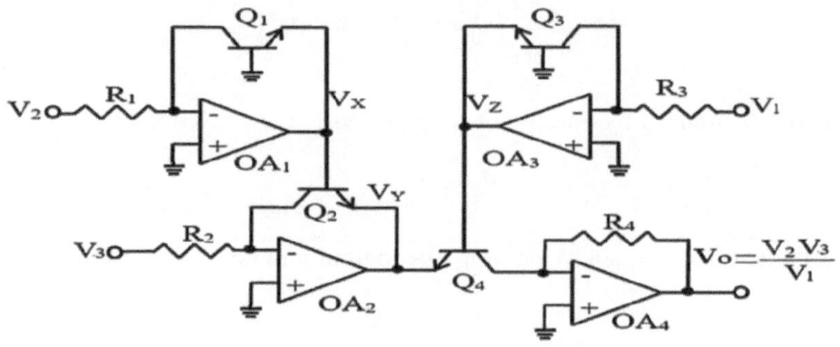

FIGURE 17.11 Log-antilog multiplier-cum-divider – type I.

$$V_{BEQ1} = 0 - V_X = V_{TQ1} \ln \frac{I_{CQ1}}{I_{SQ1}} = V_{TQ1} \ln \frac{V_2}{R_1 I_{SQ1}} \quad (17.32)$$

Considering log-amp OA_3, the collector current of Q_3 will be the current through the resistor R_3.

$$I_{CQ3} = \frac{V_1}{R_3}$$

$$V_{BEQ3} = 0 - V_Z = V_{TQ3} \ln \frac{I_{CQ3}}{I_{SQ3}} = V_{TQ3} \ln \frac{V_1}{R_3 I_{SQ3}} \quad (17.33)$$

Assume the transistors Q_1 and Q_3 are a matched pair such that $V_{TQ1}=V_{TQ3}= V_{T1}$ and $I_{SQ1} = I_{SQ3} = I_{S1}$, then the above Eqs. (17.32) and (17.33) become

$$-V_X = V_{T1} \ln \frac{V_2}{R_1 I_{S1}} \quad (17.34)$$

$$-V_Z = V_{T1} \ln \frac{V_1}{R_3 I_{S1}} \quad (17.35)$$

Equations (17.34) and (17.35) give

$$-V_X + V_Z = V_{T1} \ln \frac{V_2}{R_1 I_{S1}} - V_{T1} \ln \frac{V_1}{R_3 I_{S1}}$$

$$V_Z - V_X = V_{T1} \ln \frac{V_2}{V_1} \frac{R_3}{R_1} \quad (17.36)$$

Considering the log-amp OA_2, the collector current of Q_2 will be the current through the resistor R_2.

$$I_{CQ2} = \frac{V_3}{R_2}$$

$$V_{BEQ2} = V_X - V_Y = V_{TQ2} \ln \frac{I_{CQ2}}{I_{SQ2}} = V_{TQ2} \ln \frac{V_3}{R_2 I_{SQ2}} \tag{17.37}$$

Considering the antilog amplifier OA_4, the collector current of Q_4 will be the current through the resistor R_4.

$$I_{CQ4} = \frac{V_O}{R_4}$$

$$V_{BEQ4} = V_Z - V_Y = V_{TQ4} \ln \frac{I_{CQ4}}{I_{SQ4}} = V_{TQ4} \ln \frac{V_O}{R_4 I_{SQ4}} \tag{17.38}$$

Assume transistors Q_2 and Q_4 are matched such that $V_{TQ2} = V_{TQ4} = V_{T2}$ and $I_{SQ4} = I_{SQ2} = I_{S2}$

$$V_X - V_Y = V_{T2} \ln \frac{V_3}{R_2 I_{S2}} \tag{17.39}$$

$$V_Z - V_Y = V_{T2} \ln \frac{V_O}{R_4 I_{S2}} \tag{17.40}$$

Equations (17.40) and (17.39) give

$$V_Z - V_Y - V_X + V_Y = V_{T2} \ln \frac{V_O}{R_4 I_{S2}} \frac{R_2 I_{S2}}{V_3}$$

$$V_Z - V_X = V_{T2} \ln \frac{V_O}{V_3} \frac{R_2}{R_4} \tag{17.41}$$

Further assume that all transistors are matched and are kept adjacent to each other such that $V_{T2} = V_{T1} = V_T$, $I_{S2} = I_{S1} = I_S$ and assume $R_1 = R_2 = R_3 = R_4 = R$. Then, Eqs. (17.36) and (17.41)

$$V_{T1} \ln \frac{V_2}{V_1} \frac{R_3}{R_1} = V_{T2} \ln \frac{V_O}{V_3} \frac{R_2}{R_4} \tag{17.42}$$

$$\ln\frac{V_2}{V_1} = \ln\frac{V_O}{V_3} \tag{17.43}$$

$$V_O = \frac{V_2V_3}{V_1} \tag{17.44}$$

17.8 LOG-ANTILOG MULTIPLIER-CUM-DIVIDER – TYPE II

The circuit diagram of MCD type II is shown in Figure 17.12.

The logarithmic operation of bipolar transistor is

$$V_{BE} = V_T \ln\frac{I_C}{I_S} \tag{17.45}$$

where V_{BE} = base emitter voltage, V_T = KT/Q, I_C = collector current, and I_S = emitter saturation current.

In Figure 17.12, let us assume all transistors Q_1, Q_2, Q_3 and Q_4 are identical and packed in a single monolithic chip. Hence, $V_{TQ1} = V_{TQ2} = V_{TQ3} = V_{TQ4} = V_T$ and $I_{SQ1} = I_{SQ2} = I_{SQ3} = I_{SQ4} = I_S$.

Logging transistor Q_3,

$$V_A - V_X = V_T \ln\frac{I_3}{I_S} \tag{17.46}$$

Logging transistor Q_1,

$$0 - V_X = V_T \ln\frac{I_1}{I_S} \tag{17.47}$$

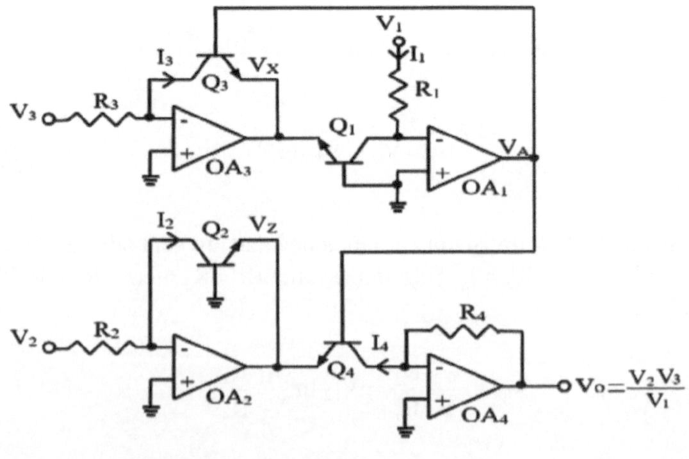

FIGURE 17.12 Log-antilog multiplier-cum-divider – type II.

Equations (17.46) and (17.47) give

$$V_A = V_T \ln \frac{I_3}{I_1} \tag{17.48}$$

Logging transistor Q_2,

$$0 - V_Z = V_T \ln \frac{I_2}{I_S} \tag{17.49}$$

Logging transistor Q_4,

$$V_A - V_Z = V_T \ln \frac{I_4}{I_S} \tag{17.50}$$

Equations (17.50) and (17.49) give

$$V_A = V_T \ln \frac{I_4}{I_2} \tag{17.51}$$

From Eqs. (17.48) and (17.51),

$$V_T \ln \frac{I_3}{I_1} = V_T \ln \frac{I_4}{I_2}$$

$$\frac{I_3}{I_1} = \frac{I_4}{I_2} \tag{17.52}$$

It is observed from Figure 17.12 that $I_1 = V_1/R_1$, $I_2 = V_2/R_2$, $I_3 = V_3/R_3$ and $I_4 = V_O/R_4$.

$$\frac{V_3R_1}{R_3V_1} = \frac{V_OR_2}{R_4V_2} \tag{17.53}$$

Let us assume $R_1 = R_2 = R_3 = R_4 = R$, then

$$\frac{V_3}{V_1} = \frac{V_O}{V_2}$$

$$V_O = \frac{V_2V_3}{V_1} \tag{17.54}$$

17.9 MULTIPLIER-CUM-DIVIDER USING FIELD EFFECT TRANSISTORS

The circuit diagram of MCD using FETs is shown in Figure 17.13. FET is to be used in a triode region as a voltage variable resistor (VVR). The two FETs Q_1 and Q_2 are

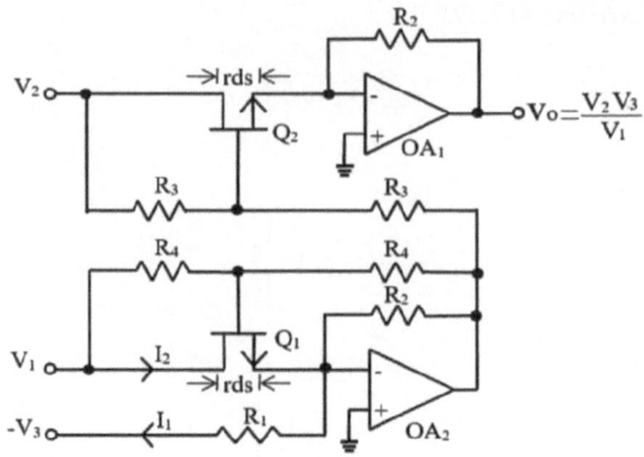

FIGURE 17.13 FET-based multiplier-cum-divider.

identical such that for the same V_{GS}, r_{ds} of both FETs are the same. R_3 are linearizing resistors of Q_2, and R_4 are linearizing resistors of Q_1.

The output of inverting amplifier OA_1 will be

$$V_O = -\frac{R_2}{r_{ds}} V_2 \tag{17.55}$$

The current through R_1 is given as

$$I_1 = -\frac{V_3}{R_1} \tag{17.56}$$

The current through FET Q_1 is

$$I_2 = \frac{V_1}{r_{ds}}$$

$$I_1 = I_2$$

$$-\frac{V_3}{R_1} = \frac{V_1}{r_{ds}}$$

$$r_{ds} = -\frac{V_1 R_1}{V_3} \tag{17.57}$$

Equation (17.57) in (17.55) gives

$$V_O = \frac{R_2 V_2 V_3}{V_1 R_1}$$

If $R_1 = R_2$, then

$$V_O = \frac{V_2 V_3}{V_1} \tag{17.58}$$

17.10 MULTIPLIER-CUM-DIVIDER USING MOSFETS

The MCD using MOSFETs is shown in Figure 17.14. MOSFETs Q_1 and Q_2 are an identically matched pair and work as voltage controlled resistors. When the voltage between gate and drain exceeds the threshold voltage, i.e., $V_{GD} > V_{TH}$, the drain to source current is given as

$$I_{DS} = K\left[2\left(V_{GS} - V_T\right)V_{DS} - V_{DS}^2\right]$$

$$I_{DS} \simeq 2K\left(V_{GS} - V_T\right)V_{DS} \tag{17.59}$$

The input voltages V_1 and V_2 are positive and hence the source terminals of MOSFETs are at virtual ground by the op-amps OA_1 and OA_2. The current through R_1 will be

$$I_1 = -\frac{V_3}{R_1} \tag{17.60}$$

$$I_1 = I_2 = I_{DS2} = 2K\left(V_{GS2} - V_{T2}\right)V_{DS2} \tag{17.61}$$

$$V_{GS2} = V_G \tag{17.62}$$

$$V_{DS2} = V_1 \tag{17.63}$$

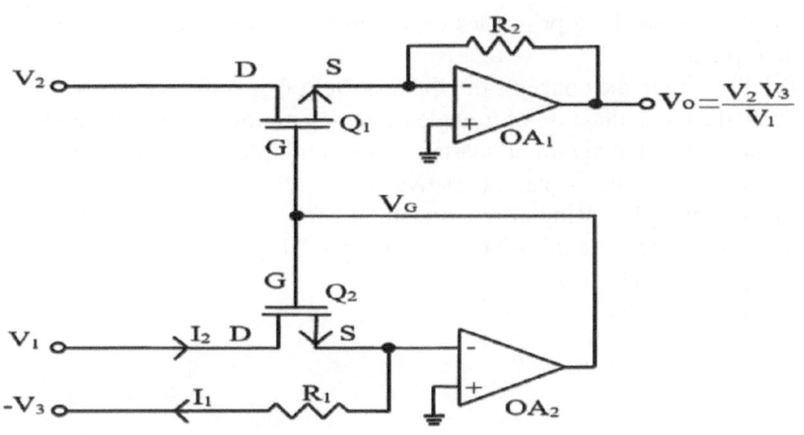

FIGURE 17.14 Multiplier-cum-divider using MOSFETs.

$$V_G - V_{T2} = -\frac{V_3}{2KR_1V_1} \tag{17.64}$$

Considering the other MOSFET Q_1, we have

$$I_{DS1} = 2K\left(V_{GS1} - V_{T1}\right)V_{DS1} \tag{17.65}$$

$$V_{GS1} = V_G \tag{17.66}$$

$$V_{DS1} = V_2 \tag{17.67}$$

Equations (17.66) and (17.67) in (17.65) give

$$I_{DS1} = -\frac{V_2V_3}{R_1V_1} \tag{17.68}$$

The output voltage V_O will be

$$V_O = -I_{DS1}R_2 \tag{17.69}$$

Let $R_1 = R_2 = R$.
Equation (17.68) in (17.69) gives

$$V_O = \frac{V_2V_3}{V_1} \tag{17.70}$$

TUTORIAL EXERCISES

1. Design a multiplier using diode-based log-antilog amplifiers.
2. Design a multiplier using FETs.
3. Explain the working principles of a transistorized variable transconductance multiplier.
4. Briefly explain the concepts of Gilbert's multiplier cell.
5. Draw the block diagram of triangle averaging multiplier and explain briefly.
6. Draw the block diagram of quarter-squarer multiplier and explain neatly.
7. You are given a two squarer (positive and negative squarer). Design a quarter-squarer multiplier using differential amplifiers.
8. (a) Design 3 log–1 antilog MCD. (b) Design 2 log–2 antilog MCD.
9. Design an MCD using FETs.
10. Design an MCD using MOSFETS.

18 Conversion of Function Circuits

All function circuits are interlinked. If one can get a multiplier, then it is easy to convert that multiplier into a divider, squarer, square rooter, MCD, VMC, etc. Similarly if one can get a square rooter, it is easy to convert this square rooter into a divider, multiplier, squarer, MCD, VMC, etc. This chapter deals with how a multiplier can be converted into a squarer, divider, and square rooter, and a how a divider can be converted into a multiplier and square rooter. A multiplier-cum-divider can be converted into multiplier, divider, squarer, square rooter, and vector magnitude circuits. Multipliers-cum-dividers can be obtained from multipliers and dividers. Details of the conversion of function circuits are described in this chapter.

18.1 MULTIPLIER TO SQUARER

The multiplier output is given as

$$V_O = \frac{V_1 V_2}{V_R} \tag{18.1}$$

Let $V_2 = V_1$, then

$$V_O = \frac{V_1^2}{V_R} \tag{18.2}$$

Realizing the operation of a squarer.

DESIGN EXERCISES

1. The sawtooth wave-based time division multiplier shown in Figure 12.3(b) is converted into a squarer and shown in Figure 18.1. (i) Explain the working operation of a squarer shown in Figure 18.1, (ii) draw waveforms at appropriate places, and (iii) deduce expression for its output voltage. Verify this squarer by experimenting with the AFC trainer kit.
2. Similarly, convert all multipliers explained in Chapters 11 and 12 into squarers. (i) Draw circuit diagrams, (ii) explain their working operations, (iii) draw waveforms at appropriate places, and (iv) deduce expression for their output voltage. Verify these squarers by experimenting with the AFC trainer kit.

DOI: 10.1201/9781003221449-22

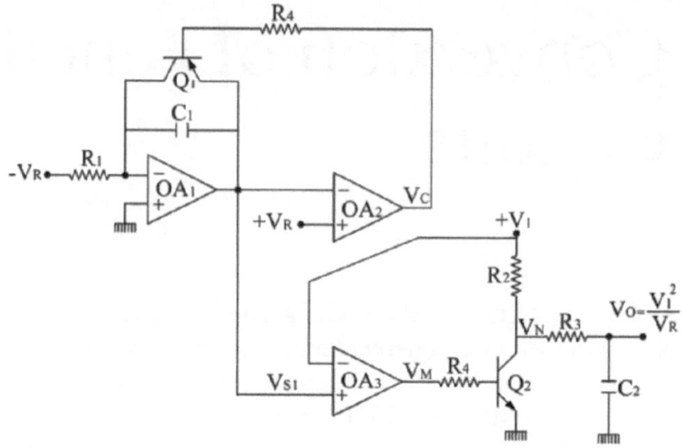

FIGURE 18.1 Sawtooth wave-based time division squarer.

18.2 MULTIPLIER TO DIVIDER

A multiplier can be converted into a divider as shown in Figure 18.2. Figure 18.2(a) shows type I methods, and Figure 18.2(b) shows type II methods. The multiplier output is given as

$$V_X = \frac{V_1 V_O}{V_R} \tag{18.3}$$

where V_R is the multiplier constant.

In Figure 18.2(a),

$$\text{The current through resistor } R_1; I_1 = \frac{V_X}{R_1} \tag{18.4}$$

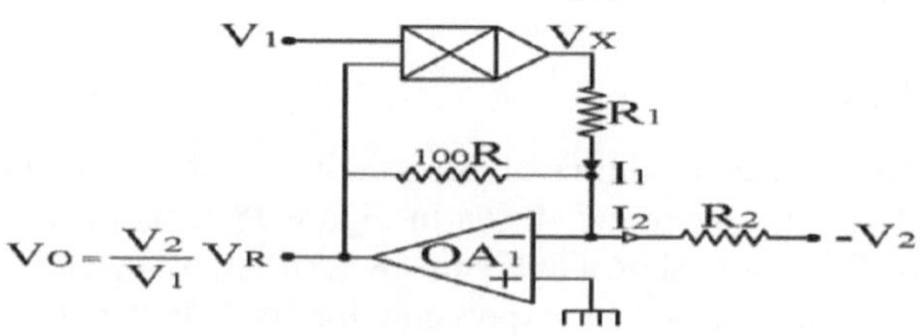

FIGURE 18.2(A) Multiplier to divider – type I.

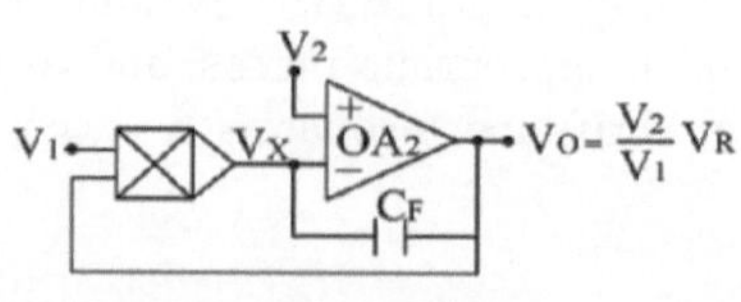

FIGURE 18.2(B) Multiplier to divider – type II.

$$\text{The current through resistor } R_2;\ I_2 = \frac{V_2}{R_2} \tag{18.5}$$

Since the op-amp has high input impedance, current I_1 will not enter into op-amp, and hence $I_1 = I_2$.

$$\frac{V_X}{R_1} = \frac{V_2}{R_2} \tag{18.6}$$

Let us assume $R_1 = R_2 = R$.

$$V_X = V_2 \tag{18.7}$$

From Eqs. (18.3) and (18.7)

$$V_O = \frac{V_2}{V_1} V_R \tag{18.8}$$

In the circuit shown in Figure 18.2(b), the op-amp is at a negative closed loop feedback and a positive DC voltage is ensured in the feedback loop. Hence, its non-inverting terminal voltage is equal to its inverting terminal voltage.

$$V_2 = V_X \tag{18.9}$$

From Eqs. (18.3) and (18.9)

$$V_O = \frac{V_2}{V_1} V_R \tag{18.10}$$

DESIGN EXERCISES

1. The sawtooth wave-based multiplier shown in Figure 12.3(b) is converted into a divider by using a type II method as shown in Figure 18.3. (i) Explain

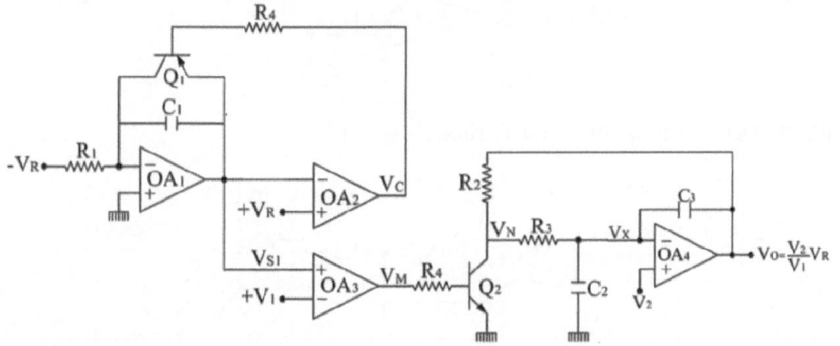

FIGURE 18.3 Sawtooth wave-based time division divider from multiplier.

the working operation of the divider shown in Figure 18.3 (ii) Draw waveforms at appropriate places, and (iii) deduce expression for the output voltage. Verify this divider by experimenting with the AFC trainer kit.

2. Similarly, convert all multipliers explained in Chapters 11 and 12 into dividers by using both type I and type II methods. In each, (i) draw circuit diagrams, (ii) explain their working operations, (iii) draw waveforms at appropriate places, and (iv) deduce expression for their output voltage. Verify these dividers by experimenting with the AFC trainer kit.

18.3 MULTIPLIER TO SQUARE ROOTER

A multiplier can be converted into a square rooter as shown in Figure 18.4. Figure 18.4(a) shows type I methods, and Figure 18.4(b) shows type II methods.

The multiplier output is given as

$$V_X = \frac{{V_O}^2}{V_R} \tag{18.11}$$

where V_R is the multiplier constant.

In Figure 18.4(a),

$$\text{The current through resistor } R_1; I_1 = \frac{V_X}{R_1} \tag{18.12}$$

$$\text{The current through resistor } R_2; I_2 = \frac{V_1}{R_2} \tag{18.13}$$

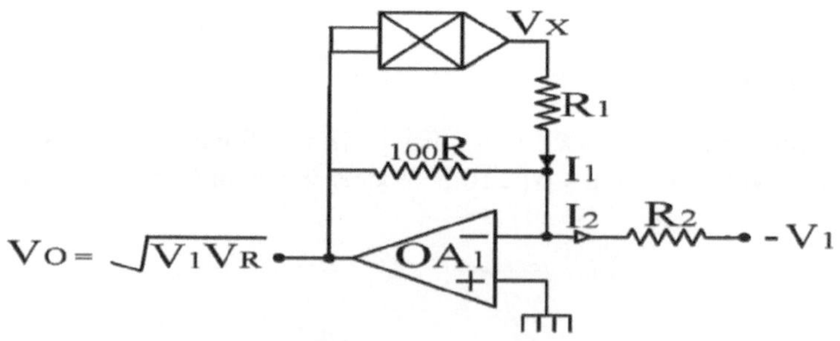

FIGURE 18.4(A) Multiplier to square rooter – type I.

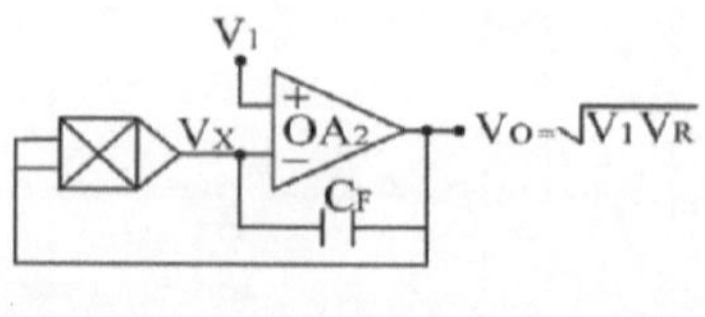

FIGURE 18.4(B) Multiplier to square rooter – type II.

Since the op-amp has high input impedance, current I_1 will not enter into op-amp and hence $I_1 = I_2$.

$$\frac{V_X}{R_1} = \frac{V_1}{R_2} \tag{18.14}$$

Let us assume $R_1 = R_2 = R$.

$$V_X = V_1 \tag{18.15}$$

From Eqs. (18.11) and (18.15)

$$V_O = \sqrt{V_1 V_R} \tag{18.16}$$

In the circuit shown in Figure 18.4(b), the op-amp is at a negative closed loop feedback and a positive DC voltage is ensured in the feedback loop. Hence, its non-inverting terminal voltage is equal to its inverting terminal voltage.

$$V_1 = V_X \tag{18.17}$$

From Eqs. (18.11) and (18.17)

$$V_O = \sqrt{V_1 V_R} \tag{18.18}$$

DESIGN EXERCISES

1. The sawtooth wave-based time division multiplier shown in Figure 12.3(b) is converted into a square rooter by using type II method and shown in Figure 18.5. (i) Explain the working operation of square rooter shown in Figure 18.5, (ii) draw waveforms at appropriate places, and (iii) deduce expression for its output voltage. Verify this square rooter by experimenting with the AFC trainer kit.

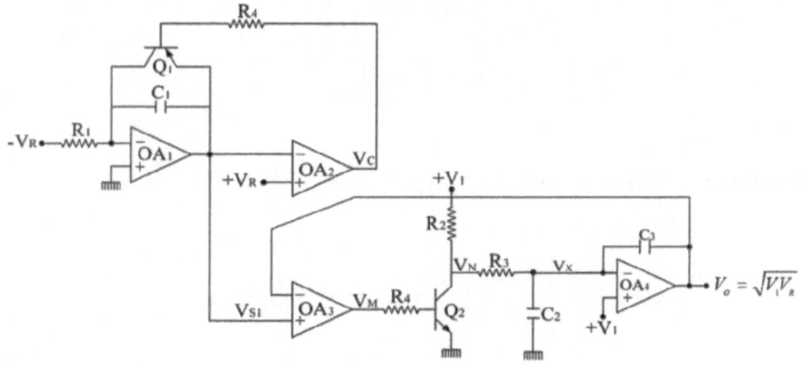

FIGURE 18.5 Sawtooth wave-based time division square rooter.

2. Similarly, convert all multipliers explained in Chapters 11 and 12 into square rooters by both type I and type II methods. In each, (i) draw circuit diagrams, (ii) explain their working operations,, (iii) draw waveforms at appropriate places and (iv) deduce expression for their output voltage. Verify these square rooters by experimenting with the AFC trainer kit.

18.4 DIVIDER TO MULTIPLIER

A divider can be converted into a multiplier as shown in Figure 18.6. Figure 18.6(a) shows type I methods, and Figure 18.6(b) shows type II methods.

The divider output is given as

$$V_X = \frac{V_O}{V_1} V_R \tag{18.19}$$

where V_R is the divider constant.

In Figure 18.6(a),

$$\text{The current through resistor } R_1; I_1 = \frac{V_X}{R_1} \tag{18.20}$$

$$\text{The current through resistor } R_2; I_2 = \frac{V_2}{R_2} \tag{18.21}$$

Since the op-amp has high input impedance, current I_1 will not enter into op-amp, and hence $I_1=I_2$.

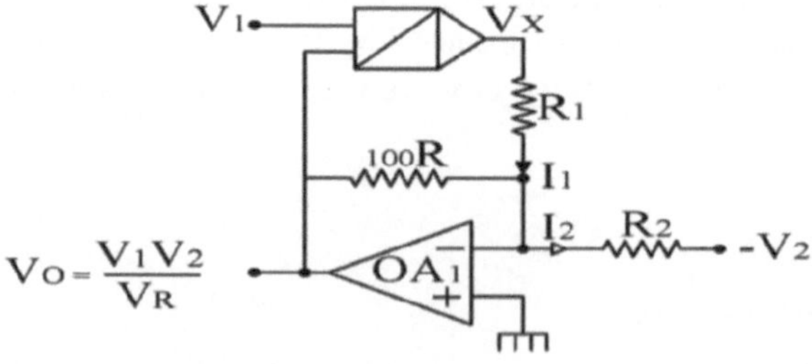

FIGURE 18.6(A) Divider to multiplier – type I.

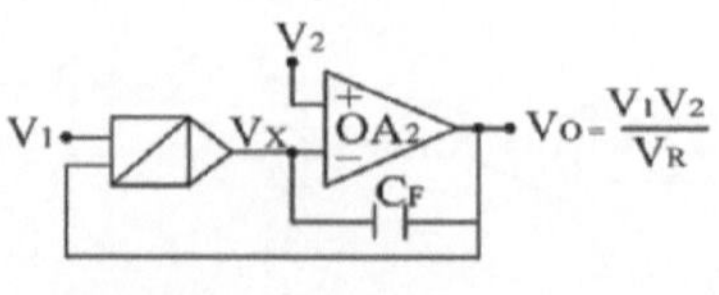

FIGURE 18.6(B) Divider to multiplier – type II.

$$\frac{V_X}{R_1} = \frac{V_2}{R_2} \tag{18.22}$$

Let us assume $R_1 = R_2 = R$.

$$V_X = V_2 \tag{18.23}$$

From Eqs. (18.19) and (18.23)

$$V_O = \frac{V_1 V_2}{V_R} \tag{18.24}$$

In the circuit shown in Figure 18.6(b), the op-amp is at a negative closed loop feedback and a positive DC voltage is ensured in the feedback. Hence, its non-inverting terminal voltage will be equal to its inverting terminal voltage.

$$V_2 = V_X \tag{18.25}$$

From Eqs. (18.19) and (18.25)

$$V_O = \frac{V_1 V_2}{V_R} \tag{18.26}$$

DESIGN EXERCISES

1. The time division divider with no reference shown in Figure 13.1(a) is converted into a multiplier type II method and shown in Figure 18.7. (i) Explain the working operation of the multiplier shown in Figure 18.7, (ii) draw waveforms at appropriate places, and (iii) deduce expression for the output voltage. Verify this multiplier by experimenting with the AFC trainer kit.
2. Similarly convert all dividers explained in Chapter 13 into multipliers by using type I and type II methods. In each, (i) draw circuit diagrams, (ii) explain their working operations, (iii) draw waveforms at appropriate places, and (iv) deduce expression for their output voltage. Verify these multipliers by experimenting with the AFC trainer kit.

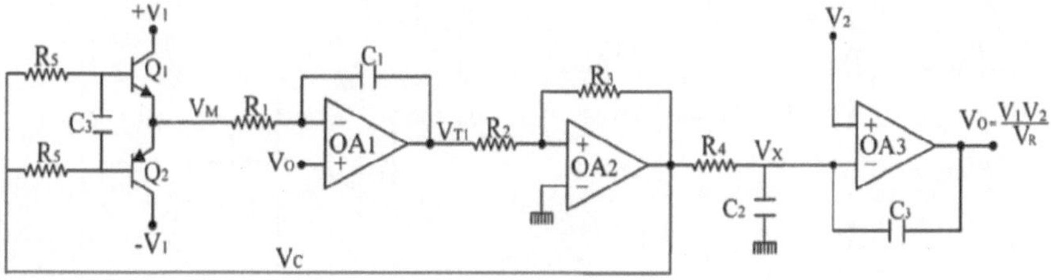

FIGURE 18.7 Time division multiplier from divider.

18.5 DIVIDER TO SQUARE ROOTER

The divider output is given as

$$V_O = \frac{V_2}{V_1} V_R \tag{18.27}$$

In the above equation, if $V_1 = V_O$, $V_2 = V_1$, then the equation will become

$$V_O = \sqrt{V_1 V_R} \tag{18.28}$$

DESIGN EXERCISES

1. The divider circuit shown in Figure 13.1(a) is converted into a square rooter and shown in Figure 18.8. (i) Explain the working operation of the square rooter shown in Figure 18.8, (ii) draw waveforms at appropriate places, and (iii) deduce expression for the output voltage. Verify this square rooter by experimenting with the AFC trainer kit.
2. Similarly, convert all dividers explained in Chapter 13 into square rooters. (i) Draw circuit diagrams, (ii) explain their working operations, (iii) draw waveforms at appropriate places, and (iv) deduce expression for their output voltage. Verify these square rooters by experimenting with the AFC trainer kit.

18.6 MULTIPLIER-CUM-DIVIDER TO MULTIPLIER

The MCD output is given as

$$V_O = \frac{V_2 V_3}{V_1} \tag{18.29}$$

In the above equation, if we keep $V_1 = V_R$, $V_2 = V_1$ and $V_3 = V_2$, then the output will become

$$V_O = \frac{V_1 V_2}{V_R} \tag{18.30}$$

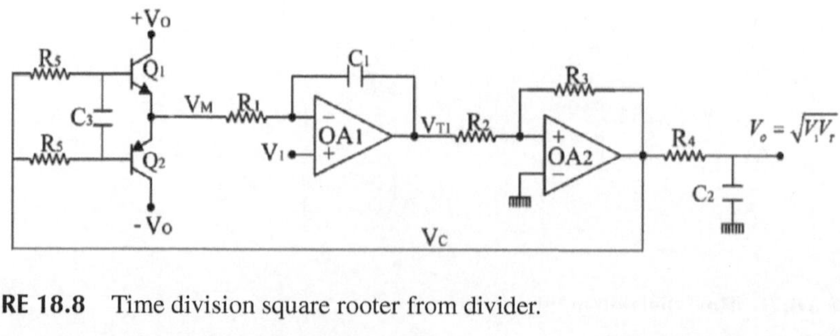

FIGURE 18.8 Time division square rooter from divider.

DESIGN EXERCISES

1. The double single slope peak responding multiplier-cum-divider shown in Figure 16.1 is converted into a multiplier and shown in Figure 18.9. (i) Explain the working operation of the multiplier shown in Figure 18.9, (ii) draw waveforms at appropriate places, and (iii) deduce expression for its output voltage. Verify this multiplier by experimenting with the AFC trainer kit.
2. Similarly, convert all multipliers-cum-dividers explained in Chapters 14–16 into multipliers. (i) Draw circuit diagrams, (ii) explain their working operations, (iii) draw waveforms at appropriate places, and (iv) deduce expression for their output voltage. Verify these multipliers by experimenting with the AFC trainer kit.

18.7 MULTIPLIER-CUM-DIVIDER TO DIVIDER

The MCD output is given as

$$V_O = \frac{V_2 V_3}{V_1} \tag{18.31}$$

In the above equation, if we keep $V_3 = V_R$, then the output will become

$$V_O = \frac{V_2}{V_1} V_R \tag{18.32}$$

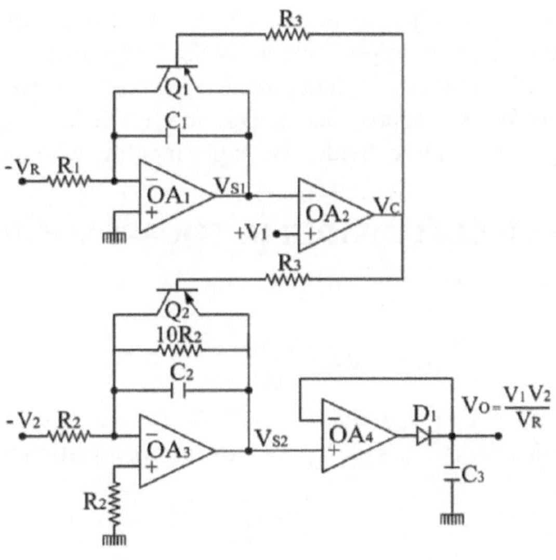

FIGURE 18.9(A) Double single slope peak detecting multiplier from MCD.

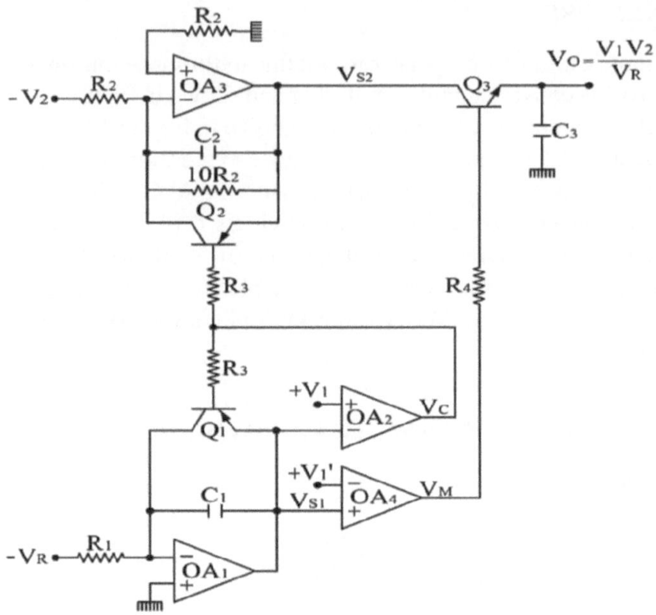

FIGURE 18.9(B) Double single slope peak sampling multiplier from MCD.

DESIGN EXERCISES

1. The double single slope peak responding multipliers-cum-dividers shown in Figure 16.1 are converted into a divider and shown in Figure 18.10. (i) Explain the working operation of the divider shown in Figure 18.10, (ii) draw waveforms at appropriate places, and (iii) deduce expression for the output voltage. Verify this divider by experimenting with the AFC trainer kit.
2. Similarly, convert all multipliers-cum-dividers explained in Chapters 14–16 into dividers. (i) Draw circuit diagrams, (ii) explain their working operations, (iii) draw waveforms at appropriate places, and (iv) deduce expression for their output voltage. Verify these dividers by experimenting with the AFC trainer kit.

18.8 MULTIPLIER-CUM-DIVIDER TO SQUARE ROOTER

The MCD output is given as

$$V_O = \frac{V_2 V_3}{V_1} \tag{18.33}$$

If we keep $V_3 = V_R$, $V_1 = V_O$, $V_2 = V_1$, then the above equation will become

$$V_O = \sqrt{V_1 V_R} \tag{18.34}$$

Realizing the operation of a square rooter.

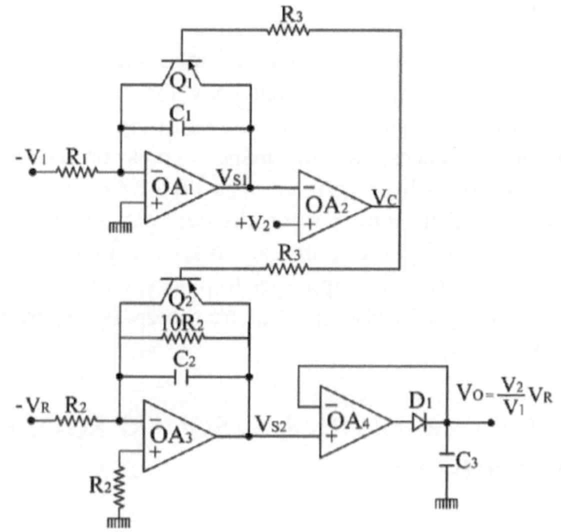

FIGURE 18.10(A) Double single slope peak detecting divider.

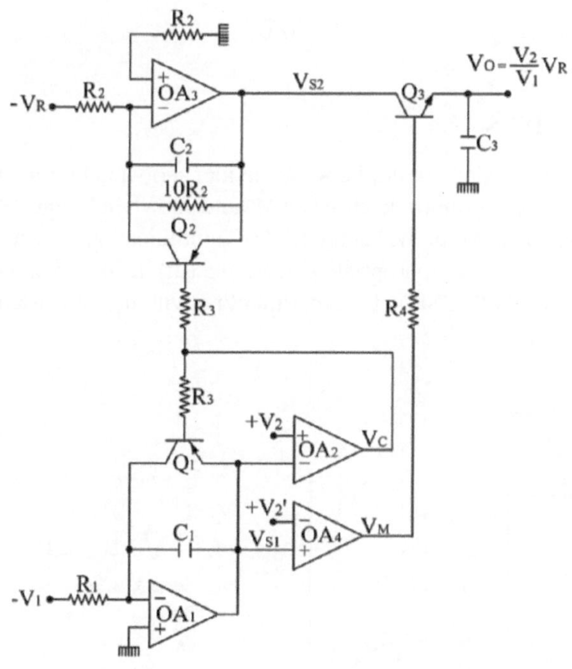

FIGURE 18.10(B) Double single slope peak sampling divider.

DESIGN EXERCISES

1. The sawtooth wave-based time division multiplier-cum-divider shown in Figure 15.1 is converted into a square rooter and shown in Figure 18.11. (i) Explain the working operation of the square rooter shown in Figure 18.11, (ii) draw waveforms at appropriate places, and (iii) deduce expression for the output voltage. Verify this square rooter by experimenting with the AFC trainer kit.
2. Similarly, convert all multipliers-cum-dividers explained in Chapters 14–16 into a square rooter. (i) Draw circuit diagrams, (ii) explain their working operations, (iii) draw waveforms at appropriate places, and (iv) deduce expression for their output voltage. Verify these square rooters by experimenting with the AFC trainer kit.

18.9 MULTIPLIER-CUM-DIVIDER TO SQUARE ROOT OF MULTIPLICATION

The MCD output is given as

$$V_O = \frac{V_2 V_3}{V_1} \tag{18.35}$$

In the above equation, if $V_1 = V_O$, $V_2 = V_1$ and $V_3 = V_2$, then the output will become

$$V_O = \sqrt{V_1 V_2} \tag{18.36}$$

DESIGN EXERCISES

1. The double single slope multiplier-cum-divider shown in Figure 16.1 is converted into a square root of multiplication (SRM) and shown in Figure 18.12. (i) Explain the working operation of the square root of multiplication shown in Figure 18.12, (ii) draw waveforms at appropriate places, and (iii) deduce expression for the output voltage. Verify this SRM by experimenting with the AFC trainer kit.

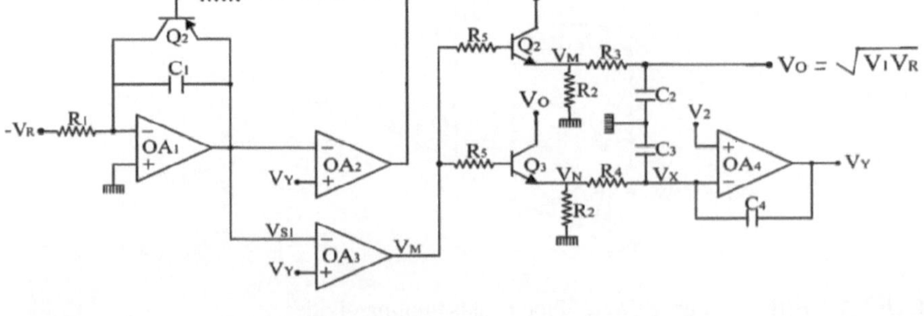

FIGURE 18.11 Sawtooth wave-based square rooter.

2. Similarly, convert all multipliers-cum-dividers explained in Chapters 14–16 into a square root of multiplication (SRM). (i) Draw circuit diagrams, (ii) explain their working operations, (iii) draw waveforms at appropriate places, and (iv) deduce expression for their output voltage. Verify these SRMs by experimenting with the AFC trainer kit.

18.10 MULTIPLIER-CUM-DIVIDER TO SQUARING AND DIVIDING

The MCD output is given as

$$V_O = \frac{V_2 V_3}{V_1} \tag{18.37}$$

If $V_3 = V_2$, then the output will become

$$V_O = \frac{V_2^2}{V_1} \tag{18.38}$$

DESIGN EXERCISES

1. The double single slope peak responding multiplier-cum-divider shown in Figure 15.1 is converted into squaring and dividing (SAD) and is shown in Figure 18.13. (i) Explain the working operation of squaring and dividing

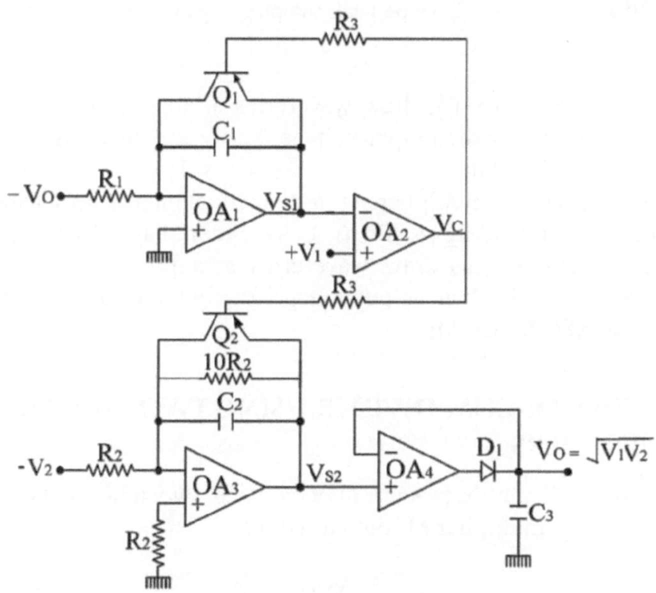

FIGURE 18.12(A) Double single slope peak detecting square root of multiplication from multiplier-cum-divider.

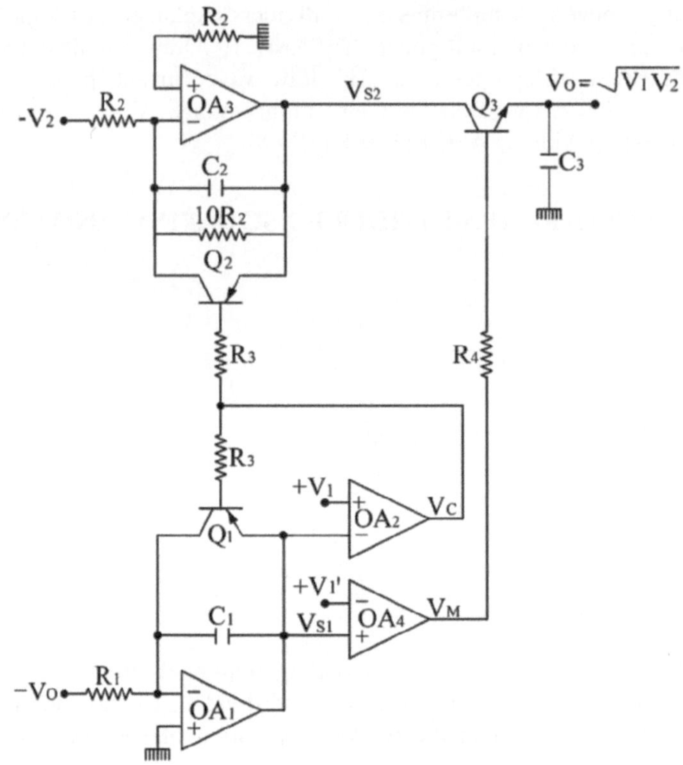

FIGURE 18.12(B) Double single slope peak sampling square root of multiplication from multiplier-cum-divider.

shown in Figure 18.13, (ii) draw waveforms at appropriate places, and (iii) deduce expression for the output voltage. Verify this SAD by experimenting with the AFC trainer kit.

2. Similarly, convert all multipliers-cum-dividers explained in Chapters 14–16 into squaring and dividing (SAD). (i) Draw circuit diagrams, (ii) explain their working operations, (iii) draw waveforms at appropriate places, and (iv) deduce expression for their output voltage. Verify these SADs by experimenting with the AFC trainer kit.

18.11 MULTIPLIER-CUM-DIVIDER USING TWO ANALOG MULTIPLIERS

The block diagram of multiplier-cum-divider using two multipliers is shown in Figure 18.14. The first multiplier M_1 output will be

$$V_A = \frac{V_2 V_3}{V_R} \tag{18.39}$$

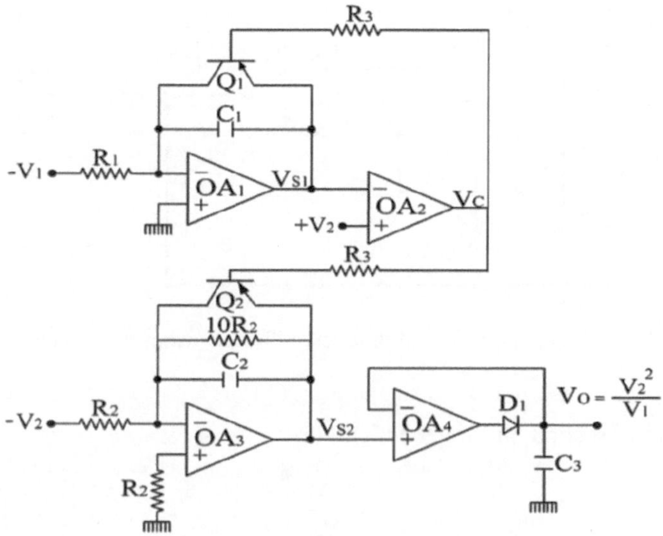

FIGURE 18.13(A) Double single slope peak detecting squaring and dividing from multiplier-cum-divider.

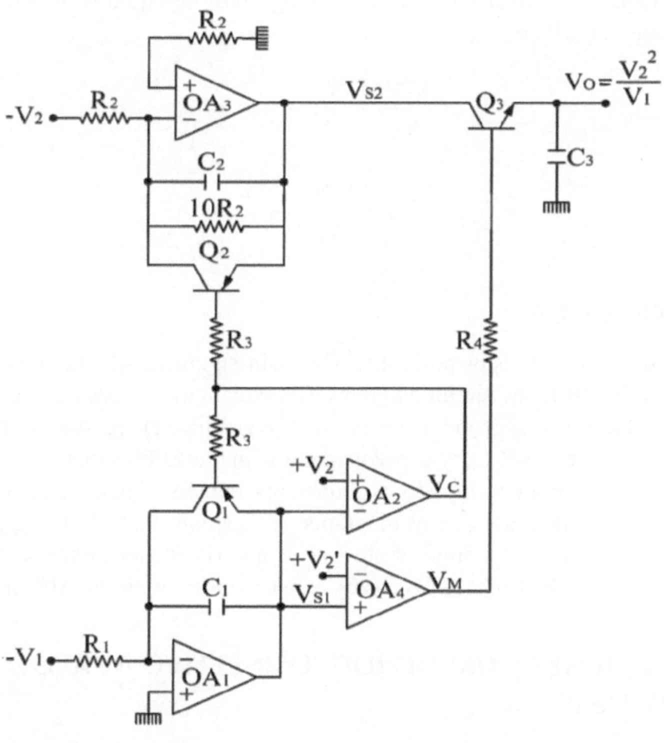

FIGURE 18.13(B) Double single slope peak sampling squaring and dividing from multiplier-cum-divider.

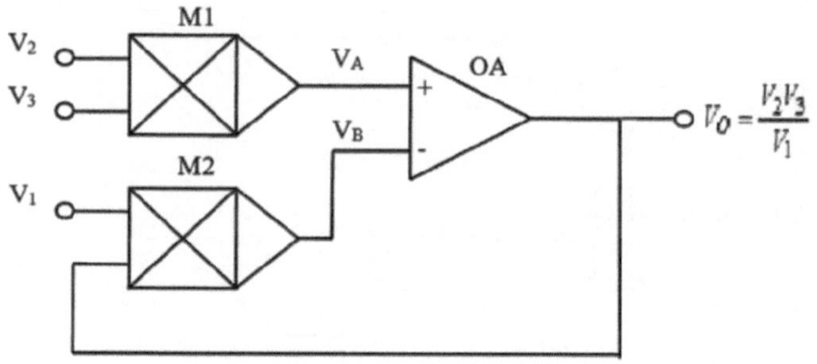

FIGURE 18.14 Block diagram multiplier-cum-divider using two multipliers and an op-amp.

The output of second multiplier M_2 will be

$$V_B = \frac{V_1 V_O}{V_R} \tag{18.40}$$

In Eqs. (18.39) and (18.40), V_R are the multiplier constants. The op-amp OA is connected in a negative feedback loop and a positive DC voltage is ensured in the closed loop. Hence, its inverting terminal voltage will be equal to its non-inverting terminal voltage, i.e., $V_A = V_B$.

$$\frac{V_2 V_3}{V_R} = \frac{V_1 V_O}{V_R} \tag{18.41}$$

$$V_O = \frac{V_2 V_3}{V_1} \tag{18.42}$$

DESIGN EXERCISES

1. By using two time division multipliers shown in Figure 12.1(b), obtain a multiplier-cum- divider. (i) Draw circuit diagrams, (ii) explain their working operations, (iii) draw waveforms at appropriate places, and (iv) deduce expression for their output voltage. Verify this MCD by experimenting with the AFC trainer kit.
2. Similarly, convert all multipliers explained in Chapters 11 and 12 into multipliers-cum-dividers. (i) Draw circuit diagrams, (ii) explain their working operations, (iii) draw waveforms at appropriate places, and (iv) deduce expression for their output voltage. Verify these MCDs by experimenting with the AFC trainer kit.

18.12 MULTIPLIER-CUM-DIVIDER USING TWO ANALOG DIVIDERS

The block diagram of multiplier-cum-divider using two dividers is shown in Figure 18.15. The first divider D_1 output will be

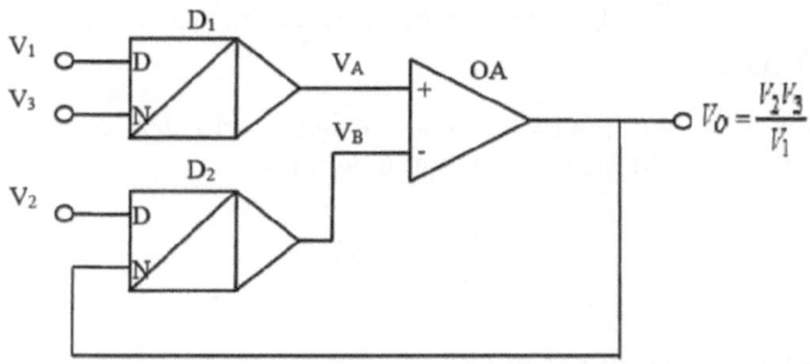

FIGURE 18.15 Block diagram of multiplier-cum-divider using two analog dividers.

$$V_A = \frac{N}{D} = \frac{V_3}{V_1} V_R \tag{18.43}$$

The second divider D_2 output will be

$$V_B = \frac{N}{D} = \frac{V_O}{V_2} V_R \tag{18.44}$$

In Eqs. (18.43) and (18.44), V_R is the divider constant. The op-amp OA is configured in a negative closed loop configuration and a positive DC voltage is ensured in the feedback. Hence, the voltage at its positive input voltage must be equal to the voltage at its negative input voltage.

$$V_A = V_B \tag{18.45}$$

$$\frac{V_3}{V_1} = \frac{V_O}{V_2}$$

$$V_O = \frac{V_2 V_3}{V_1} \tag{18.46}$$

DESIGN EXERCISES

1. By using two time division dividers shown in Figure 13.1(a), obtain a multiplier-cum-divider. (i) Draw circuit diagrams, (ii) explain their working operations, (iii) draw waveforms at appropriate places, and (iv) deduce expression for their output voltage. Verify this MCD experimenting with the AFC trainer kit.
2. Similarly, convert all dividers explained in Chapter 13 into MCDs. (i) Draw circuit diagrams, (ii) explain their working operations, (iii) draw waveforms at appropriate places, and (iv) deduce expression for their output voltage. Verify these MCDs by experimenting with the AFC trainer kit.

18.13 MULTIPLIER-CUM-DIVIDER USING TWO ANALOG DIVIDERS IN CASCADE

The block diagram of multiplier-cum-divider using two dividers in cascade is shown in Figure 18.16(a). The first divider D_1 output will be

$$V_X = \frac{N}{D} = \frac{V_1}{V_2} V_R \quad (18.47)$$

The second divider D_2 output will be

$$V_O = \frac{N}{D} = \frac{V_3}{V_X} V_R \quad (18.48)$$

In Eqs. (18.47) and (18.48), V_R is the divider constant.

$$V_O = \frac{V_3 V_R}{V_1 V_R} V_2 \quad (18.49)$$

$$V_O = \frac{V_2 V_3}{V_1} \quad (18.50)$$

DESIGN EXERCISES

1. By using two time division dividers shown in Figure 13.1(a), obtain a multiplier-cum-divider. (i) Draw circuit diagrams, (ii) explain their working operations, (iii) draw waveforms at appropriate places, and (iv) deduce expression for their output voltage. Verify this MCD by experimenting with the AFC trainer kit.

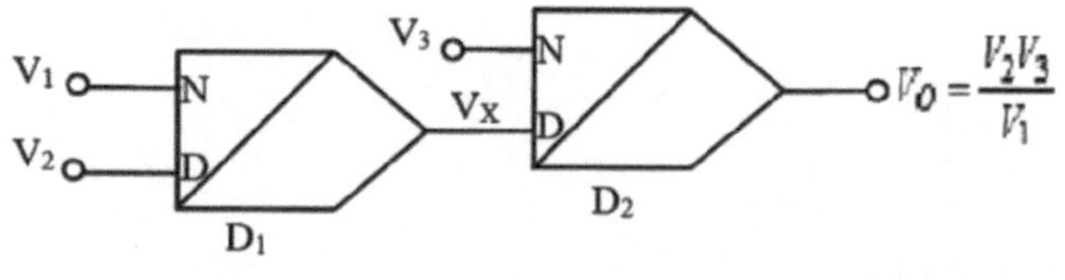

FIGURE 18.16(A) Block diagram of multiplier-cum-divider using dividers.

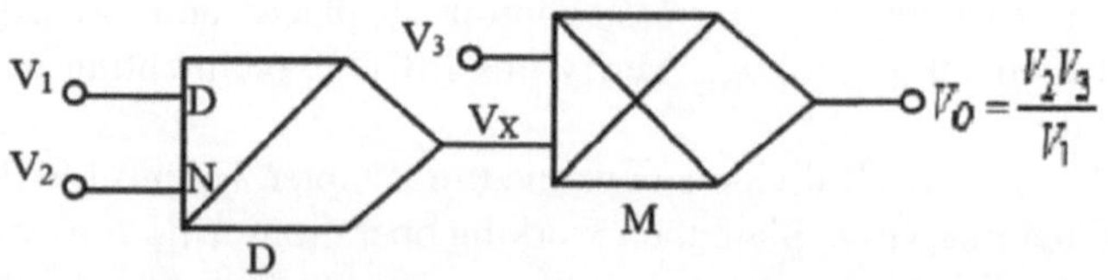

FIGURE 18.16(B) Block diagram of multiplier-cum-divider using divider and multiplier.

2. Similarly, convert all dividers explained in Chapter 13 into MCDs. (i) Draw circuit diagrams, (ii) explain their working operations, (iii) draw waveforms at appropriate places, and (iv) deduce expression for their output voltage. Verify these MCDs by experimenting with the AFC trainer kit.

18.14 MULTIPLIER-CUM-DIVIDER USING A DIVIDER AND A MULTIPLIER

The block diagram of a proposed multiplier-cum-divider using a divider and a multiplier is shown in Figure 18.16(b). The divider D output will be

$$V_X = \frac{N}{D} = \frac{V_2}{V_1} V_R \tag{18.51}$$

The multiplier M output will be

$$V_O = \frac{V_X V_3}{V_R} \tag{18.52}$$

In Eqs. (18.51) and (18.52), V_R are the divider and multiplier constants.

$$V_O = \frac{V_2 V_R}{V_1 V_R} V_3 \tag{18.53}$$

$$V_O = \frac{V_2 V_3}{V_1} \tag{18.54}$$

DESIGN EXERCISES

1. By using the time division multiplier in Figure 12.1(a) and the time division divider in Figure 13.1(a), obtain a multiplier-cum-divider. (i) Draw circuit diagrams, (ii) explain their working operations, (iii) draw waveforms at appropriate places, and (iv) deduce expression for their output voltage. Verify this MCD by experimenting with the AFC trainer kit.
2. Similarly, convert all multipliers and dividers explained in Chapters 11–13 into MCDs. (i) Draw circuit diagrams, (ii) explain their working operations, (iii) draw waveforms at appropriate places, and (iv) deduce expression for their output voltage. Verify these MCDs by experimenting with the AFC trainer kit.

Part E

Miscellaneous Function Circuits

19 Vector Magnitude Circuits

A vector magnitude circuit (VMC) receives two input voltages and produces an output voltage. The output voltage is the square root of the sum of two squared input voltages. There are several applications of this circuit in measurements, instrumentation, and communication systems. A few examples are (i) impedance and power measurements, (ii) construction of signal-envelope circuits, (iii) diversity combiner circuits, (iv) radiometer circuits, and (v) other applications in which the power of separate waveforms must be added. VMCs are obtained from multipliers, dividers, and multipliers-cum-dividers. Various VMCs using these function circuits are explained in this chapter.

19.1 VECTOR MAGNITUDE CIRCUIT USING TWO MULTIPLIERS

The block diagram of VMC using two multipliers is shown in Figure 19.1. One multiplier M_1 is at the non-inverting terminal path, and another multiplier M_2 is at the negative feedback path of the op-amp OA. The multiplier M_1 output is

$$V_A = \frac{V_1^2}{V_R} \tag{19.1}$$

The multiplier M_2 output will be

$$V_B = \frac{V_Y V_X}{V_R} \tag{19.2}$$

In the above Eqs. (19.1) and (19.2), V_R is the multiplier constant. As the op-amp OA is connected in the negative feedback loop, its non-inverting terminal voltage will be equal to its inverting terminal voltage. Hence, $V_A = V_B$.

From Eqs. (19.1) and (19.2),

$$V_X = \frac{V_1^2}{V_Y} \tag{19.3}$$

The unity gain summer S2 output will be

$$V_Y = V_O - V_2 \tag{19.4}$$

DOI: 10.1201/9781003221449-24

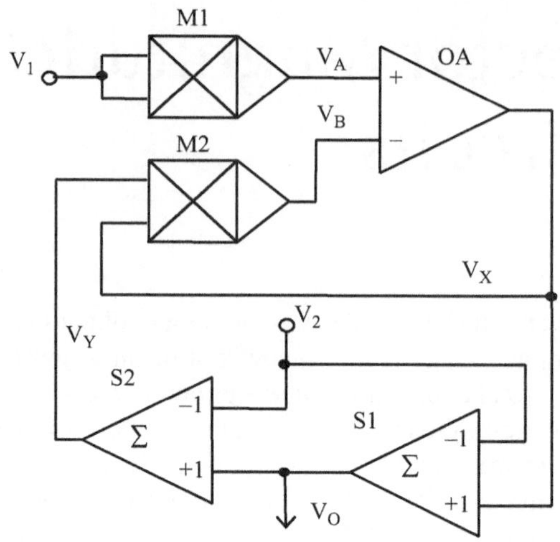

FIGURE 19.1 Block diagram of vector magnitude circuit using two multipliers.

The unity gain summer S1 output will be

$$V_O = V_X - V_2 \tag{19.5}$$

From Eqs. (19.3), (19.4), and (19.5)

$$V_O = \sqrt{V_1^{\,2} + V_2^{\,2}} \tag{19.6}$$

The circuit diagram of the VMC is shown in Figure 19.2, and its associated waveforms are shown in Figure 19.3. One time division multiplier is realized by op-amp OA_4, switch S_3, and R_2C_2 lowpass filter, and the other multiplier is realized by op-amp OA_3, switch S_2, and R_1C_1 lowpass filter. For both multipliers, a sawtooth wave is applied as reference clock. An asymmetrical rectangular waveform V_{P2} is generated by comparing the input voltage V_1 with the reference sawtooth wave V_{S1} of peak value V_R, time period 'T' by comparator OA_4. The ON time δ_{T1} of the V_{P2} is

$$\delta_{T1} = \frac{V_1}{V_R} T \tag{19.7}$$

Another asymmetrical rectangular waveform V_{P1} is generated by comparing the voltage V_Y with the reference sawtooth wave V_{S1} of peak value V_R, time period 'T' by the comparator OA_3. The ON time δ_{T2} of the V_{P1} is

$$\delta_{T2} = \frac{V_Y}{V_R} T \tag{19.8}$$

The switch S_2 connects voltage V_X during the ON time δ_{T2} and zero volts during OFF time of V_{P1}, to the R_1C_1 lowpass filter. A pulse waveform V_M with maximum

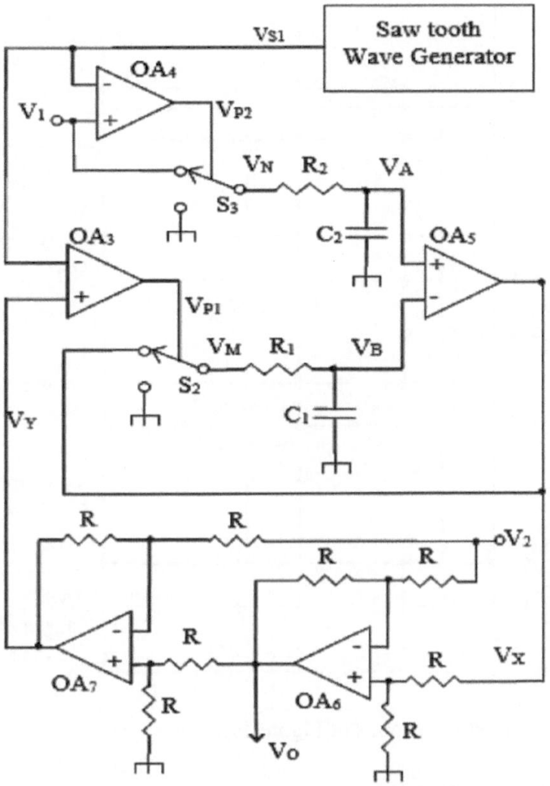

FIGURE 19.2 Circuit diagram of proposed vector magnitude circuit.

value of V_X is generated at the output of switch S_2. The R_1C_1 lowpass filter gives the average value of the pulse train V_M and is given by

$$V_B = \frac{1}{T}\int_0^{\delta_{T2}} V_X\,dt = \frac{V_X}{T}\delta_{T2} = \frac{V_X V_Y}{V_R} \tag{19.9}$$

The switch S_3 connects input voltage V_1 during the ON time δ_{T1} and zero volts during OFF time of pulse V_{P2}, to the R_2C_2 lowpass filter. A pulse waveform V_N with maximum value of V_1 is generated at the output of switch S_3. The R_2C_2 lowpass filter gives the average value of the pulse train V_N and is given by

$$V_A = \frac{1}{T}\int_0^{\delta_{T1}} V_1\,dt = \frac{V_1}{T}\delta_{T1} = \frac{V_1^2}{V_R} \tag{19.10}$$

In Figure 19.2, since the op-amp OA_5 is at negative feedback loop, its inverting terminal voltage will be equal to its non-inverting voltage. Hence, $V_A = V_B$.

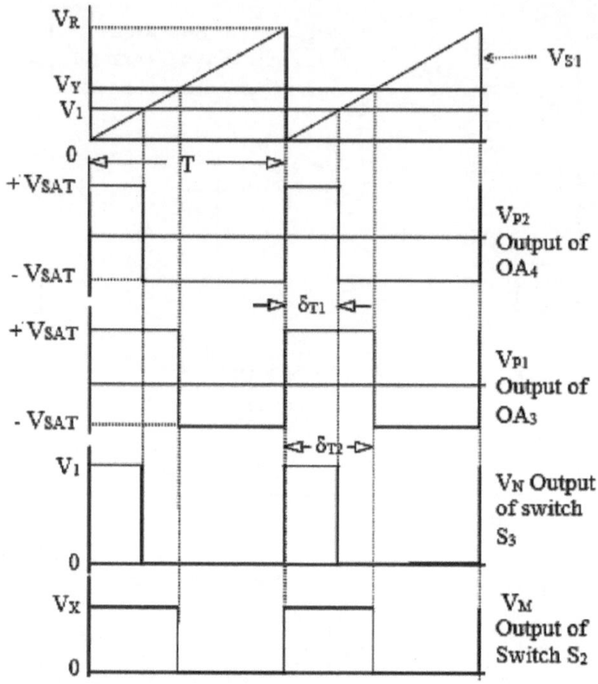

FIGURE 19.3 Associated waveforms of Figure 19.2.

From Eqs. (19.9) and (19.10)

$$V_X = \frac{V_1^2}{V_Y} \tag{19.11}$$

The output of the differential amplifier OA_7 is given as

$$V_Y = V_O - V_2 \tag{19.12}$$

The output of the differential amplifier OA_6 is given as

$$V_O = V_X - V_2 \tag{19.13}$$

From Eqs. (19.11), (19.12), and (19.13)

$$V_O = \sqrt{V_1^2 + V_2^2} \tag{19.14}$$

19.2 VECTOR MAGNITUDE CIRCUIT USING THREE MULTIPLIERS

The block diagram of VMC using three multipliers is shown in Figure 19.4. The multiplier M_1 output will be

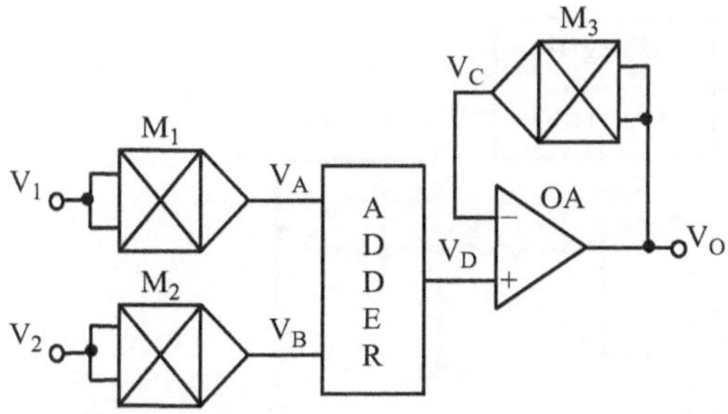

FIGURE 19.4 Block diagram of vector magnitude circuit using three multipliers.

$$V_A = \frac{V_1^2}{V_T} \tag{19.15}$$

The multiplier M_2 output will be

$$V_B = \frac{V_2^2}{V_T} \tag{19.16}$$

The multiplier M_3 output will be

$$V_C = \frac{V_O^2}{V_T} \tag{19.17}$$

In Eqs. (19.15)–(19.17) V_T are the multiplier constants. The output of the adder will be

$$V_D = \frac{V_1^2 + V_2^2}{V_T} \tag{19.18}$$

The op-amp OA is at a negative feedback loop and a positive DC voltage is ensured in the feedback. Hence, its non-inverting terminal voltage will be equal to its inverting terminal voltage, i.e., $V_C = V_D$.

From Eqs. (19.17) and (19.18)

$$V_O^2 = V_1^2 + V_2^2 \tag{19.19}$$

$$V_O = \sqrt{V_1^2 + V_2^2} \tag{19.20}$$

19.3 VECTOR MAGNITUDE CIRCUIT USING THREE TRIANGULAR WAVE–BASED TIME DIVISION MULTIPLIERS

The VMC diagram using three triangular wave–based TDMs is shown in Figure 19.5, and its associated waveforms are shown in Figure 19.6. The triangular wave V_{T1} is

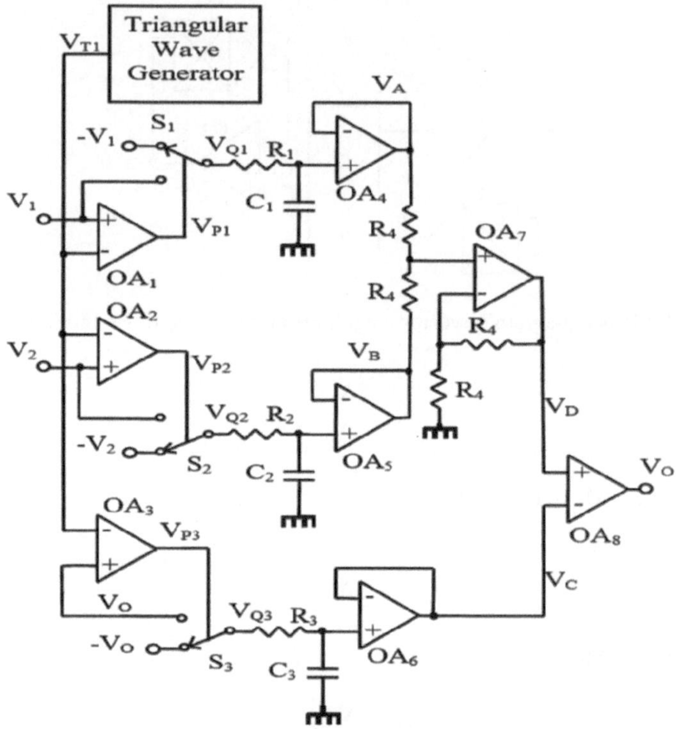

FIGURE 19.5 Vector magnitude circuit using triangular wave-based multipliers.

compared with one input voltage V_1 by comparator OA_1. An asymmetric rectangular pulse waveform V_{P1} with peak-to-peak of $\pm V_{SAT}$ is generated at the output of OA_1. From the waveforms shown in Figure 19.6,

$$T_2 = \frac{V_T + V_1}{2V_T} T \tag{19.21}$$

$$T_1 = \frac{V_T - V_1}{2V_T} T \tag{19.22}$$

$$T = T_1 + T_2 \tag{19.23}$$

The switch S_1 connects V_1 during ON time of pulse V_{P1} and $-V_1$ during the OFF time of pulse V_{P1} at its output V_{Q1}. This pulse V_{Q1} is another asymmetric rectangular waveform with peak-to-peak value of $\pm V_1$ as shown in Figure 19.6. The pulse V_{Q1} is given to the R_1C_1 lowpass filter. The output of this lowpass filter will be

$$V_A = \frac{1}{T}\left[\int_0^{T_2} (V_1)\,dt + \int_{T_2}^{T_1+T_2} (-V_1)\,dt\right]$$

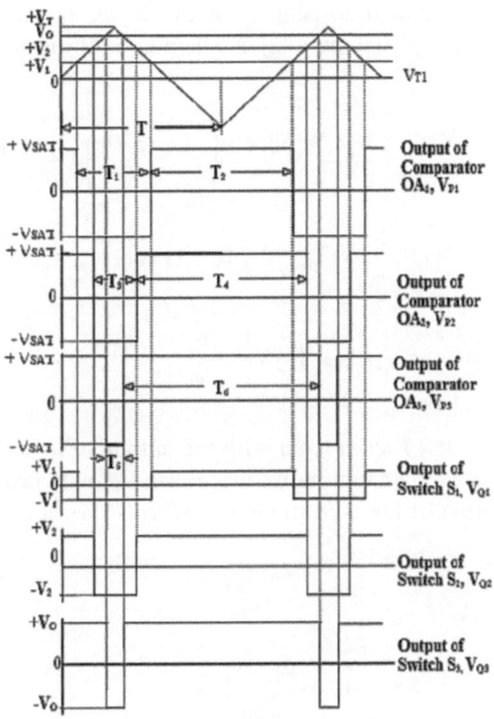

FIGURE 19.6 Associated waveforms of Figure 19.5.

$$V_A = \frac{1}{T}\left[V_1T_2 - V_1\left(T_1 + T_2 - T_2\right)\right]$$

$$= \frac{1}{T}\left(T_2 - T_1\right)V_1 = \frac{V_1^2}{V_T} \tag{19.24}$$

The triangular wave V_{T1} is compared with the other input voltage V_2 by comparator OA_2. An asymmetric rectangular pulse waveform V_{P2} with peak-to-peak of $\pm V_{CC}$ is generated at the output of OA_2. From the waveforms shown in Figure 19.6,

$$T_4 = \frac{V_T + V_2}{2V_T}T \tag{19.25}$$

$$T_3 = \frac{V_T - V_2}{2V_T}T \tag{19.26}$$

$$T = T_3 + T_4 \tag{19.27}$$

The switch S_2 connects V_2 during ON time of the pulse V_{P2}, and $-V_2$ during the OFF time of the pulse V_{P2} at its output V_{Q2}. This pulse V_{Q2} is another asymmetric rectangular waveform with peak-to-peak value of $\pm V_2$ as shown in Figure 19.6. The pulse V_{Q2} is given to the R_2C_2 lowpass filter. The output of this lowpass filter will be

$$V_B = \frac{1}{T}\left[\int_0^{T_4}(V_2)dt + \int_{T_4}^{T_3+T_4}(-V_2)dt\right]$$

$$V_B = \frac{1}{T}\left[V_2T_4 - V_2(T_3 + T_4 - T_4)\right]$$

$$= \frac{1}{T}(T_4 - T_3)V_2 = \frac{V_2^2}{V_T} \tag{19.28}$$

The triangular wave V_{T1} is compared with the output voltage V_O by the comparator OA_3. An asymmetric rectangular pulse waveform V_{P3} with peak-to-peak of $\pm V_{SAT}$ is generated at the output of OA_3. From the waveforms shown in Figure 19.6,

$$T_6 = \frac{V_T + V_O}{2V_T}T \tag{19.29}$$

$$T_5 = \frac{V_T - V_O}{2V_T}T \tag{19.30}$$

$$T = T_5 + T_6 \tag{19.31}$$

The switch S_3 connects V_O during ON time of the pulse V_{P3} and $-V_O$ during the OFF time of the pulse V_{P3} at its output V_{Q3}. This pulse V_{Q3} is another asymmetric rectangular waveform with peak-to-peak value of $\pm V_O$ as shown in Figure 19.6. The pulse V_{Q3} is given to the R_3C_3 lowpass filter. The output of this lowpass filter will be

$$V_C = \frac{1}{T}\left[\int_0^{T_6}(V_O)dt + \int_{T_6}^{T_5+T_6}(-V_O)dt\right]$$

$$V_C = \frac{1}{T}\left[V_OT_6 - V_O(T_5 + T_6 - T_6)\right]$$

$$= \frac{1}{T}(T_6 - T_5)V_O = \frac{V_O^2}{V_T} \tag{19.32}$$

The op-amp OA_7 is configured as an adder. Its output will be

$$V_D = V_A + V_B$$

$$V_D = \frac{V_1^2}{V_T} + \frac{V_2^2}{V_T} \tag{19.33}$$

The op-amp OA_8 is at a negative feedback loop (output voltage is given as feedback input to switch S3 and comparator OA_3) and a positive DC voltage is ensured in the feedback.

Hence, its non-inverting terminal voltage will be equal to its inverting terminal voltage, i.e., $V_C = V_D$.

From Eqs. (19.32) and (19.33)

$$V_O = \sqrt{V_1^2 + V_2^2} \tag{19.34}$$

19.4 VECTOR MAGNITUDE CIRCUIT USING THREE SAWTOOTH WAVE-BASED TIME DIVISION MULTIPLIERS

The circuit diagram of a VMC using three sawtooth wave-based TDMs is shown in Figure 19.7, and its associated waveforms are shown in Figure 19.8. A sawtooth

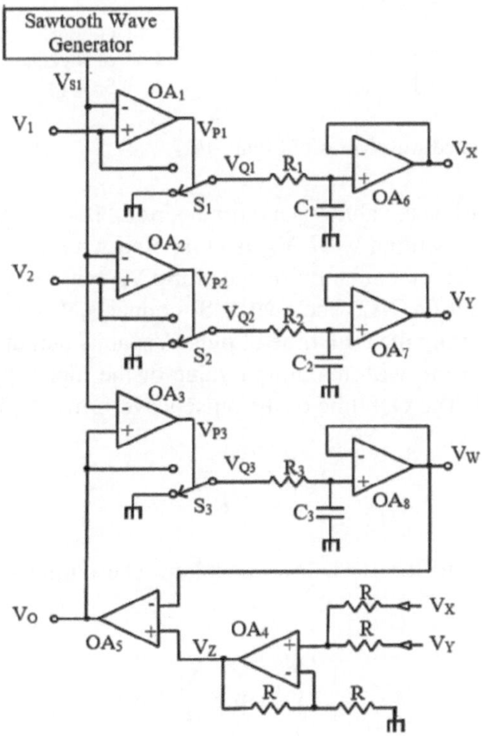

FIGURE 19.7 Circuit diagram of proposed vector magnitude circuit.

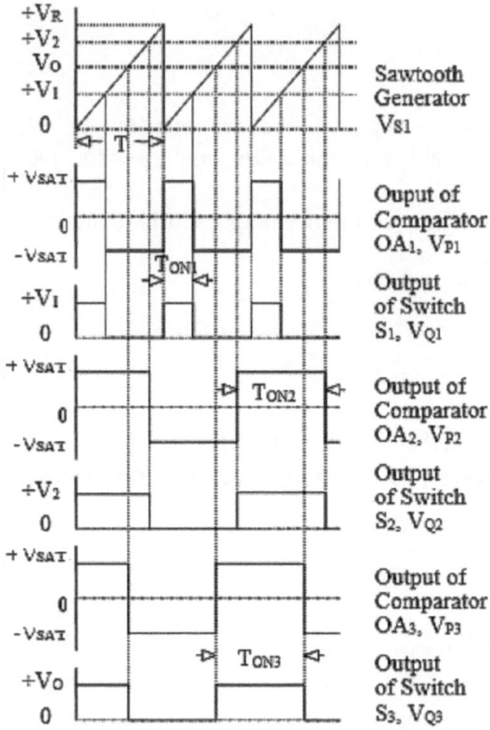

FIGURE 19.8 Associated waveforms of Figure 19.7.

wave, marked as V_{S1} of peak value V_R and time period 'T' is generated by a sawtooth wave generator. This sawtooth wave V_{S1} is compared with one input voltage V_1 by the comparator OA_1. A rectangular pulse waveform V_{P1} with peak-to-peak of ±Vcc is generated at the output of OA_1. The switch S_1 connects V_1 during ON time of the pulse V_{P1} and GND during the OFF time of pulse V_{P1} at its output V_{Q1}. This pulse V_{Q1} is a rectangular waveform with maximum value of the input applied voltage V_1 as shown in Figure 19.8. The ON time of the pulse waveforms V_{Q1} is given by

$$T_{ON1} = \frac{V_1}{V_R} T \tag{19.35}$$

Pulse V_{Q1} is given to the R_1C_1 lowpass filter. The output of the lowpass filter will be

$$V_X = \frac{1}{T} \int_0^{T_{ON1}} V_1 \, dt = \frac{V_1}{T} T_{ON1}$$

$$V_X = \frac{V_1^2}{V_R} \tag{19.36}$$

The sawtooth wave V_{S1} is compared with another input voltage V_2 by comparator OA_2.

A rectangular pulse waveform V_{P2} with peak-to-peak of ±Vcc is generated at the output of OA_2. The switch S_2 connects V_2 during ON time of the pulse V_{P2} and GND during the OFF time of the pulse V_{P2} at its output V_{Q2}. This pulse V_{Q2} is a rectangular waveform with maximum value of the input voltage V_2 as shown in Figure 19.8. The ON time of the pulse waveforms V_{Q2} is given by

$$T_{ON2} = \frac{V_2}{V_R} T \tag{19.37}$$

The pulse V_{Q2} is given to the R_2C_2 lowpass filter. The output of the lowpass filter will be

$$V_Y = \frac{1}{T} \int_0^{T_{ON2}} V_2 \, dt = \frac{V_2}{T} T_{ON2} = \frac{V_2^{\,2}}{V_R} \tag{19.38}$$

The sawtooth wave V_{S1} is compared with the voltage V_O by comparator OA_3. A rectangular pulse waveform V_{P3} with peak-to-peak of ±Vcc is generated at the output of OA_3. The switch S_3 connects V_O during ON time of the pulse V_{P3} and GND during the OFF time of the pulse V_{P3} at its output V_{Q3}. This pulse V_{Q3} is a rectangular waveform with maximum value of the voltage V_O as shown in Figure 19.8. The ON time of the pulse waveform V_{Q3} is given by

$$T_{ON3} = \frac{V_O}{V_R} T \tag{19.39}$$

The pulse V_{Q3} is given to the R_3C_3 lowpass filter. The output of the lowpass filter will be

$$V_W = \frac{1}{T} \int_0^{T_{ON3}} V_O \, dt = \frac{V_O}{T} T_{ON3}$$

$$V_W = \frac{V_O^{\,2}}{V_R} \tag{19.40}$$

The output of an adder realized with op-amp OA_4 will be

$$V_Z = V_X + V_Y = \frac{V_1^{\,2} + V_2^{\,2}}{V_R} \tag{19.41}$$

The op-amp OA_5 is at negative feedback loop, and a positive DC voltage is ensured in the feedback. Hence, its non-inverting terminal voltage will be equal to its inverting terminal voltage, i.e., $V_W = V_Z$.

From Eqs. (19.40) and (19.41)

$$V_O = \sqrt{V_1^2 + V_2^2} \quad (19.42)$$

19.5 VECTOR MAGNITUDE CIRCUIT USING TWO DIVIDERS

The block diagram of VMC using two dividers is shown in Figure 19.9. The output of divider D1 will be

$$V_Z = \frac{V_Y}{V_1} V_R \quad (19.43)$$

The output of divider D2 will be

$$V_X = \frac{V_1}{V_Z} V_R = \frac{V_1^2}{V_Y} \quad (19.44)$$

In Eqs. (19.43) and (19.44), V_R is the divider constant. The difference amplifier S2 output will be

$$V_Y = V_O - V_2 \quad (19.45)$$

The difference amplifier S1 output will be

$$V_O = V_X - V_2 \quad (19.46)$$

From Eqs. (19.45) and (19.46)

$$V_O = \frac{V_1^2}{V_Y} - V_2 = \frac{V_1^2 - V_2 V_Y}{V_Y} \quad (19.47)$$

The above equation can be simplified as

$$V_O = \sqrt{V_1^2 + V_2^2} \quad (19.48)$$

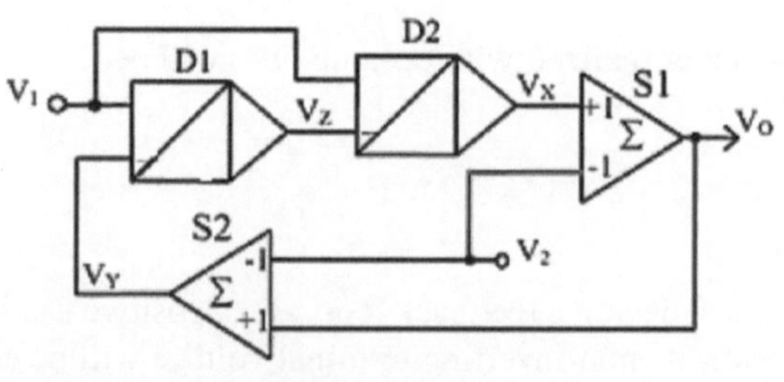

FIGURE 19.9 Block diagram of the proposed method.

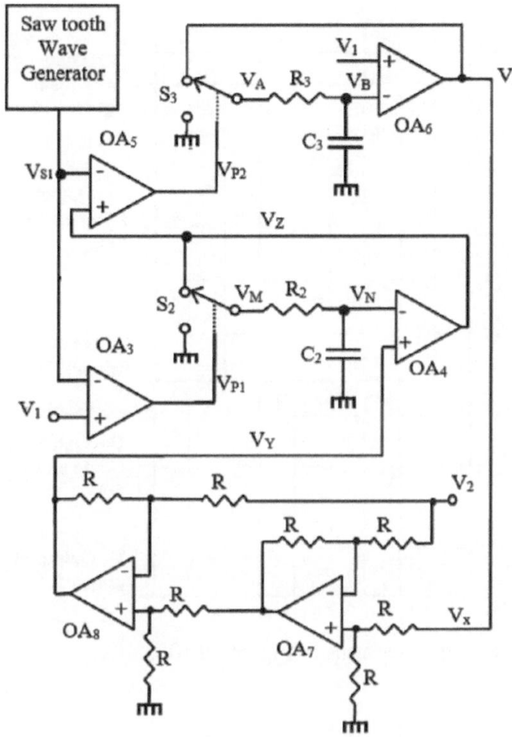

FIGURE 19.10 Circuit diagram of proposed vector magnitude circuit using two analog dividers.

The vector magnitude circuit diagram is shown in Figure 19.10, and its associated waveforms are shown in Figure 19.11. The comparator OA_3, switch S_2, R_2C_2 lowpass filter, and op-amp OA_4 form one divider D1. The comparator OA_5, switch S_3, R_3C_3 lowpass filter and op-amp OA_6 form another divider D_2. The comparator OA_3 compares the sawtooth wave V_{S1} of peak value V_R and time period 'T', with the one input voltage V_1. An asymmetrical rectangular wave V_{P1} is generated at the output of comparator OA_3 with a $\pm V_{SAT}$ as peak-to-peak value, where $\pm V_{CC}$ is the power supply to the circuit. Its ON time (δ_{T1}) will be

$$\delta_{T1} = \frac{V_1}{V_R} T \tag{19.49}$$

V_{P1} controls the switch S_2. The switch connects V_Z to the R_2C_2 lowpass filter during the ON time δ_{T1} of V_{P1} and zero volts during the OFF time of V_{P1}. The pulse train V_M of same time period 'T' and ON time δ_{T1} with maximum value of V_Z is generated at the output of switch S_2. The average value of this pulse train V_M will be

$$V_N = \frac{1}{T} \int_0^{\delta_{T1}} V_Z \, dt \tag{19.50}$$

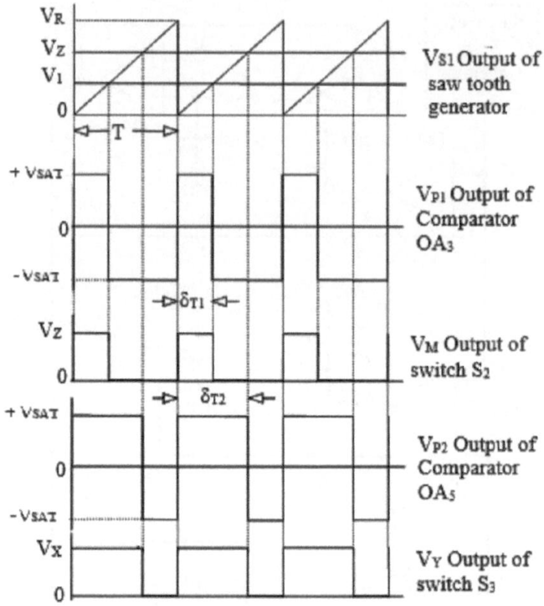

FIGURE 19.11 Associated waveforms of Figure 19.10.

$$V_N = \frac{1}{T} V_Z \delta_{T1}$$

$$V_N = \frac{V_Z V_1}{V_R} \tag{19.51}$$

The op-amp OA_4 is connected in a negative closed loop feedback, and a positive DC voltage is ensured in the feedback loop. Hence, voltage at its non-inverting terminal will be equal to voltage at its inverting terminal, i.e., $V_Y = V_N$.

$$V_Y = \frac{V_Z V_1}{V_R} \tag{19.52}$$

$$V_Z = \frac{V_Y}{V_1} V_R \tag{19.53}$$

Comparator OA_5 compares the same sawtooth wave V_{S1} of peak value V_R and time period 'T', with the voltage V_Z. An asymmetrical rectangular wave V_{P2} is generated at the output of comparator OA_5 with a $\pm V_{SAT}$ as peak-to-peak value. Its ON time (δ_{T2}) will be

$$\delta_{T2} = \frac{V_Z}{V_R} T \tag{19.54}$$

V_{P2} controls the switch S_3. The switch connects V_X to the R_3C_3 lowpass filter during the ON time δ_{T2} of V_{P2} and zero volt during the OFF time of V_{P2}. Another pulse train V_A of the same time period 'T' and ON time δ_{T2} with maximum value of V_X is generated at the output of switch S_3. The average value of this pulse train V_A will be

$$V_B = \frac{1}{T}\int_0^{\delta_{T2}} V_X\, dt = \frac{1}{T} V_X \delta_{T2} = \frac{V_Z V_X}{V_R} \tag{19.55}$$

The op-amp OA_6 is connected in a negative closed loop feedback, and a positive DC voltage is ensured in the feedback loop. Hence, the voltage at its non-inverting terminal will be equal to the voltage at its inverting terminal, i.e., $V_1 = V_B$.

$$V_1 = \frac{V_Z V_X}{V_R} \tag{19.56}$$

$$V_X = \frac{V_1 V_R}{V_Z} = V_1 V_R \frac{V_1}{V_Y V_R}$$

$$V_X = \frac{V_1^2}{V_Y} \tag{19.57}$$

The output of the differential amplifier OA_8 is given by

$$V_Y = V_O - V_2 \tag{19.58}$$

The output of the differential amplifier OA_7 is given by

$$V_O = V_X - V_2 = \frac{V_1^2}{V_Y} - V_2 = \frac{V_1^2 - V_2 V_Y}{V_Y} \tag{19.59}$$

The above equation can be simplified as

$$V_O = \sqrt{V_1^2 + V_2^2} \tag{19.60}$$

19.6 VECTOR MAGNITUDE CIRCUIT USING MULTIPLIER-CUM-DIVIDER

A VMC can be obtained from an MCD as shown in Figure 19.12. In this figure

$$V_B = V_2 + V_O \tag{19.61}$$

$$V_A = \frac{V_X V_Y}{V_Z} = \frac{V_1^2}{V_B} \tag{19.62}$$

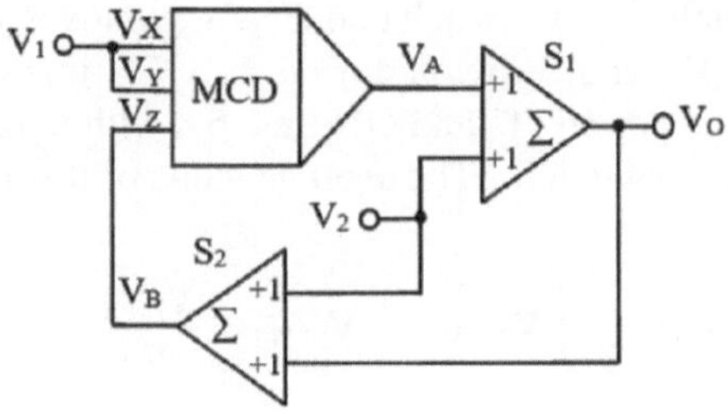

FIGURE 19.12 Block diagram of vector magnitude circuit using multiplier-cum-divider.

$$V_O = V_A + V_2 \tag{19.63}$$

Solving the above Eqs. (19.61)–(19.63),

$$V_O = \sqrt{V_1^2 + V_2^2} \tag{19.64}$$

DESIGN EXERCISES

1. Convert all MCDs described in Chapters 14–16 into VMC. (i) Explain their working operations, (ii) draw waveforms at appropriate places, and (iii) deduce expression for the output voltage. Verify this VMC by experimenting with the AFC trainer kit.

20 Multifunction Converters

In manipulating signals derived from a physical process, it is necessary to realize an output signal that is a non-linear function of an input. Thus, there is a need for a circuit whose non-linear transfer function can be tailored arbitrarily. Function circuits are suited for such a purpose. A multifunction converter (MFC) is a function circuit which receives three input voltages V_1, V_2, V_3 and produces an output voltage $V_O = V_2\left(\frac{V_3}{V_1}\right)^m$ where 'm' is an integer. Figure 20.1 shows the symbol of an MFC. MFC can be converted into multiplier, divider, squarer, square rooter, MCD, and VMC. The logarithmic characteristics of transistors are to perform log and antilog amplifiers. These log and antilog amplifiers are used to develop multifunction converters and are described in this chapter.

20.1 LOG-ANTILOG MULTIFUNCTION CONVERTERS

20.1.1 Log-Antilog MFC For m = 1

The log-antilog MFC for m = 1 works as an MCD. Figure 20.2 shows the circuit diagram of MFC for m = 1.

The logarithmic operation of bipolar transistor is

$$V_{BE} = V_T \ln \frac{I_C}{I_S} \tag{20.1}$$

where V_{BE} = base emitter voltage, V_T = KT/Q, I_C = collector current, and I_S = emitter saturation current.

In Figure 20.2, let us assume all transistors Q_1, Q_2, Q_3 and Q_4 are identical and packed in a single monolithic chip. Hence, $V_{TQ1} = V_{TQ2} = V_{TQ3} = V_{TQ4} = V_T$ and $I_{SQ1} = I_{SQ2} = I_{SQ3} = I_{SQ4} = I_S$.

Logging transistor Q_3,

$$V_A - V_X = V_T \ln \frac{I_3}{I_S} \tag{20.2}$$

Logging transistor Q_1,

$$0 - V_X = V_T \ln \frac{I_1}{I_S} \tag{20.3}$$

DOI: 10.1201/9781003221449-25

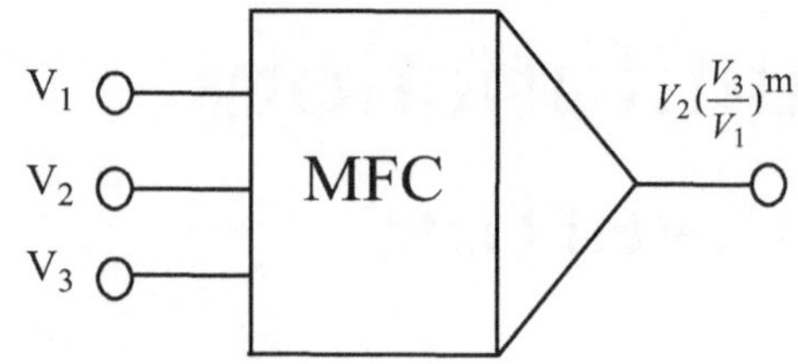

FIGURE 20.1 Symbol of a multifunction converter.

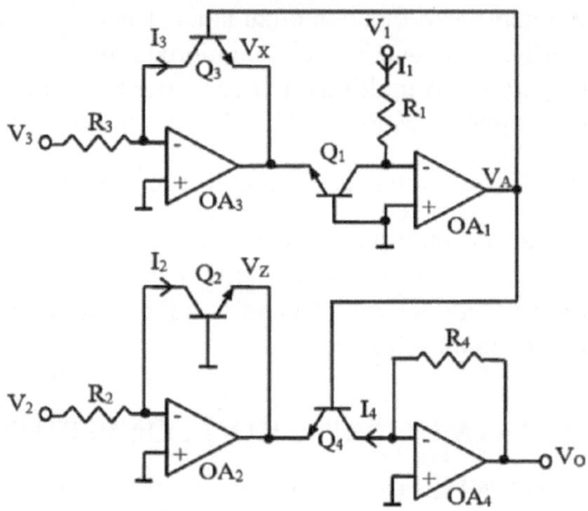

FIGURE 20.2 Circuit diagram of MFC for m = 1.

Equations (20.2) and (20.3) give

$$V_A = V_T \ln \frac{I_3}{I_1} \tag{20.4}$$

Logging transistor Q_2,

$$0 - V_Z = V_T \ln \frac{I_2}{I_S} \tag{20.5}$$

Logging transistor Q_4,

$$V_A - V_Z = V_T \ln \frac{I_4}{I_S} \tag{20.6}$$

Equations (20.6)–(20.5) give

$$V_A = V_T \ln \frac{I_4}{I_2} \tag{20.7}$$

From Eqs. (20.4) and (20.7)

$$V_T \ln\frac{I_3}{I_1} = V_T \ln\frac{I_4}{I_2}$$

$$\frac{I_3}{I_1} = \frac{I_4}{I_2}$$

It is observed from Figure 20.2 that $I_1 = V_1/R_1$, $I_2 = V_2/R_2$, $I_3 = V_3/R_3$ and $I_4 = V_O/R_4$.

$$\frac{V_3R_1}{R_3V_1} = \frac{V_OR_2}{R_4V_2}$$

Let us assume $R_1 = R_2 = R_3 = R_4 = R$, then

$$\frac{V_3}{V_1} = \frac{V_O}{V_2}$$

$$V_O = V_2\left(\frac{V_3}{V_1}\right)^m \tag{20.8}$$

where m = 1.

20.1.2 Log-Antilog MFC for m < 1

The log-antilog MFC for m < 1 is shown in Figure 20.3.

The logarithmic operation of bipolar transistor is

$$V_{BE} = V_T \ln\frac{I_C}{I_S} \tag{20.9}$$

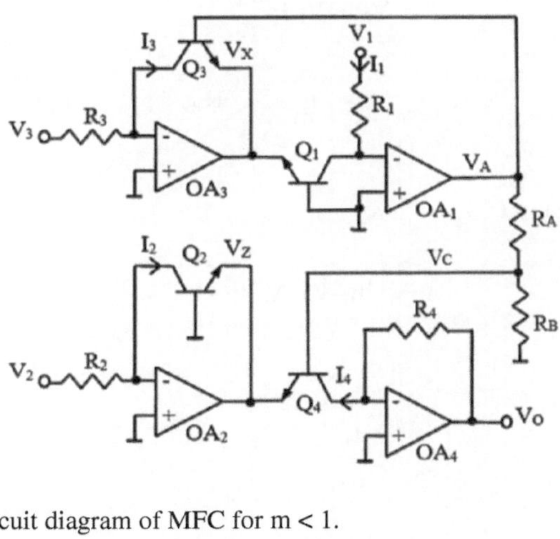

FIGURE 20.3 Circuit diagram of MFC for m < 1.

where V_{BE} = base emitter voltage, $V_T = KT/Q$, I_C = collector current, and I_S = emitter saturation current. In Figure 20.3, let us assume all transistors Q_1, Q_2, Q_3 and Q_4 are identical and packed in a single monolithic chip. Hence, $V_{TQ1} = V_{TQ2} = V_{TQ3} = V_{TQ4} = V_T$ and $I_{SQ1} = I_{SQ2} = I_{SQ3} = I_{SQ4} = I_S$.

Logging transistor Q_3,

$$V_A - V_X = V_T \ln\frac{I_3}{I_S} \tag{20.10}$$

Logging transistor Q_1,

$$0 - V_X = V_T \ln\frac{I_1}{I_S} \tag{20.11}$$

Equations (20.10)–(20.11) give

$$V_A = V_T \ln\frac{I_3}{I_1} \tag{20.12}$$

Logging transistor Q_2,

$$0 - V_Z = V_T \ln\frac{I_2}{I_S} \tag{20.13}$$

Logging transistor Q_4,

$$V_C - V_Z = V_T \ln\frac{I_4}{I_S} \tag{20.14}$$

Equations (20.14)–(20.13) give

$$V_C = V_T \ln\frac{I_4}{I_2} \tag{20.15}$$

$$V_C = mV_A, \quad m = \frac{R_B}{R_A + R_B} \tag{20.16}$$

Equations (20.12) in (20.16) give

$$mV_T \ln\frac{I_3}{I_1} = V_T \ln\frac{I_4}{I_2}$$

$$\left(\frac{I_3}{I_1}\right)^m = \frac{I_4}{I_2}$$

It is observed from Figure 20.3 that $I_1 = V_1/R_1$, $I_2 = V_2/R_2$, $I_3 = V_3/R_3$ and $I_4 = V_O/R_4$.

$$\left(\frac{V_3R_1}{R_3V_1}\right)^m = \frac{V_OR_2}{R_4V_2}$$

Let us assume $R_1 = R_2 = R_3 = R_4 = R$, then

$$\left(\frac{V_3}{V_1}\right)^m = \frac{V_O}{V_2}$$

$$V_O = V_2\left(\frac{V_3}{V_1}\right)^m \tag{20.17}$$

where $m < 1$.

20.1.3 Log-Antilog MFC for m > 1

The log-antilog MFC for $m > 1$ is shown in Figure 20.4.

Logging transistor Q_3,

$$V_B - V_X = V_T \ln\frac{I_3}{I_S} \tag{20.18}$$

Logging transistor Q_1,

$$0 - V_X = V_T \ln\frac{I_1}{I_S} \tag{20.19}$$

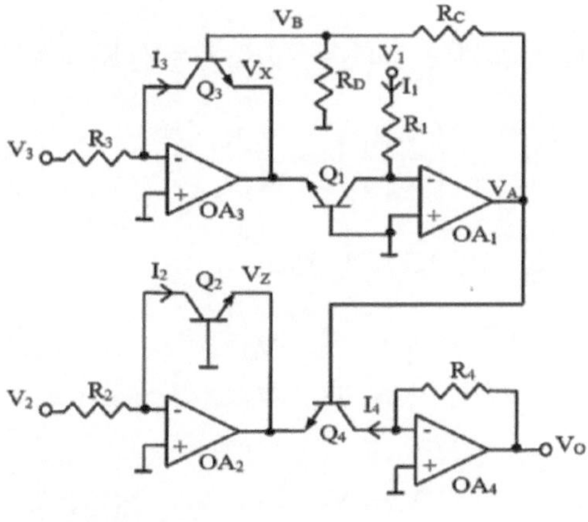

FIGURE 20.4 Circuit diagram of MFC for m > 1.

Equations (20.18)–(20.19) give

$$V_B = V_T \ln \frac{I_3}{I_1} \tag{20.20}$$

Logging transistor Q_2,

$$0 - V_Z = V_T \ln \frac{I_2}{I_S} \tag{20.21}$$

Logging transistor Q_4,

$$V_A - V_Z = V_T \ln \frac{I_4}{I_S} \tag{20.22}$$

Equations (20.22)–(20.21) give

$$V_A = V_T \ln \frac{I_4}{I_2} \tag{20.23}$$

$$V_A = mV_B, \quad m = \frac{R_C + R_D}{R_D} \tag{20.24}$$

Equations (20.20) in (20.24) give

$$mV_T \ln \frac{I_3}{I_1} = V_T \ln \frac{I_4}{I_2}$$

$$\left(\frac{I_3}{I_1}\right)^m = \frac{I_4}{I_2}$$

It is observed from Figure 20.4 that $I_1 = V_1/R_1$, $I_2 = V_2/R_2$, $I_3 = V_3/R_3$ and $I_4 = V_O/R_4$.

$$\left(\frac{V_3R_1}{R_3V_1}\right)^m = \frac{V_OR_2}{R_4V_2}$$

Let us assume $R_1 = R_2 = R_3 = R_4 = R$, then

$$\left(\frac{V_3}{V_1}\right)^m = \frac{V_O}{V_2}$$

$$V_O = V_2\left(\frac{V_3}{V_1}\right)^m \tag{20.25}$$

where $m > 1$.

20.2 MULTIFUNCTION CONVERTERS USING SUBTRACTOR, ADDER, LOG, AND ANTILOG AMPLIFIERS

An MFC using subtractor, adder, log, and antilog amplifiers is shown in Figure 20.5.

The log amplifiers' outputs are given as

$$V_X = \log V_1 \tag{20.26}$$

$$V_Y = \log V_2 \tag{20.27}$$

$$V_Z = \log V_3 \tag{20.28}$$

The output of differential amplifier OA_1 will be

$$V_M = \frac{R_2}{R_1}\left(V_Y - V_X\right) \tag{20.29}$$

Equations (20.26) and (20.27) in (20.29) give

$$V_M = m\left(\log V_2 - \log V_1\right)$$

where $m = R_2/R_1$

$$V_M = m\log\left(\frac{V_2}{V_1}\right) = \log\left(\frac{V_2}{V_1}\right)^m$$

The output of adder OA_2 will be

$$V_N = V_M + V_Z = \log\left(\frac{V_2}{V_1}\right)m + \log V_3$$

$$V_N = \log\left[V_3\left(\frac{V_2}{V_1}\right)^m\right] \tag{20.30}$$

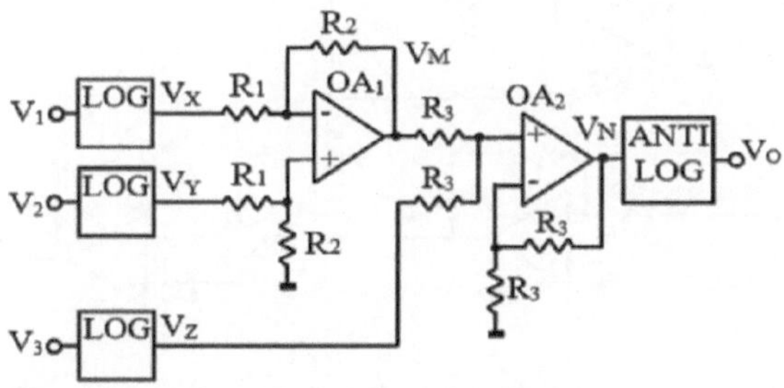

FIGURE 20.5 MFC using subtractor, adder, log, and antilog amplifiers.

The output voltage V_O is the antilog of V_N. Hence,

$$V_O = V_3\left(\frac{V_2}{V_1}\right)^m \tag{20.31}$$

20.3 MULTIFUNCTION CONVERTERS USING THREE LOG AND ONE ANTILOG AMPLIFIERS

An MFC circuit using three log and one antilog amplifiers is shown in Figure 20.6.

Let us assume all transistors Q_1, Q_2, Q_3 and Q_4 are identical and packed in a single monolithic chip. Hence, $V_{TQ1} = V_{TQ2} = V_{TQ3} = V_{TQ4} = V_T$ and $I_{SQ1} = I_{SQ2} = I_{SQ3} = I_{SQ4} = I_S$.

Logging transistor Q_1,

$$0 - V_D = V_T \ln\frac{I_1}{I_S} \tag{20.32}$$

Logging transistor Q_3,

$$V_A - V_D = V_T \ln\frac{I_3}{I_S} \tag{20.33}$$

Equations (20.32)–(20.33) give

$$-V_A = V_T \ln\frac{I_1}{I_3} \tag{20.34}$$

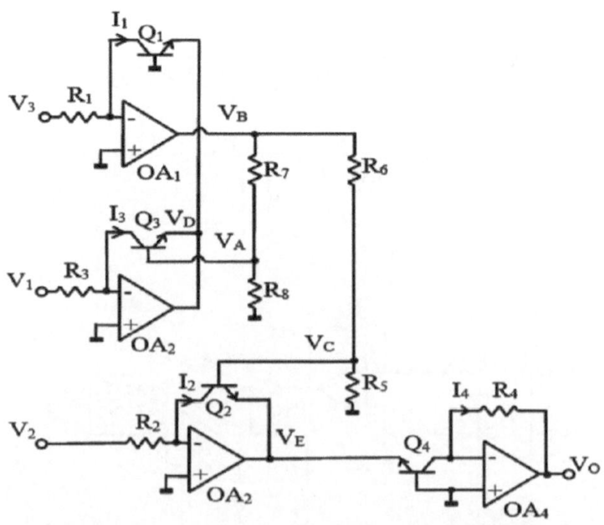

FIGURE 20.6 Multifunction converter using three log and one antilog amplifiers.

Logging transistor Q_2,

$$V_C - V_E = V_T \ln \frac{I_2}{I_S} \tag{20.35}$$

Logging transistor Q_4,

$$0 - V_E = V_T \ln \frac{I_4}{I_S} \tag{20.36}$$

Equations (20.36)–(20.35) give

$$-V_C = V_T \ln \frac{I_4}{I_2} \tag{20.37}$$

From Figure 20.6

$$V_C = \frac{V_B}{R_6 + R_5} R_5 \tag{20.38}$$

$$V_A = \frac{V_B}{R_7 + R_8} R_8 \tag{20.39}$$

From Eq. (20.38)

$$V_B = \frac{R_5 + R_6}{R_5} V_C \tag{20.40}$$

From Eq. (20.39)

$$V_B = \frac{R_7 + R_8}{R_8} V_A \tag{20.41}$$

From Eqs. (20.40) and (20.41)

$$V_C = \frac{R_5\left(R_7 + R_8\right)}{R_8\left(R_5 + R_6\right)} V_A$$

$$-V_C = -mV_A \tag{20.42}$$

$$\text{where } m = \frac{R_5\left(R_7 + R_8\right)}{R_8\left(R_5 + R_6\right)}$$

Equations (20.34) and (20.37) in (20.42) give

$$mV_T \ln \frac{I_1}{I_3} = V_T \ln \frac{I_4}{I_2} \tag{20.43}$$

$$\left(\frac{I_1}{I_3}\right)^m = \frac{I_4}{I_2} \tag{20.44}$$

It is observed from Figure 20.6 that $I_1 = V_3/R_1$, $I_2 = V_2/R_2$, $I_3 = V_1/R_3$ and $I_4 = V_O/R_4$.

$$\left(\frac{V_3R_3}{R_1V_1}\right)^m = \frac{V_OR_2}{R_4V_2} \tag{20.45}$$

Let us assume $R_1 = R_2 = R_3 = R_4 = R$, then

$$\left(\frac{V_3}{V_1}\right)^m = \frac{V_O}{V_2}$$

$$V_O = V_2\left(\frac{V_3}{V_1}\right)^m \tag{20.46}$$

20.4 MULTIFUNCTION CONVERTER APPLICATIONS

20.4.1 Sine Function Converter

The approximate sine function can be expressed with only two terms as:

$$\sin x = x - \frac{x^{2.827}}{6.28} \tag{20.47}$$

The error curve plot is shown in Figure 20.7.

If 90° corresponds to 10 V, the above equation can be written as

$$V_O = 10 \sin 9V_I$$

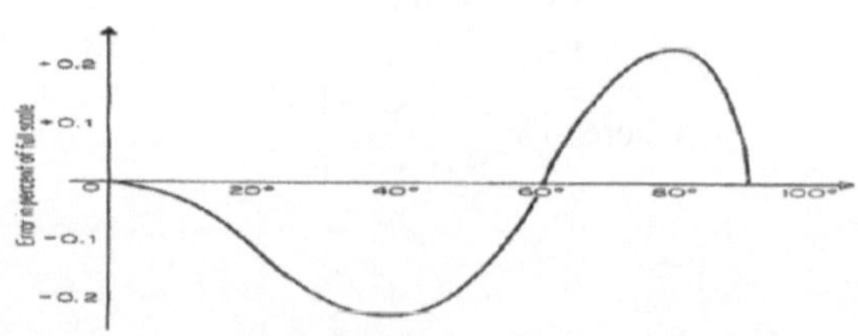

FIGURE 20.7 Error plot.

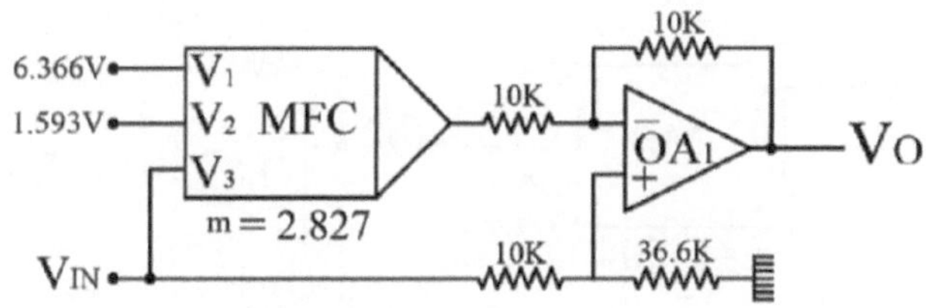

FIGURE 20.8 Sine function converter.

This can be approximated by

$$V_O \approx 1.5708V_I - 1.5924\left(\frac{V_I}{6.366}\right)^{2.827} \tag{20.48}$$

The simulation of the above Eq. (20.48) using an MFC in conjunction with an op-amp is shown in Figure 20.8.

20.4.2 Cosine Function Generator

The approximate cosine function can be expressed as

$$\cos x = 1 - \frac{x^2}{2!} + \frac{x^4}{4!} - \frac{x^6}{6!} + \ldots\ldots\ldots \tag{20.49}$$

The above equation can be implemented with three terms which will have an error of 2% using two multipliers, or an MFC and an op-amp. The cosine function can be represented as

$$\cos x = 1 + 0.235x - \frac{x^{1.504}}{1.445} \tag{20.50}$$

The error curve plot is shown in Figure 20.9.

The actual transfer function is optimized if it is scaled such that

$$V_O = 10\cos 9V_I \tag{20.51}$$

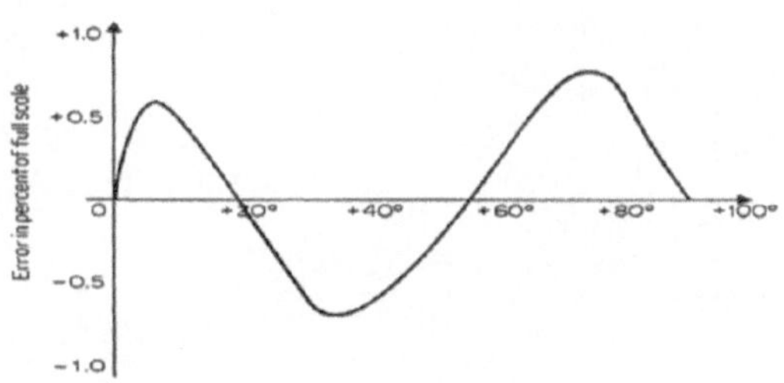

FIGURE 20.9 Error plot.

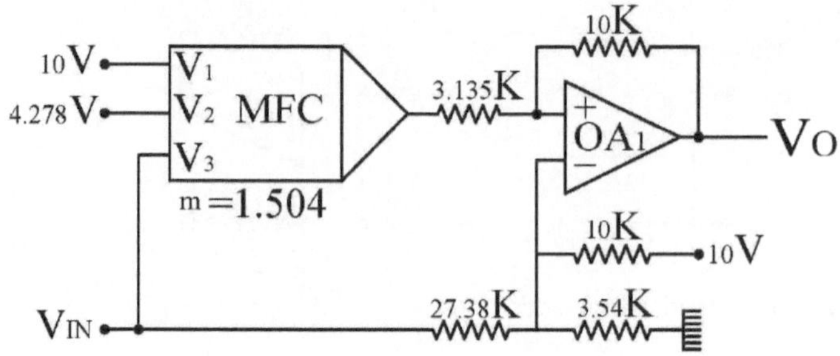

FIGURE 20.10 Cosine function converter.

From Eqs. (20.49) and (20.50)

$$V_O = 10 + 0.3652V_I - 0.4276V_I^{1.504} \tag{20.52}$$

Equation (20.52) can be implemented using an MFC and an op-amp and is shown in Figure 20.10.

20.4.3 Arctangent Function Generator

The approximate $\tan^{-1}$ 1 function can be expressed as:

$$\tan^{-1} x = \frac{\pi}{2} \frac{x^{1.2125}}{1 + x^{1.2125}} \tag{20.53}$$

Figure 20.11 shows the error curve.

$$V_O = 9\tan^{-1}\left(V_X/V_Y\right)$$

0 V < V_O < 9 V corresponding to 0–90°. This can be approximated to

$$V_O = \frac{9\left(Vx/Vz\right)^{1.2125}}{1 + \left(Vx/Vz\right)^{1.2125}}\left(1V = 10°\right) \tag{20.54}$$

The circuit implementation of Eq. (20.54) is shown in Figure 20.12.

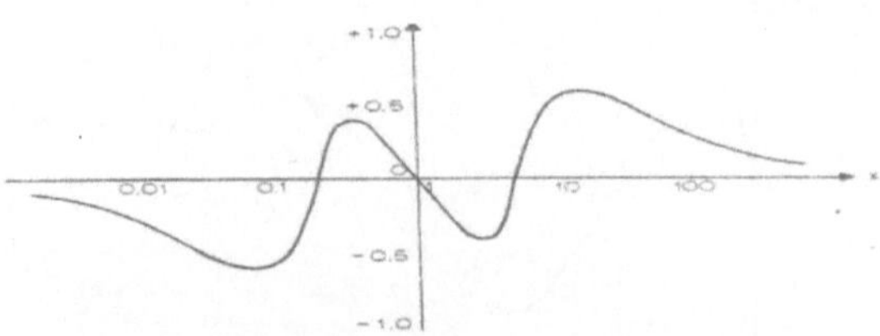

FIGURE 20.11 Error plot.

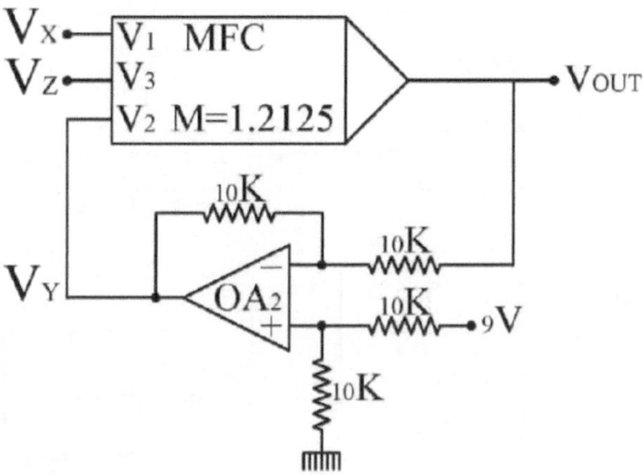

FIGURE 20.12 Arctangent function convertor.

20.5 MULTIFUNCTION CONVERTER CONVERSIONS

The MFC output is

$$V_O = V_2\left(\frac{V_3}{V_1}\right)^m$$

Let $V_1 = V_R$, $V_2 = V_1$, $V_3 = V_2$ and m = 1, then

$V_O = \frac{V_1V_2}{V_R}$; a multiplier operation can be obtained

Let $V_2 = V_R$, $V_3 = V_2$ and m = 1, then

$V_O = \frac{V_2}{V_1}V_R$; a divider operation can be obtained

Let m = 1, then

$V_O = \frac{V_2V_3}{V_1}$; a multiplier-cum-divider operation can be obtained.

Let $V_1 = V_O$, $V_2 = V_1$, $V_3 = V_R$ and m = 1, then

$V_O = \sqrt{V_1V_R}$; a square rooter operation can be obtained.

Let $V_1 = V_O$, $V_2 = V_1$, $V_3 = V_2$ and m = 1, then

$V_O = \sqrt{V_1V_2}$; a square root of multiplication operation can be obtained.

The MFC circuit shown in Figure 20.2 is converted into a multiplier and is shown in Figure 20.13. The MFC circuit shown in Figure 20.5 is converted into a divider and is shown in Figure 20.14. The MFC circuit shown in Figure 20.6 is converted into a square rooter and is shown in Figure 20.15. The MFC circuit shown in Figure 20.2 is converted into a square root of multiplication and is shown in Figure 20.16.

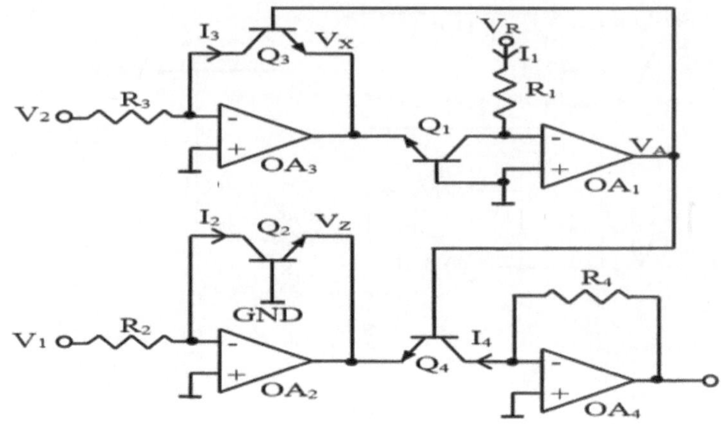

FIGURE 20.13 Multiplier from multifunction converter.

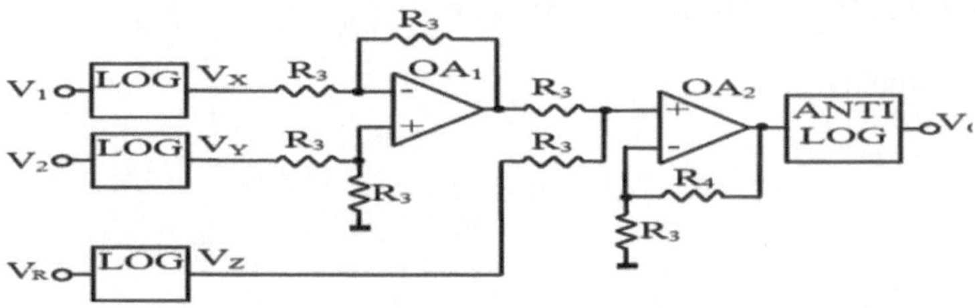

FIGURE 20.14 Divider from multifunction converter.

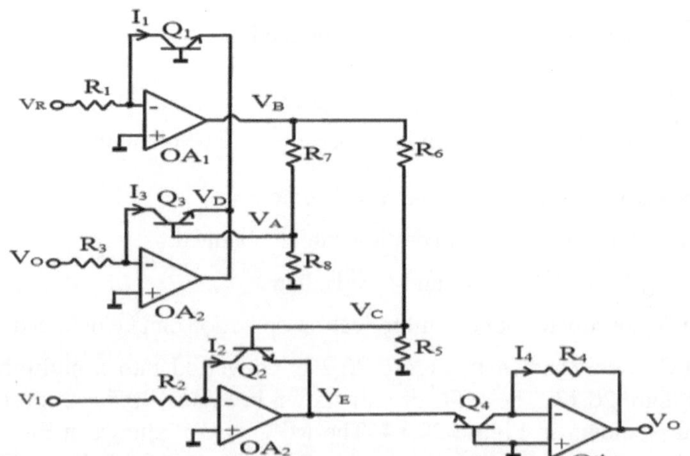

FIGURE 20.15 Square rooter from multifunction converter.

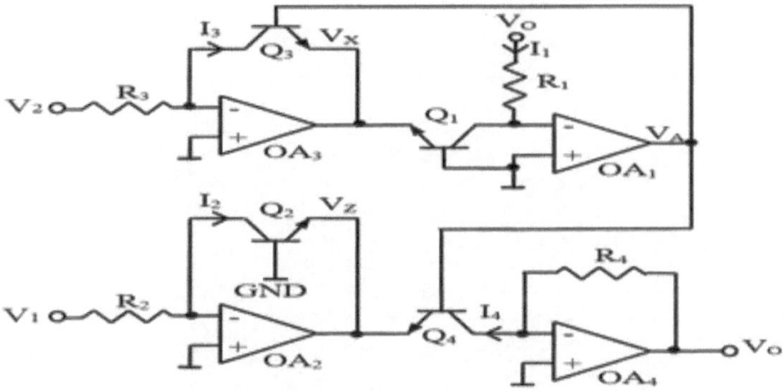

FIGURE 20.16 Square root of multiplication from multifunction converter.

20.6 EXPONENTIATOR

An exponentiator is a function circuit which accepts an input voltage V_1 and produces an output as

$$V_O = V_1^{\,m}$$

If we keep $V_2 = V_1 = V_R$ and $V_3 = V_1$ in an MFC circuit, then this will work as an exponentiator.

TUTORIAL EXERCISES

1. Design a function circuit to produce output $V_O = V_2\left(\frac{V_3}{V_1}\right)^5$ for given input voltages of V_1, V_2 and V_3.
2. Design a function circuit to produce output voltage $V_O = \left(\frac{V_1}{2}\right)^2\left(\frac{V_1}{V_1}\right)^2$.
3. Design a function circuit to produce output voltage $V_O = \left(\frac{V_1}{10}\right)^{0.76}$.
4. Design a multifunction converter circuit using log and antilog amplifiers.
5. How will you simulate a sine function using a multifunction converter?
6. How will you simulate a cosine function using a multifunction converter?

21 Phase Sensitive Detectors

A phase sensitive detector (PSD) is a function circuit which accepts two sinusoidal signals V_S and V_R of the same frequency with phase difference Ø and provides two DC output voltages in which one DC output voltage V_Q is proportional to the quadrature component SinØ and the another DC output voltage V_I is proportional to the in-phase component CosØ. Figure 21.1 shows the symbol of a PSD.

PSDs are used in (1) impedance measurement in polar and rectangular forms, (2) active and reactive power measurement, (3) phase and power factor measurement, and (4) test instrument for current transformer errors. There are several conventional types of PSDs: (1) Multiplying, (2) switching, (3) sampling, (4) switching-sampling, and (5) position-sampled. All these PSDs are described in this chapter.

21.1 MULTIPLYING PHASE SENSITIVE DETECTORS

A PSD using multipliers is shown in Figure 21.2. The reference signal V_2 of peak value V_{S2} and the phase-shifted signal V_1 of peak value V_{S1} are given to multiplier M_1.

$$V_1 = V_{S1}\sin(\omega t+\theta) \tag{21.1}$$

$$V_2 = V_{S2}\sin\omega t \tag{21.2}$$

The output of multiplier M_1 will be

$$V_X = \frac{V_{S1}V_{S2}\sin(\omega t+\theta)\sin\omega t}{V_R}$$

$$V_X = \frac{V_{S1}V_{S2}}{2V_R}\left[\cos\theta-\cos(2\omega t+\theta)\right]$$

where V_R is the multiplier constant. The lowpass filter at the output of multiplier M_1 removes AC component and allows only DC term. The in-phase output voltage V_I is

$$V_I = \frac{V_{S1}V_{S2}}{2V_R}\cos\theta \tag{21.3}$$

The 90° phase-shifted reference signal V_2' and the phase-shifted signal V_1 are applied to another multiplier M_2. Its output V_Y will be

DOI: 10.1201/9781003221449-26

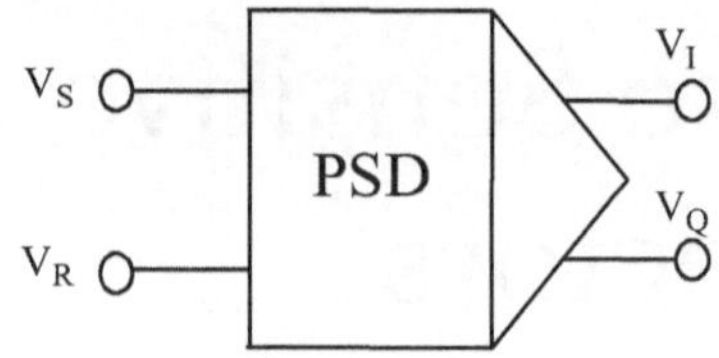

FIGURE 21.1 Symbol of phase sensitive detector.

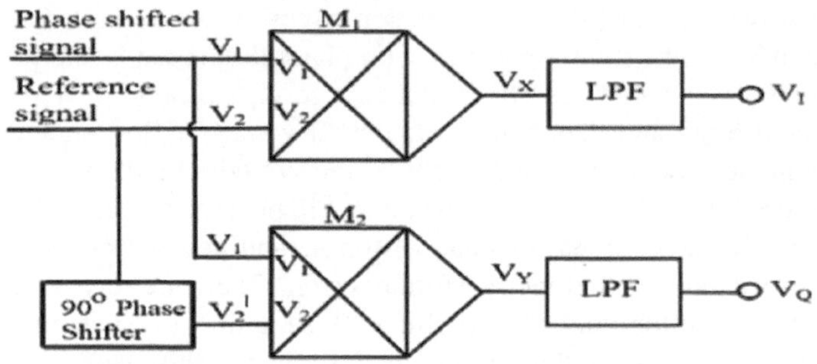

FIGURE 21.2 Phase sensitive detector using multipliers.

$$V_Y = \frac{V_{S1}V_{S2}\sin(\omega t + 90°)\sin(\omega t + \theta)}{V_R}$$

$$= \frac{V_{S1}V_{S2}}{2V_R}\left[\sin\theta + \sin(2\omega t + \theta)\right]$$

where V_R is the multiplier constant. The lowpass filter at the output of multiplier M_2 removes AC component and allows only DC term. The quadrature output voltage V_Q is

$$V_Q = \frac{V_{S1}V_{S2}}{2V_R}\sin\theta \tag{21.4}$$

21.2 SWITCHING PHASE SENSITIVE DETECTORS

The switching PSD is shown in Figure 21.3, and its associated waveforms are shown in Figure 21.4. During positive half-cycle of reference signal, the input signal is connected to the switch SW_2 output and during negative half-cycle of the reference signal, inverted input signal is connected to the switch SW_2 output. The switch SW_2 output is given to the lowpass filter II to get in-phase component V_I. Similarly during positive half-cycle of reference signal the 90° phase-shifted input signal is connected to the switch SW_1 output and during negative half-cycle of reference signal inverted 90° phase-shifted signal is connected to switch SW_1. The switch SW_1 output is given to the lowpass filter I to get quadrature component V_Q.

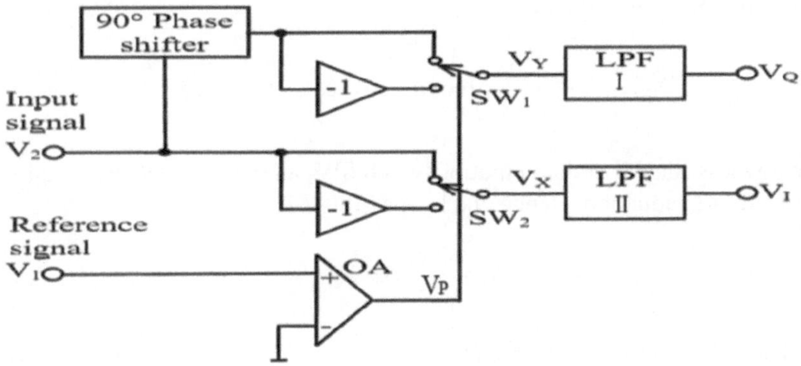

FIGURE 21.3 Switching PSD.

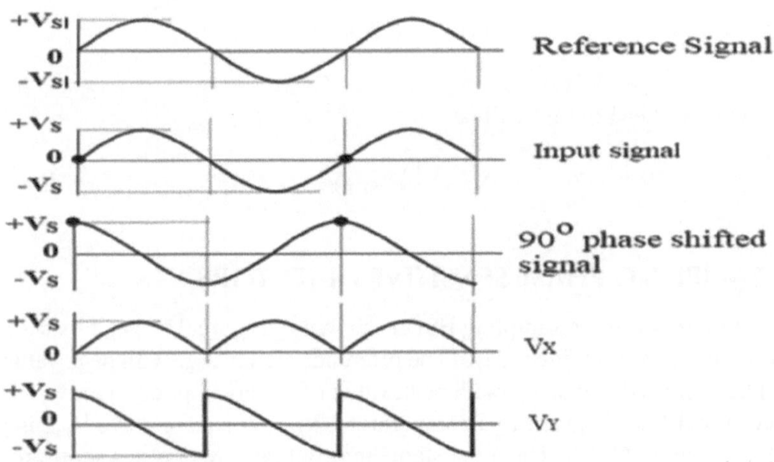

FIGURE 21.4 Associated waveforms of Figure 21.3.

Let V_1 be the reference signal with peak value of V_{S1}, and the input signal be V_2 with peak value of V_S.

$$V_1 = V_{S1} \sin \omega t$$

$$V_2 = V_S \sin(\omega t - \theta)$$

The reference signal is given to the zero crossing detector OA and the resulted square wave V_F can expanded by Fourier serial as

$$V_P = \frac{4}{\pi}\left[\sin\omega t - \frac{1}{3}\sin 3\omega t + \ldots\ldots\ldots\right]$$

The switch SW_2 output will be

$$V_X = V_P V_2 = \frac{2V_S}{\pi}\left[\cos\theta - \cos\left(2\omega t - \theta + ...\right]\right.$$

The lowpass filter II at the output of switch SW_2 allows only DC term and eliminates AC term in equation. Hence, the lowpass filter II output is

$$V_I = \frac{2V_S}{\pi}\cos\theta \tag{21.5}$$

Similarly the output of switch SW_1 will be

$$V_Y = V_P\left(V_2 + 90°\right) = V_P V_S \cos\left(\omega t - \theta\right)$$

$$V_Y = \frac{2V_S}{\pi}\left[\sin\theta + \sin\left(2\omega t - \theta + ...\right]\right.$$

The lowpass filter I output will be

$$V_Q = \frac{2V_S}{\pi}\sin\theta \tag{21.6}$$

21.3 SAMPLING PHASE SENSITIVE DETECTORS

The circuit diagram of the sampling PSD is shown in Figure 21.5, and its associated waveforms are shown in Figure 21.6. The reference signal V_{REF} with peak value of V_S and the phase-shifted signal V_{IN} with peak value of V_{SI} are applied to the circuit. The reference signal V_{REF} is given to a comparator CMP_1. A square wave V_{CP1} is generated at the output of CMP_1. The monostable multivibrator M_1 gives a short pulse V_{M1} during rising edge of its input pulse V_{CP1}. This short pulse V_{M1} is acting as a sampling pulse to the sample and hold circuit realized by the switch S_1, capacitor C_1, and op-amp OA_1. The output of op-amp OA_1 is the quadrature component and is given as

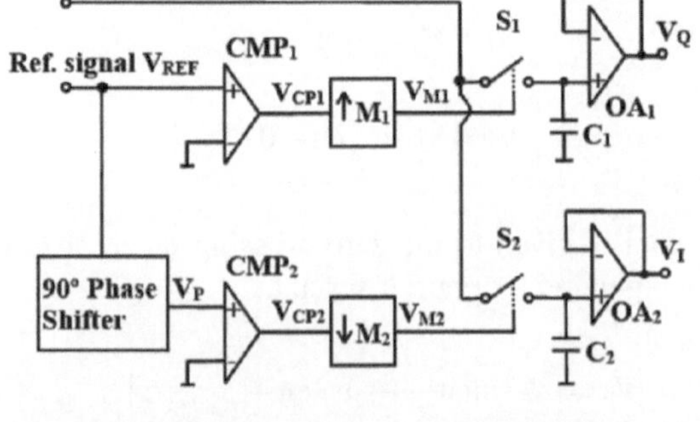

FIGURE 21.5 Circuit diagram of sampling phase sensitive detector.

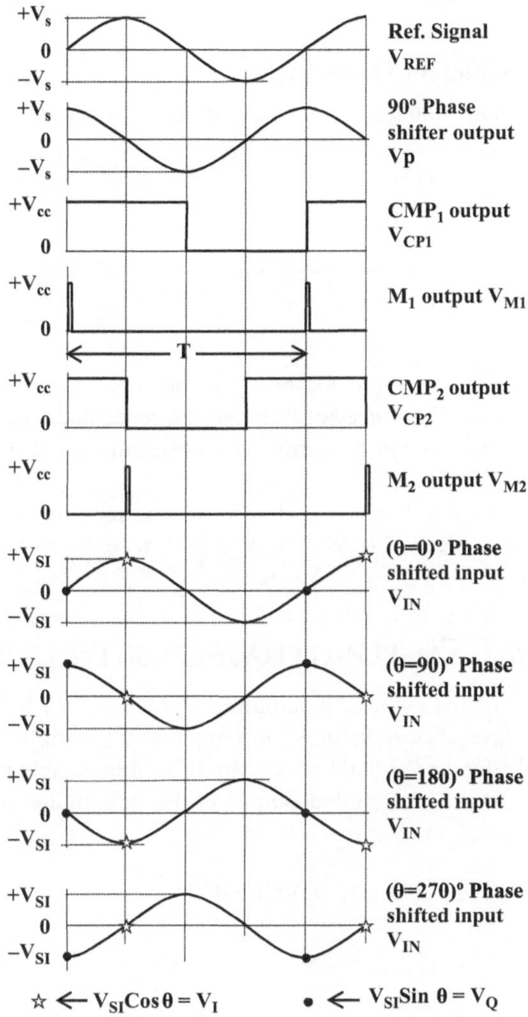

FIGURE 21.6 Associated waveforms of Figure 21.5 for different phase angles.

$$V_Q = Vs_I \sin\theta \tag{21.7}$$

The 90° phase-shifted reference signal V_P with peak value of V_S is given to another comparator CMP_2. Another square wave V_{CP2} is generated at the output of CMP_2. The monostable multivibrator M_2 gives a short pulse V_{M2} during falling edge of its input pulse V_{CP2}. This short pulse V_{M2} is acting as a sampling pulse to the sample and hold circuit realized by the switch S_2, capacitor C_2, and op-amp OA_2. The output of op-amp OA_2 is the in-phase component and is given as

$$V_I = Vs_I \cos\theta \tag{21.8}$$

TABLE I
Outputs for Different Phase Angles

Serial No.	Phase Difference θ	V_I (V_{SI} Cosθ)	V_Q (V_{SI} Sinθ)
1	0°	V_{SI}	0
2	90°	0	V_{SI}
3	180°	$-V_{SI}$	0
4	270°	0	$-V_{SI}$

In the above Eqs. (21.7) and (21.8), the V_{SI} is the peak value of the phase-shifted signal, and **θ** is the phase difference between the reference and phase-shifted signals. Equations (21.7) and (21.8) can be verified by observing the waveforms shown in Figure 21.6.

The waveforms in Figure 21.6 are used to show how sampling points are achieved for a given phase differences of 0, 90, 180 and 270° as examples. The observations are given in Table I.

21.4 SWITCHING-SAMPLING PHASE SENSITIVE DETECTORS

The circuit diagram of the switching-sampling PSD is shown in Figure 21.7. The phase-shifted sine wave of peak value V_S is sampled by a sampling pulse generated during +ve edge of reference signal by the switch S_1, monostable multivibrator, and op-amps OA_1 and OA_3. The sampled output is the quadrature component and is given by

$$V_Q = Vs \operatorname{Sin} \theta \tag{21.9}$$

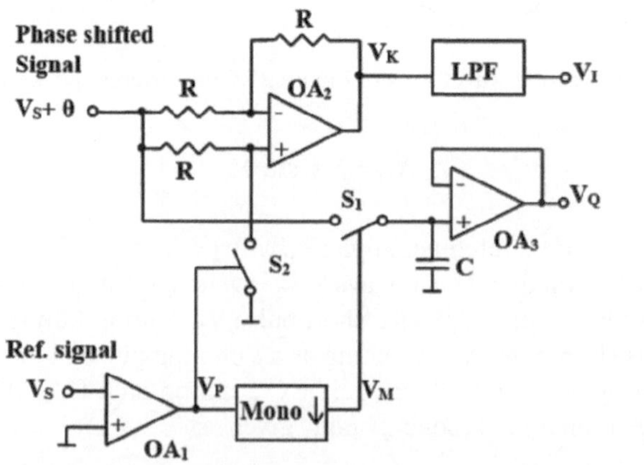

FIGURE 21.7 Circuit diagram of switching-sampling phase sensitive detector.

This can be graphically verified by the waveforms shown in Figure 21.8 for difference phase angles of 0, 90, 180, and 270°, respectively, as shown in Table II. The phase-shifted sine wave Vs+θ is given to a controlled amplifier OA_2 which is controlled by the reference signal through zero crossing detector OA_1 and switch S_2. During +ve half-cycle of reference signal, the switch S_2 is open, and the control amplifier OA_2 will work as a non-inverting amplifier. During –ve half-cycle, the switch S_2 shorts, and the control amplifier OA_2 will work as an inverting amplifier. Let reference sine wave = $V1(t) = V_S \sin \omega t$ and phase-shifted sine wave be $V_2(t) = V_s$

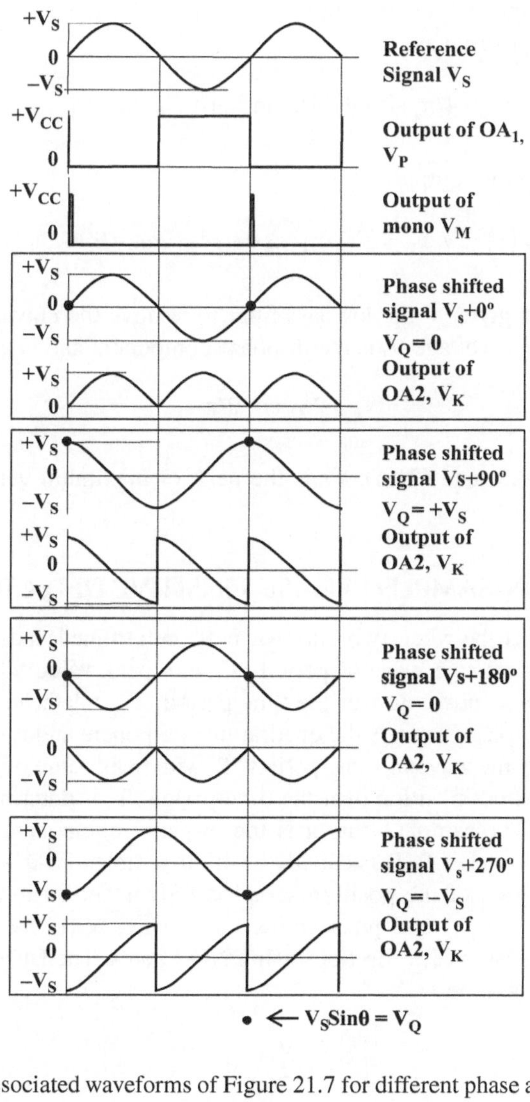

FIGURE 21.8 Associated waveforms of Figure 21.7 for different phase angles of 0, 90, 180, and 270°.

TABLE II
Outputs for Different Phase Angles

Serial No.	Phase Difference θ	V_I (V_S Cosθ)	V_Q (V_S Sinθ)
1	0°	$2V_S/\pi$	0
2	90°	0	V_S
3	180°	$-2V_S/\pi$	0
4	270°	0	$-V_S$

sin (ωt+θ). Since the output of comparator OA_1 is a square wave V_p, it can be expanded by Fourier serial as,

$$V_p = 4/\pi\left[\sin\omega t - 1/3\sin 3\omega t + \ldots\ldots\ldots\ldots\right] \quad (21.10)$$

The control amplifier output will be

$$V_K(t) = V_p V_2 = 2V_S/\pi[\cos\theta - \cos(2\omega t + \theta + \ldots] \quad (21.11)$$

The signal V_K is given to the lowpass filter to remove the unwanted AC components. The DC output voltage V_I is the in-phase component and is given by

$$V_I = 2V_S \cos\theta/\pi \quad (21.12)$$

In both Eqs. (21.3) and (21.6), Vs is the peak or maximum value of the phase-shifted signal.

21.5 POSITION-SAMPLED PHASE SENSITIVE DETECTORS

The phase difference between two sine waves is determined and converted into a voltage V_θ. A constant sine wave of period 'T' with peak value V_S is sampled by a sampling pulse whose position over the time period 'T' is determined by the phase voltage V_θ. The sampled output is the quadrature component and is given by $V_Q = V_S$ Sinθ. A constant cosine wave of same period 'T' with peak value of V_S is sampled by a sampling pulse whose position over the time period 'T' is determined as the same phase voltage V_θ. The sampled output is the in-phase component and is given by $V_I = V_S$ Cosθ. This type of PSD is called "position-sampled PSD."

The circuit diagram of the position-sampled PSD is shown in Figure 21.9. The reference sine wave $V_1(t)$ and phase-shifted sine wave $V_2(t)$ are given the phase detector which consists of RS flip flops using NOR gates. The DC output voltage of this phase detector will be

$$V_\theta = \frac{V_{SAT}}{360}\theta \quad (21.13)$$

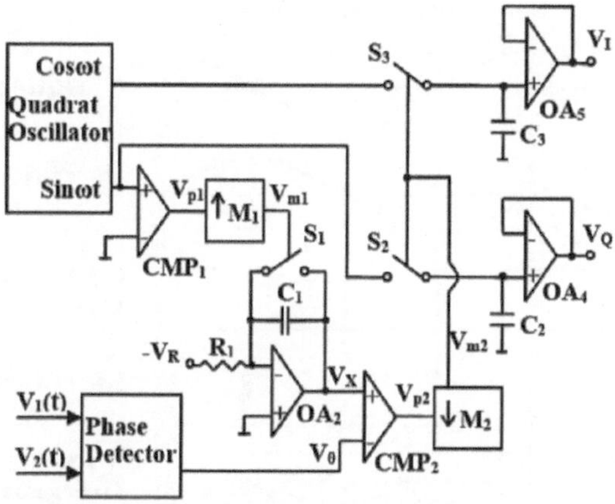

FIGURE 21.9 Circuit diagram of position-sampled phase sensitive detector.

where Vcc is the power supply voltage, and θ is the phase difference between the input sine waves $V_1(t)$ and $V_2(t)$.

From Eq. (21.1), the phase voltages V_θ for different phase angles are given in Table III.

The quadrature oscillator gives standard reference sine and cosine waves. The sine wave is converted into a square wave V_{P1} by comparator CMP_1. A short pulse V_{M1} is obtained during rising edge of this square wave V_{P1}. This short pulse V_{M1} shorts capacitor C_1 in integrator OA_2. A reference voltage $-V_R$ given to the integrator OA_2 and its output will be

$$V_X = \frac{1}{RC}\int V_R\,dt = \frac{V_R}{RC}t \qquad (21.14)$$

A sawtooth waveform is generated at the output of integrator OA_2. The associated waveforms are shown in Figure 21.10. From the waveforms shown and from Eq. (21.14) at t = T, $V_X = V_T$ where V_T is the peak value of the sawtooth wave. Let V_T = Vcc, and it is given by

$$V_T = \frac{V_R}{RC}T = V_{SAT}$$

TABLE III
Phase Detector Output for Different Phase Angles

Serial No.	Phase Difference θ	Phase Detector Output V_θ
1	90°	$V_{SAT}/4$
2	180°	$V_{SAT}/2$
3	270°	$3V_{SAT}/4$

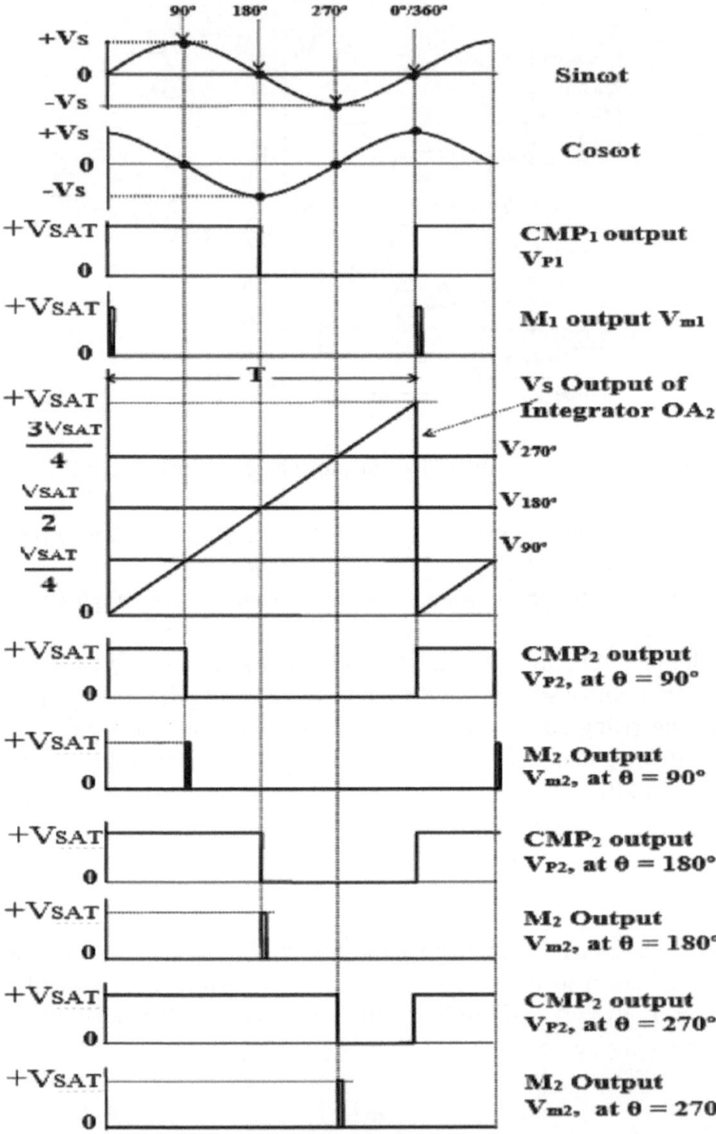

FIGURE 21.10 Associated waveforms of Figure 21.9.

The phase voltage V_θ is compared with this sawtooth waveform by comparator CMP_2. A pulse waveform V_{P2} is generated at the output of CMP_2. A short pulse V_{M2} is generated during falling edge of this pulse train V_{P2}. The short pulse V_{M2} is acting as a sampling pulse to sample and hold circuits realized with op-amps (OA_4 and OA_5), capacitors (C_2 and C_3), and switches (S_2 and S_3). It samples sine wave for quadrature and cosine wave for in-phase components. It is observed from the

TABLE IV
Outputs for Different Phase Angles

Serial No.	Phase θ	V_S Cosθ	V_S Sinθ
1	90°	0	V_S
2	180°	$-V_S$	0
3	270°	0	$-V_S$

waveforms of Figure 21.10 for different phase angles of 90, 180, and 270° that the output of OA_4 and OA_5 will be

$$V_Q = V_S \operatorname{Sin}\theta \tag{21.15}$$

$$V_I = V_S \operatorname{Cos}\theta \tag{21.16}$$

The observations are given in Table IV. V_S is the peak value of the quadrature oscillator output signals.

21.6 PHASE SENSITIVE DETECTORS USING 4046 PLL IC

The switching PSD using 4046 PLL is shown in Figure 21.11.

The 4046 IC is configured such that it maintains 90° phase shift between signal input and its VCO output. The switch S1, op-amp OA1, and lowpass filter 1 constitutes switching phase sensitive detector for in-phase component. As discussed in Section 21.2, the output of lowpass filter 1 will be

$$V_I = \frac{2V_S}{\pi} \operatorname{Cos}\theta \tag{21.17}$$

The switch S2, 4046 as 90° phase shifter (T/4 delay generator) and lowpass filter 2 constitutes another phase sensitive detector for quadrature component. As discussed in Section 21.2, the output of lowpass filter 2 will be

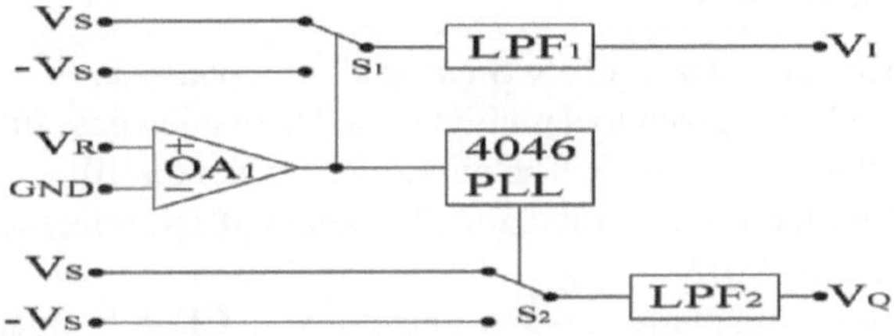

FIGURE 21.11 Switching phase sensitive detector with 4046 PLL.

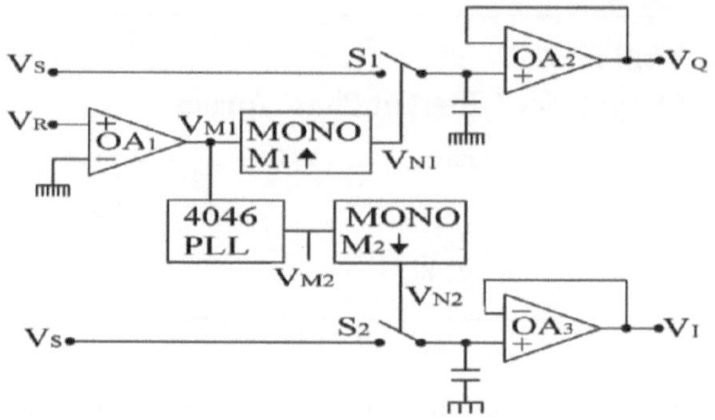

FIGURE 21.12 Sampling phase sensitive detector with 4046 PLL.

$$V_Q = \frac{2V_S}{\pi} \sin\theta \qquad (21.18)$$

The sampling PSD using 4046 PLL is shown in Figure 21.12.

The peak value of the signal is V_{S1}. The reference signal V_R is converted into square wave V_{M1} by comparator OA_1. The monostable multivibrator M1 gives a short spike V_{N1} during every rising edge of the square wave V_{M1}. The spike V_{N1} is acting as a sampling pulse to the sample and hold circuit realized by switch S_1, capacitor C_1, and op-amp OA_2. The sample and hold output is given as

$$V_Q = V_{S1} \sin\theta \qquad (21.19)$$

The 4046 IC is configured such that it maintains 90° phase shift between signal input and VCO output. The 90° phase- shifted output (or time delay T/4) from VCO of 4046 PLL is given to another monostable multivibrator M2 to give a spike V_{N2} every falling edge of V_{M2}. The spike V_{N2} is acting as a sampling pulse to the sample and hold circuit realized by switch S_2, capacitor C_2, and op-amp OA_3. The sample and hold output is given as

$$V_I = V_{S1} \cos\theta \qquad (21.20)$$

TUTORIAL EXERCISES

1. A reference sine wave of 5 Sin∂t and the phase-shifted sine wave of 2 Sin(∂t + 60°) are given to a multiplier and then a lowpass filter. Find output voltage of the lowpass filter. The multiplier constant is 10.
2. What will be the output voltage to Problem 1 if the reference sine wave is phase-shifted by 90°?
3. Design a switching phase sensitive detector with CD 4053 analog multiplexer IC.
4. What is the advantage of a sampling phase sensitive detector over other phase sensitive detectors?

5. A reference sine wave of $1V_P$ and $2V_P$, 60° phase-shifted sine wave are given to a switching phase sensitive detector. Find out its in-phase and quadrature output voltages.
6. A 2 V peak and 140° phase-shifted signal is sampled by a sampling pulse generated during positive zero crossing of the reference signal. What will be the sampled output?

Part F

Applications of Function Circuits

22 Applications of Analog Multipliers

Analog multipliers are used to get voltage-tunable active filters. An analog multiplier as a non-linear network element is realizing the transfer characteristics active filters.

Voltage-mode universal filters using two current conveyors can realize lowpass, bandpass, highpass, notch, and allpass filters. They have the following advantages: (i) orthogonal control of the natural frequency and quality factor by grounded resistors, (ii) easy conversion into a voltage-controlled filter, (iii) low component spread, (iv) minimum active and passive components, (v) lacking match components except for allpass applications, and (vi) low active and passive sensitivities.

New active only current-mode integrator and differentiator with electronically tunable time constants are composed of one operational amplifier (OA) and two operational transconductance amplifiers (OTAs), and they are suitable for monolithic implementation either with CMOS or bipolar technologies. No realizability conditions are imposed for the proposed circuits and all of the active sensitivities are low.

Tunable active filters are adjusted manually using potentiometers. Electronic tuning provides for automatic adjustment and such tuning can be accomplished with analog multipliers in active filters. By multiplying the signal impressed on a resistor, the resulting current is increased as though the resistance were divided by the same factor. In other words, a change in the effective time constant can be achieved. With this ability to control time constants, very rapid adjustments to filter characteristics can be made by varying the control voltages on multiplier input. This technique is described in this chapter for basic first-order lowpass, highpass, and bandpass filters.

22.1 VOLTAGE-TUNABLE HIGHPASS FILTERS

The circuit diagram is shown in Figure 22.1. The output of op-amp OA_1 is given by

$$V_X = -\left(V_I + V_O\right)10R_1C_1s \tag{22.1}$$

$$V_O = \frac{V_X V_{C1}}{10} \tag{22.2}$$

From Eqs. (22.1) and (22.2)

$$\frac{V_O(s)}{V_I(s)} = -\frac{R_1C_1sV_{C1}}{1+R_1C_1sV_{C1}} \tag{22.3}$$

DOI: 10.1201/9781003221449-28

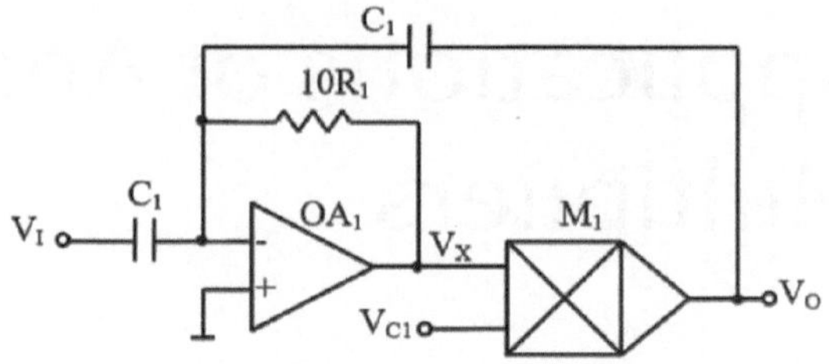

FIGURE 22.1 Circuit diagram of voltage-tunable highpass filter.

Converting the equation from ‘s’ notation to ‘jω’, we get

$$\frac{V_O(j\omega)}{V_I(j\omega)} = -\frac{j\omega R_1 C_1 V_{C1}}{1 + j\omega R_1 C_1 V_{C1}} = -\frac{j2\pi f R_1 C_1 V_{C1}}{1 + j2\pi R_1 C_1 V_{C1}} \tag{22.4}$$

A highpass filter will have transfer function like the below Eq. (22.5)

$$\frac{V_O(j\omega)}{V_I(j\omega)} = AF \frac{j\left(\frac{f}{fC}\right)}{1 + j\left(\frac{f}{fC}\right)} \tag{22.5}$$

From Eqs. (22.4) and (22.5), the circuit shown in Figure 22.1 is proved to be a highpass filter and its cut-off frequency

$$f_C = \frac{1}{V_{C1} R_1 C_1} \tag{22.6}$$

And highpass filter gain is

$$AF = -1 \tag{22.7}$$

The control voltage V_{C1} will have only positive polarity.

22.2 VOLTAGE-TUNABLE LOWPASS FILTERS

The circuit diagram is shown in Figure 22.2. A multiplier M is connected in a second feedback loop of an integrator.

As the non-inverting amplifier of integrator OA_1 is grounded, the inverting terminal voltage is zero volts and no current flows into op-amp OA_1. The output of the integrator will be in Laplace transform

$$V_X = -\frac{10}{R_2 C_2 s}(V_I + V_O) \tag{22.8}$$

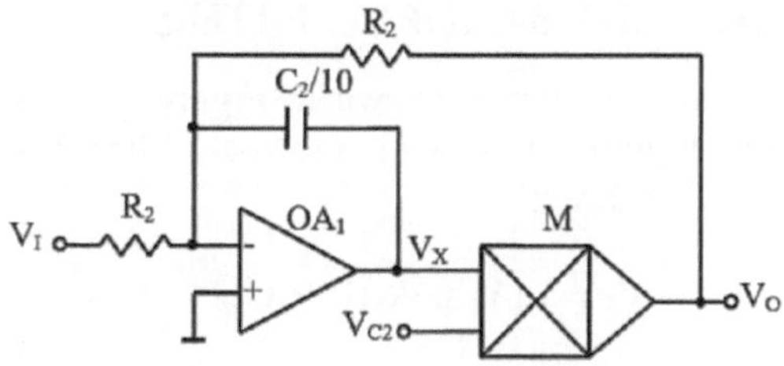

FIGURE 22.2 Circuit diagram of voltage-tunable lowpass filter.

$$V_O = \frac{V_X V_{C2}}{10} \tag{22.9}$$

where 10 is the multiplier constant.

From Eqs. (22.8) and (22.9)

$$\frac{V_O(s)}{V_I(s)} = -\frac{V_{C2}}{V_{C2} + R_2C_2s} = -\frac{1}{1 + R_2C_2s/V_{C2}} \tag{22.10}$$

Converting the 's' notation to 'jω' notation

$$\frac{V_O(j\omega)}{V_I(j\omega)} = -\frac{1}{1 + R_2C_2j\omega/V_{C2}} = -\frac{1}{1 + j\dfrac{2\pi fR_2C_2}{V_{C2}}} \tag{22.11}$$

If any first-order filter circuit in the form of equation,

$$\frac{V_O(j\omega)}{V_I(j\omega)} = \frac{AF}{1 + j\left(\dfrac{f}{f_C}\right)} \tag{22.12}$$

it is called a "lowpass filter." Considering Eqs. (22.11) and (22.12), the cut-off frequency is given as

$$f_C = \frac{V_{C2}}{2\pi R_2C_2} \tag{22.13}$$

And lowpass filter gain will be

$$AF = -1 \tag{22.14}$$

Hence, the filter cut-off frequency is controlled by a control voltage V_{C2} and it should also have positive voltage only.

22.3 VOLTAGE-TUNABLE BANDPASS FILTERS

The voltage-tunable bandpass filter is shown in Figure 22.3. A bandpass filter is obtained by cascading a highpass filter with a lowpass filter. The output expression for Figure 22.3 is given as follows:

$$V_X = -\left(V_I + V_Y\right)10R_1C_1s \tag{22.15}$$

$$V_Y = \frac{V_X V_{C1}}{10} \tag{22.16}$$

From Eqs. (22.15) and (22.16)

$$V_Y = -\frac{R_1C_1sV_{C1}}{1+R_1C_1sV_{C1}}V_I \tag{22.17}$$

$$V_Z = -\frac{10}{R_2C_2s}\left(V_Y + V_O\right) \tag{22.18}$$

$$V_O = \frac{V_Z V_{C2}}{10} \tag{22.19}$$

From Eqs. (22.18) and (22.19)

$$V_O = -\frac{V_{C2}}{V_{C2}+R_2C_2s}V_Y = -\frac{1}{1+R_2C_2s/V_{C2}}V_Y \tag{22.20}$$

Equation (22.17) in (22.20) gives

$$\frac{V_O(s)}{V_I(s)} = \frac{R_1C_1sV_{C1}}{\left(1+R_1C_1sV_{C1}\right)\left(1+\dfrac{R_2C_2s}{V_{C2}}\right)} \tag{22.21}$$

Converting the above equation into jω notation

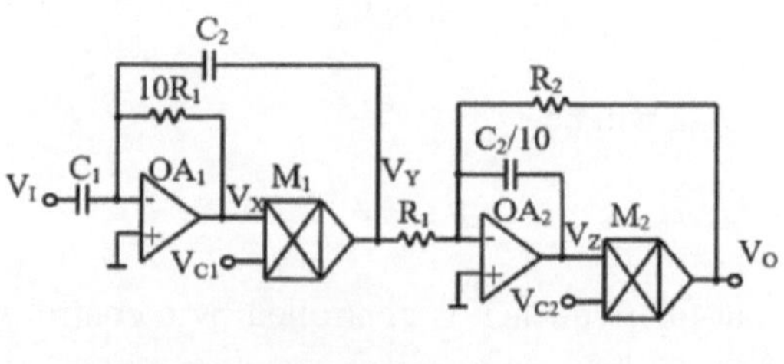

FIGURE 22.3 Circuit diagram of bandpass filter.

$$\frac{V_O(j\omega)}{V_I(j\omega)} = \frac{j\omega R_1C_1V_{C1}}{\left(1 + j\omega R_1C_1V_{C1}\right)\left(1 + \dfrac{j\omega R_2C_2}{V_{C2}}\right)} \tag{22.22}$$

A filter circuit is said to be "bandpass" if it follows the below Eq. (22.23)

$$\frac{V_O(j\omega)}{V_I(j\omega)} = \frac{j\omega/\omega_L}{\left(1 + j\omega/\omega_L\right)\left(1 + j\omega/\omega_H\right)} \tag{22.23}$$

On seeing Eqs. (22.22) and (22.23), it is proved that Figure 22.3 works as a bandpass filter whose lower and higher cut-off frequencies are given by

$$f_L = \frac{1}{2\pi R_1C_1V_{C1}} \tag{22.24}$$

$$f_H = \frac{V_{C2}}{2\pi R_2C_2} \tag{22.25}$$

22.4 VOLTAGE-TUNABLE BANDSTOP FILTERS

A voltage-tunable bandstop filter can be obtained from a voltage-tunable bandpass filter as shown in Figure 22.4. The input signal is passed through the voltage-tunable bandpass filter and is subtracted from the input signal.

The cut-off frequency is given by

$$f_O = \sqrt{f_L f_H} \tag{22.26}$$

22.5 VOLTAGE-TUNABLE UNIVERSAL ACTIVE FILTERS

Adding two multipliers to the universal active filter, a voltage-tunable universal active filter can be obtained. Its frequency is controllable by a control voltage and provides simultaneous lowpass, highpass, and bandpass outputs. Figure 22.5 shows

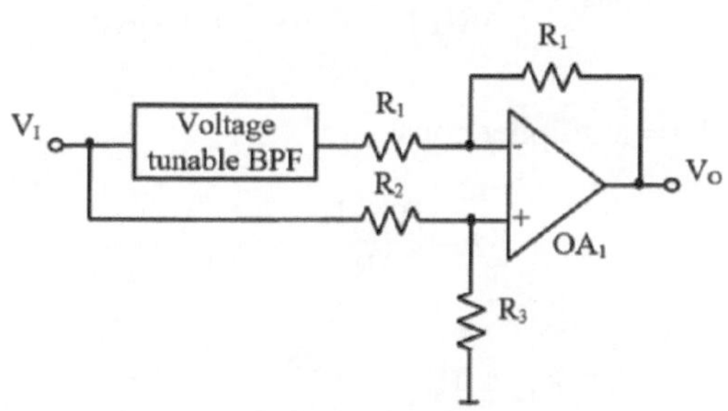

FIGURE 22.4. Circuit of voltage-tunable bandstop filter.

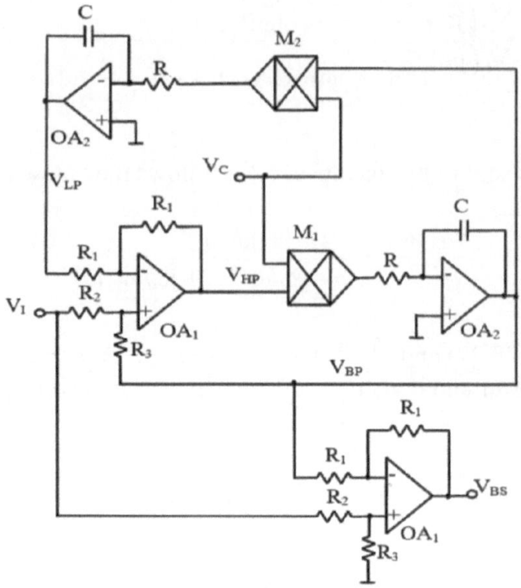

FIGURE 22.5 Voltage-tunable universal active filter.

the circuit diagram of voltage tunable universal active filter. The voltage transfer function between respective outputs and the input are as follows:

$$\text{Lowpass} = \frac{V_{LP}(s)}{Vi(s)} = \frac{K_{LP}\omega_O^2}{s^2 + s\omega_O/Q + \omega_O^2} \tag{22.27}$$

$$\text{Highpass} = \frac{V_{HP}(s)}{Vi(s)} = \frac{K_{HP}s^2}{s^2 + s\omega_O/Q + \omega_O^2} \tag{22.28}$$

$$\text{Bandpasss} = \frac{V_{BP}(s)}{Vi(s)} = \frac{K_{BP}s\omega_O/Q}{s^2 + s\omega_O/Q + \omega_O^2} \tag{22.29}$$

where, $\omega_O = \frac{V_C}{10RC}$, 10 = multiplier constant

$$Q = \frac{R_2 + R_3}{2R_2} \tag{22.30}$$

$$K_{LP} = K_{HP} = \frac{2R_3}{R_2 + R_3}$$

$$K_{BP} = \frac{R_3}{R_2}$$

The control voltage V_C is to be positive DC voltage, and hence a two-quadrant multiplier is to be used.

22.6 BALANCED MODULATORS

Modulation and frequency doubling are basically processes of multiplication. Modulation is a process by which some characteristics of one wave called a "carrier" are varied in accordance with some characteristics of another wave called a "modulating signal." Multipliers are used for modulators and are described below.

The multiplier as a balanced modulator or suppressed-carrier double sideband modulation is shown in Figure 22.6.

V_1 is the modulating signal: $V_1 = E_m \text{Sin}\omega_m t$
V_2 is the carrier signal: $V_2 = E_c \text{Sin}\omega_c t$
V_1 and V_2 are applied to the multiplier. Its output will be

$$V_O = \frac{V_1 V_2}{V_R} = \frac{(E_m \text{ Sin } \omega_m t)(E_c \text{ Sin } \omega_c t)}{V_R} \tag{22.31}$$

$$V_O = \frac{E_m E_c}{2V_R}\left[\cos(\omega_c - \omega_m)t - \cos(\omega_c + \omega_m)t\right] \tag{22.32}$$

In the above equation, the V_R is a constant value and usually is 10. Also in the above equation, the carrier frequency term does not appear and hence the term "suppressed carrier" is obtained. Balanced modulators are widely used in communication systems, measurements and instrumentation, and control systems. They have the advantage over other modulation schemes that the carrier is suppressed and does not appear in the output, and hence power consumption is reduced. Modulating circuits have their spectrum centered about the second harmonic of the carrier frequency or any multiple of it, and they require complex filters to eliminate the unwanted frequencies.

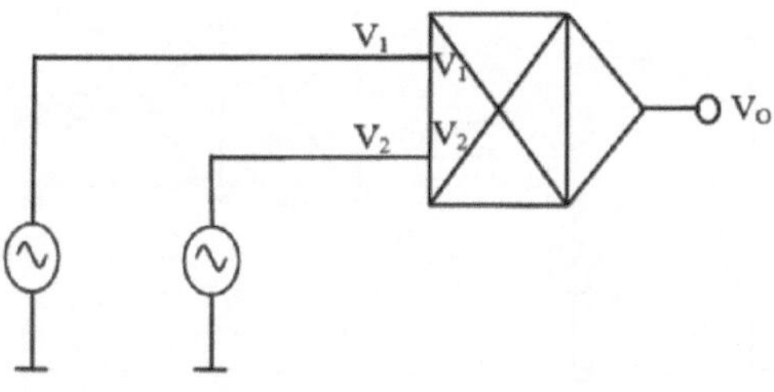

FIGURE 22.6 Multiplier as balanced modulator.

A common problem in communication is to extract information from single sideband (ss) signals received. The ss signal can be written as

$$e_{ss} = K\sin\left(\omega_m + \omega_c\right)t \tag{22.33}$$

If e_{ss} is multiplied by an appropriate carrier signal of a $\sin\omega_c t$, the resulting output will be

$$V_O = \frac{\left[K\sin\left(\omega_m + \omega_c\right)t\right]\left(A\sin\omega_c t\right)}{V_R}$$

$$V_O = \frac{KA}{2V_R}\left[\cos\omega_m t - \cos\left(\omega_m + 2\omega_c\right)t\right] \tag{22.34}$$

In the above equation, the term $KA/V_R \cos \omega_m t$ can be extracted by using a simple filter to remove the second high frequency term.

22.7 AMPLITUDE MODULATORS

Similar to a balanced modulator, when a DC voltage is added to the modulating signal, the multiplier performs amplitude modulation. In this, the carrier is passed through the multiplier when the modulating signal is zero.

From Figure 22.7

$$V_1 = E_m + mE_m \sin\omega_m t = E_m\left(1 + m\right)\sin\omega_m t$$

$V_2 = E_c \sin \omega_c t$

$$V_O = \frac{\left(E_m + mE_m \sin\omega_m t\right)\left(E_c \sin\omega_c t\right)}{V_R} \tag{22.35}$$

$$V_O = \frac{E_m E_c}{V_R}[\sin\omega_c t + \frac{m}{2}\cos\left(\omega_c - \omega_m\right)t - \frac{m}{2}\cos\left(\omega_c + \omega_m\right)t \tag{22.36}$$

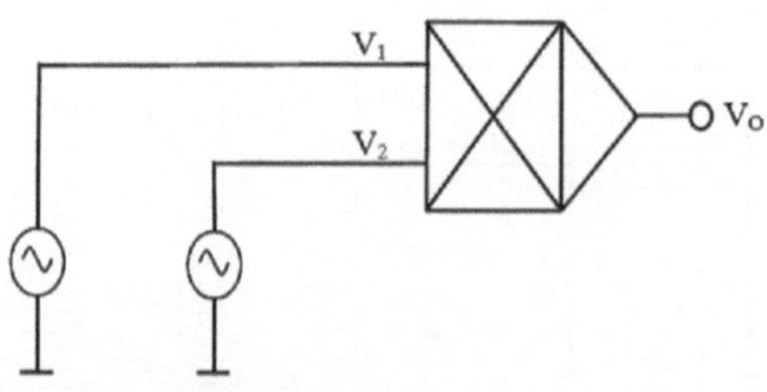

FIGURE 22.7 Multiplier as amplitude modulator.

where m = modulation index. If we keep the peak amplitude of the modulating wave equal to the DC offset voltage, then 100% modulation can be achieved.

22.8 FREQUENCY DOUBLERS

The frequency doubler operation using multiplier is shown in Figure 22.8.

Let $V_1 = V_2 = A\sin\omega t$

$$V_O = \frac{\left(A\sin\omega t\right)^2}{V_R} \tag{22.37}$$

$$V_O = \frac{A^2}{2V_R}\left(1-\cos 2\omega t\right) \tag{22.38}$$

The multiplier output V_O contains a DC voltage in association with the second harmonic of the input signal, and this DC voltage can be removed through AC coupling.

22.9 PHASE ANGLE DETECTORS

Figure 22.9 shows a multiplier in connection with a lowpass filter to determine phase difference between two sinusoidal signals of the same frequency.

The principle of the circuit is based on the trigonometric identity

$$\sin\alpha\sin\beta = \frac{1}{2}\left[\cos\left(\alpha-\beta\right)-\cos\left(\alpha+\beta\right)\right]$$

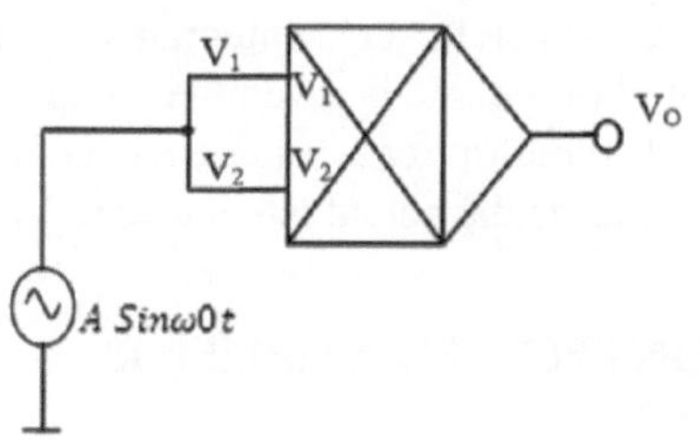

FIGURE 22.8 Frequency doubler using multiplier.

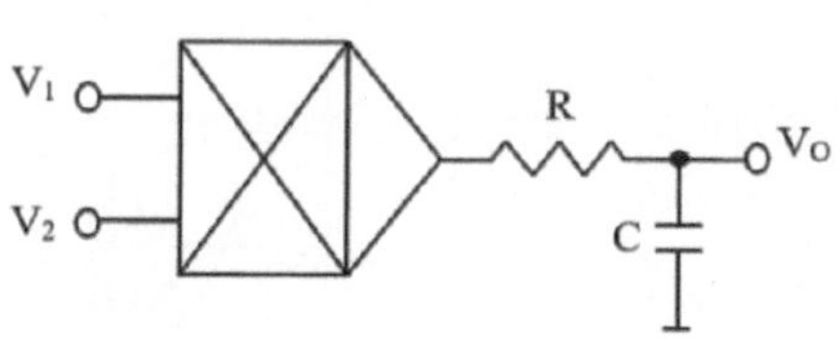

FIGURE 22.9 Phase detector using multiplier.

The waveform at the output of multiplier V_X consists of an AC voltage superimposed on a DC level. The frequency of the AC term is twice that of input signals, and the DC voltage level is proportional to the phase difference between the two inputs.

Let $V_1 = A \sin \omega t$

And $V_2 = B \sin(\omega t + \theta)$

$$V_X = \frac{AB}{2V_R}\left[\cos\theta - \cos(2\omega t + \theta)\right] \tag{22.39}$$

The RC lowpass filter eliminates the AC term and allows only the DC voltage V_O, and it is given as

$$V_O = \frac{AB}{2V_R}\cos\theta \tag{22.40}$$

$$\theta = \cos^{-1}\frac{2V_R V_O}{AB}$$

22.10 ROOT MEAN SQUARE DETECTORS

Figure 22.10 shows a root mean square (RMS) detector using a multiplier.

The RMS value of a signal V_I is given by

$$V_{RMS} = \sqrt{\frac{1}{T}\int_0^T (V_I)^2\, dt} \tag{22.41}$$

The input signal V_I is squared by the multiplier M_1, and the squared signal V_X is given to integrator OA_1. The integrator OA_1 integrates V_X and gives V_Y as the integrated output. The integrated output V_Y is given the square rooter realized by multiplier M_2 and op-amp OA_2. The multiplier as square rooter is discussed in Chapter 18. The output of the square rooter is the actual RMS output V_O.

22.11 VOLTAGE-CONTROLLED AMPLIFIERS

The multiplier output is given as

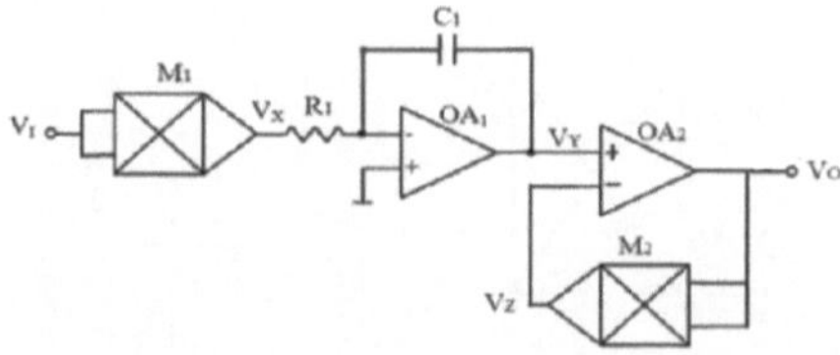

FIGURE 22.10 Circuit diagram of root mean square detector.

$$V_O = \frac{V_1 V_2}{V_R}$$

V_1 and V_2 are input voltages and V_R is the constant voltage.
Let $V_1 = V_I$ (Input voltage)
$V_2 = V_C$ (Control voltage)

The multiplier output will be

$$V_O = V_I \frac{V_C}{V_R} \tag{22.42}$$

Thus, the multiplier amplifies the input voltage V_I with a gain V_C/V_R.

22.12 RECTIFIERS

The full wave rectifier using a multiplier is shown in Figure 22.11. The op-amp OA acts as a zero crossing detector and converts input sine wave into square wave.

The multiplier is a four-quadrant multiplier whose output is configured as always positive. The multiplier output is the full wave rectified output as shown in Figure 22.12.

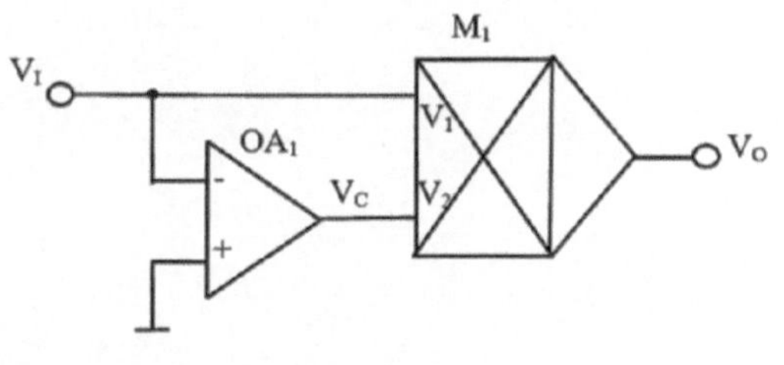

FIGURE 22.11 Multiplier as full wave rectifier.

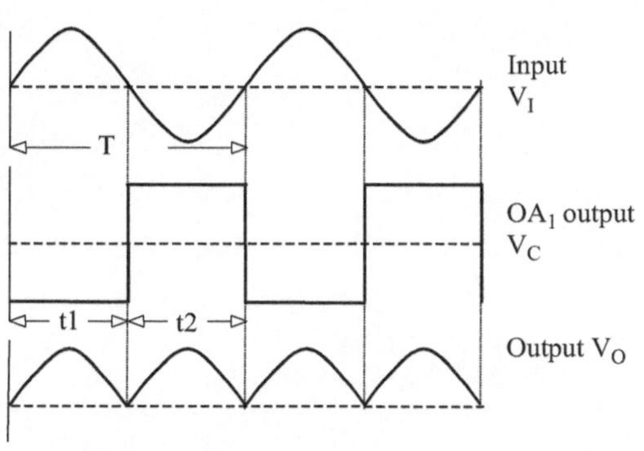

FIGURE 22.12 Associated waveforms of Figure 22.11.

TUTORIAL EXERCISES

1. A voltage-tunable highpass filter has R_1 and C_1 values as $R_1 = 1\ K\Omega$, $C_1 = 0.1\ \mu F$. 5 V is applied as its control voltage. Find the cut-off frequency of the highpass filter.
2. In Problem 1, if the R_1C_1 parts are interchanged, what is the function of the circuit? Find its cut-off frequency.
3. Design voltage-tunable bandpass filter for $f_L = 100\ H_Z = f_H = 500\ H_Z$ with control voltages of 1 and 5 V, respectively.
4. The voltage-tunable bandpass filter of Problem 3 is to be converted into voltage-tunable bandstop filter. Draw the circuit diagram and find its cut-off frequency.
5. Two input signals $V_1 = 5 \sin \omega t$ and $V_2 = 9 \sin (\omega t + 25°)$ are applied to a multiplier. Find the DC output voltage of the multiplier. Ensure that multiplier constant = 10.
6. Draw output waveforms of a full wave rectifier with multiplier of Figure 22.11 if the op-amp OA input terminals (–) and (+) are interchanged.

23 Impedance Measurements

A wide range of electrical and electronic instruments are used in engineering, science, and other non-scientific fields. The three passive circuit elements—resistors, inductors, and capacitors—are essentially building these instruments.

There are several methods for measuring impedance:

1. Null method: The unknown impedance is computed from the values of known elements after the reduction of an output voltage or current to zero.
2. Deflection method: The unknown impedance value is indicated on an instrument. Examples are an ohmmeter, vector impedance meter, Q meter, etc.
3. Absolute method: It requires no standard impedance elements for comparison.
4. Comparison method: Bridge and potentiometric techniques are used to measure unknown impedance. Bridges have the unknown impedance connected to other known elements in the form of a four-arm bridge. The Wheatstone and Kelvin bridges are primarily meant for the measurement of resistors under DC excitation. Maxwell, Wien, Hay, Anderson, and Schering bridges are a few examples of widely used AC bridges for measurement of inductors and capacitors. Though these bridges can provide better accuracy compared to deflection methods, obtaining of the null is often a cumbersome and time consuming process. Automatic balancing would simplify this task but with increased cost.

Some other techniques for measurement of impedance using function circuits are described in this chapter.

23.1 BASIC IMPEDANCE MEASUREMENT

Figure 23.1 shows the block diagram of the impedance measurement. The current 'I' is $I = V_S/R_S$.

The voltage across the unknown impedance 'Z' is $V_Z = IZ$

$$Z = \frac{V_Z}{I} = V_Z \frac{R_S}{V_S}$$

If we keep R_S and V_S constant, then the impedance is proportional to the impedance voltage V_Z. The phase sensitive detector (PSD) extracts the impedance voltage into quadrature component V_Q and in-phase component V_I. At this stage, the impedance voltage can be represented as

$$V_Z = V_I + jV_Q \tag{23.1}$$

DOI: 10.1201/9781003221449-29

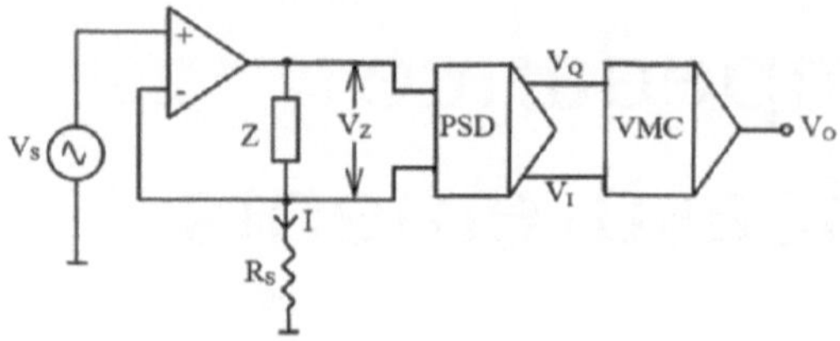

FIGURE 23.1 Block diagram of impedance measurement.

in rectangular form.

However, to represent impedance voltage in polar form, a vector magnitude circuit (VMC) is used. Impedance can also be given by

$$(\text{Voltage})V_Z = V_O\underline{|\theta} \tag{23.2}$$

in polar form where $\mathbf{V_O} = \sqrt{V_I^2 + V_Q^2}$ and θ is the phase difference between the exciting voltage V_S and the impedance voltage V_Z.

23.2 INDUCTANCE MEASUREMENT BY MAGNITUDE RESPONSE

23.2.1 L Measurement in Rectangular Form

Figure 23.2 shows inductance measurement in rectangular form using a PSD. The voltage across the unknown inductor L_X is given to PSD as its input signal through the instrumentation amplifier (IA). The source voltage V_S is also given to the PSD as a reference signal.

As discussed in Chapter 21, PSD gives two outputs: in-phase component V_I and quadrature component V_Q. Let us assume that sampling PSD is used here. Hence,

$$V_I = V_P \cos\theta \tag{23.3}$$

$$V_Q = V_P \sin\theta \tag{23.4}$$

where V_P is the peak value of inductor voltage V_X.

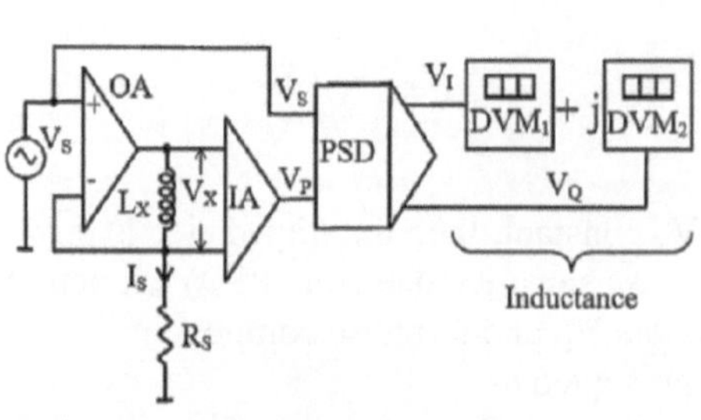

FIGURE 23.2 Measurement of L in rectangular form.

From Figure 23.2

$$V_X = I_S L_X = \frac{V_S}{R_S} L_X \tag{23.5}$$

Let us keep V_S and R_S as standard constants. Hence,

$$V_X \propto L_X \tag{23.6}$$

The two digital voltmeters as shown in Figure 23.2 read V_X in the form $V_X = V_I + jV_Q$. DVM_1 reads in-phase component V_I, DVM_2 reads quadrature component V_Q, and both simultaneously display V_X (or in other words L_X) in rectangular form.

DESIGN EXERCISES

1. Use switching, multiplying, sampling, and switching-sampling PSDs in the circuit shown in Figure 23.2. In each, (i) draw circuit diagrams and (ii) explain their working operation.

23.2.2 L Measurement in Polar Form

The inductance measurement in polar form is shown in Figure 23.3. As discussed in the previous section, the inductor voltage is proportional to the unknown inductance. The reference sine wave V_S and the inductor voltage V_X are given to a phase detector which consists of zero crossing detectors OA_1, OA_2, Ex-OR gate G, and RC lowpass filter. The phase detector output V_{O2} is the phase angle between reference and inductor voltage and is displayed in DVM_2.

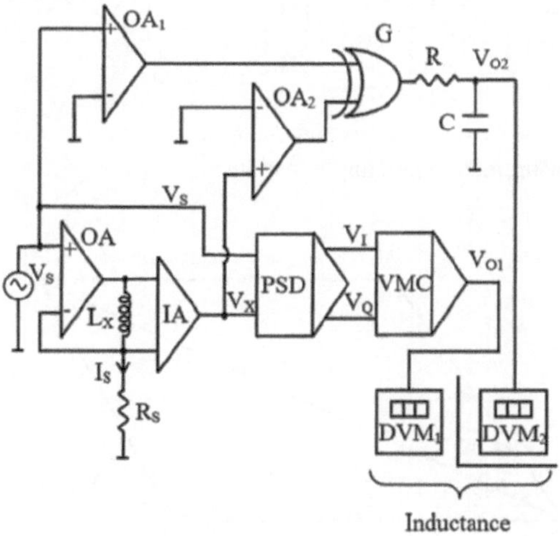

FIGURE 23.3 L measurement in polar form.

The reference and inductor voltages are given to a PSD to get in-phase component V_I and quadrature component V_Q of the inductor voltage. The outputs of PSD are given to a VMC. As discussed in Chapter 19, the VMC output will be

$$V_{O1} = \sqrt{V_I^2 + V_Q^2} \tag{23.7}$$

The output of VMC V_{O1} is given to the DVM_1 to display the magnitude of inductor voltage, which in turn displays the value of inductance. Both DVMs read the inductor voltage V_X and display in turn unknown inductance in polar form.

DESIGN EXERCISES

1. Use switching, multiplying, sampling, and switching-sampling PSDs in the circuit shown Figure 23.3. In each, (i) draw circuit diagrams and (ii) explain their working operation.

23.3 INDUCTANCE MEASUREMENT BY PHASE ANGLE RESPONSE

The block diagram of inductance measurement by phase angle response is shown in Figure 23.4. The unknown inductor is connected in series with a known standard resistance R_S and the combination is passed by a constant current source I_S of constant frequency. The phasors of the voltages V_S, V_R, V_L are indicated in Figure 23.5, where Vs is the source voltage, V_R is the voltage across the resistor, and V_L is the voltage across the unknown inductor.

It can be deduced from Figure 23.5 that

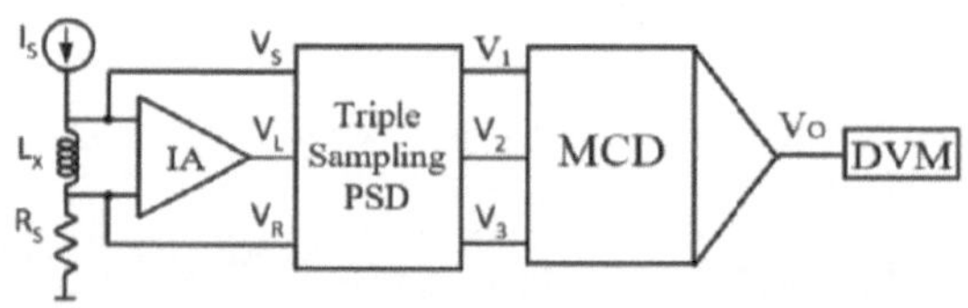

FIGURE 23.4 L measurement phase angle response.

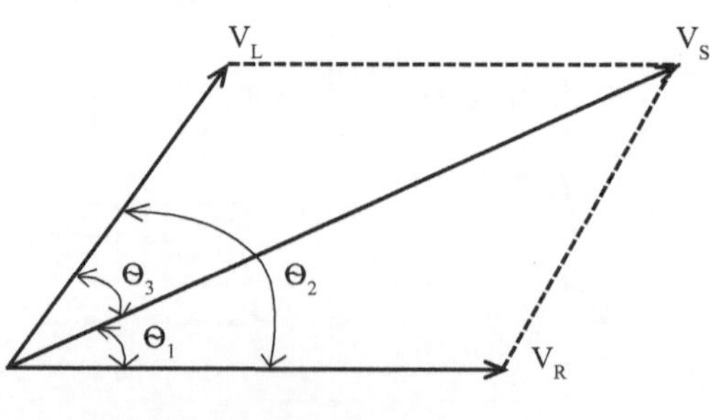

FIGURE 23.5 Phasor diagram for inductive impedance.

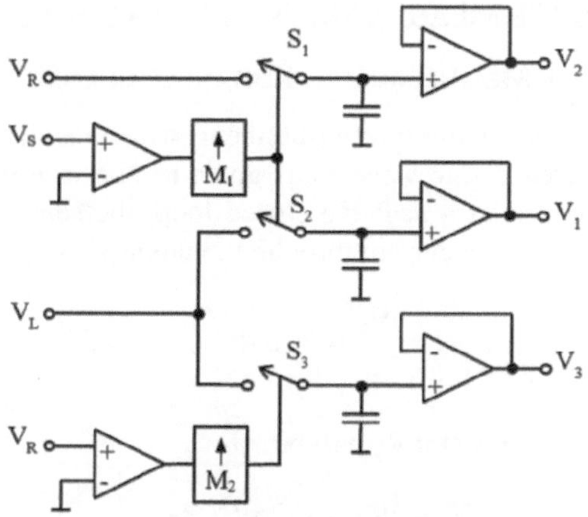

FIGURE 23.6 Circuit diagram of triple sampling type phase sensitive detectors.

$$L = \frac{R}{\omega} \frac{\text{Sin}\,\theta_1\,\text{Sin}\,\theta_2}{\text{Sin}\,\theta_3} \tag{23.8}$$

The triple sampling phase sensitive detectors are shown in Figure 23.6. As discussed in Chapter 21, a phase-shifted signal V_R is sampled by a sampling pulse generated during positive zero crossings of the reference signal V_S. Then the sampled output will be

$$V_2 = V_R\,\text{Sin}\,\theta_1 \tag{23.9}$$

where V_R is the peak value of the phase-shifted signal, and θ_1 is the phase difference between V_S and V_R. Similarly,

$$V_1 = V_L\,\text{Sin}\,\theta_3 \tag{23.10}$$

$$V_3 = V_L\,\text{Sin}\,\theta_2 \tag{23.11}$$

In Figure 23.4, the MCD output will be

$$V_O = \frac{V_2 V_3}{V_1} = V_R \frac{\text{Sin}\,\theta_1\,\text{Sin}\,\theta_2}{\text{Sin}\,\theta_3} \tag{23.12}$$

On comparing Eqs. (23.8) and (23.12), MCD output is proportional to unknown inductance and is displayed in the digital voltmeter (DVM).

DESIGN EXERCISES

1. Use switching, multiplying, sampling, and switching-sampling phase sensitive detectors in the circuit shown in Figure 23.4. In each, (i) draw circuit diagrams and (ii) explain their working operation.

23.4 CAPACITANCE MEASUREMENT BY MAGNITUDE RESPONSE

23.4.1 Capacitance Measurement in Rectangular Form

The block diagram of capacitance measurement in rectangular form is shown in Figure 23.7. The reference sine wave V_S is given to non-inverting terminal of op-amp. Since the op-amp is at a negative closed loop, the same V_S appeared on its inverting terminal. Hence, the current through the unknown capacitor C_X will be

$$I = \frac{V_S}{X_C} = V_S 2\pi f C_X \tag{23.13}$$

The voltage across the resistor R_S will be

$$V_R = IR_S = V_S 2\pi f C_X R_S \tag{23.14}$$

If we keep V_S, f, and R_S as constants

$$V_R \propto C_X \tag{23.15}$$

V_R is measured and displayed in terms of C_X. The voltage V_R is given to PSD through an Instrumentation Amplifier IA. The reference signal V_S is also given to PSD. The PSD gives two outputs: in-phase component V_I and quadrature component V_Q. Let us assume switching-sampling PSD is used here. As discussed in Section 21.4 the outputs of switching-sampling PSD are

$$V_I = \frac{2V_P}{\pi} \cos\theta \tag{23.16}$$

$$V_Q = V_P \sin\theta \tag{23.17}$$

where V_P is the peak value of V_R, and θ is the phase difference between V_R and V_S. The V_R can be represented as

$$V_R = V_I + jV_Q \tag{23.18}$$

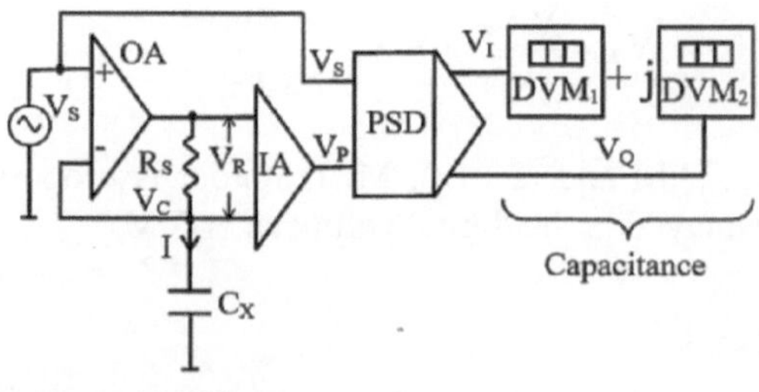

FIGURE 23.7 Capacitance measurement in rectangular form.

DVM_1 reads the in-phase component V_I, and DVM_2 reads quadrature component V_Q. Both DVMs simultaneously indicate the unknown capacitance C_X in rectangular form.

DESIGN EXERCISES

1. Use switching, multiplying, and sampling phase sensitive detectors in the circuit shown Figure 23.7. In each, (i) draw circuit diagrams and (ii) explain their working operation.

23.4.2 Capacitance Measurement in Polar Form

The capacitance measurement in polar form is shown in Figure 23.8. The reference sine wave V_S and the resistor voltage V_R which is proportional to unknown capacitance 'C' are given to a phase detector which consists of zero crossing detectors OA_1, OA_2, Ex-OR gate G, and RC lowpass filter. The phase detector output V_{O2} is the phase angle between reference and capacitor voltage and is displayed in DVM_2.

The reference voltage V_S and resistor voltage V_R are given to a PSD to get in-phase component V_I and quadrature component V_Q of the capacitor voltage. The outputs of PSD are given to a VMC. As discussed in Chapter 19, the VMC output will be

$$V_{O1} = \sqrt{V_I^2 + V_Q^2} \tag{23.19}$$

The output of VMC V_{O1} is given to the DVM_1 to display magnitude of resistor voltage, which in turn displays the value of capacitance. Both DVMs read the resistor voltage V_R and display in turn the unknown capacitance in polar form.

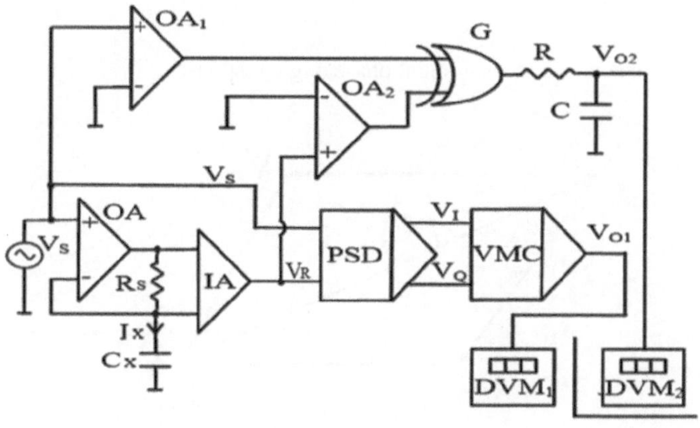

FIGURE 23.8 Capacitance measurement in polar form.

DESIGN EXERCISES

1. Use switching, multiplying, sampling, and switching-sampling PSDs in the circuit shown Figure 23.8. In each, (i) draw circuit diagrams and (ii) explain their working operation.

23.5 CAPACITANCE MEASUREMENT BY PHASE ANGLE RESPONSE

The capacitance measurement by phase angle response is given in Figure 23.9, the unknown capacitor is connected in series with a known standard resistance R_S and the combination is passed by a constant current source I_S of constant frequency. The phasors of the voltages V_S, V_R, V_C are indicated in Figure 23.10, where Vs is the source voltage, V_R is the voltage across the resistor, and V_C is the voltage across the unknown capacitor.

It can be deduced from Figure 23.10 that

$$C_P = \frac{1}{R\omega}\frac{\sin\theta_3 \sin\theta_2}{\sin\theta_1} \tag{23.20}$$

The triple sampling PSDs are shown in Figure 23.11. As discussed in Section 21.3, a phase-shifted signal V_R is sampled by a sampling pulse generated during positive zero crossings of the reference signal V_S. Then the sampled output will be

$$V_2 = V_R \operatorname{Sin}\theta_1 \tag{23.21}$$

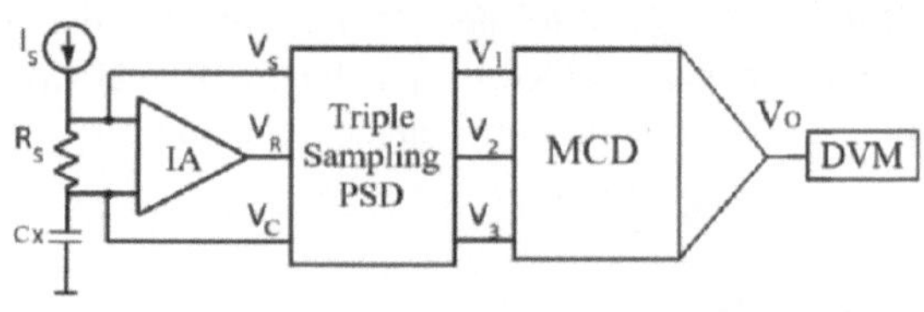

FIGURE 23.9 Capacitance measurement phase angle response.

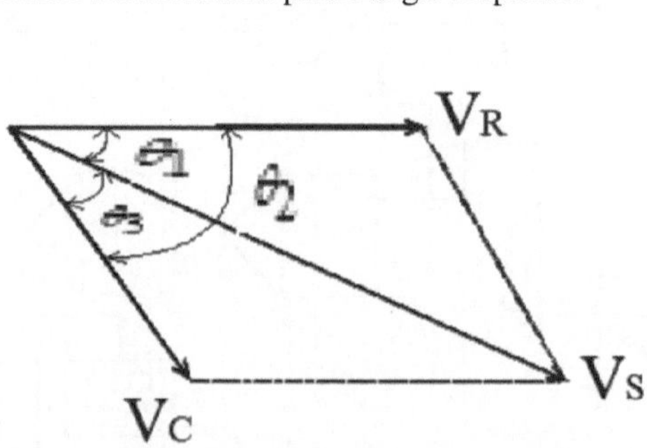

FIGURE 23.10 Phasor diagram of capacitive impedance.

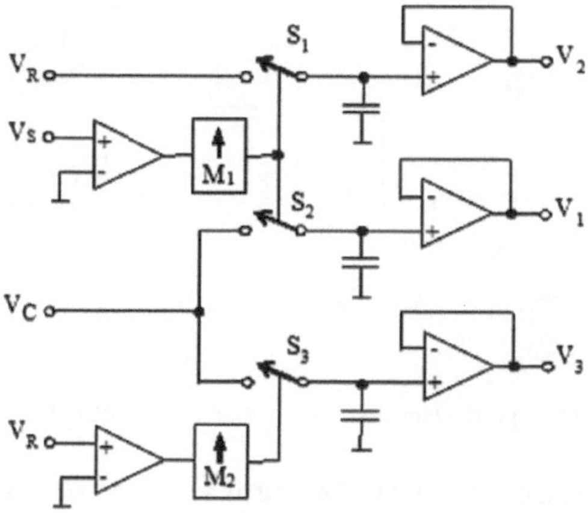

FIGURE 23.11 Triple sampling PSDs.

where V_R is the peak value of the phase-shifted signal, and θ_1 is the phase difference between V_S and V_R. Similarly

$$V_1 = V_C \operatorname{Sin} \theta_3 \tag{23.22}$$

$$V_3 = V_C \operatorname{Sin} \theta_2 \tag{23.23}$$

In Figure 23.9, the MCD output is given as

$$V_O = \frac{V_2 V_3}{V_1} = V_R \frac{\operatorname{Sin} \theta_1 \operatorname{Sin} \theta_2}{\operatorname{Sin} \theta_3} \tag{23.24}$$

On comparing Eqs. (23.20) and (23.24), the MCD output voltage is proportional to the unknown capacitance and is displayed in the DVM.

DESIGN EXERCISES

1. Use switching, multiplying, sampling, and switching-sampling PSDs in the circuit shown Figure 23.9. In each, (i) draw circuit diagrams and (ii) explain their working operation.

23.6 CAPACITANCE MEASUREMENT BY COMPARISON METHOD

The capacitance measurement by comparison method is shown in Figure 23.12. The unknown capacitor C_X can be represented by a parallel combination of a capacitor C_P and a resistance R_P. For an excitation voltage V sin ωt, the current I would be

$$I_C = V_S \left(G_P \sin \omega t + \omega C_P \cos \omega t \right) \tag{23.25}$$

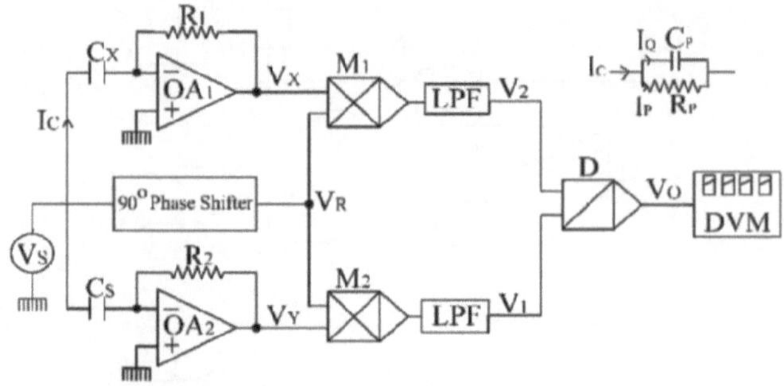

FIGURE 23.12 Capacitance measurement by comparison method.

where G_P is the equivalent conductance and is equal to $1/R_P$. Hence, current I_C is made of a component

$$I_P = V_S G_P \sin \omega t \tag{23.26}$$

which is in-phase with the applied voltage and a component

$$I_Q = V_S \omega C_P \cos \omega t \tag{23.27}$$

C_P and G_P can be determined in quadratic phase by separating the in-phase and quadrature components of I_C with respect to V as

$$C_P = \frac{I_Q}{\omega} V_S \tag{23.28}$$

$$G_P = \frac{I_P}{V_S} \tag{23.29}$$

where I_P, I_Q and V are respective peak values. The current through unknown capacitor C_X excited by the sinusoidal source V_S is converted to a proportional voltage by the op-amp OA_1. Similarly, standard capacitance C_S is also excited by the same voltage V_S by the op-amp OA_2. The output of OA_1 will be

$$V_X = -V_S R_1 \left(Gp \sin \omega t + \omega Cp \cos \omega t \right) \tag{23.30}$$

And output of OA_2 will be

$$V_Y = -V_S R_2 \left(G_S \sin \omega t + \omega C_S \cos \omega t \right) \tag{23.31}$$

The output of OA_1 V_X and 90° phase-shifted source voltage are given to multiplier M_1. The multiplier M_1 output will be

$$V_2 = \frac{V_S V_T}{2 V_R} R_1 \omega C_P \tag{23.32}$$

A 90° phase-shifted source signal and output of OA_2 V_Y are given to multiplier M_2 and its output will be

$$V_1 = \frac{V_S V_T}{2V_R} R_2 \omega C_S \tag{23.33}$$

In Eqs. (23.32) and (23.33) Vs is the peak value of the excitation signal, V_T is the peak value of 90° phase-shifter and V_R is the multiplier constant. The voltages V_1 and V_2 are given to a divider and the divider output V_O is given by

$$V_O = \frac{V_2}{V_1}$$

$$V_O = \frac{C_P}{C_S}\frac{R_1}{R_2}$$

$$V_O = KC_P \tag{23.34}$$

The divider output voltage is proportional to unknown capacitance C_X and is displayed in the DVM.

DESIGN EXERCISES

1. Replace multiplying PSD in Figure 23.12 with switching, sampling, and switching-sampling PSDs. In each, (i) draw circuit diagrams and (ii) explain their working operation.

23.7 MEASUREMENT OF Q FACTOR

Figure 23.13 shows Q factor measurement using a PSD. The voltage across the unknown inductor L_X is given to PSD as its input signal through the IA. The source voltage V_S is also given to the PSD as a reference signal.

As discussed in Chapter 21, PSD gives two outputs: in-phase component V_I and quadrature component V_Q. Let us assume that sampling PSD is used here. Hence,

$$V_I = V_P \cos\theta \tag{23.35}$$

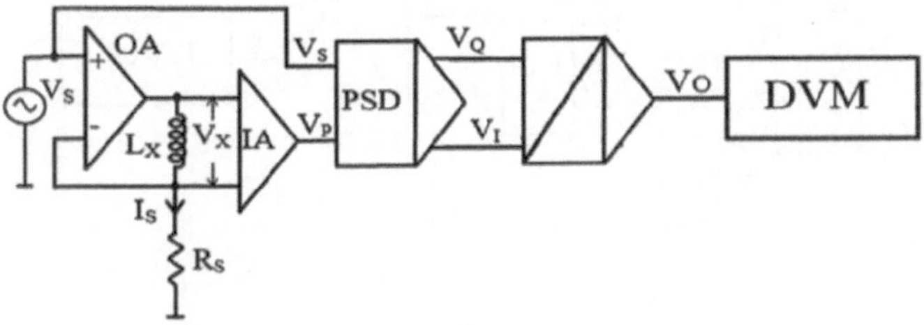

FIGURE 23.13 Measurement of Q factor.

$$V_Q = V_P \sin\theta \tag{23.36}$$

where V_P is the peak value of inductor voltage V_X.

$$Q = \frac{\text{Quadrature component}}{\text{Inphase component}} = \frac{V_Q}{V_I} \tag{23.37}$$

The output of divider is proportional to Q factor of inductor and is displayed in DVM.

DESIGN EXERCISES

1. Use switching, multiplying, sampling, and switching-sampling PSDs in the circuit shown in Figure 23.13. In each, (i) draw circuit diagrams and (ii) explain their working operation.

23.8 MEASUREMENT OF TAN δ

The Tan δ measurement is shown in Figure 23.14. The reference sine wave V_S is given to non-inverting terminal of op-amp. The voltage V_R is given to PSD through IA. The reference signal V_S is also given to PSD. The PSD gives two outputs: in-phase component V_I and quadrature component V_Q. Let us assume switching-sampling type PSD is used here. As discussed in Section 21.4 the outputs of switching-sampling PSD are

$$V_I = \frac{2V_P}{\pi}\cos\theta \tag{23.38}$$

$$V_Q = V_P \sin\theta \tag{23.39}$$

where V_P is the peak value of V_R and θ is the phase difference between V_R and V_S.

$$\text{TAN}\,\delta = \frac{\text{Inphase component}}{\text{Quadrature component}} = \frac{V_I}{V_Q} \tag{23.40}$$

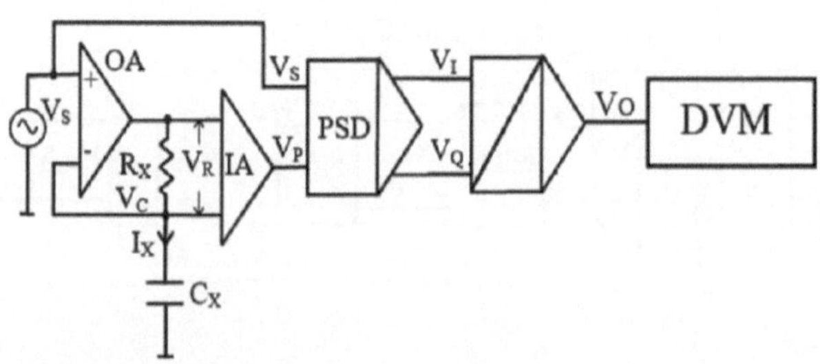

FIGURE 23.14 Measurement of Tan δ.

The output of divider is proportional to Tan δ of capacitor and is displayed in DVM.

DESIGN EXERCISES

1. Use switching, multiplying, sampling, and switching-sampling PSDs in the circuit shown in Figure 23.4. In each, (i) draw circuit diagrams and (ii) explain their working operation.

24 Power and Power Factor Measurements

An electric power is the product of voltage (V) and current (I). Let the phase difference between voltage and current be Ø. Electric power is classified into (i) apparent power, (ii) active power, and (iii) reactive power. Apparent power is defined as voltage-to-current (V/I), active power is defined as VIcosØ, and reactive power is defined as VIsinØ. In the early days, analog watt meters were used, but nowadays digital watt meters replace analog watt meters because of accuracy, compactness, cost, etc. All digital watt meters use analog multipliers to multiply voltage and current signals and display in a display device. In this chapter, we will see how function circuits are used to measure power and power factor.

The power and power factor measurement setup for a single phase is shown in Figure 24.1(a), and for a three-phase it is shown in Figure 24.1(b). In Figure 24.1(a), the source voltage is reduced to a required voltage level through a potential transformer (PT). The reduced voltage is called V_1, which means a voltage representing source voltage V. The current is passed from source to load through a current transformer (CT). The secondary terminal of CT is shorted through a burden resistor R_B and another voltage V_2 is existing across the burden resistance R_B. This V_2 is proportional to current I. V_2 means a voltage representing current I.

In Figure 24.1(b), the R-B voltage is reduced to a voltage V_{X1} by the potential transformer PT_X and the R phase current is converted to a voltage V_{X2} with current transformer CT_X and burden resistor R_{BX}. The Y-B voltage is reduced to a voltage V_{Y1} by the potential transformer PT_Y and the Y phase current is converted to a voltage V_{Y2} with current transformer CT_Y and burden resistor R_{BY}.

In Figure 24.1(c), the R-N voltage is reduced to a voltage V_{X1} by the potential transformer PT_X and the R phase current is converted to a voltage V_{X2} with current transformer CT_X and burden resistor R_{BX}. The Y-N voltage is reduced to a voltage V_{Y1} by the potential transformer PT_Y and the Y phase current is converted to a voltage V_{Y2} with current transformer CT_Y and burden resistor R_{BY}. The B-N voltage is reduced to a voltage V_{Z1} by the potential transformer PT_Z and the B phase current is converted to a voltage V_{Z2} with current transformer CT_Z and burden resistor R_{BZ}.

24.1 ACTIVE POWER MEASUREMENTS

Figure 24.2(a) shows an active power measurement system in a single phase, Figure 24.2(b) shows an active power measurement in three phases (2 watt meter method), and Figure 24.2(c) shows an active power measurement in three phases (3 watt meter method).

DOI: 10.1201/9781003221449-30

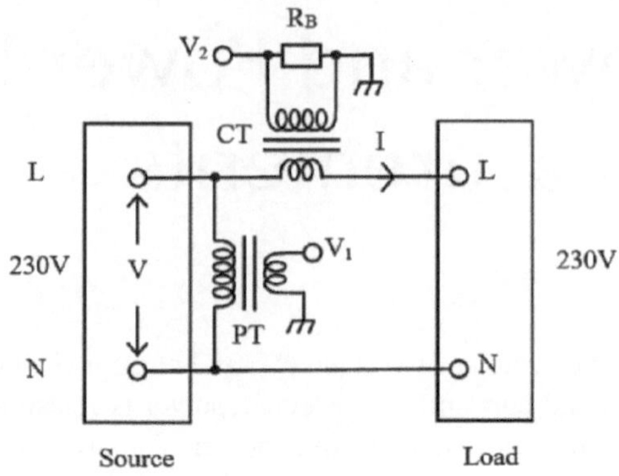

FIGURE 24.1(A) Power measurement setup in single phase.

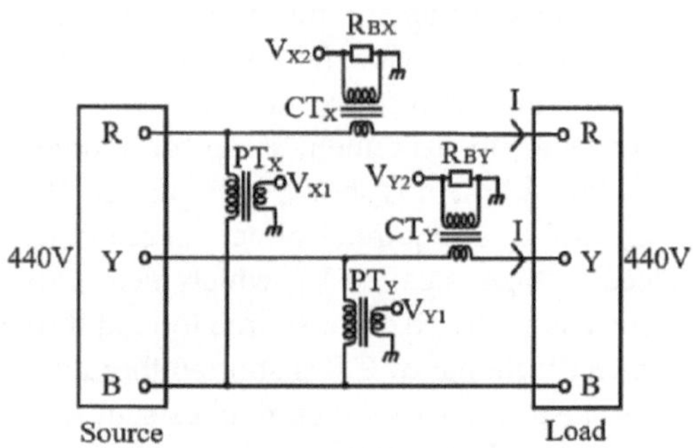

FIGURE 24.1(B) A 2 watt meter power measurement setup in three phases.

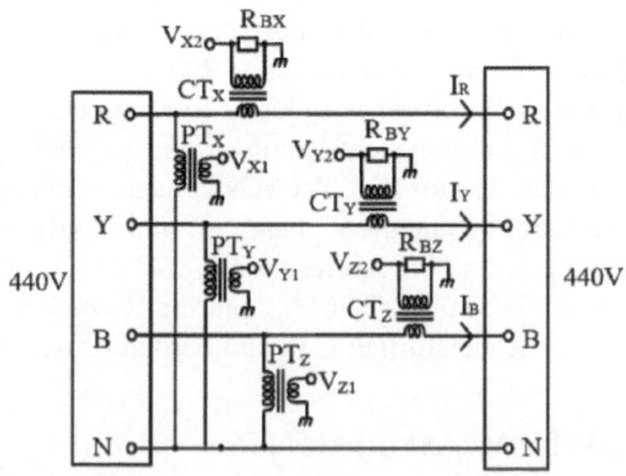

FIGURE 24.1(C) A 3 watt meter power measurement setup in three phases.

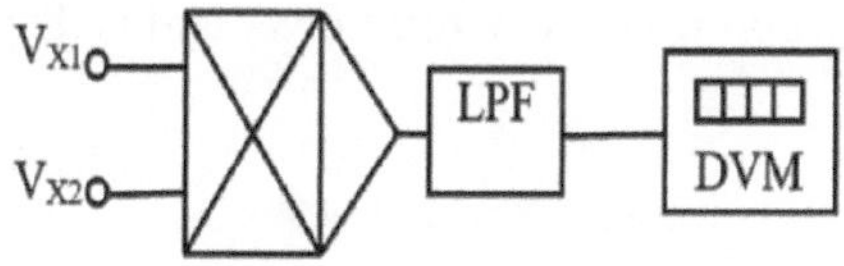

FIGURE 24.2(A) Active power measurement in single phase.

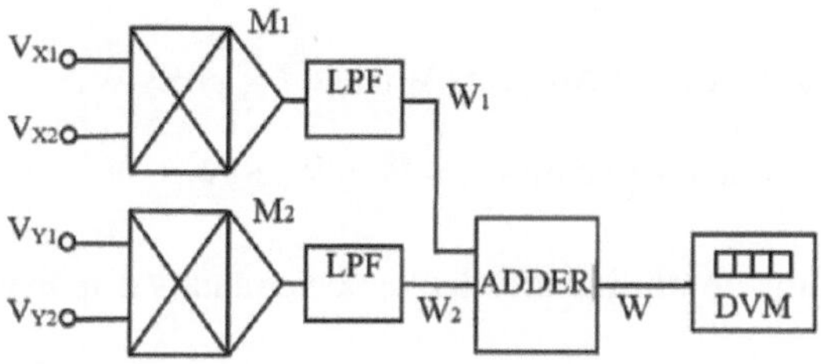

FIGURE 24.2(B) Power measurement in three-phase 2 watt meter method.

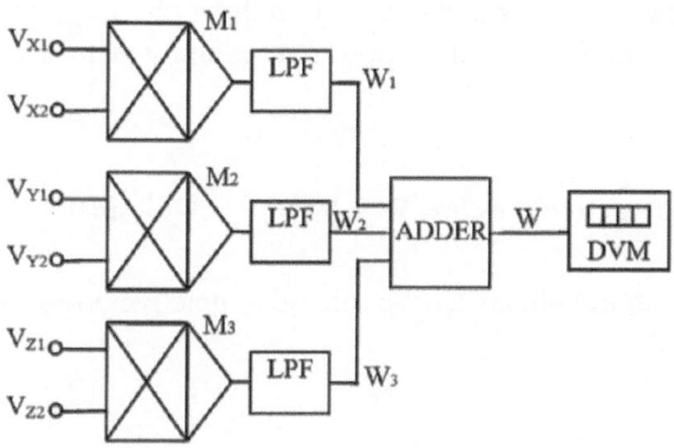

FIGURE 24.2(C) Active power measurement in three-phase 3 watt meter method.

In Figure 24.2(a),

$$\text{Active power} = W = VI\cos\theta = V_1V_2\cos\theta \tag{24.1}$$

The multiplier output voltage is proportional to active power and is displayed in the digital voltmeter (DVM).

In Figure 24.2(b),

$$W_1 = V_{X1}V_{X2}\cos\theta \tag{24.2}$$

$$W_2 = V_{Y1}V_{Y2}\cos\theta \tag{24.3}$$

The multiplier M_1 gives W_1, and M_2 gives W_2.

$$\text{Actual active power} = W = W_1 + W_2 \tag{24.4}$$

The adder output is proportional to active power and is displayed in the DVM.

In Figure 24.2(c),

$$W_1 = V_{X1}V_{X2}\cos\theta \quad (24.5)$$

$$W_2 = V_{Y1}V_{Y2}\cos\theta \quad (24.6)$$

$$W_3 = V_{Z1}V_{Z2}\cos\theta \quad (24.7)$$

The multiplier M_1 gives W_1, M_2 gives W_2, and M_3 gives W_3.

$$\text{Actual active power} = W = W_1 + W_2 + W_3 \quad (24.8)$$

The adder output is proportional to active power and is displayed in the DVM.

24.2 REACTIVE POWER MEASUREMENTS

Figure 24.3(a) shows a reactive power measurement system in a single phase, Figure 24.3(b) shows a reactive power measurement in three phases (2 watt meter method), and Figure 24.3(c) shows a reactive power measurement in three phases (3 watt meter method).

In Figure 24.3(a),

$$\text{Reactive power} = W = VI\sin\theta = V_1V_2\sin\theta \quad (24.9)$$

The multiplier output voltage is proportional to reactive power and is displayed in the DVM.

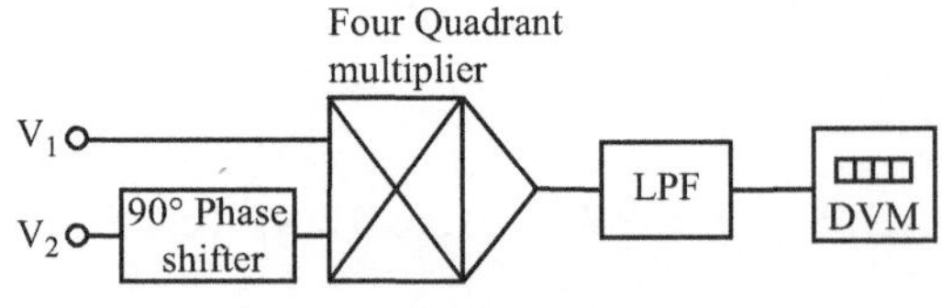

FIGURE 24.3(A) Reactive power measurement in single phase.

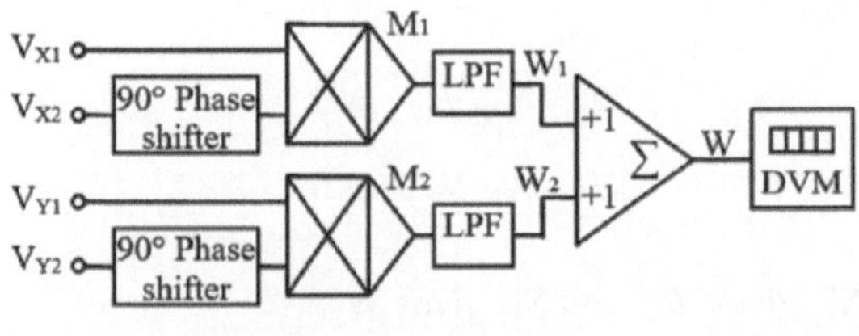

FIGURE 24.3(B) Reactive power measurement in three-phase 2 watt meter method.

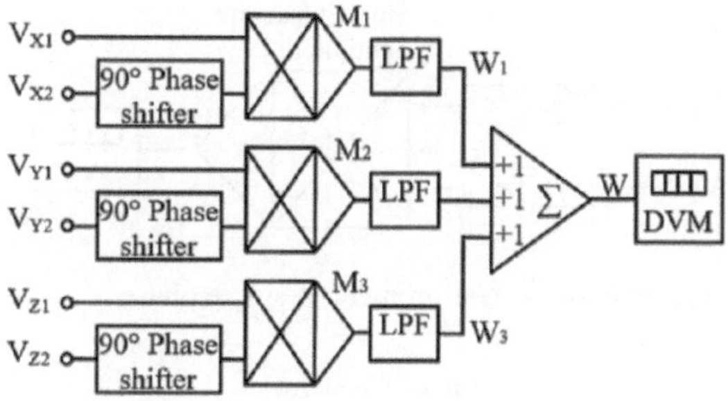

FIGURE 24.3(C) Reactive power measurement in three-phase 3 watt meter method.

In Figure 24.3(b),

$$W_1 = V_{X1}V_{X2}\sin\theta \tag{24.10}$$

$$W_2 = V_{Y1}V_{Y2}\sin\theta \tag{24.11}$$

The multiplier M_1 gives W_1, and M_2 gives W_2.

$$\text{Actual reactive power} = W = W_1 + W_2 \tag{24.12}$$

The adder output is proportional to reactive power and is displayed in the DVM.
In Figure 24.3(c),

$$W_1 = V_{X1}V_{X2}\sin\theta \tag{24.13}$$

$$W_2 = V_{Y1}V_{Y2}\sin\theta \tag{24.14}$$

$$W_3 = V_{Z1}V_{Z2}\sin\theta \tag{24.15}$$

The multiplier M_1 gives W_1, M_2 gives W_2, and M_3 gives W_3.

$$\text{Actual reactive power} = W = W_1 + W_2 + W_3 \tag{24.16}$$

The adder output is proportional to the reactive power and is displayed in the DVM.

24.3 MEASUREMENTS OF APPARENT POWER

Figure 24.4(a) shows an apparent power measurement system, Figure 24.4(b) shows an apparent power measurement in three phases (2 watt meter method), and Figure 24.4(c) shows an apparent power measurement in three phases (3 watt meter method).

In Figure 24.4(a),

$$\text{Apparent power} = W = VI = V_1V_2 \tag{24.17}$$

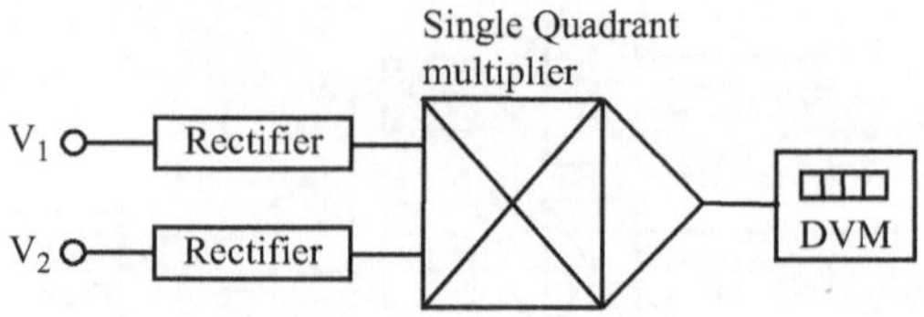

FIGURE 24.4(A) Apparent power measurement in single phase.

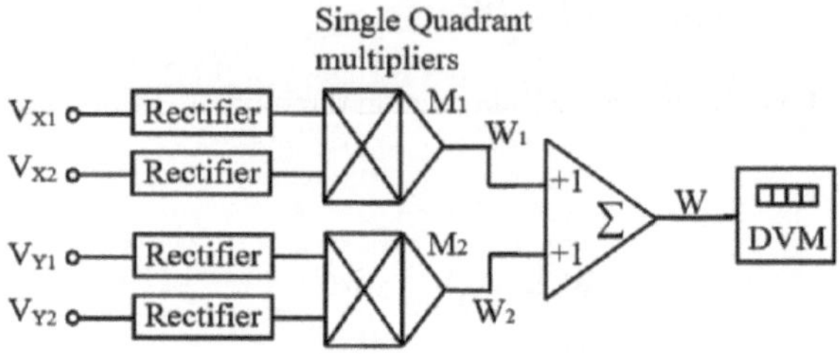

FIGURE 24.4(B) Apparent power measurement in three-phase 2 watt meter method.

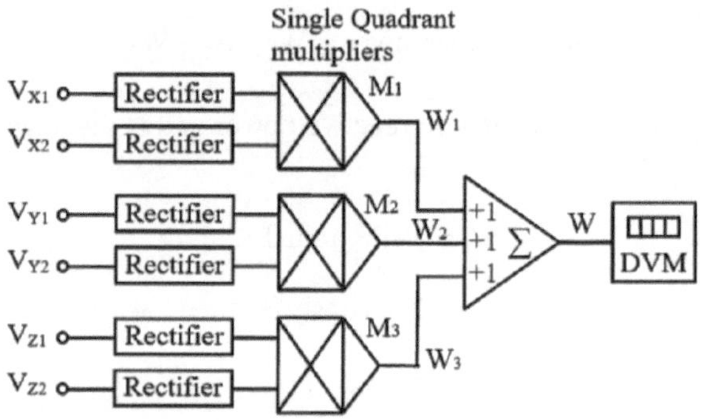

FIGURE 24.4(C) Apparent power measurement in three-phase 3 watt meter method.

The multiplier output voltage is proportional to the apparent power and is displayed in the DVM.

In Figure 24.4(b),

$$W_1 = V_{X1}V_{X2} \tag{24.18}$$

$$W_2 = V_{Y1}V_{Y2} \tag{24.19}$$

The multiplier M_1 gives W_1, and M_2 gives W_2.

$$\text{Actual apparent power} = W = W_1 + W_2 \tag{24.20}$$

The adder output is proportional to apparent power and is displayed in the DVM.

In Figure 24.4(c),

$$W_1 = V_{X1}V_{X2} \tag{24.21}$$

$$W_2 = V_{Y1}V_{Y2} \tag{24.22}$$

$$W_3 = V_{Z1}V_{Z2} \tag{24.23}$$

The multiplier M_1 gives W_1, M_2 gives W_2, and M_3 gives W_3.

$$\text{Actual apparent power} = W = W_1 + W_2 + W_3 \tag{24.24}$$

The adder output is proportional to apparent power and is displayed in the DVM.

24.4 MEASUREMENTS OF POWER FACTOR

$$\text{Power factor} = \cos\varnothing = \frac{\text{Active power}}{\text{Apparent power}} = \frac{VI\cos\theta}{VI} \tag{24.25}$$

Figure 24.5 shows a block diagram of power factor measurement, Figure 24.5(a) shows one for a single phase, and Figure 24.5(b) shows one for a three-phase measurement.

In Figure 24.5(a), the multiplier M_1 and lowpass filter (LPF) output is given as

$$V_X = V_1V_2\cos\theta = VI\cos\theta \tag{24.26}$$

The multiplier M_2 output will be

$$V_Y = V_1V_2 = VI \tag{24.27}$$

The IC7107 DVM reads its two input voltages, V_X and V_Y, and displays a reading which is corresponding to

$$\text{DVM reading} = \frac{V_{IN}}{V_{REF}} = \frac{V_X}{V_Y} = \cos\theta \tag{24.28}$$

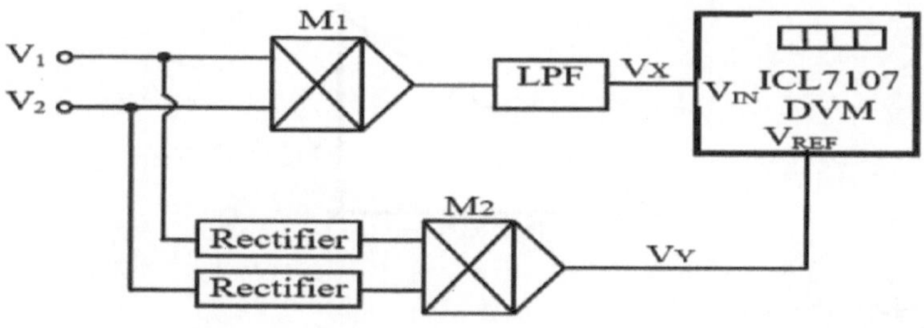

FIGURE 24.5(A) Power factor measurement in single phase.

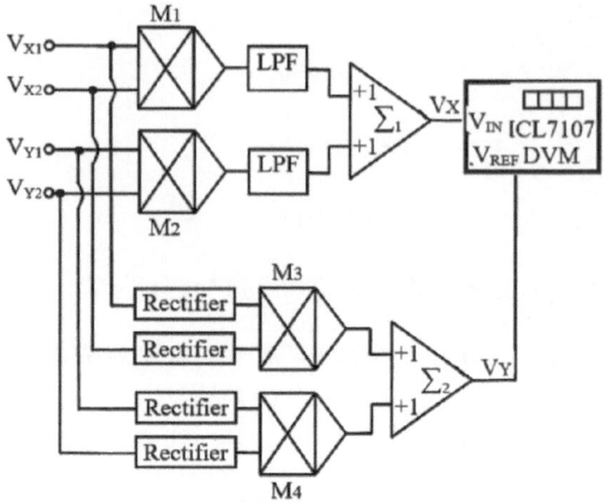

FIGURE 24.5(B) Power factor measurement in three phases.

In Figure 24.5(b), the multipliers M_1 and M_2, LPF, and adder 1 constitute an active power measurement system. The multipliers M_3, M_4, and adder 2 constitute an apparent power measurement system. The reading in IC7107 is proportional to power factor cos Ø.

24.5 TRIVECTOR POWER MEASUREMENTS

A trivector meter is a power meter which indicates apparent power, active power, and reactive power in a power system. Figure 24.6 shows a power triangle.

$$\text{Apparent Power} = S = \text{Volt Amps} = VI$$

$$\text{Active Power} = P = \text{Watts} = VI\cos\varnothing$$

$$\text{Reactive Power} = Q = \text{Volt Amps Reactive} = VI\sin\varnothing$$

$$\text{Power Factor} = PF = \cos\varnothing = P/S$$

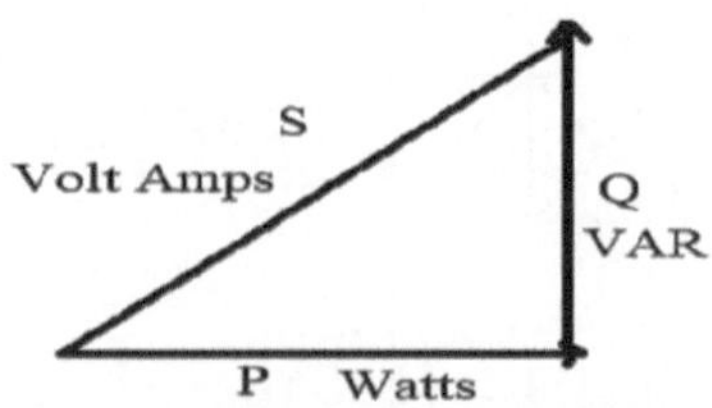

FIGURE 24.6 Power triangle.

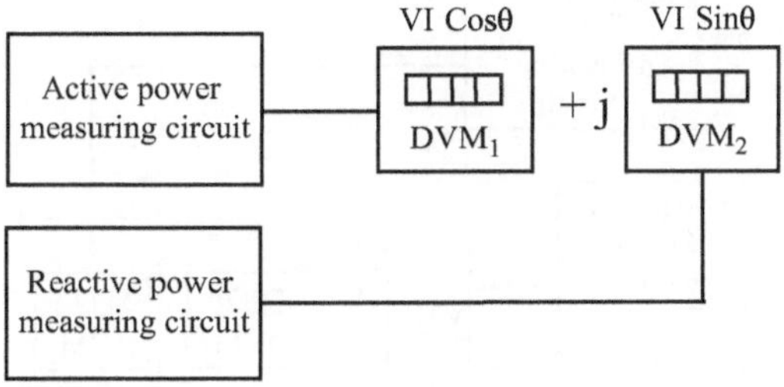

FIGURE 24.7 Power representation in rectangular form.

24.5.1 Power Measurement in Rectangular Form

Figure 24.7 shows power measurement in a rectangular form. The power can be represented in rectangular form as

$$S = P + jQ \tag{24.29}$$

As shown in Figure 24.7, the active power P is measured in DVM_1, which is called "in-phase component," and the reactive power Q is measured and displayed in DVM_2, which is the "quadrature component." The two DVMs together represent apparent power in a rectangular form.

Figure 24.8 shows a circuit diagram of power measurement in a rectangular form. The PSD outputs are given as

$$V_I = K_1 I \cos \varnothing$$

$$V_Q = K_2 I \sin \varnothing$$

where K_1 and K_2 are constants.

The multiplier M_1 output will be proportional to $VI \cos Ø$ and is displayed in DVM_1. The multiplier M_2 output will be proportional to $VI \sin Ø$ and is displayed in DVM_2.

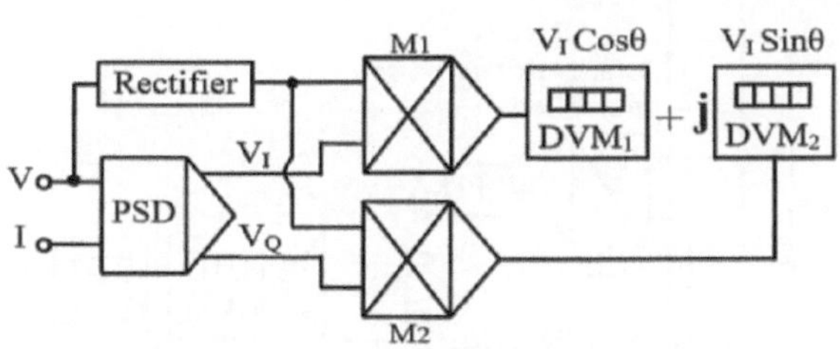

FIGURE 24.8 Power measurement in rectangular form.

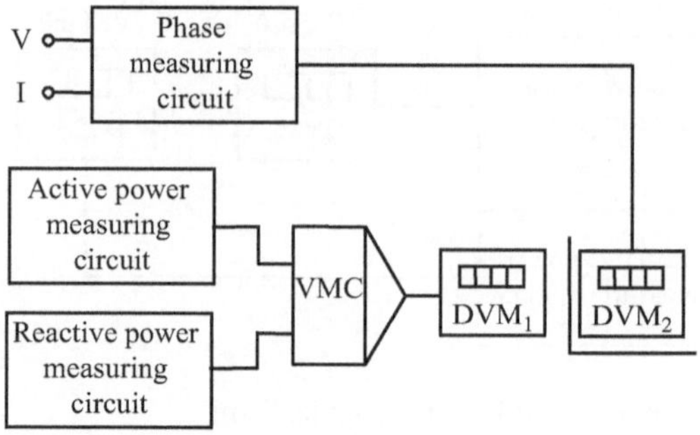

FIGURE 24.9 Block diagram of power measurement in polar form.

24.5.2 Power Measurement in Polar Form

Figure 24.9 shows power measurement in a polar form. Power 'S' can be represented in polar form as

$$S = \left[\sqrt{V_P^2 + V_Q^2} \right] \underline{|\varnothing} \quad (24.30)$$

where Ø is the phase difference between voltage and current.

The block diagram in Figure 24.9 is implemented using multipliers and is shown in Figure 24.10.

The multiplier M_1 output is proportional to active power 'P', and multiplier M_2 output is proportional to reactive power 'Q'. The VMC output will be

$$V_O = \sqrt{{V_P}^2 + {V_Q}^2} \quad (24.31)$$

This is called "magnitude" and is displayed in DVM_1. The phase angle between voltage and current is measured by an EX-OR gate and is displayed in another DVM_2. The two DVMs together display power in a polar form.

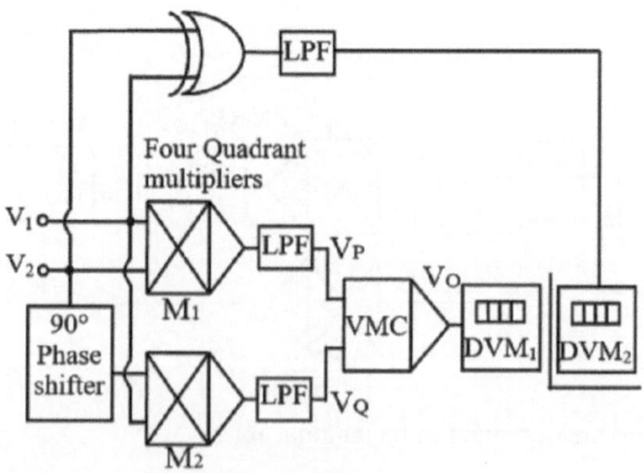

FIGURE 24.10 Power measurement in polar form using multipliers.

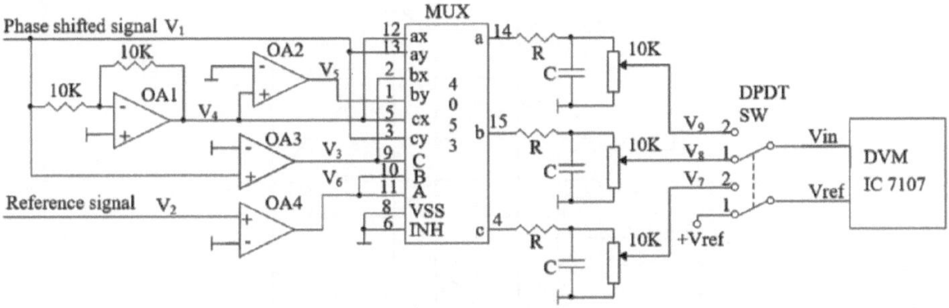

FIGURE 24.11 Circuit diagram of measurement of power factor and phase angle.

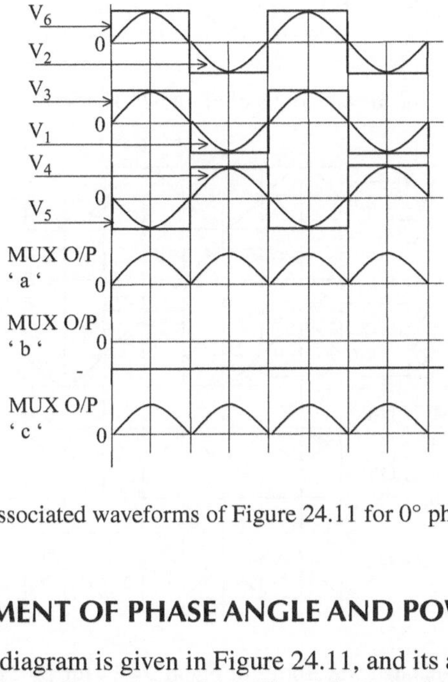

FIGURE 24.12(A) Associated waveforms of Figure 24.11 for 0° phase angle.

24.6 MEASUREMENT OF PHASE ANGLE AND POWER FACTOR

The proposed circuit diagram is given in Figure 24.11, and its associated waveforms in Figures 24.12(a), 24.12(b), and 24.12(c). The reference signal V_2 and the phase-shifted signal V_1 are given to zero crossing detectors OA_4 and OA_3 to get square waveforms V_6 and V_3, respectively. The phase-shifted signal V_1 is inverted by using an inverting amplifier OA_1 and then given to zero crossing detector OA_2 to get square waveform V_5.

The multiplexer IC4053 has three, 2 to 1 multiplexers. Its operation is shown in Table I. For Multiplexer 1, the input 'ay' is connected to output 'a' if control pin 'A' is at HIGH, and the input 'ax' is connected to output 'a' if control pin 'A' is at LOW. For Multiplexer 2, the input 'by' is connected to output 'b' if control pin 'B' is at HIGH, and the input 'bx' is connected to output 'b' if control pin 'B' is at LOW. For Multiplexer 3, the input 'cy' is connected to output 'c' if control pin 'C' is at HIGH, and the input 'cx' is connected to output 'c' if control pin 'C' is at LOW.

The operation of each multiplexer is explained below:

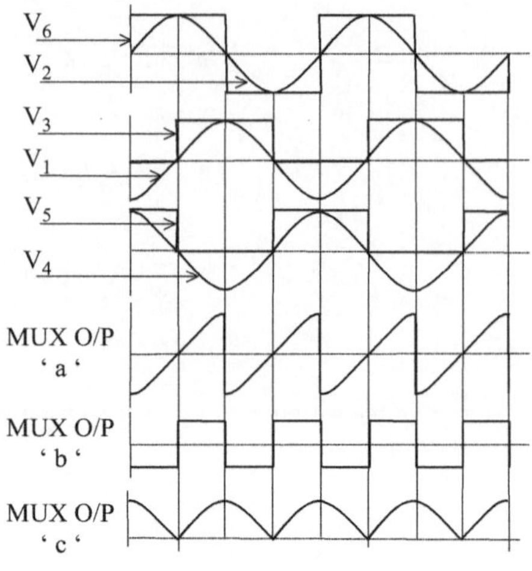

FIGURE 24.12(B) Associated waveforms of Figure 24.11 for 90° phase angle.

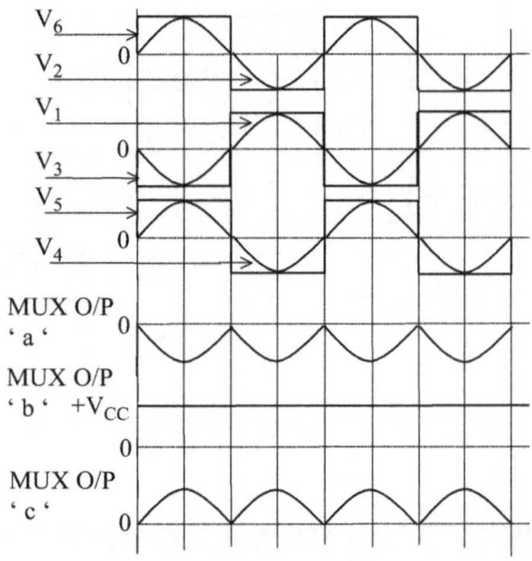

FIGURE 24.12(C) Associated waveforms of Figure 24.11 for 180° phase angle.

Multiplexer 1:
This multiplexer performs the operation $V_1 \cos \varnothing$

$$\text{Let } V_2(t) = V_2 \sin \omega t$$

$$V_1(t) = V_1 \sin(\omega t + \varnothing)$$

Since V_6 is a square wave, it can be expanded by Fourier series as,

$$V_6 = 4/\pi\left[\sin \omega t - 1/3 \sin 3\omega t + \cdots\right]$$

TABLE I
Multiplexer Operation

Multiplexer Operation

Input Status			Output Status		
C	B	A	c	b	a
0	0	0	Cx	bx	ax
0	0	1	Cx	bx	Ay
0	1	0	Cx	by	Ax
0	1	1	Cx	by	Ay
1	0	0	Cy	bx	Ax
1	0	1	Cy	bx	Ay
1	1	0	Cy	by	Ax
1	1	1	Cy	by	Ay

The multiplexer's output at 'a' will be

$$a(t) = V_6 V_1 = 2V_1/\pi[\cos\varnothing - \cos(2\omega t + \varnothing + \ldots] \tag{24.32}$$

The RC lowpass filter removes the unwanted AC components, and the DC output voltage V_9 is given by

$$V_9 = 2V_1 \cos\varnothing/\pi \tag{24.33}$$

Multiplexer 2:
This multiplexer works as an EX-OR gate. The average of its output is proportional to the phase difference between the input pulses. The characteristics of EX-OR gate are shown in Figure 24.12(d). From this figure, it is observed that the maximum DC output occurs when the phase difference is π radians. The slope of the curve between 0 and π radians is the conversion gain Kp of the phase detector. If IC has +Vcc = 5 V, then Kp = 5/180°. The output of multiplexer 2 at 'b' is fed to an RC lowpass filter to

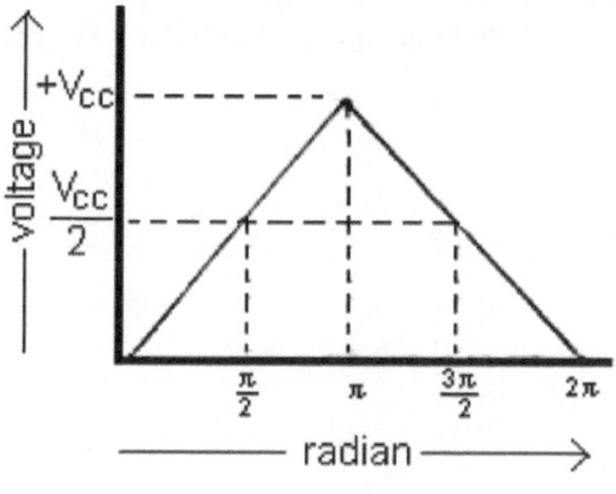

FIGURE 24.12(D) Characteristics of EX-OR gate.

remove high frequency components and to get DC voltage V_8 which is proportional to the phase difference Ø.

Multiplexer 3:
This works as a full wave rectifier. From Figure 24.2, it is observed that the Multiplexer 3 output at 'c' is a full wave rectified waveform, and this can be expanded by Fourier series as,

$$c(t) = 2V_1/\pi - 4V_1 \cos 2\omega t/3\pi - 4V_1 \cos 4\omega t/15\pi - \cdots$$

The output of RC lowpass filter is

$$V_7 = 2V_1/\pi \quad (24.34)$$

Digital Voltmeter:
The conventional digital voltmeter (DVM) using IC7107 is used here to display phase angle and power factor. The reading of DVM is proportional to V_{in}/V_{ref}. When the double pole double throw (DPDT) switch is in position '1', the DVM reads phase angle for a constant reference voltage $+V_{ref}$. When the DPDT switch is changed to position '2', the DVM reads $2V_1$ Cos Ø/$2V_1$ = Cos Ø (Power Factor).

TUTORIAL EXERCISES

1. In the active power measurement system described in Section 24.1, a multiplier is used as a phase sensitive detector (PSD). Use other PSDs described in Chapter 21 and implement them for action power measurement.
2. Replace the multiplier with a 90° phase shifter in Figure 24.3(a) with a sampling PSD.
3. Design a power factor meter using a (i) switching PSD, (ii) switching-sampling PSD, and (iii) sampling PSD. In each, draw circuit diagrams, waveforms, and working of their operation.
4. Design a three-phase volt-amp-reactive (VAR) meter using sampling PSDs.
5. Design a single-phase watt meter using switching PSDs.
6. Design a power factor meter using a (i) multiplier, (ii) switching PSD, and (iii) sampling PSD.

25 Miscellaneous Applications of Function Circuits

Function circuits are finding application in signal processing, data acquisition, communications, controls, instrumentation, and analog computations. Numerous useful applications of function circuits include performing as automatic gain control, voltage-controlled waveform generator, voltage-controlled quadrature oscillator, voltage-controlled exponentiator, mass gas flow measurement, etc. A function circuit is an input–output block whose output voltage is a prescribed function of the input signal. The function may be given in an equation form, graphical form, or a table of values. In this chapter, we will discuss a systematic method for approximating a given function and then modeling it with analog function circuits.

25.1 AUTOMATIC GAIN CONTROL CIRCUIT

The block diagram of an automatic gain control (AGC) circuit is shown in Figure 25.1. A multiplier is used as a voltage-controlled linear amplifier.

The amplitude of the output signal V_O is rectified and then filtered to get DC component only. The DC voltage at the output of the lowpass filter (LPF) is compared with a reference voltage through the integrator to generate an error signal. The error signal is integrated into the high gain integrator. When the DC component of the output voltage V_O is equal to the reference voltage, the integrator input is zero and the output of the integrator is steady. The output of the integrator is multiplied by the input signal V_I and as varying the gain.

The amplitude stability of the output signal depends on the stability of the DC reference voltage and the integrator gain. The rate at which the loop can realize after sudden changes in input level depends on the cut-off frequency of the LPF and the integrator time constant. AGC circuits are widely used to stabilize the signal amplitude of oscillation to keep a signal's amplitude constant, while its phase angle is varied by filtering and for various other purposes.

25.2 AUTOMATIC GAIN CONTROL CIRCUITS USING ANALOG DIVIDERS

The divider is used in an AGC system as shown in Figure 25.2 for positive reference voltage, and Figure 25.3 for negative reference voltage.

The numerator of the divider accepts bipolar voltages, and the denominator accepts only negative voltages. The output voltage V_O is a half-wave rectified to a DC reference by an integrator OA_2. The integrator output is given the denominator of

DOI: 10.1201/9781003221449-31

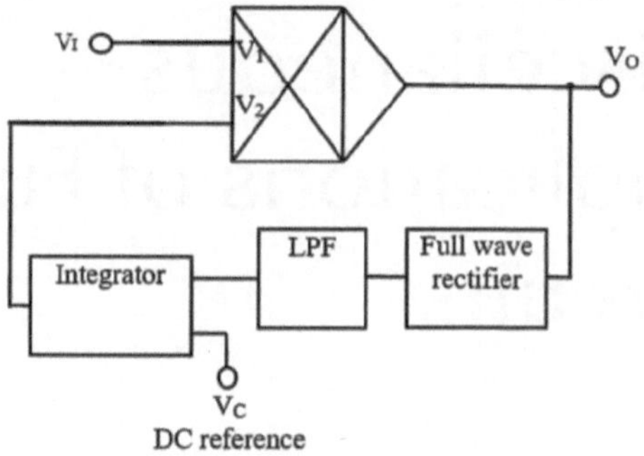

FIGURE 25.1 Automatic gain control circuit using a multiplier.

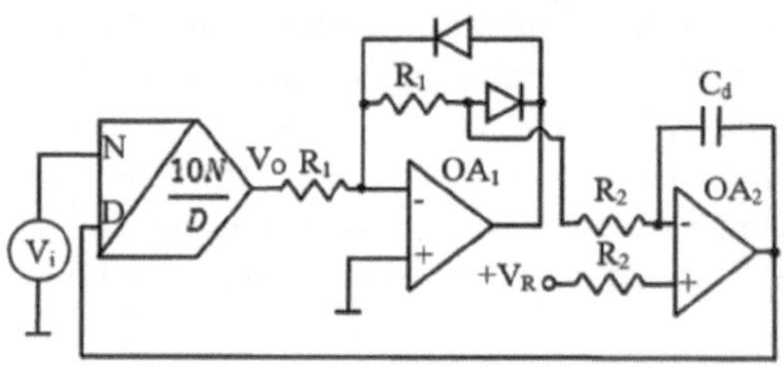

FIGURE 25.2 Automatic gain control using divider for positive reference voltage.

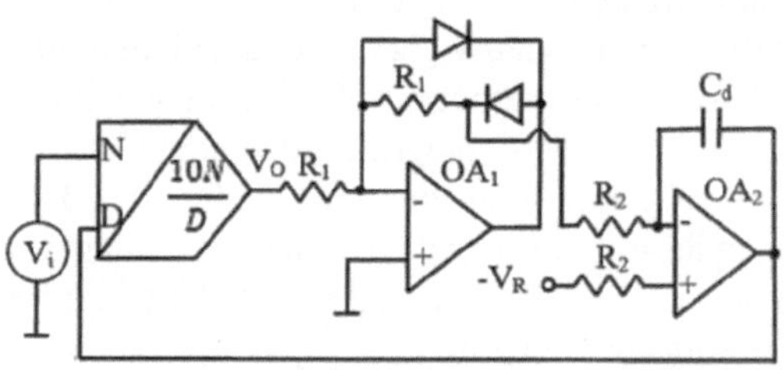

FIGURE 25.3 Automatic gain control using divider for negative reference voltage.

divider as a control signal. Therefore, variations in the output V_O caused by the input voltage changes will create a control signal that corresponds to the input voltage changes. Under a balanced condition,

$$\frac{2V_P}{\pi} = V_R \tag{25.1}$$

where V_P = peak value of input signal, and V_R = reference voltage.

25.3 VOLTAGE-CONTROLLED WAVEFORM GENERATOR

Figure 25.4 shows the voltage-controlled waveform generator using a multiplier. It provides simultaneous square and triangular waveforms. Op-amp OA_1 works as

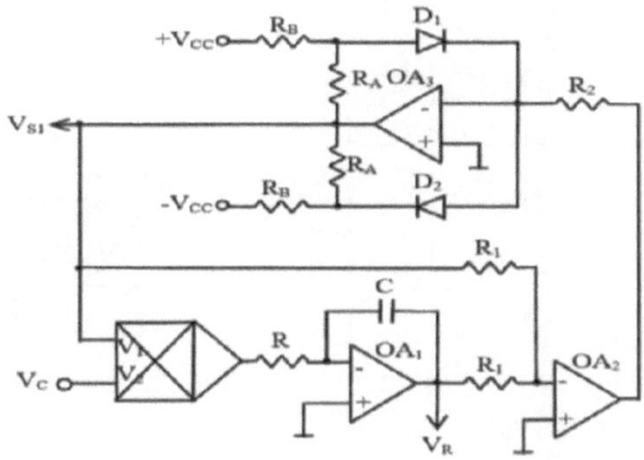

FIGURE 25.4 Voltage-controlled waveform generator using multiplier.

integrator; OA_2 and OA_3 work as comparators. The positive feedback around the comparator provides fast, uniform switching when the oscillation frequency is changing. The simple diode limiter (R_A, R_B, D_1, and D_2) around OA_3 sets the amplitude of square wave output. A multiplier is connected in the feedback loop, and control voltage V_C tunes the amplitude of square wave feeding back to the integrator.

The multiplier acts as a frequency modulator in the circuit. The frequency f_O of the output square wave and triangular wave is given as

$$f_O = \frac{V_C}{4V_R RC} \tag{25.2}$$

where V_C is the control voltage and V_R is the multiplier constant.

25.4 VOLTAGE-CONTROLLED QUADRATURE OSCILLATOR

In general, a quadratic oscillator consists of two integrators and some sort of amplitude limiting circuits. With addition of two multipliers, the frequency of oscillation can be linearly proportional to a control voltage V_C. Figure 25.5 shows such a voltage-controlled quadrature oscillator.

Its oscillation frequency f_O is given as

$$f_O = \frac{V_C}{V_R 2\pi RC} \tag{25.3}$$

where V_R is the multiplier constant.

25.5 VOLTAGE-CONTROLLED EXPONENTIATOR

A multifunction converter (MFC) discussed in Chapter 20 is connected to a voltage-controlled exponentiator with a multiplier, as shown in Figure 25.6.

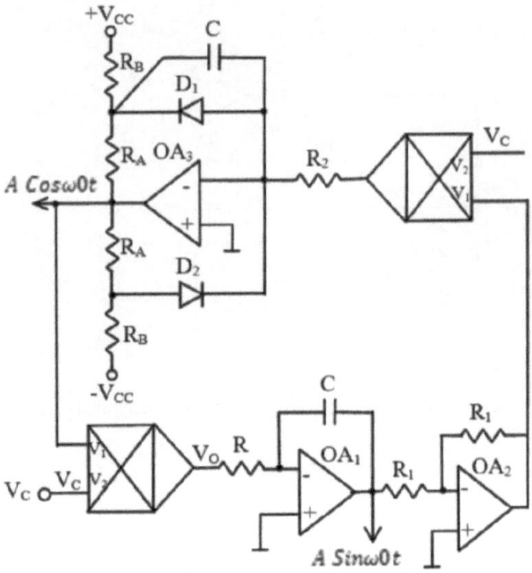

FIGURE 25.5 Voltage-controlled quadrature oscillator using multiplier.

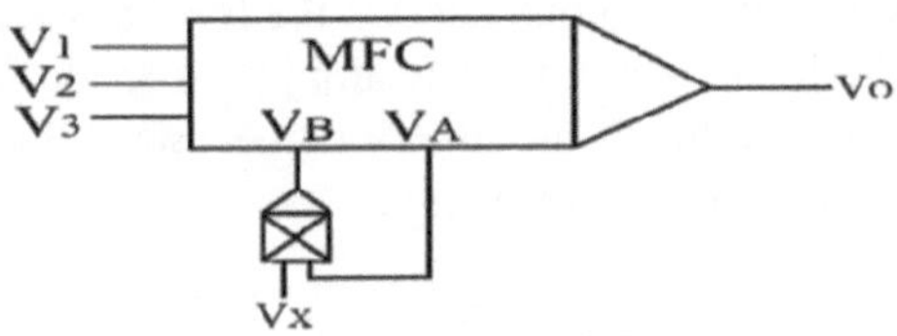

FIGURE 25.6 Voltage-controlled exponentiator.

The MFC output is given as

$$V_O = V_2\left(\frac{V_3}{V_1}\right)^m \tag{25.4}$$

where V_1, V_2, and V_3 are input voltages and 'm' is an arbitrary exponent (m > 1) determined by two external resistors. Using a multiplier in the place of these two resistors, exponent 'm' can be made voltage-tunable. The output of Figure 25.6 is given as

$$V_O = V_2\left(\frac{V_3}{V_1}\right)^{V_R/V_X} \tag{25.5}$$

where $V_X > 0$, V_R is a constant voltage.

25.6 MASS GAS FLOW MEASUREMENT

Analog function circuits are used in industrial automation and process control applications.

The equation of mass gas flow is given as

$$V_O(\text{flow}) = \sqrt{\frac{P\Delta P}{T}}K \tag{25.6}$$

where ΔPis the differential pressure across the orifice, P is the absolute pressure, T is the absolute temperature, and K is constant scaled to utilize full output range.

The equation is realized by using two multifunction converters (MFCs) in Figure 25.7 and by using two multiplier-cum-dividers (MCDs) in Figure 25.8.

In Figure 25.7, the first MFC output will be

$$V_W = V_2\left(\frac{V_3}{V_1}\right)^{m=1} = \frac{V_2V_3}{V_1} = \frac{P\Delta P}{T} \tag{25.7}$$

The second MFC output is

$$V_O = \frac{V_2V_3}{V_1} = \sqrt{\frac{P\Delta P}{T}}K \tag{25.8}$$

In Figure 25.8, the first MCD output will be

$$V_W = \frac{V_2V_3}{V_1} = \frac{P\Delta P}{T} \tag{25.9}$$

The second MCD output is

$$V_O = \frac{V_2V_3}{V_1} = \sqrt{\frac{P\Delta P}{T}}K \tag{25.10}$$

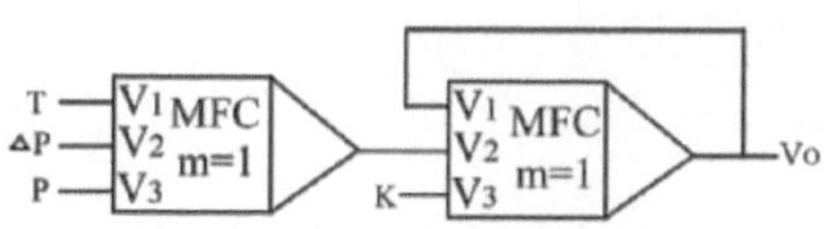

FIGURE 25.7 Mass gas flow measurement with two multifunction converters.

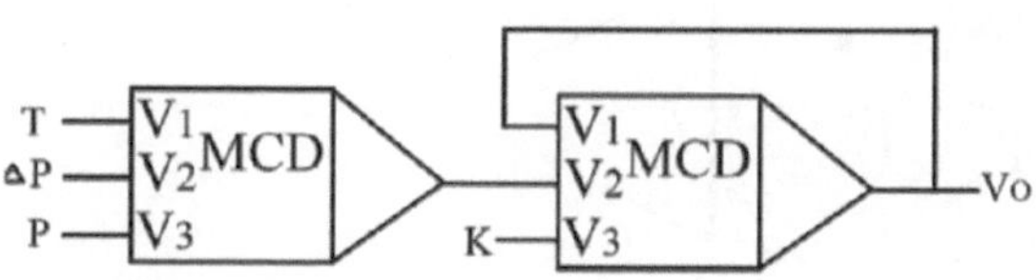

FIGURE 25.8 Mass gas flow measurement with two multipliers-cum-dividers.

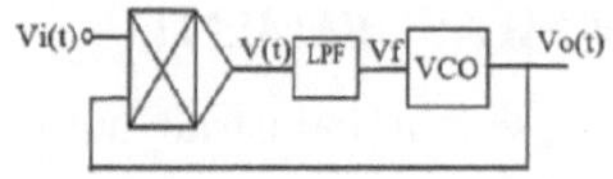

FIGURE 25.9 Phase locked loop using multiplier.

25.7 PHASE LOCKED LOOP

The phase locked loop (PLL) using a multiplier is shown in Figure 25.9. The multiplier is used here as a phase detector. Let initially the input voltage Vi(t) be grounded, then the multiplier output and the lowpass filter (LPF) output Vf be zero. The voltage controlled oscillator at present will work as a free-running multivibrator and produces free-running frequency fo determined by its internal RC values. When an input signal Vi(t) is applied to the multiplier, the multiplier compares phase and frequencies of input signal Vi(t) and VCO signal output Vo(t) and produces an error signal V(t). The LPF extracts the DC component of V(t) and is applied to VCO. If the frequency of Vi(t) is in the vicinity of Vo, the feedback loop will force the VCO to synchronize or lock on the input signal. Then the frequency of VCO will be the same as the frequency of the input signal except for a finite phase difference. The PLL is used as a demodulator following phase or frequency modulation. PLL is used as a matched filter operating as a coherent detector.

25.8 SIMULATION OF EQUATIONS

Function circuits can be used to simulate equations. For example, Figure 25.10(a) shows simulation for equation $V_O = 6V + 5V_1$, Figure 25.10(b) shows simulation for equation $V_O = 3V - 6V_1 + 9V_1^2$, Figure 25.10(c) shows simulation for equation $V_O = 2V + 5V_1 + 6V_1^2 + 4V_1^3$, Figure 25.10(d) shows simulation for equation

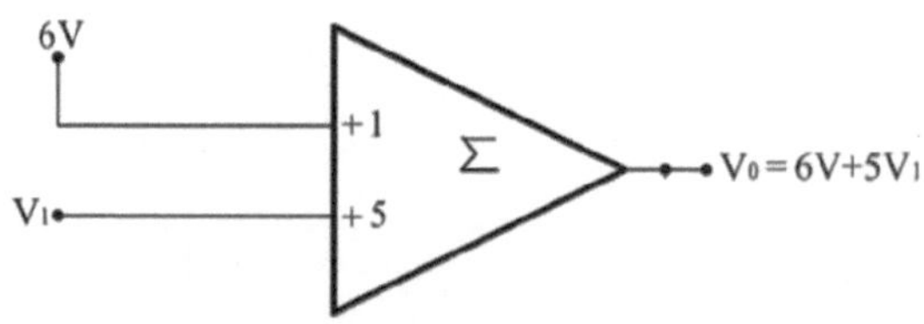

FIGURE 25.10(A) Simulation of equation $V_O = 6V+5V_1$.

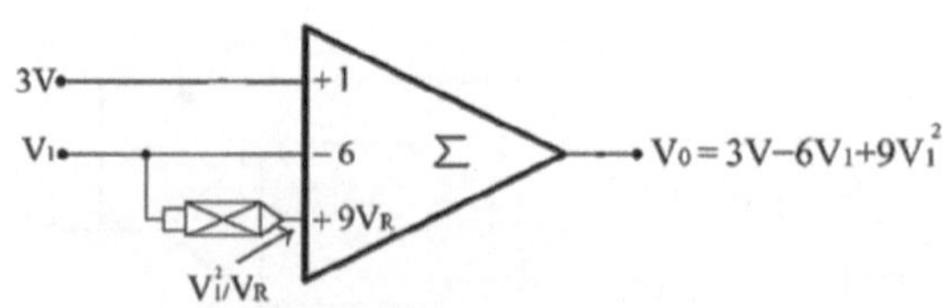

FIGURE 25.10(B) Simulation of equation $V_O = 3V-6V_1+9V_1^2$.

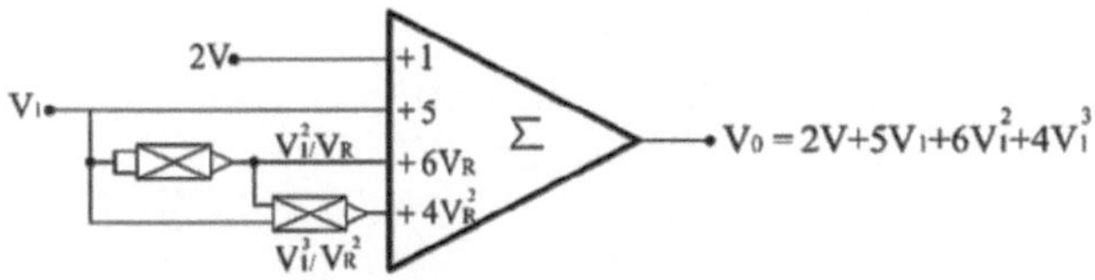

FIGURE 25.10(C) Simulation of equation $V_O = 2V+5V_1+6V_1^2+4V_1^3$.

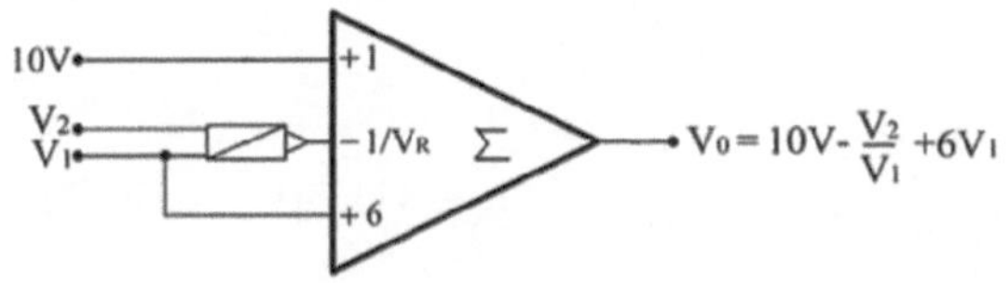

FIGURE 25.10(D) Simulation of equation $V_O = 10V-V_2/V_1+6V_1$.

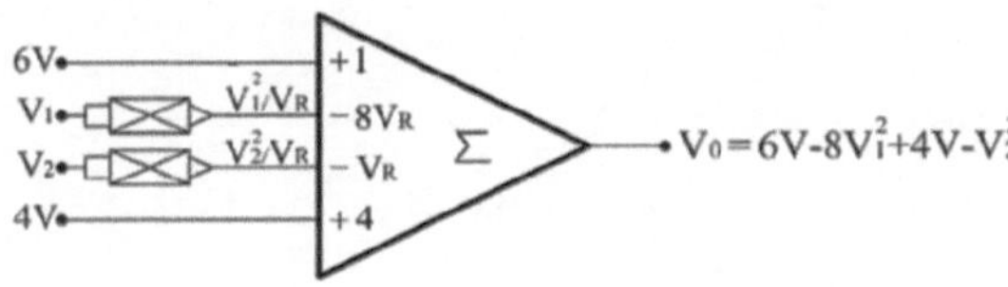

FIGURE 25.10(E) Simulation of equation $V_O = 6V-8V_1^2+4V-V_2^2$.

$V_O = 10V - V_2/V_1 + 6V_1$, and Figure 25.10(e) shows simulation for equation $V_O = 6V - 8V_1^2 + 4V - V_2^2$. All parts of Figure 25.10 are self-explanatory.

TUTORIAL EXERCISES

1. Design a voltage-controlled function generator using a multiplier. The frequency of oscillation is 10 KHz and control voltage is 1 V.
2. Design a quadrature oscillator using a multiplier with oscillation frequency of 1 KHz.
3. Draw a circuit diagram of the simulation shown in Figure 25.10(a) using an op-amp.
4. Simulate the function $V_O = V_1^2 - V_2^2 + V_3^2 - V_4^2$.

Appendix A
Analog Function Circuits Tutorial Kit

The author designed and developed an "Analog Function Circuits Tutorial Kit," also referred to in the book as an "AFC Trainer Kit," to practically verify his concepts and theory on function circuits. The kit has all the components of function circuits as individual blocks. The respective blocks can be connected by external wires or patch cards to form a required function circuit. The kit has internally generated input voltages V_1, V_2, and V_3. Multimeters, power supplies, and oscilloscopes are required externally to experiment with the function circuit. On reading this book, the reader is able to verify the analog function circuits by experimenting with this trainer kit. This experimental kit is very useful, and it is recommended for use in every electronics laboratory in the world. The photograph of this kit is given below.

For further details and to get this kit, contact: kcselvam@ee.iitm.ac.in

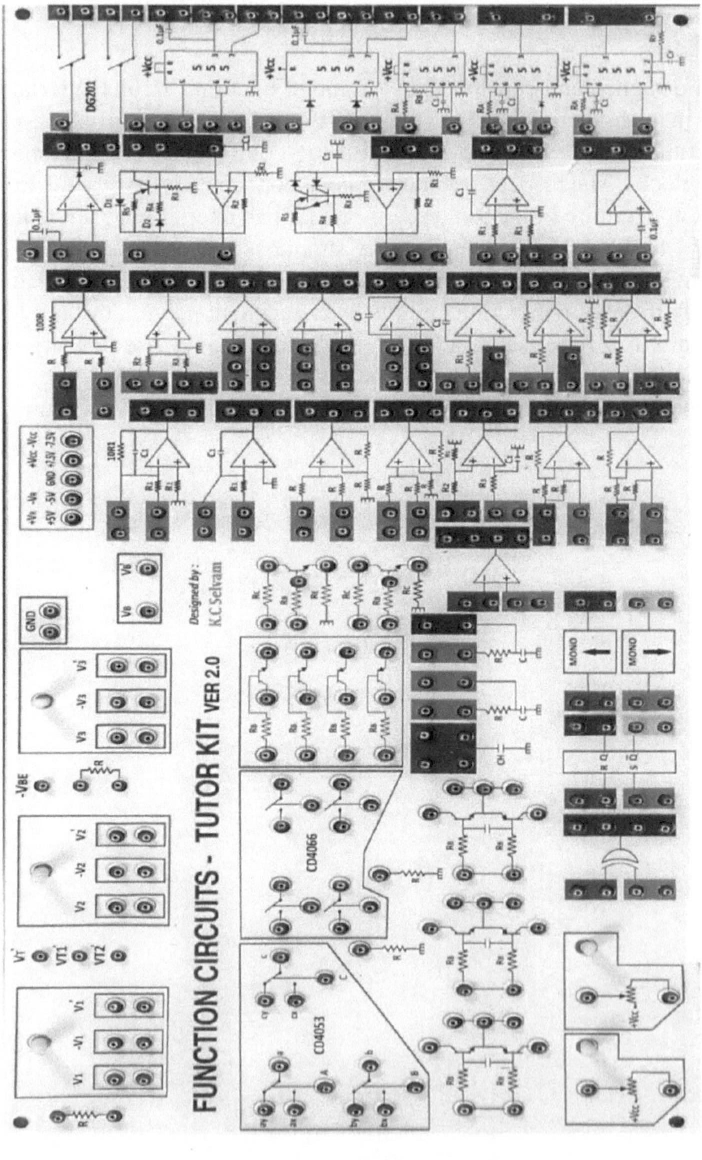
FUNCTION CIRCUITS - TUTOR KIT VER 2.0
Designed by :
K.C.Selvam
CD4053
CD4066
DG201
GND
MONO
MONO

Index

W

Z